Primate Craniofacial Function and Biology

DEVELOPMENTS IN PRIMATOLOGY: PROGRESS AND PROSPECTS
Series Editor: Russell H. Tuttle, University of Chicago, Chicago, Illinois

This peer-reviewed book series melds the facts of organic diversity with the continuity of the evolutionary process. The volumes in this series exemplify the diversity of theoretical perspectives and methodological approaches currently employed by primatologists and physical anthropologists. Specific coverage includes: primate behavior in natural habitats and captive settings; primate ecology and conservation; functional morphology and developmental biology of primates; primate systematics; genetic and phenotypic differences among living primates; and paleoprimatology.

PRIMATE BIOGEOGRAPHY
    Edited by Shawn M. Lehman and John Fleagle

REPRODUCTION AND FITNESS IN BABOONS: BEHAVIORAL, ECOLOGICAL,
AND LIFE HISTORY PERSPECTIVES
    Edited By Larissa Swedell and Steven R. Leigh

RINGTAILED LEMUR BIOLOGY: LEMUR CATTA IN MADAGASCAR
    Edited by Alison Jolly, Robert W. Sussman, Naoki Koyama, and Hantanirina Rasamimanana

PRIMATES OF WESTERN UGANDA
    Edited by Nicholas E. Newton-Fisher, Hugh Notman, James D. Paterson, and Vernon Reynolds

PRIMATE ORIGINS: ADAPTATIONS AND EVOLUTION
    Edited by Matthew J. Ravosa and Marian Dagosto

LEMURS: ECOLOGY AND ADAPTATION
    Edited by Lisa Gould and Michelle L. Sauther

PRIMATE ANTI-PREDATOR STRATEGIES
    Edited by Sharon L. Gursky and K.A.I. Nekaris

CONSERVATION IN THE 21ST CENTURY: GORILLAS AS A CASE STUDY
    Edited by T.S. Stoinski, H.D. Steklis, and P.T. Mehlman

ELWYN SIMONS: A SEARCH FOR ORIGINS
    Edited by John G. Fleagle and Christopher C. Gilbert

THE BONOBOS: BEHAVIOR, ECOLOGY, AND CONSERVATION
    Edited by Takeshi Furuichi and Jo Thompson

PRIMATE CRANIOFACIAL FUNCTION AND BIOLOGY
    Edited by Chris Vinyard, Matthew J. Ravosa, and Christine E. Wall

Chris Vinyard · Matthew J. Ravosa · Christine Wall
Editors

Primate Craniofacial Function and Biology

Springer

Editors
Chris Vinyard
Department of Anatomy
Northeastern Ohio University
 College of Medicine
4209 Cpy Route 44
Rootstown, OH 44272-0095
USA
cvinyard@neoucom.edu

Matthew J. Ravosa
University of Missouri School of Medicine
Columbia, MO 65212
USA
ravosam@health.missouri.edu

Christine Wall
Department of Biological
 Anthropology and Anatomy
Duke University
Box 90383
Durham, NC 27708
USA
christine_wall@baa.mc.duke.edu

ISBN: 978-0-387-76584-6   e-ISBN: 978-0-387-76585-3
DOI: 10.1007/978-0-387-76585-3

Library of Congress Control Number: 2008931523

© 2008 Springer Science+Business Media, LLC
All rights reserved. This work may not be translated or copied in whole or in part without the written permission of the publisher (Springer Science+Business Media, LLC, 233 Spring Street, New York, NY 10013, USA), except for brief excerpts in connection with reviews or scholarly analysis. Use in connection with any form of information storage and retrieval, electronic adaptation, computer software, or by similar or dissimilar methodology now known or hereafter developed is forbidden.
The use in this publication of trade names, trademarks, service marks, and similar terms, even if they are not identified as such, is not to be taken as an expression of opinion as to whether or not they are subject to proprietary rights.

Printed on acid-free paper

springer.com

Preface

William Hylander is synonymous with primate craniofacial function. For the first three-quarters of the 20th century, studies of the primate masticatory apparatus typically inferred function by examining form. William Hylander revolutionized studies of the primate masticatory apparatus through his use of in vivo techniques to quantify bone strain, jaw-muscle activity, and jaw movements in living primates while they chewed. His direct measures of jaw, tooth, and jaw-muscle function during chewing are the empirical cornerstone that many biological anthropologists build upon today. We dedicate this volume surveying recent developments in primate craniofacial function and biology to William Hylander and his lifelong contribution to biological anthropology.

Today, the amount of craniofacial research on primates is immense. Functional studies alone range from in vivo analyses of living animals to morphological and finite element explorations of extinct primate cranial remains. The results of these research efforts have been fundamental in developing our understanding of primate biology and evolution. The sheer magnitude of craniofacial studies affirm that furthering our knowledge of primate craniofacial biology is one of the most important research agendas in modern biological anthropology.

Outside of primatology, many mammalian biologists have provided key insights into primate form and function through their comparative analyses of mammalian clades. Some biological anthropologists have continued a tradition of studying non-primate mammals as model taxa, as alternative functional designs, or as comparative radiations for exploring form–function associations observed in primates. We make a concerted effort to include this broader mammalian perspective to build on its fundamental contribution to our understanding of primate craniofacial biology.

It is impossible to adequately incorporate current research on primate craniofacial function and biology into a single volume. Our strategy was to put together a volume in honor of William Hylander, which will give readers an overview of what is current in a number of different research areas. Because several of the contributors worked closely with Hylander throughout his career, we have the deepest coverage in topics focusing on craniofacial function during feeding. This having been said, we attempted to provide a wide range of current research. We hope that readers will be able to capitalize on this approach to integrate otherwise disparate ideas and methodologies for their own work. The cost of our approach is that good scientists and good science were left out. If by bringing together this wide range of researchers we can help catalyze future work on primate craniofacial biology and function, then we feel this cost will have been worth it.

Several of the chapters in this volume were initially presented at a 2005 symposium entitled: "Primate Craniofacial Function and Biology: A Symposium in Honor of William L. Hylander" during the American Association of Physical Anthropology Meetings in Milwaukee, WI. Building from this initial group of presentations, we had the good fortune of adding several chapters to the current volume. In the end, we were able to include twenty chapters in five sections that broadly explore different approaches to studying the skulls of primates and other mammals.

This volume would not have been possible without the advice and assistance of numerous individuals. Specifically, thanks to Andrea Macaluso, Tom Brazda, Lisa Tenaglia, Krista Zimmer, Melanie Wilichinsky, and Russell Tuttle for guiding us through the editorial process. Diana Dillon and Marie Dockery provided invaluable administrative assistance. Several external reviewers provided insightful comments that advanced the scholarship of these contributions. We thank the American Association of Physical Anthropologists for hosting and supporting the 2005 symposium in honor of William Hylander, which was the catalyst for this volume. Kirk Johnson has made one of the biggest contributions to the studies of primate feeding. Throughout his more than 25 years of work in the field, each of us became indebted to Kirk for his guidance and friendship – thank you.

Finally, we wanted to express our utmost thanks to Bill Hylander. There is no end to our appreciation of your friendship, advice, and commitment to biological anthropology and experimental biology.

Rootstown, OH, USA	Christopher J. Vinyard
Columbus, MO, USA	Matthew J. Ravosa
Durham, NC, USA	Christine E. Wall

Contents

Part I Historical Perspective on Experimental Research in Biological Anthropology

1. Experimental Comparative Anatomy in Physical Anthropology: The Contributions of Dr. William L. Hylander to Studies of Skull Form and Function 3
   Daniel Schmitt, Christine E. Wall, and Pierre Lemelin

Part II In Vivo Research into Masticatory Function

2. A Nonprimate Model for the Fused Symphysis: In Vivo Studies in the Pig 19
   Susan W. Herring, Katherine L. Rafferty, Zi Jun Liu, and Zongyang Sun

3. Symphyseal Fusion in Selenodont Artiodactyls: New Insights from In Vivo and Comparative Data 39
   Susan H. Williams, Christine E. Wall, Christopher J. Vinyard, and William L. Hylander

4. Does the Primate Face Torque? 63
   Callum F. Ross

5. Motor Control of Masticatory Movements in the Southern Hairy-Nosed Wombat *(Lasiorhinus latifrons)* 83
   Alfred W. Crompton, Daniel E. Lieberman, Tomasz Owerkowicz, Russell V. Baudinette*, and Jayne Skinner

6. Specialization of the Superficial Anterior Temporalis in Baboons for Mastication of Hard Foods 113
   Christine E. Wall, Christopher J. Vinyard, Susan H. Williams, Kirk R. Johnson, William L. Hylander

Part III Modeling Masticatory Apparatus Function

7 Effects of Dental Alveoli on the Biomechanical Behavior of the Mandibular Corpus 127
David J. Daegling, Jennifer L. Hotzman, and Andrew J. Rapoff

8 Surface Strain on Bone and Sutures in a Monkey Facial Skeleton: An In Vitro Approach and its Relevance to Finite Element Analysis . 149
Qian Wang, Paul C. Dechow, Barth W. Wright, Callum F. Ross, David S. Strait, Brian G. Richmond, Mark A. Spencer, and Craig D. Byron

9 Craniofacial Strain Patterns During Premolar Loading: Implications for Human Evolution 173
David S. Strait, Barth W. Wright, Brian G. Richmond, Callum F. Ross, Paul C. Dechow, Mark A. Spencer, and Qian Wang

Part IV Jaw-Muscle Architecture

10 Scaling of Reduced Physiologic Cross-Sectional Area in Primate Muscles of Mastication 201
Fred Anapol, Nazima Shahnoor, and Callum F. Ross

11 Scaling of the Chewing Muscles in Prosimians 217
Jonathan M.G. Perry and Christine E. Wall

12 The Relationship Between Jaw-Muscle Architecture and Feeding Behavior in Primates: Tree-Gouging and Nongouging Gummivorous Callitrichids as a Natural Experiment 241
Andrea B. Taylor and Christopher J. Vinyard

Part V Bone and Dental Morphology

13 Relationship Between Three-Dimensional Microstructure and Elastic Properties of Cortical Bone in the Human Mandible and Femur 265
Paul C. Dechow, Dong Hwa Chung, and Mitra Bolouri

14 Adaptive Plasticity in the Mammalian Masticatory Complex: You Are What, and How, You Eat 293
Matthew J. Ravosa, Elisabeth K. Lopez, Rachel A. Menegaz, Stuart R. Stock, M. Sharon Stack, and Mark W. Hamrick

15 **Mandibular Corpus Form and Its Functional Significance: Evidence from Marsupials** 329
Aaron S. Hogue

16 **Putting Shape to Work: Making Functional Interpretations of Masticatory Apparatus Shapes in Primates** 357
Christopher J. Vinyard

17 **Food Physical Properties and Their Relationship to Morphology: The Curious Case of *kily*** 387
Nayuta Yamashita

18 **Convergence and Frontation in Fayum Anthropoid Orbits** 407
Elwyn L. Simons

19 **What Else Is the Tall Mandibular Ramus of the Robust Australopiths Good For?** 431
Yoel Rak and William L. Hylander

20 **Framing the Question: Diet and Evolution in Early *Homo*** 443
Susan C. Antón

Index 483

Contributors

Fred Anapol
Department of Anthropology, University of Wisconsin-Milwaukee, Milwaukee, WI 53211, fred@uwm.edu

Susan C. Antón
Center for the Study of Human Origins, Department of Anthropology, New York University, New York, NY 10003, susan.anton@nyu.edu

Russell V. Baudinette
School of Earth and Environmental Sciences, University of Adelaide, Adelaide, SA 5005, Australia

Mitra Bolouri
Department of Biomedical Sciences, Baylor College of Dentistry, Texas A&M Health Science Center, Dallas, TX 75254

Craig D. Byron
Department of Biology, Mercer University, Macon, GA 31207

Dong Hwa Chung
Department of Orthodontics, School of Dentistry, Dankook University, Chung Nam, South Korea

Alfred W. Crompton
Museum of Comparative Zoology, Harvard University, Cambridge, MA 02138, acrompton@oeb.harvard.edu

David J. Daegling
Department of Anthropology, University of Florida, Gainesville, FL 32611–7305, daegling@anthro.ufl.edu

Paul C. Dechow
Department of Biomedical Sciences, Baylor College of Dentistry, Texas A&M Health Science Center, Dallas, TX 75254, pdechow@bcd.tamhsc.edu

Mark W. Hamrick
Department of Cellular Biology and Anatomy, Medical College of Georgia, Augusta, GA 30912

Susan W. Herring
Department of Orthodontics, University of Washington, Seattle, WA 98195–7446, herring@u.washington.edu

Aaron S. Hogue
Department of Biological Sciences, Salisbury University, Salisbury, MD 21801, ashogue@salisbury.edu

Jennifer L. Hotzman
Department of Anthropology, University of Florida, Gainesville, FL 32611–7305

William L. Hylander
Department of Biological Anthropology and Anatomy, Duke University Lemur Center, Duke University, Durham, NC 27710

Kirk R. Johnson
Department of Biological Anthropology and Anatomy, Duke University, Durham, NC 27710, USA

Pierre Lemelin
Division of Anatomy, Faculty of Medicine and Dentistry, University of Alberta, Edmonton, Alberta, Canada T6G 2H7

Daniel E. Lieberman
Peabody Museum, Harvard University, Cambridge, MA 02138

Zi Jun Liu
Department of Orthodontics, University of Washington, Seattle, WA 98195–7446

Elisabeth K. Lopez
Department of Cell and Molecular Biology, Northwestern University Feinberg School of Medicine, Chicago, IL 60611–3008

Rachel A. Menegaz
Department of Pathology and Anatomical Sciences, University of Missouri School of Medicine, Columbia, MO 65212

Tomasz Owerkowicz
School of Earth and Environmental Sciences, University of Adelaide, Adelaide, SA 5005, Australia

Jonathan M.G. Perry
Department of Biological Anthropology and Anatomy, Duke University Medical Center, Durham, NC 27710, jmp20@duke.edu

Katherine L. Rafferty
Department of Orthodontics, University of Washington, Seattle, WA 98195–7446

Yoel Rak
Department of Anatomy and Anthropology, Sackler Faculty of Medicine, Tel Aviv University, Tel Aviv 69978, Israel, yoelrak@post.tau.ac.il

Contributors

Andrew J. Rapoff
Department of Mechanical Engineering, Union College, Schenectady, NY 12308–3107

Matthew J. Ravosa
Department of Pathology and Anatomical Sciences, University of Missouri School of Medicine, Columbia, MO 65212, ravosam@missouri.edu

Brian G. Richmond
Center for the Advanced Study of Hominid Paleobiology, Department of Anthropology, The George Washington University, Washington DC 20052

Callum F. Ross
Department of Organismal Biology & Anatomy, University of Chicago, Chicago, IL 60637, rossc@uchicago.edu

Daniel Schmitt
Department of Evolutionary Anthropology, Duke University Box 90383, Durham, NC 27708, USA, Daniel.Schmitt@duke.edu

Nazima Shahnoor
Department of Anthropology, University of Wisconsin-Milwaukee, Milwaukee, WI 53211

Elwyn L. Simons
Duke Primate Center, Division of Fossil Primates, Department of Biological Anthropology and Anatomy, Duke University, Durham, NC 27705, esimons@duke.edu

Jayne Skinner
School of Earth and Environmental Sciences, University of Adelaide, Adelaide, SA 5005, Australia

Mark A. Spencer
School of Human Evolution and Social Change, Institute of Human Origins, Arizona State University, Tempe, AZ 85287

M. Sharon Stack
Department of Pathology and Anatomical Sciences, University of Missouri School of Medicine, Columbia, MO 65212

Stuart R. Stock
Department of Molecular Pharmacology and Biological Chemistry, Northwestern University Feinberg School of Medicine, Chicago, IL 60611–3008

David S. Strait
Department of Anthropology, University at Albany, Albany, NY 12222, dstrait@albany.edu

Zongyang Sun
Department of Orthodontics, University of Washington, Seattle, WA 98195–7446

Andrea B. Taylor
Doctor of Physical Therapy Division, Department of Community and Family Medicine, Duke University School of Medicine and Department of Evolutionary Anthropology, Duke University, Durham, NC 27708, andrea.taylor@duke.edu

Christopher J. Vinyard
Department of Anatomy, Northeastern Ohio Universities College of Medicine, Rootstown, OH 44272, cvinyard@neoucom.edu

Christine E. Wall
Department of Evolutionary Anthropology, Duke University, Durham, NC 27708, USA, christine_wall@baa.mc.duke.edu

Qian Wang
Division of Basic Medical Sciences, Mercer University School of Medicine, Macon, GA 31207, WANG_Q2@Mercer.edu.

Susan H. Williams
Department of Biomedical Sciences, Ohio University, Athens, OH 45701, willias7@ohio.edu

Barth W. Wright
Department of Anatomy, Kansas City University of Medicine and Biosciences, Kansas City, MO 64106–1453

Nayuta Yamashita
Department of Anthropology, University of Southern California, Los Angeles, CA 90089–1692, nayutaya@usc.edu

List of Figures

2.1	Theoretical patterns of deformation of the pig symphysis	22
2.2	Pig mandibles, showing methodology used	25
2.3	Examples of raw recordings of mastication	27
3.1	Summary data of peak masseter activity in primates and selenodont artiodactyls	42
3.2	Summary data of in vitro symphyseal strains along the labial (*left*) and lingual surfaces of the alpaca symphysis during simulating lateral transverse bending, or wishboning	43
3.3	Drawing of an alpaca mandible sectioned in the sagittal plane through the symphysis	46
3.4	Principal strains and the direction of maximum principal tension recorded from the symphyses of anesthetized alpacas during simulated manual reverse wishboning	47
3.5	Expected patterns of strain during reverse wishboning and wishboning of the alpaca symphysis	47
3.6	Raw and transformed symphyseal strains during mastication by Alpaca 4	48
3.7	Summary strain data from the symphysis of alpacas during left and right chews	50
3.8	Theoretical stresses and their expected patterns of strain during dorsoventral shear, anteroposterior shear, and transverse twisting of the symphysis	53
3.9	Graphs showing the relationship between the principal strains, direction of $\varepsilon_1(\alpha)$ and integrated masseter EMG from a single power stroke during chewing by an alpaca	56
3.10	Symphyses of four species of unfused selenodont artiodactyls	57
3.11	The symphyseal region in South American camelids	58
4.1	Diagram of rostral and lateral views of *Eulemur fulvus* skull illustrating the principal external forces hypothesized to impose twisting moments (torques) on the facial skeleton	65
4.2	Diagrams of cylinders fixed at one end and subjected to opposite torques at the free end	67

12.3	Superior view of the internal architecture of marmoset masseter and temporalis	249
12.4	Schematic of a bipinnate muscle depicting the measurements taken in this study	251
12.5	Box plots comparing masseter and temporalis muscle fiber lengths	252
12.6	Box plots comparing relative masseter and temporalis muscle fiber lengths	253
12.7	Box plots comparing relative masseter and temporalis PCSAs	253
12.8	Box plots comparing the ratio of total tendon length to fasciculus + tendon length	254
12.9	Bivariate plot demonstrating the architectural trade-off between muscle force and muscle excursion for masseter and temporalis muscles	255
12.10	The PCSA of the masseter and temporalis, expressed as a percentage of the combined PCSA for both muscles	257
13.1	Region of bone sampling for human mandibular bone specimens	268
13.2	Graphical explanation of material anisotropy	268
13.3	Hypothetical Haversian canal segment within a box oriented to the orthogonal material axes of the cortical bone specimen	272
13.4	Ultrasonic longitudinal velocities (10 MHz) versus wave direction in the femoral and mandibular specimens used in the μCT portion of the study	273
13.5	Images of mandibular cortical bone reconstructed from confocal microscope scans	274
13.6	Compilations of collapsed confocal images to show entire cross-sections of the bone specimen in the 2-3 plane or perpendicular to the 3 or longitudinal axis	276
13.7	The number of nodes correlates with the longitudinal length of the osteonal segments	277
13.8	Example of a splitting pattern in a complex osteonal segment	277
13.9	Rosette histograms of angles of all osteonal segments in all 95 reconstructed confocal volumes	278
13.10	Reconstructions of cortical bone volumes from μCT scans	279
13.11	Rosette histograms of the orientations of all osteonal segments per micron length relative to the plane of projection	280
14.1	MicroCT of symphysis proportions and biomineralization	300
14.2	MicroCT of TMJ biomineralization	301
14.3	Symphyseal cortical bone thickness	304
14.4	Symphyseal proteoglycan content	305
14.5	TMJ proteoglycan content	306
14.6	Symphyseal type II collagen	306
14.7	TMJ type II collagen	307
14.8	Symphyseal apoptosis	307
14.9	TMJ apoptosis	308
14.10	TMJ anatomy and proportions	310

14.11	TMJ proteoglycan content	312
14.12	TMJ type II collagen	312
14.13	Symphyseal proteoglycan content	314
14.14	TMJ proteoglycan content	315
14.15	Symphyseal type II collagen	316
14.16	TMJ type II collagen	317
14.17	Symphyseal apoptosis	317
14.18	TMJ apoptosis	318
15.1	Schematic depiction of the principle loading regimes experienced by the mandibular corpus	333
15.2	Phylogeny of the 65 marsupial species in this sample	343
15.3	Plot of ACA ln M_2/M_3 corpus I_x versus ACA ln mandibular length	345
15.4	Plot of ACA M_2/M_3 corpus I_x LS residuals versus the proportion of foliage or hard, strong, or tough foods in the diet	346
15.5	Plot of standardized contrasts of M_2/M_3 corpus I_x LS residuals versus standardized contrasts of the proportion of foliage or hard, strong, or tough foods in the diet	347
15.6	Plot of ACA ln M_2/M_3 corpus K versus ACA ln mandibular length	347
15.7	Plot of ACA M_2/M_3 corpus K LS residuals versus the proportion of foliage, vertebrates, or products of tree gouging in the diet	348
15.8	Plot of standardized contrasts of M_2/M_3 corpus K LS residuals versus standardized contrasts of the proportion of foliage, vertebrates, or products of gouging in the diet	349
16.1	A diagrammatic representation of the mechanical advantage of the jaw-closing muscles for producing vertical bite force at the incisors	362
16.2	Examination of the functional consequences and scaling of mechanical advantage for the jaw-closing muscles across 45 haplorhine primates	363
16.3	A cartoon example of changing jaw shapes with similar functional consequences	365
16.4	Box plots comparing corporal depth of the mandible at M_1 in papionins and colobines relative to body mass$^{1/3}$ and jaw length	369
16.5	Box Plots of relative jaw length among galagids created using five different shape ratios	376
17.1	Unripe kily fruit and seeds dropped by sifakas	391
17.2	Force–displacement graph of scissors cutting test of ripe kily fruit shell and unripe kily fruit	392
17.3	Time spent feeding on kily parts throughout year compared to total annual feeding time on same food parts	395
17.4	Logged toughness (R) values of phenophases of *Tamarindus indica*	396
17.5	Toughness (R) comparisons of entire *kily* diet of two lemur species	396
17.6	Toughness (R) comparisons of individual *kily* plant parts between the two lemur species	397
17.7	Comparisons of total crest lengths between ringtailed lemurs and sifakas using within-family residuals	398

17.8	Comparisons of protocone/talonid radii between ringtailed lemurs and sifakas using raw data values	399
17.9	Comparisons of talonid basin area between ringtailed lemurs and sifakas using within-family residuals	399
17.10	Average cusp radius and height between ringtailed lemurs and sifakas using within-family residuals. Comparisons are not significant (P = 0.062 and 0.846, respectively). See Fig. 17.5 caption for explanation of symbols	400
18.1	*Parapithecus grangeri* and *Aegyptopithecus zeuxis* crania	412
18.2	*Proteopithecus sylviae* CGM 42214 and *Catopithecus browni* DPC 11594 show convergence angles measured from the degree symbols in individuals of these two species	415
18.3	Crania of *Aegyptopithecus zeuxis* showing age-related changes in convergence angles	416
18.4	Convergence angles for *Apidium bowni* DPC 5264, *Arsinoea kallimos* DPC 11434, and *Proteopithcus sylviae* DPC 14095	417
19.1	Occlusal topography of *Australopithecus africanus* (*left*) and *A. robustus*	436
19.2	A comparison of the effect of a short mandibular ramus and a tall mandibular ramus on the movement of the lower teeth relative to the upper teeth in the final stages of occlusion	438
19.3	A schematic representation of the effect of anterior condylar translation on the height of the axis of rotation and the magnitude of the anterior movement of the lower teeth	438
19.4	The glenoid fossa and articular eminence in a chimpanzee and *A. boisei* specimen KNM-ER 23000.	439
20.1	Relationship between body size (femur length and stature) and brain size	455
20.2	Dentognathic summary statistics by species group	457
20.3	Dentognathic summary statistics for *H. habilis s.l.*, African *H. erectus* (including Dmanisi), Sangiran, Ternifine, and Zhoukoudian *H. erectus*	458
20.4	Bivariate plots among molar dimensions, corpus size, and symphyseal size	460
20.5	Molar dimensions, symphysis, and corpus size	461
20.6	Bivariate relationship between molar dimensions and body size (femur length and stature)	463
20.7	Bivariate relationship between jaw dimensions and body size (femur length and stature)	465
20.8	Molar and jaw dimensions relative to cranial capacity	466
20.9	Molar and jaw dimensions through geological time	468

List of Tables

2.1	Subjects and instrumentation	24
2.2	Mandibular distortion in the transverse direction during chewing	28
2.3	Magnitude ($\mu\varepsilon \pm$ s.d.) and orientation of principal strains during mastication	29
2.4	Condition of the symphyseal suture in 115 pig mandibles as a function of age	31
3.1	Summary data of symphyseal strains at 25% loading, peak, and 25% unloading of the power stroke of mastication	49
4.1	Descriptive statistics for maximum principal strain magnitudes and directions recorded from the postorbital bars during experiments 72, 73, and 74	74
4.2	Descriptive statistics for maximum principal strain magnitudes and directions recorded from the dorsal orbital and dorsal interorbital regions during experiments 76, 78, and 79	75
5.1	Weight and percentage of the total weight of the adductor muscles, in *Lasiorhinus latifrons*	93
5.2	Data collected on nine hairy-nosed wombats	95
5.3	Summary of cycle length and percentage of cycle with active adductors in six wombats feeding on different foods	99
5.4	Working-to-balancing side ratios of adductor activity during the power stroke in different adductors of *Lasiorhinus latifrons*	100
5.5	Descriptive statistics for peak principal strains and angle of peak tensile strain on the ventral margin of the right hemi-mandible and ventral surface of the mandibular symphysis	105
6.1	Fiber type proportions in the deep and superficial anterior temporalis based on myosin ATPase reactivity	116
6.2	EMG amplitude during hard- and soft-object feeding in the deep anterior temporalis and the superficial anterior temporalis in adult male and adult female *Papio*	119
6.3	Hard/soft ratio in the working- and balancing-side deep anterior temporalis and superficial anterior temporalis in adult male and female *Papio*	120
7.1	Effects of tooth removal on mandibular bone strain	137

7.2	Effect of indenter orientation on hardness	141
8.1	Strain gage sites, orientations, and descriptions	154
8.2	In vitro strain measurements when a load of 130 N was placed on right and left central incisors	158
8.3	In vitro strains measurements when a load of 195 N was placed on right and left fourth premolars	161
8.4	vitro strains measurements when a load of 260 N was placed on right and left third molars	163
9.1	Muscle forces applied to finite element model	177
9.2	Elastic properties employed in finite element analysis	180
9.3	Validation of FE model	182
9.4	Comparisons of maximum shear strains between experiments	189
9.5	Strains in the anterior rostrum	191
10.1	Mean variables of species used in the regressions	205
10.2	Pearson's product–moment correlation coefficients between body weight$^{1/3}$ (M) × craniobasal length (CBL), RPCA$^{1/2}$ × body weight$^{1/3}$, and RPCA$^{1/2}$ × craniobasal length	206
10.3	Regression statistics for linear regressions of craniobasal length on body weight$^{1/3}$, RPCA$^{1/2}$ on body weight$^{1/3}$, and RPCA$^{1/2}$ on craniobasal length	207
10.4	Review of previous published scaling studies relevant to the data presented in this paper	212
11.1	Specimens included in this study	219
11.2	Synonyms for masticatory muscle terminology used here	222
11.3	Values of physiological cross-sectional area for sixteen species of prosimians	226
11.4	Values of reduced physiological cross-sectional area for twelve species of prosimians	227
11.5	Muscle mass for sixteen species of prosimians	228
11.6	Fiber length for sixteen species of prosimians	229
11.7	Pinnation values, PCSA values relative to body mass, and RPCSA values relative to body mass	229
11.8	Results of reduced major axis regressions of muscle dimensions on body mass	231
12.1	Muscle fiber architecture variables and predicted differences between tree-gouging marmosets and a nongouging tamarin	250
12.2	Comparison of masseter and temporalis fiber architecture between tree-gouging marmosets and a non-gouging tamarin	252
14.1	Dietary properties of rabbit experimental foods	298
14.2	Comparison of size-adjusted means for load-resisting and force-generating elements between cohorts of 10 subjects each	303
14.3	Comparison of rabbit symphyseal mean biomineralization levels between cohorts of seven subjects each	303

14.4	Comparison of rabbit TMJ mean biomineralization levels between cohorts of eight subjects each	304
14.5	Comparison of size-adjusted means of load-resisting and force-generating structures between loading cohorts	309
14.6	DFA comparison of mouse TMJ biomineralization levels	310
14.7	DFA of mouse symphysis bone-density levels that are similar to patterns for the TMJ	311
14.8	Comparison of rabbit symphyseal mean mineral density levels between cohorts of three subjects each	313
14.9	Comparison of rabbit TMJ mean biomineralization levels between cohorts of three subjects each	314
15.1	Marsupial sample, body mass, dietary classification, and number of specimens	337
15.2	Regression equations for ACA ln corpus I_x and K versus ACA ln mandibular length	344
15.3	Correlation of ACA corpus I_x and K residuals with ACA diet proportions	345
15.4	Kruskal-Wallis One-Way ANOVA results contrasting ACA corpus I_x and K residuals between diet categories	346
15.5	Slopes of LS regression lines and correlation coefficients for standardized contrasts of corpus I_x and K residuals versus standardized contrasts of diet proportions	346
16.1	The six denominators used in shape variable comparisons	371
16.2	Morphological predictions used to examine how shape variable denominators influence functional hypothesis tests	373
16.3	The five-level ranking scale for assessing similarity among hypothesis test results	373
16.4	Percentage similarity of ranks based on statistical outcomes from hypothesis tests using various shape variable denominators.	374
17.1	Lemur species included in regressions	393
17.2	Comparisons between lemur species of toughness of kily parts eaten	397
17.3	Comparisons of tooth features between the two lemur species[a]	398
18.1	Convergence angles calculated for Fayum primates who belong in diverse taxonomic groups	414
18.2	Frontation angles for Fayum primates	414
20.1	Body size by geographic region in *H. habilis* and *H. erectus*	446
20.2	Dentognathic summary statistics for *H. habilis* and *H. erectus* (worldwide) and for *H. erectus* by region/locality	451
20.3	Individuals and chimera-used in body size-brain size scaling relationships	452
20.4	Regression results for the relationship between body size and cranial capacity	455
20.5	Results of ordinary least-squares regression of dentognathic height and breadth dimensions in fossil *Homo*	459

20.6	Results of ordinary least-squares regression of dental variables versus jaw size	462
20.7	Results of ordinary least-squares regression of dentognathic variables versus body size estimates	464
20.8	Results of ordinary least-squares regression of dentognathic size versus cranial capacity	467
20.9	Results of ordinary least-squares regression of dentognathic size and time (in Ma)	469
20.10	Samples and chimeras used in analyses.	474

Part I
Historical Perspective on Experimental Research in Biological Anthropology

Anthropologists have committed countless hours to understanding how the skull works. These endeavors have been fundamental in shaping biological anthropology over the generations. For the past 35 years, William Hylander has been a leader in this effort by furthering our understanding of primate skull function. The core of Hylander's contribution focuses on his rigorous attempts to document how muscles are recruited and how bones respond when alert primates chew and bite. These in vivo, experimental data provide the empirical baseline for numerous studies examining skull form, feeding behaviors, and craniofacial adaptations in living and fossil primates.

Schmitt, Lemelin, and Wall begin the volume with a historical review of William Hylander's contribution to experimental research in biological anthropology. They initially place Hylander's work in the broader context of the "new physical anthropology" championed by Washburn in the 1950s, which advocated an experimental comparative approach. Schmitt et al. proceed by discussing three of Hylander's key contributions and noting how each paper has fundamentally advanced biological anthropology. They conclude by discussing how Hylander's collective work will benefit the future of primate craniofacial research.

Chapter 1
Experimental Comparative Anatomy in Physical Anthropology: The Contributions of Dr. William L. Hylander to Studies of Skull Form and Function

Daniel Schmitt, Christine E. Wall, and Pierre Lemelin

Contents

1.1 Introduction	3
1.2 Experimental Comparative Anatomy in Physical Anthropology	4
1.3 The Experimental Approach and Contributions of Bill Hylander	7
1.3.1 Bone Mechanical Properties	8
1.3.2 Scaling in Craniofacial Structures	10
1.3.3 EMG of Mastication	10
1.4 Conclusion	11
References	12

1.1 Introduction

Two important events occurred during the 74th Annual Meeting of the American Association of Physical Anthropology (AAPA) held in Milwaukee, WI (USA), in the spring of 2005. On the one hand, it was the 75th anniversary of the AAPA and a special symposium was held discussing the scientific impact of its founders and early contributors. On the other hand, colleagues and friends of Professor William Hylander gathered to present their most recent work and pay tribute to Dr. Hylander's lifetime contribution to studies of skull functional morphology in primates. Although these two important events were coincidental, they both reminded us of a critical event in the history of physical anthropology: that is, the rise of what Washburn defined as the "New Physical Anthropology" (Washburn, 1951a, b), which advocated for the use of an experimental approach in physical anthropology.

D. Schmitt
Department of Evolutionary Anthropology, Duke University Box 90383, Durham, NC 27708, USA
e-mail: Daniel.Schmitt@duke.edu

Our goal in this chapter is to briefly review the historical context that led to Washburn's important, albeit at the time controversial, call for a comparative experimental anatomy and to show how that call was answered by the functional morphologists interested in the evolution of the skull and feeding apparatus of primates and other mammals. In particular, the contribution of Bill Hylander to this important area of research will be reviewed and some directions for the future, which build on Bill's pioneering work, will be discussed.

1.2 Experimental Comparative Anatomy in Physical Anthropology

In 1951, Sherwood Washburn published two seminal essays in which he explicitly urged physical anthropologists to adopt a laboratory-based methodology to address questions of human evolution (Washburn, 1951a, b). The view espoused in these papers is considered a paradigm shift from the typological approach that characterized physical anthropology at the time to a more multidisciplinary approach where experiments played a major role (Stini, 2005). Washburn's (1947; Mednick and Washburn, 1956) approach using laboratory-based data to resolve conflicts in scenarios of human evolution was not always well received by his peers at the time. Indeed, one famous contemporary of Washburn questioned him about the validity of using rats to understand the human phenomenon (Washburn, 1983; Marks, 2000). Moreover, many of his peers saw the experimental method as a major threat because "they thought it was destroying the evidence" (DeVore, 1992:417).

At the time Washburn made the call for a "modern, experimental, comparative anatomy" (Washburn, 1951a:67), there was little activity of that type in the field of physical anthropology. The pioneering studies of Hildebrand (1931) and Elftman and Manter (1935) represent some of the few laboratory-based studies of mastication and locomotion, respectively, completed at the time. Despite the scarcity of such studies in 1951, Washburn foresaw the relevance of these types of data for the evaluation of functional and evolutionary hypotheses in physical anthropology. In addition, he understood the burgeoning technological revolution that would make the experimental approach possible.

The term "experimental comparative anatomy" is cumbersome and on the surface seems too vague to have any meaning. But what Washburn meant is very clear. He meant to expand the available tools for assessing the functional significance of a given morphological trait by controlled, laboratory-based testing of hypotheses derived from comparative anatomy. He wanted anthropologists to go beyond the "opinions," theoretical models, and simple correlations that dominated comparative anatomy at the time.

By the late 1960s and 1970s, there was a rapidly growing trend in the use of laboratory-based methods to study the functional morphology of primates. Much of this was directed toward a better understanding of the primate postcranium and began incorporating the newest techniques available (see Fleagle, 1979; Jouffroy, 1989; Churchill and Schmitt, 2003; Schmitt, 2003; Lemelin and Schmitt, 2007 for historical accounts). But the focus of this chapter is the parallel development of laboratory-based studies of primate craniofacial biology.

In order to fully explore the ways in which laboratory-based studies have influenced our understanding of primate craniofacial anatomy and evolution, it is worth beginning by discussing the foundations of experimental comparative anatomy. In a seminal paper in 1977, Rich Kay and Matt Cartmill formalized the methodological principles that had guided studies of functional morphology for many decades. They laid out four clear steps for making inferences about behavior in fossil animals. Kay and Cartmill (1977) stressed that any trait which we wished to use as a surrogate for a behavior had to be present in extant animals and its function had to be known. But they did not discuss in detail how to determine the function of a trait.

Treated simply, at least two comparative approaches can be used to assess function based on studying morphological traits: (1) an approach capitalizing on convergence or divergence to relate organismal form with presumed functions (Brooks and McLennan, 1991) and (2) an "argument from design" that applies theoretical principles of engineering and design to interpret organismal function based on form (Rudwick, 1964; Lauder, 1996). Both approaches have their strengths and flaws that have been reviewed elsewhere (Bock and von Walhert, 1965; Bock, 1977; Fleagle, 1979; Homberger, 1988; Lauder, 1995, 1996), so we will just briefly touch on them here. The central point here is that conclusions derived from traditional comparative anatomy or engineering-style "argument from design" studies must be viewed not as end-results, but rather as hypotheses to be tested. In many instances, those tests are best accomplished in the laboratory.

Laboratory-based hypothesis testing is exactly what Washburn had in mind and what Hylander and his contemporaries have been pursuing for the past 30 years. But they represent the leaders of a small group within physical anthropology. In spite of the theoretical strengths and observed success of an experimental comparative approach, few physical anthropologists test their functional models with experimental data. There are a lot of reasons for the lack of rigorous testing using laboratory data. Many anthropologists misunderstand how the experimental approach can be used to test functional hypotheses. Too often, criticisms are made about small sample sizes, unnatural laboratory conditions, and the highly technical aspect of the methods used in the laboratory. These concerns inhibit the willingness of physical anthropologists to collect experimental data and the acceptance of such data when they are presented. In the absence of experimental data, confirmation of a functional model can only be achieved via traditional comparative anatomy (e.g., the prediction that long legs are mechanically critical for leaping primates is confirmed by the observation that other leaping animals have long legs). This mode of checking functional models may lead to correct conclusions; but as Bock (1977), Homberger (1988), and others have noted, this is not always the case. Lauder (1996:56) points out that such conclusions are based on untested assumptions and that:

> ...in our desire to draw conclusions about biological design and to support theoretical views of how organism are built, we have been too willing to make assumptions about the relationship between structure and mechanical function...[and]... we have not often

conducted the mechanical and performance tests needed to assess the average quality of organismal design.

Similarly, Fleagle (1979:316) noted that "Regardless of how mechanically plausible and convincing [functional] explanations may be, they can be rigorously tested only by in vivo studies". The work of Bill Hylander and his colleagues represents an ideal combination of basic design principles (e.g., beam theory to describe the jaw) and experimental analyses. His approach has led to a better understanding of both structure–function relationships and the evolution of primate craniofacial function and is a model for how to proceed in testing morphological hypotheses with laboratory data.

There are three possible outcomes when using an experimental approach to determine structure–function relationships. The first possibility is that a given hypothesis developed from comparative data or engineering principles will be supported. This is a common, although not universal, outcome.

A second possible outcome is that a mechanical hypothesis relating a specific structure to a particular function may be found to be in error; however, the underlying correlation between structure and behavior still exists. In this case, experimental studies can yield a better understanding of the underlying causal relationship between a trait and its function, and may lead to novel areas for investigation. One good example of how laboratory studies may clarify or change conclusions based on the measurements of morphology is the debate over adaptations to tree gouging in primates. Prior to any laboratory work, researchers measuring jaws and teeth, especially the highly modified incisors of many gouging primates, argued that tree gouging induced large forces on the masticatory system (e.g., Szalay and Seligsohn, 1977; Rosenberger, 1992; Dumont, 1997; Spencer, 1999). This conclusion seems logical, but in vivo work on marmosets has shown that this is not the case and that, in fact, the forces generated during gouging can be seen as being relatively low (Vinyard et al., 2001, in press). This does not invalidate the correlation between chisel-like incisors and gouging behavior in callitrichids. But the finding of relatively low bite forces during gouging in callitrichids is unexpected and it opens up new areas of research into structural modifications of the bones and muscles of the skull that may not be related to force production but instead allow for the use of large gapes in primates (Williams et al., 2002; Taylor and Vinyard, 2004; Perry, 2006).

A final potential outcome is that experimental data change our perception of both structure–function correlations and causal relationships. Understanding the mechanics of phase II during the power stroke of mastication, which began with Hiiemae and Kay (1972, 1974a, b) and was further investigated by Hylander and Crompton in the late 1980s (Hylander and Crompton, 1986; Hylander et al., 1987), represents a perfect example. Prior to the work of Hylander and Crompton, it was assumed that powerful grinding occurs during phase II. This is a heavily embedded assumption in comparative studies of the primate molar dentition. For example, certain occlusal facets on the molar teeth are identified as "phase II facets," and pitting and abrasions observed on the occlusal surfaces have been linked to forceful contact between the teeth and food during phase II (Kay, 1977, 1987; Grine, 1986). More recently, in a combined video and electromyographic study designed to confirm Hylander's earlier work, Wall et al. (2006) found in baboons that there is negligible force produced by the major jaw adductors during phase II and that mandibular movements were minimal during phase II as compared to phase I. Their results showed that phase II movement is likely trivial for breaking down food. Instead, most food breakdown on the phase II facets probably takes place late in phase I movement, in association with crushing of the food object (see also Hiiemae, 1984; Teaford and Walker, 1984; Teaford, 1985). Additional work to characterize phase II jaw movements in other primate species is underway, but the experimental data clearly suggest the need for an alternative functional explanation for the structure and microwear patterns on the phase II facets of primates.

1.3 The Experimental Approach and Contributions of Bill Hylander

Clearly, the experimental approach is a critical tool for physical anthropologists. Despite the obvious relevance of this approach to the study of primate craniofacial biomechanics, little laboratory-based research was conducted before Bill Hylander began his career. Bill was a graduate student at the University of Chicago in the 1960s. He returned to graduate school after practicing dentistry for several years. His goal was to combine his knowledge of the functional anatomy of the teeth and jaws with the study of the evolution of the primate skull. In the 1970s, he pioneered the experimental approach to understanding the functional anatomy of the skull in primates.

As a graduate student, Bill was interested in developing biomechanical models explaining craniofacial form and function that could ultimately be tested with experimental data. In his thesis, Bill evaluated competing hypotheses to explain the functional anatomy of the Eskimo skull (Hylander, 1972). Specifically, he wanted to know which of the features seen in the cranium and mandible were related to the aspects of tooth use and diet and which could be explained as cold adaptations. This work was eventually published in an edited volume (Hylander, 1977a) in the same year that Bill published his first in vivo bone strain paper (Hylander, 1977b). Bill developed a biomechanical model to explain such features as a robust mandible, high temporal lines, and large bicondylar dimensions as structures that aid in the generation and dissipation of high vertical bite forces.

In another study that demonstrates his knack for critically evaluating competing hypotheses, Bill published a seminal analysis of the human mandible as a lever system (Hylander, 1975). A number of workers had claimed that the jaw functioned as a mechanical link, and thus generated no reaction force at the temporomandibular joint (TMJ). Bill showed that the mandible does function as a third-class lever, and that many aspects of mandibular morphology can be explained in this context.

Since then Bill has made an enormous and fundamental contribution to studies of primate skulls and teeth. Bill has published over 75 research articles, including papers in many edited volumes. A Science Citation search indicates that since 1980, Bill's work has been cited more than 933 times in published articles. His research has been continuously funded by the National Science Foundation, National Institutes of Health, and private foundations for more than 30 years.

Although Bill is probably best known for his work documenting and interpreting the functional significance of in vivo patterns of bone strain in the primate skull (Hylander, 1977b, 1979a, b, c, 1984; Hylander and Johnson, 1992, 1997a, b; Hylander and Ravosa, 1992; Hylander et al., 1991a, b, 1998; Ravosa et al., 2000; Ross and Hylander, 1996 and many others), he has made seminal contributions in the areas of theoretical jaw mechanics (Hylander, 1975b), jaw kinematics (Hylander, 1978; Hylander and Crompton, 1986; Hylander et al., 1987), bone biology (Bouvier and Hylander, 1981, 1996; Hylander et al., 1991b; Hylander and Johnson, 2002), muscle function (Hylander and Johnson, 1985, 1994; Hylander et al., 2000, 2005), and comparative morphometrics (Hylander, 1972, 1975a, 1977, 1985, 1988; Kay and Hylander, 1978). Along the way, he has trained a number of graduate and post-doctoral students who have published extensively in these areas.

Because of the massive volume of Bill's work, it is impossible to review it entirely in this short contribution. Furthermore, we feel that simply reiterating all that Bill has done cannot do justice to how influential his thinking has been on the field of primate craniofacial function and biology. As an alternative approach, we consider what we think are three key articles and discuss how they have molded thinking on mammalian craniofacial biology and the evolution of the primate masticatory system. Other authors might have chosen different articles, but to our way of thinking the work we describe below exemplifies not only the scope of Bill's career but also raise important issues that will remain at the forefront of research on craniofacial functional anatomy for decades to come. We will discuss the implications of Bill's pioneering studies on understanding bone mechanical properties, the scaling of craniofacial structures in primates, and muscle activity during mastication.

1.3.1 Bone Mechanical Properties

One area that Bill focused on is the mechanical determinants of bone size and shape. In 2002, he published a paper with his long-time associate Kirk Johnson, which summarizes his research on this topic and places these results within the context of the functional adaptation/optimal strain environment model of bone remodeling (Hylander and Johnson, 2002; see also Bassett, 1968; Hylander and Johnson, 1997a, b; Lanyon and Rubin, 1985; Pauwels, 1980; Rubin, 1984). Briefly, the phrase "functional adaptation" describes the supposition that normal, day-to-day loading conditions play an important role in maintaining bone mass and geometry (e.g., Hert et al., 1971; Bouvier and Hylander, 1981; Lanyon and Rubin, 1985;

Carter, 1987). Lanyon and Rubin, (1985) argue that functional adaptation maintains bone strain magnitudes within a physiological range of values (usually understood to be between 2,000 and 3,000 $\mu\varepsilon$), in order to maintain this optimal strain environment. Under this model, bone models or remodels in response to altered loads in order to maintain the optimal strain environment. One prediction based on the observation of functional adaptation to a narrow range of physiological strains in long bones is that bones should be optimized for resisting routine, physiological loads and should, by definition, exhibit a maximum of strength with a minimum of material for load-bearing purposes. This putative relationship is widely accepted among physical anthropologists (see Lanyon and Rubin, 1985 and Churchill and Schmitt, 2003 for a review of engineering models and bone strength in general biology and anthropology, respectively), but has rarely been tested experimentally. In several studies where it has been tested in the experimentally, significant questions have arisen (Lanyon and Rubin, 1985; Demes et al., 2001).

Hylander and Johnson (2002) bring an enormous amount of data on strain patterns within the jaws and face of primates to test the predictions of the functional adaptation hypothesis. Their evidence unequivocally shows that functional adaptation is not the only mechanism acting to determine the size and shape of bones in the skull. Although some regions (e.g., the anterior root of the zygoma) probably respond to the functional loads incurred during mastication and other feeding behaviors, many bony regions appear grossly over-designed or under-designed for their purported load-resisting function during feeding. The problem that Hylander and Johnson (2002) lay out with respect to the distribution of bone and bone strains is that it is not possible to predict strain from bone geometry (see also Daegling, 1993). The good news is that high strain regions are generally correlated with strong bone (either due to type, amount, or distribution). The bad news is that it is difficult to predict high strain regions simply by proximity to muscle or teeth. Furthermore, low strains are a bad indicator of underlying bone geometry. Several implications follow from these profound findings. Most critically for physical anthropologists, the data suggest that bone size and shape within the skull is not "largely or exclusively determined by or associated with routine and habitual forces associated with mastication, incision, or isometric biting"... and that "reconstructing the masticatory behavior and biomechanics of primates from the fossilized remains of the craniofacial skeleton is extremely problematic, particularly when done in the complete absence of a detailed understanding of the biomechanical environment of the craniofacial region of living primates" (Hylander and Johnson, 2002:43–44). Hylander and Johnson (2002) go on to cite numerous studies published since 1955, which have made this unwarranted assumption. It would seem that though Washburn's call for an experimental approach was heeded by a number of anthropologists that went on to collect in vivo data, those data are not always incorporated into comparative studies.

The importance of the results presented in Hylander and Johnson (2002) is not just about the inability to link bone structure and function in the absence of experimental data. They also provide a testable alternative to explain bone deposition and maintenance in regions where strains do not match bone mass or geometry. They reason that heredity must play an important role in maintaining bone mass in such regions as the browridge.

1.3.2 Scaling in Craniofacial Structures

The second paper we want to review is entitled "Mandibular Function and Biomechanical Stress and Scaling," and was published by Bill Hylander over 20 years ago (Hylander, 1985). This paper is important because it lays out a very explicit methodology for examining the relationship between mechanical loads and the form of the masticatory apparatus across several primate species. Bill makes the important observation that animals do not walk on their faces, and that therefore gravity or a surrogate for gravity, such as body mass, is not a directly relevant independent variable in the functional analyses of size and shape of the masticatory apparatus. There is no a priori reason to predict geometric similarity of masticatory structures. There is, however, ample experimental evidence that predicts strong mechanical influences on the size and shape of masticatory structures. Hylander's (1985) call to develop mechanical predictions of how various features of the masticatory apparatus should scale had a strong influence on studies of craniofacial biomechanics (Daegling, 1993; Daegling and McGraw, 2001; Ravosa, 1996a, b; Ross, 2001; Vinyard and Ravosa, 1998). Here Hylander develops independent variables with specific mechanical relevance to the question at hand, rather than relying on the ones that may be less relevant but are easy to measure (e.g., mandibular length is not always the most relevant mechanical variable).

Moreover, Hylander (1985) advice is much more far-reaching. He pointed out the need for more data on muscle cross-sectional area and to develop explicit models of bone size and scaling under masticatory stress regimes. More recently, Lucas (2004) has developed a set of scaling predictions based on food material properties. Future work should focus on testing various aspects of Lucas's (2004) model, and should incorporate appropriate independent variables into analyses. For example, analysis of moment arms may reveal interesting results about bite force generation when scaled relative to some other estimate of muscle force (e.g., mass or cross-sectional area). These muscle data are now available for many primate species (Anapol et al., Chapter 10; Perry and Wall, Chapter 11) and are relatively easy to collect for others that remain undocumented.

1.3.3 EMG of Mastication

The third paper, Hylander et al. (2005), summarizes much of what we know about the activity of the temporalis muscle during mastication in primates and represents close to 15 years of EMG data collection by Hylander and his colleagues. This paper, in addition to Hylander and Johnson (1994), Hylander et al. (2000), Vinyard et al. (2005, 2006), and Wall et al. (2006, Chapter 6), provide up-to-date information on timing differences and working-side versus balancing-side muscle activity for a wide array of primates and treeshrews as an outgroup species. One important focus of Bill's EMG work (Hylander and Johnson, 1994; Hylander et al., 2000, 2004) has been to evaluate the link between the late peak in activity of the balancing-side deep masseter and wishboning of the mandibular symphysis in anthropoids. Research is underway to extend our understanding of interspecific variation in EMG patterns both by studying other primate species (e.g., *Propithecus verreauxi*, *Eulemur fulvus*, *Saimiri sciureus*, and *Cebus apella*) and looking at specific regions within the temporalis, medial pterygoid, and masseter muscles in representative species. One basic comparison that would be useful for understanding the relationship between EMG, muscle force, and interspecific variation is to compare variability in EMG variables during incision, puncture-crushing, and rhythmic chewing (i.e., tooth-tooth contact) cycles to look for patterns with mechanical significance. For example, do animals avoid unpredictable loads of the symphysis and corpus during the early part of food breakdown (puncture-crushing) as compared to the later part (rhythmic chewing)? If they do, we would expect animals to avoid high peaks in favor of multiple, lower-amplitude peaks.

1.4 Conclusion

Bill's career is hardly at an end. He is an emeritus professor and is currently working on generating testable hypotheses about important features of the primate masticatory apparatus. One of his current projects is to quantify the relationship between canine length, mandible length, maximum passive gape, and muscular development in catarrhine primates. Hylander and Vinyard (2006) find that short canines are correlated with small gape. They suggest that canine reduction, such as that seen in the earliest hominins, is a functional requirement to minimize canine interference associated with decreased gape.

In this chapter, we tried to highlight the role that Bill, his colleagues, and his students have played in advancing our understanding of the biomechanics and evolution of craniofacial structures in primates. The field of physical anthropology owes Bill a tremendous debt of gratitude not only for the specific data he generated concerning primate skull anatomy but also for the model he provides as an experimentalist and for the platform from which much current and future work is and will be built. It is often tedious, difficult, and time-consuming to collect in vivo data, especially on primates. Bill Hylander spent more than 30 years collecting the bone strain, kinematic, and EMG data that have so enriched our understanding of the function, biology, and evolution of the primate craniofacial region. This work represents, we hope, only a beginning for our field. The students that Bill trained and, in turn, the students that they will train need to keep on building on the remarkable dataset that Bill created, because it is the only way to fully understand craniofacial morphology in primates and other mammals.

Acknowledgments We want to thank Bill Hylander for his mentorship and friendship to all three of us over the past 15 years. We thank Chris Vinyard and Matt Ravosa for their insightful comments on this manuscript. This manuscript also benefited from many helpful conversations with Matt Cartmill.

References

Bassett, C. A. L. (1968). Biological significance of piezoelectricity. *Calcif. Tissue Res.* 1:252–272.
Bock, W. J. (1977). Adaptation and the comparative method. In: Hecht, M. K., Goody, P. G., and Hecht, B. M. (eds.), *Major Patterns in Vertebrate Evolution*. Plenum Press, New York, pp. 57–82.
Bock, W. J., and von Walhert, G. (1965). Adaptation and the form-function complex. *Evol.* 19:269–299.
Bouvier, M., and Hylander, W. L. (1981). Effect of bone strain on cortical bone structure in macaques (*Macaca mulatta*). *J. Morphol.* 167:1–12.
Bouvier, M., and Hylander, W. L. (1996). The function of secondary osteonal bone: mechanical or metabolic? *Arch. Oral Biol.* 41:941–950.
Brooks, D. R., and McLennan, D. A. (1991). *Phylogeny, Ecology, and Behavior*, Chicago, The University of Chicago Press.
Carter, D. R. (1987). Mechanical loading history and skeletal biology. *J. Biomech.* 20:1095–1109.
Churchill, S. E., and Schmitt, D. (2003). Biomechanics in paleoanthropology: engineering and experimental approaches to the investigation of behavioral evolution in the genus *Homo*. In: Harcourt, C., and Crompton, R. (eds.), *New Perspectives in Primate Evolution and Behavior*. Linnaean Society, London, pp. 59–90.
Daegling, D. J. (1993) The relationship of *in vivo* bone strain to mandibular corpus morphology in *Macaca fascicularis*. *J. Hum. Evol.* 25:247–269.
Daegling, D. J., and McGraw, W. S. (2001). Feeding, diet, and jaw form in West African *Colobus* and *Procolobus*. *Int. J. Primatol.* 22:1033–1055.
Demes, B., Qin, Y. -X., Stern, J. T., Jr., Larson, S. G., and Rubin, C. T. (2001). Patterns of strain in the macaque tibia during functional activity. *Am. J. Phys. Anthropol.* 116:257–265.
DeVore, I. (1992). An interview with Sherwood Washburn. *Curr. Anthropol.* 33:411–423.
Dumont, E. R. (1997) Cranial shape in fruit, nectar, and exudate feeders: implications for interpreting the fossil record. *Am. J. Phys. Anthropol.* 102:187–202.
Elftman, H., and Manter, J. (1935). Chimpanzee and human feet in bipedal walking. *Am. J. Phys. Anthropol.* 20:69–79.
Fleagle, J. G. (1979). Primate positional behavior and anatomy: naturalistic and experimental approaches. In: Morbeck, M. E., Preuschoft, H., and Gomberg, N. (eds.), *Environment, Behavior, and Morphology: Dynamic Interactions in Primates*. Gustav Fischer, New York, pp. 313–326.
Grine, F. E. (1986). Dental evidence for dietary differences in *Australopithecus* and *Paranthropus*: A quantitative analysis of permanent molar microwear. *J. Hum. Evol.* 15:783–822.
Hert, J., Liskova, M., and Landa, J. (1971). Reaction of bone to mechanical stimuli. 1. Continuous and intermittent loading of tibia in rabbit. *Folia Morphol. (Praha)* 19:290–300.
Hiiemae, K. M. (1984). Functional aspects of primate jaw morphology. In: Chivers, D. J., Wood, B. A., and Bilsborough, A. (eds.), *Food Acquisition and Processing in Nonhuman Primates*. Academic Press, London, pp. 359–281.
Hiiemae, K. M., and Kay, R. F. (1972). Trends in the evolution of primate mastication. *Nature* 240:486–487.
Hildebrand, G. Y. (1931). Studies in the masticatory movements of the lower jaw. *Scand. Arch. Physiol.* Suppl. 61:1–190.
Homberger, D. G. (1988). Models and tests in functional morphology: The significance of description and integration. *Am. Zool.* 28:217–229.

Hylander, W. L. (1972). The Adaptive Significance of Eskimo Craniofacial Morphology. Unpublished Ph. D. thesis, University of Chicago.

Hylander, W. L. (1975a). Incisor size and diet in anthropoids with special reference to Cercopithecidae. *Science* 189:1095–1098.

Hylander, W. L. (1975b). The human mandible: Lever or link? *Am. J. Phys. Anthropol.* 43:227–242.

Hylander, W. L. (1977a). The adaptive significance of Eskimo craniofacial morphology. In: Dahlberg, A. A., and Graber, T. M. (eds.), *Orofacial Growth and Development*. Mouton, The Hague, pp. 129–169.

Hylander, W. L. (1977b). In vivo bone strain in the mandible of *Galago crassicaudatus*. *Am. J. Phys. Anthropol.* 46:309–326.

Hylander, W. L. (1978). Incisal bite force direction in humans and the functional significance of mammalian mandibular translation. *Am. J. Phys. Anthropol.* 48:1–8.

Hylander, W. L. (1979a) The functional significance of primate mandibular form. *J. Morphol.* 160:223–240.

Hylander, W. L. (1979b). Mandibular function in *Galago crassicaudatus* and *Macaca fascicularis*: An *in vivo* approach to stress analysis of the mandible. *J. Morphol.* 159:253–296.

Hylander, W. L. (1979c). An experimental analysis of temporomandibular joint reaction force in macaques. *Am. J. Phys. Anthropol.* 51:433–456.

Hylander, W. L. (1984). Stress and strain in the mandibular symphysis of primates: A test of competing hypotheses. *Am. J. Phys. Anthropol.* 64:1–46.

Hylander, W. L. (1985). Mandibular function and biomechanical stress and scaling. *Am. Zool.* 25:315–330.

Hylander, W. L. (1988). Implications of *in vivo* experiments for interpreting the functional significance of "robust" australopithecine jaws. In: Grine, F. E. (ed.), *Evolutionary History of the "Robust" Australopithecines*, Aldine de Gruyter, New York, U.S.A. pp. 55–83.

Hylander, W. L., and Crompton, A. W. (1986). Jaw movements and patterns of mandibular bone strain during mastication in the monkey *Macaca fascicularis*. *Arch. Oral Biol.* 31:841–848.

Hylander, W. L., and Johnson, K. R. (1985). Temporalis and masseter muscle function during incision in macaques and humans. *Int. J. Primatol.* 6:289–322.

Hylander, W. L., and Ravosa, M. J. (1992). An analysis of the supraorbital region of primates: a morphometric and experimental approach. In: Smith, P., and Tchernov, E. (eds.), *Structure, Function, and Evolution of Teeth*. Freund, Tel Aviv, pp. 233–255.

Hylander, W. L., and Johnson, K. R. (1992). Strain gradients in the craniofacial region of primates. In: Davidovitch, Z. (ed.), *The Biological Mechanisms of Tooth Movement and Craniofacial Adaptation*. Ohio State University College of Dentistry, Columbus, Ohio, U.S.A., pp. 559–569.

Hylander, W. L., and Johnson, K. R. (1994). Jaw muscle function and wishboning of the mandible during mastication in macaques and baboons. *Am. J. Phys. Anthropol.* 94:523–547.

Hylander, W. L., and Johnson, K. R. (1997a). *In vivo* bone strain patterns in the zygomatic arch of macaques and the significance of these patterns for functional interpretations of cranial form. *Am. J. Phys. Anthropol.* 102:203–232.

Hylander, W. L., and Johnson, K. R. (1997b). *In vivo* bone strain patterns in the craniofacial region of primates. In: McNeill, C., Hannam, A., and Hatcher, D. (eds.), *Occlusion: Science and Practice*. Quintessence Publishing Co., Chicago, Illinois, U.S.A., pp. 165–178.

Hylander, W. L., and Johnson, K. R. (2002). Functional morphology and *in vivo* bone strain patterns in the craniofacial region of primates: Beware of biomechanical stories about fossil bones. In: Plavcan, J. M., Kay, R. F., Jungers, W. L., and van Schaik, C. P. (eds.), *Reconstructing Behavior in the Primate Fossil Record*. Kluwer Academic/Plenum Publishers, New York, pp. 43–72.

Hylander, W. L., and Vinyard, C. J. (2006). The evolutionary significance of canine reduction in hominins: Functional links between jaw mechanics and canine size. *Am. J. Phys. Anthropol.* 129:S42:107.

Hylander, W. L., Johnson, K. R., and Crompton, A. W. (1987). Loading patterns and jaw movements during mastication in *Macaca fascicularis*: A bone-strain, electromyographic, and cineradiographic analysis. *Am. J. Phys. Anthropol.* 72:287–314.

Hylander, W. L., Picq, P. G., and Johnson, K. R. (1991a). Masticatory stress hypotheses and the supraorbital region of primates. *Am. J. Phys. Anthropol.* 86:1–36.

Hylander, W. L., Rubin, C. T., Bain, S. D., and Johnson, K. R. (1991b). Correlation between haversian remodeling and strain magnitude in the baboon face. *J. Dent. Res.* 70:360.

Hylander, W. L., Ravosa, M. J., Ross, C. F., and Johnson, K. R. (1998). Mandibular corpus strain in Primates: Further evidence for a functional link between symphyseal fusion and jaw-adductor muscle force. *Am. J. Phys. Anthropol.* 107:257–271.

Hylander, W. L., Ravosa, M. J., Ross, C. F., Wall, C. E., and Johnson, K. R. (2000). Symphyseal fusion and jaw-adductor muscle force: An EMG study. *Am. J. Phys. Anthropol.* 112:469–492.

Hylander, W. L., Vinyard, C. J., Wall, C. E., Williams, S. H., and Johnson, K. R. (2004). Jaw adductor force and symphyseal fusion. In: Anapol, F., German, R. Z., and Jablonski, N., (eds.), *Shaping Primate Evolution*. Cambridge University Press, Cambridge, pp. 229–257.

Hylander, W. L., Wall, C. E., Vinyard, C. J., Ross, C. F., Ravosa, M. J., Williams, S. H., and Johnson, K. R. (2005). Temporalis function in anthropoids and strepsirrhines: An EMG study. *Am. J. Phys. Anthropol.* 125:35–56.

Jouffroy, F. K. (1989). Quantitative and experimental approaches to primate locomotion: A review of recent advances. In: Seth, P., and Seth, S. (eds.), *Perspectives in Primate Biology*. Today and Tomorrow's Printers and Publishers, New Delhi, pp. 47–108.

Kay, R. F. (1977). The evolution of molar occlusion in the Cercopithecidae and early catarrhines. *Am. J. Phys. Anthropol.* 46:327–352.

Kay, R. F. (1987). Analysis of primate dental microwear using image processing techniques. *Scanning Microscop.* 1:657–662.

Kay, R. F., and Cartmill, M. (1977). Cranial morphology and adaptations of *Palaeochthon nacimenti* and other Paromomyidae (Plesiadapoidea, ?Primates), with a description of a new genus and species. *J. Hum. Evol.* 6:19–35.

Kay, R. F., and Hiiemae, K. M. (1974a). Jaw movement and tooth use in Recent and fossil primates. *Am. J. Phys. Anthropol.* 40:227–256.

Kay, R. F., and Hiiemae, K. M. (1974b). Mastication in *Galago crassicaudatus*: a cinefluorographic and occlusal study. In: Martin, R. D., Doyle, G. A., and Walker, A. C. (eds.), *Prosimian Biology*. University of Pittsburgh Press, Pittsburgh, Penn., U.S.A., pp. 501–530.

Kay, R. F., and Hylander, W. L. (1978). The dental structure of mammalian folivores with special reference to primates and phalangeroids (Marsupialia). In: Montgomery, G. (ed.), *The Ecology of Arboreal Folivores*, Smithsonian Press, Washington, D.C., U.S.A., pp. 173–191.

Lanyon, L. E., and Rubin, C. T. (1985). Functional adaptation in skeletal structures. In: Hildebrand, M., Bramble, D. M., Liem, K. F., and Wake, D. B. (eds.), *Functional Vertebrate Morphology*. Harvard University Press, Cambridge, Mass., pp. 1–25.

Lanyon, L. E., and Rubin, C. T. (1985). Functional adaptation in skeletal structures. In: Hildebrand, M., Bramble, D. M., Liem, K. F., and Wake, D. B. (eds.), *Functional Vertebrate Morphology*. Harvard University Press, Cambridge, Mass., pp. 1–25.

Lauder, G. V. (1995). On the inference of function from structure. In: Thomasen, J. J. (ed.), *Functional Morphology in Vertebrate Paleontology*. Cambridge University Press, Cambridge, pp. 1–18.

Lauder, G. V. (1996). The argument from design. In: Lauder, G. V., and Rose, M. R. (eds.), *Adaptation*. Academic Press, New York, pp. 55–92.

Lemelin, P., and Schmitt, D. (2007). Origins of grasping and locomotor adaptations in primates: comparative and experimental approaches using an opossum model. In: Ravosa, M. J., and Dagosto, M. (eds.), *Primate Origins: Adaptations and Evolution*. Kluwer Academic Publishers, New York, pp. 329–380.

Lucas, P. W. (2004). *Dental Functional Morphology*. Cambridge University Press, Cambridge, U.K.

Marks, J. (2000). Sherwood Washburn, 1911–2000. *Evol. Anthropol.* 9:225–226.

Melnick, L. W., and Washburn, S. L. (1956). The role of sutures in the growth of the braincase of the infant pig. *Am. J. Phys. Anthropol.* 14:175–191.

Pauwels, F. (1980). *Biomechanics of the Locomotor Apparatus*. Springer-Verlag, Berlin.

Perry, J. M. G. (2006). Anatomy and scaling of prosimian chewing muscles. The FASEB Journal 20(5-II). p. A846.

Ravosa, M. J. (1996a). Jaw morphology and function in living and fossil Old World monkeys. *Int. J. Primatol.* 17:909–932.

Ravosa, M. J. (1996b). Mandibular form and function in North American and European Adapidae and Omomyidae. *J. Morphol.* 229:171–190.

Ravosa, M. J., Johnson, K. R., and Hylander, W. L. (2000). Strain in the galago facial skull. *J. Morphol.* 245:51–66.

Rosenberger, A. L. (1992). Evolution of feeding niches in new world monkeys. *Am. J. Phys. Anthropol.* 88:525–562.

Ross, C. F. (2001). In vivo function of the craniofacial haft: the interorbital pillar. *Am. J. Phys. Anthropol.* 116:108–121.

Ross, C. F., and Hylander, W. L. (1996). In vivo and in vitro bone strain in the owl monkey circumorbital region and the function of the post-orbital septum. *Am. J. Phys. Anthropol.* 101:183–215.

Rubin, C. T. (1984). Skeletal strain and the functional significance of bone architecture. *Calcif. Tissue Int.* 36:S11–S18.

Rudwick, M. J. S. (1964). The inference of structure from function in fossils. *Brit. J. Phil. Sci.* 15:27–40.

Schmitt, D. (2003). Insights into the evolution of human bipedalism from experimental studies of humans and other primates. *J. Exp. Biol.* 206:1437–1448.

Spencer, M. A. (1999). Constraints on masticatory system evolution in anthropoid primates. *Am. J. Phys. Anthropol.* 108:483–506.

Stini, W. A. (2005). Sherwood Washburn and "The New Physical Anthropology." *Am. J. Phys. Anthropol.* Suppl. 40:197–198.

Szalay, F. S., and Seligsohn, D. (1977). Why did the strepsirhine tooth comb evolve. *Folia Primatol.* 27:75–82.

Taylor, A. R., and Vinyard, C. J. (2004). Comparative analysis of masseter fiber architecture in tree-gouging (*Callithrix jacchus*) and nongouging (*Saguinus oedipus*) callitrichids. *J. Morphol.* 261:276–285.

Teaford, M. F. (1985). Molar microwear and diet in the genus *Cebus*. *Am. J. Phys. Anthropol.* 66:363–370.

Teaford, M. F., and Walker, A. (1984). Quantitative differences in dental microwear between primate species with different diets and a comment on the presumed diet of *Sivapithecus*. *Am. J. Phys. Anthorpol.* 64:191–200.

Vinyard, C. J., and Ravosa, M. J. (1998). Ontogeny, function and scaling of the mandibular symphysis in Papionin primates. *J. Morphol.* 235:157–175.

Vinyard, C. J., Wall, C. E., Williams, S. H., Schmitt, D., and Hylander, W. L. (2001). A preliminary report on the jaw mechanics during tree gouging in common marmosets (*Callithrix jacchus*). In: Brook, A. (ed.), *Dental Morphology 2001*. Sheffield Academic Press Ltd., Sheffield, U.K. pp. 283–298.

Vinyard, C. J., Williams, S. H., Wall, C. E., Johnson, K. R., and Hylander, W. L. (2005). Jaw-muscle electromyography during chewing in Belanger's treeshrews (*Tupaia belangeri*). *Am. J. Phys. Anthropol.* 127:26–45.

Vinyard, C. J., Wall, C. E., Williams, S. H., Mork, A. L., Brooke, A. G., De Oliveira Melo, L. C., Valenca-Montenegro, M. M., Valle, Y. B. M., Monteiro da Cruz, M. A. O., Lucas, P. W., Schmitt, D., Taylor, A. B., and Hylander, W. L. (in press). The evolutionary morphology of tree gouging in marmosets. In: Davis, L. C., Ford, S. M., and Porter L. M. (eds.), *The Smallest Anthropoids: The Marmoset/Callimico Radiation*. Springer, New York.

Vinyard, C. J., Wall, C. E., Williams, S. H., Johnson, K. R., and Hylander, W. L. (2006). Masseter electromyography during chewing in ring-tailed lemurs (*Lemur catta*). *Am. J. Phys. Anthropol.* 130:85–95.

Wall, C. E., Vinyard, C. J., Williams, S. H., Johnson, K. R., and Hylander, W. L. (2006). Phase II occlusion in relation to jaw movement and masseter muscle recruitment during chewing in *Papio anubis*. *Am. J. Phys. Anthropol.* 129:215–224.

Washburn, S. L. (1947). The relation of the temporal muscle to the form of the skull. *Anat. Rec.* 99:239–248.
Washburn, S. L. (1951a). The analysis of primate evolution with particular reference to the origin of man. *Cold Spring Harb Symp Quant Biol* 15:67–77.
Washburn, S. L. (1951b). The new physical anthropology. *Trans. NY Acad. Sci. (Series II)* 13:298–304.
Washburn, S. L. (1983). Evolution of a teacher. *Ann. Rev. Anthropol.* 12:1–24.
Williams, S. H., Wall, C. E., Vinyard, C. J., and Hylander, W. L. (2002). A biomechanical analysis of skull form in gum-harvesting galagids. *Folia Primatol.* 73:197–209.

Part II
In Vivo Research into Masticatory Function

One of Hylander's lasting contributions to primate feeding biology will be his seminal publications documenting how the bones of the face are strained and how the jaw muscles function during chewing and biting in alert animals. In Part II, four chapters discuss the in vivo patterns of craniofacial strains and jaw-muscle activity in primates, artiodactyls, and marsupials. Each of these chapters discusses the areas of research that Hylander worked on throughout his career.

Herring et al. and Williams et al. describe patterns of symphyseal strains during mastication in pigs and alpacas, respectively. Although both taxa fuse their symphyses early in ontogeny (i.e., contemporaneous with weaning), neither species routinely exhibit a significant wishboning loading regime typical of cercopithecine primates. Herring et al. generally observe tension on the lingual symphyseal surface that may be related to ramal eversion created by masseter contraction during mastication in pigs. Alpacas primarily twist their symphyses about a transverse axis during mastication. The combined results from pigs, alpacas, and cercopithecine primates suggest that symphyseal fusion early in ontogeny is not related to a single jaw-muscle activity pattern. When viewed from this broader mammalian perspective, the early fusing symphysis is likely driven by disparate loading regimes related to underlying differences in jaw form, tooth morphology, and chewing mechanics.

In chapter four, Ross documents the patterns of facial strains during chewing in brown lemurs (*Eulemur fulvus*) to address questions relating to how the primate face torques. Ross finds that twisting of the skull in brown lemurs is more similar to anthropoids than the more closely-related galagos. This result suggests that the two primate suborders do not exhibit distinct patterns of circumorbital loading. He concludes from this finding that the origin of anthropoids did not necessarily involve changes in circumorbital form related to suborder differences in loading. Moreover, Ross finds that the local geometry of the postorbital bar precludes a simple, a priori prediction of strain patterns from an idealized model such as a cylinder or beam.

The final two chapters in this part examine jaw-muscle activity patterns during chewing. A.W. Crompton, a long-time collaborator with Hylander, and several of Crompton's colleagues document a novel jaw-muscle motor pattern during mastication in the Southern hairy-nosed wombat (*Lasiorhinus latifrons*). This motor pattern is significantly different from those of placental mammals, including primates, studied to date. An important implication of this study is the realization that transverse compression on the lingual surface of the symphysis, while the labial surface was tensed. No rostro-caudal gradient of strain could be detected. This pattern of deformation suggests a tendency to rotate the mandibular bodies with eversion of the lower border. Symphyseal deformation also accompanied opening, with a somewhat similar pattern but diminished magnitude. Overall, strain levels at the symphysis were high compared to other tooth-bearing bones, but decreased in older animals with more completely fused sutures. A survey of 115 dried skulls revealed that fusion begins at the age of 2 months, which coincides with natural weaning. Fusion commences on the compressed lingual surface.

2.2 Introduction

Although pigs and peccaries (families Suidae and Tayassuidae, Suoidea, Cetartiodactyla) are only very distantly related to anthropoid primates, convergences with respect to the masticatory system have been noted by many authors (Ström et al., 1986; Bermejo et al., 1993; Herring, 1998). Similarities between anthropoids and the suoids include the general (but not the detailed) anatomy of the adductor muscles of the mandible, the morphology and movements of the temporomandibular joint, and the bunodont dentition. To a considerable extent, these similarities correspond to generalized, often tough, diets for which both groups of species have adopted comparable strategies for food reduction, consisting of strong vertical closing movements coupled with short but powerful transverse power strokes. Another notable similarity between suoids and anthropoids is the early fusing mandibular symphysis.

As the primitive mammalian state is an unfused symphysis, it is the repeated fusion that has taken place in different lineages that needs explanation. The advantage of fusion is presumed to be biomechanical and to involve stiffening and/or strengthening the joint for repetitive loading during feeding. A number of factors have been considered correlated with symphyseal fusion in primates and other taxa, including large body size, robust construction of the mandible, tough diet, isognathy, extensive recruitment of balancing-side muscles during mastication, transversely oriented occlusal plane, transversely oriented zygomaticomandibularis (ZM, also called deep masseter), and a pattern of balancing-side contraction in which the ZM is strongly active when the superficial masseter and medial pterygoid activity are diminished (Scapino, 1981; Ravosa and Hylander, 1994; Lieberman and Crompton, 2000; Hogue and Ravosa, 2001). Pigs are well developed in all of these characteristics (Herring and Scapino, 1973; Herring, 1976; Herring et al., 2001; Rafferty et al., 2003). Incisor usage, although possibly involved (Greaves, 1988), is not a reliable predictor of symphyseal fusion. Pigs, like many other fused-symphysis taxa (Ravosa and Hylander, 1994), assign difficult ingestion tasks to the premolars rather than the incisors (Herring and Scapino, 1973). Thus in many ways, pigs seem to be a typical example for the fused symphysis. However, the possession of these characteristics does not explain how the fused pig symphysis functions.

The pig symphysis is more horizontally oriented than that of most anthropoids, particularly humans. Pig lower incisors are extremely procumbent, and the long axis of the symphysis lies close to the occlusal plane, especially as viewed on the lingual surface. This orientation should very effectively stiffen the symphysis against bending in the transverse plane. Zhang has calculated that stresses and strains should be similar in the human and pig symphyses; but if the pig symphysis were oriented vertically as in humans, the greater muscle forces and moment arms of the pig jaw would increase stress and strain to injurious levels (Zhang, 2001). At the same time, the horizontally elongated pig symphysis converts the expectations of transverse bending from a difference in polarity between the labial and the lingual surfaces to a comparison of rostro-caudal position.

The literature about loading of a fused symphysis has changed over time. When DuBrul and Sicher first speculated about symphyseal stresses, their emphasis was on jaw protrusion and the action of the lateral pterygoid muscles in producing what they called wishboning, but is more accurately described as medial transverse bending (DuBrul and Sicher, 1954). This is an activity that should accompany jaw opening, rather than the power stroke of mastication (Fig. 2.1). That the primate symphysis does bend medially during jaw opening has been well documented by Hylander's strain gage studies of macaques (Hylander, 1984; Hylander and Johnson, 1994) and by repeated demonstrations in human subjects that the dental arches are closer together when the jaws are opened (e.g., Chen et al., 2000). Although there are no previous data for pig, a dynamic model of pig mastication based on direct measurements of the mandible and jaw muscles plus electromyographic data from the literature predicted that medial transverse bending during opening would reach a peak torque approximately 60% of maximum and would include a minor twisting component due to asymmetrical activation of the lateral pterygoid muscles (Zhang, 2001). However, because the opening strains on the macaque symphysis are relatively low compared to the strains observed during the power stroke (Hylander, 1985), they have received little attention in recent papers. Instead, attention has been focused on power stroke strains.

Of the multiple deformation patterns theoretically possible during the power stroke (Beecher, 1977; Beecher, 1979; Scapino, 1981; Hylander, 1984), emphasis has been placed on dorso-ventral shear and lateral transverse bending (Fig. 2.1). Dorso-ventral shear is thought to be most important early in the power stroke when the working-side teeth engage the bolus while the balancing side continues to be closed by adductor muscle contraction; dorso-ventral forces have been associated with partial symphyseal fusion in some prosimians (Beecher, 1977). In Zhang's simulation of mastication, dorso-ventral shear was the largest predicted force in the pig symphysis (Zhang, 2001).

Lateral transverse bending takes place primarily at the end of the power stroke when the lateral pull of the balancing-side ZM opposes the laterally directed working-side occlusal load. Zhang's dynamic simulation predicted that lateral transverse bending during the power stroke would produce torques at least 40% larger than those from medial transverse bending during opening (Zhang, 2001). Furthermore, Zhang found that lateral transverse bending could arise not only from

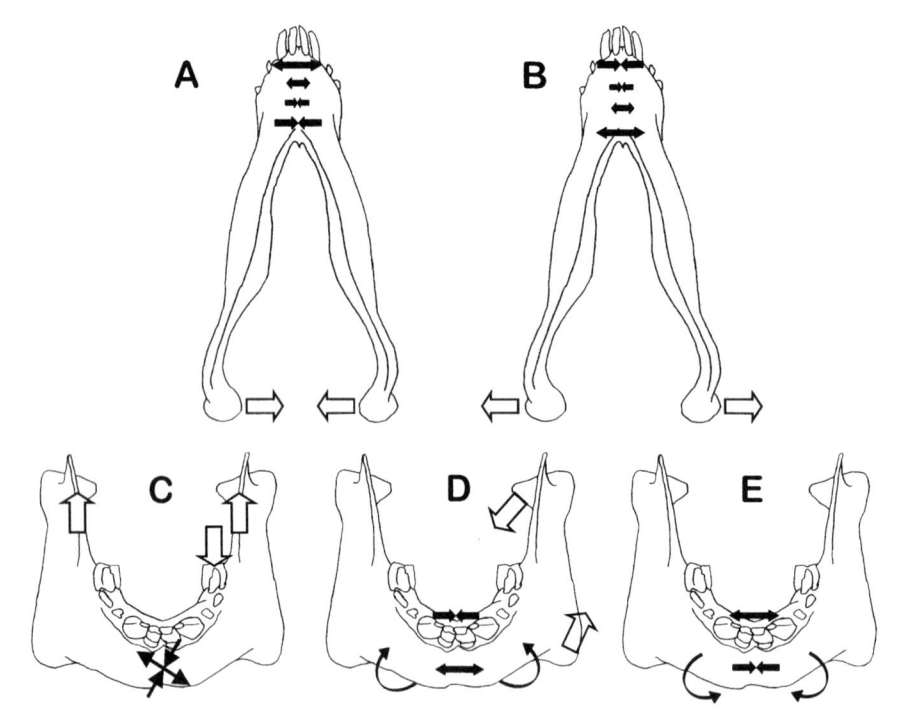

Fig. 2.1 Theoretical patterns of deformation of the pig symphysis. **A** and **B** are ventral views of the mandible and show the labial surface of the symphysis. **C–E** are anterior views and show the labial surface below and the lingual surface above the procumbent incisors. The open arrows indicate load directions. The solid double-headed arrows indicate tensile strain, while the opposed solid arrows indicate compressive strain. The thin curved arrows indicate rotation of the mandibular rami around their long axes. **A.** Medial transverse bending, such as might be caused by the medial component of lateral pterygoid muscle pull on the condyles. Both labial and lingual surfaces of the symphysis are expected to show a pattern of relative tension rostrally and relative compression caudally. **B.** Lateral transverse bending, such as might be caused by lateral pull of the ZM on the balancing side resisted by the occlusion on the working side. Both surfaces of the symphysis are expected to show a pattern of relative compression rostrally and relative tension caudally. **C.** Dorso-ventral shear, such as might be caused by a unilateral occlusal load with bilateral muscle force. Both surfaces of the symphysis are expected to show a 45° strain pattern. **D.** Rotation of mandibular rami with eversion of the ventral borders. Loads that could produce eversion include the ventral component of lateral pterygoid pull and the dorsal pull of the masseter. The symphysis is expected to be compressed lingually and tensed labially by bending in the coronal plane. **E.** Rotation of the mandibular rami with inversion of the ventral borders. The symphysis is expected to show lingual tension and labial compression

ZM contraction and molar occlusion but also from the tendency of the working-side superficial masseter to evert the mandible and from jaw joint reaction forces (Fig. 2.1). Authors who have analyzed other artiodactyls and carnivorans have also emphasized the significance of multiple transverse forces in determining symphyseal morphology (Scapino, 1981; Lieberman and Crompton, 2000; Hogue and Ravosa, 2001).

The main goal of this contribution was to ascertain whether the general conclusions about strain in the fused symphysis are valid for pigs, specifically (1) medial transverse bending occurs during opening, but is relatively small; (2) dorso-ventral shear occurs at the beginning of the power stroke; and (3) the largest strains are those of lateral transverse bending occurring at the end of the power stroke. Because of the horizontal orientation of the pig symphysis, transverse bending was expected to cause rostro-caudal differences in strain (Fig. 2.1). In particular, medial transverse bending should cause tensile strain rostrally and compressive strain caudally on both lingual and labial surfaces of the symphysis. Lateral transverse bending should cause the opposite pattern, rostral compression, and caudal tension. Dorso-ventral shear should present as a 45° orientation of strain. Figure 2.1 also depicts a third possible type of deformation caused by the rotation of each mandibular body about its own long axis, resulting in either eversion or inversion of the ventral border and bending of the symphysis in a coronal plane (terminology from Beecher, 1977). During the experiments we became aware that at least some of the animals showed incomplete surface fusion of the symphysis. Thus, a secondary goal was to establish the pattern and progress of fusion using a sample of age-known pig skulls.

2.3 Material and Methods

2.3.1 In Vivo Experiments

In vivo data on symphyseal deformation were gathered from a total of 17 young swine of various breeds and both sexes (Table 2.1). All animal procedures were reviewed and approved by the University of Washington Institutional Animal Care and Use Committee. The instrumentation varied (Fig. 2.2) and included (1) single-element strain gages (EP-08-125BT-120, Vishay Micro-Measurements, Raleigh NC) glued (cyanoacrylate) to bone in a transverse orientation across the midline; (2) stacked rosette strain gages (SK-06-030WR-120, Vishay Micro-Measurements) glued to bone either in the midline or on one side of the symphysis; and (3) pairs of 2-mm piezoelectric crystals with attachment pegs (2P-34C-40-NS, Sonometrics, London, Ont.) for digital sonomicrometry. The pegs were placed into small holes drilled in the lower borders of the left and right mandibular rami below the cheek tooth row after minor reflection of the inserting fibers of the digastric muscle. The sonomicrometry system has a theoretical resolution of 15 µm.

Before the recording session, each pig was acclimated to the laboratory environment. On the day of the recording, the animal was anesthetized by mask with isoflurane for surgical placement of transducers. The labial aspect of the symphysis and/or mandible was exposed from an extra-oral approach, whereas the lingual aspect employed an intra-oral incision through the gingival tissue. Preparation of the strain gages and bone followed standard methods (Rafferty et al., 2000). Gage orientation was measured with a protractor, and crystal locations were documented before closure of the periosteal and skin incisions. Lead wires were exited through the

The EMG recordings were used to identify the side of chewing. Pigs typically alternate the side of chewing with each stroke. Although occlusion is usually bilateral in pigs, the directionality is strongly unilateral. The side of the power stroke can be identified by late activity in the working-side masseter/medial pterygoid and balancing side temporalis/ZM. This activity pattern moves the lower incisors from the working-side past the midline to the balancing-side (Herring, 1976). Strain recordings were analyzed quantitatively. After exporting the analog signals to Excel, baseline strain was subtracted and voltages converted to microstrain. Data from the three elements of the rosette gages were used to compute the magnitude and orientation of the principal strains (Tech Note 515, Measurements Group). Peaks were identified as relative maxima of shear strain (maximum principal strain minus minimum principal strain) and were classified as "opening," "early closing," or "late closing" by their timing relative to the EMG of the masseter and other muscles. Sonometric deformation data were converted to strain by dividing the distance between pairs of crystals. Although right and left masticatory cycles (whenever they could be identified) were analyzed separately, there were no consistent differences in symphyseal strain, so data from both chewing sides were averaged.

2.3.2 Skeletal Investigation

In a sample of 115 mandibles of pigs of known age, not including any of the animals used in the present study, fusion of the symphysis was assessed by examining its outer surface under magnification. Of these mandibles, 2 were from Yucatan micropigs, 10 were from domestic farm pigs, and the remainder were from Hanford/Sinclair minipigs. To be classified as unfused, a suture had to be visible along the entire midline junction. To be classified as fused, the suture had to be visibly co-ossified everywhere, but between the incisor alveoli. All other conditions were classified as "partially fused," and the locations of the unfused portions were noted.

2.4 Results

2.4.1 Mastication

Most subjects showed at least two distinct strain peaks during chewing, one corresponding to the opening stroke and the other, usually larger, corresponding to the closing/power stroke (i.e., concurrent with masseter EMG). The exceptions were #347 (Fig. 2.3B), which lacked measurable opening strain, and #357 and #362, which had opening strains for about half of their chewing cycles. Closing peaks were seen in all animals and were often bi- or even trimodal. However, the complex closing waveforms were variable, and in most cases it was not possible to classify

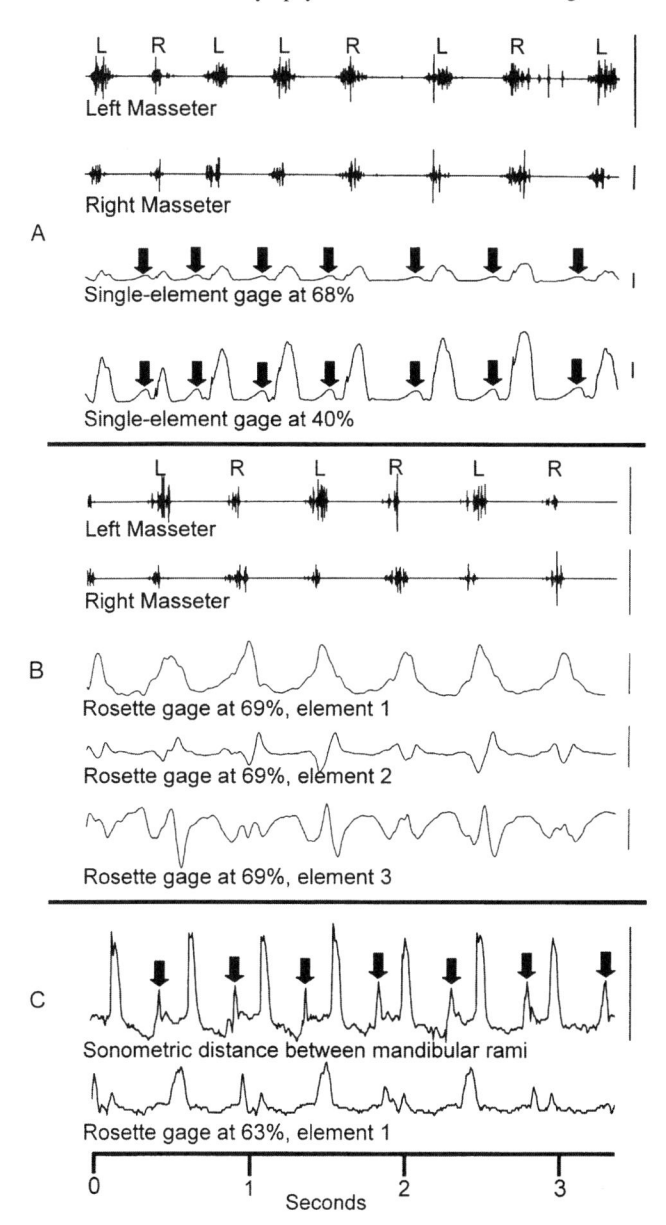

Fig. 2.3 Examples of raw recordings of mastication. Scale is indicated by the vertical bars to the right of each trace. For muscles the scale bar is 1 V, for the strain channels in **A** and **B** the bar is 1000 με, and for the sonometric trace in **C** the bar is 1 mm. The strain channel in **C** is uncalibrated. The working side is designated as left (L) or right (R) based on muscle pattern; this information is not available for **C**. **A.** #368, single-element gages at 40% and 60%. Both gages show an opening peak (arrows) and a larger, relatively simple closing peak. Strains were always tensile and always higher in the more rostral location. **B.** #347, one rosette gage 3 mm off midline at the 69% position. Element 2 was roughly aligned with the rostro-caudal axis

($n = 4$ subjects), as compared to 233 ± 214 µε ($n = 8$) for those with stabilized osteotomies.

Rostro-caudal position was not correlated with strain levels; but on the more thoroughly sampled labial side of the symphysis, the gages located at 40% (#368, Table 2.2 and Fig. 2.3A, and #359, Table 2.3) gave higher tensile strains than those located more caudally.

2.4.1.2 Closing Strains

Deformations for closing were typically much larger in magnitude than those for opening, but surprisingly similar to opening in pattern (Tables 2.2 and 2.3). The major pattern difference is that during closing, both pigs with lingual gages showed strong compression in the transverse direction. On the labial side, symphyseal strains remained predominantly tensile and between 45° and 90° for at least one phase of closing for every pig. As in opening, strain levels were not correlated with rostro-caudal position. However, compared to opening tensile orientations, closing tensile orientations were more often 45° and less often transverse (Table 2.3). There were only two subjects with distinct early and late closing peaks (#362 and #387, Table 2.3), and these suggested that tensile strain was initially transverse or 45° and then became more sagittal. As measured by sonomicrometry, the distance between the mandibular rami increased; at least in #362, the magnitude of this closing deformation was double that of opening (Fig. 2.3C).

A comparison between intact pigs and those with mandibular osteotomies using rosette data for closing strains showed that overall strain levels were higher in the osteotomized group (shear strain averaged 265 ± 63 µε for four intact mandibles and 588 ± 182 µε for eight osteotomized mandibles, $p < 0.01$). However, the osteotomized pigs were also generally younger than the intact group and open symphyses were more common (Table 2.1); for both groups combined, strain levels were found to be inversely correlated with age ($r = -0.81$, $p < 0.01$).

2.4.2 Fusion of the Mandibular Symphysis

Table 2.4 shows the results of the survey of symphyseal fusion in mandibles of known age. In very young animals (less than 1 month), the open suture was wider at the caudal margin than elsewhere. Fusion was not seen in animals less than 2 months old. By the age of 2 months, many symphyses had begun fusing, and all symphyses from pigs older than 2 months were at least partially fused. The first completely fused symphyses were also seen in the third month of life, but the time of complete fusion was variable, with partial fusion still seen in one 6-month-old pig. Of the 33 partially fused symphyses, all showed fusion on the rostral part of the lingual surface. A visible suture was most common on the labial surface, either rostral or caudal ($n = 13$), the caudal edge ($n = 11$), and the caudal part of the lingual surface ($n = 7$). Two symphyses had visible suture segments in more than one region.

Table 2.4 Condition of the symphyseal suture in 115 pig mandibles as a function of age

Age	Unfused	Partially fused	Fused
Less than 2 months	11	0	0
2 months	8	5	0
2.25–3 months	0	10	8
3.25–4 months	0	12	13
4.25–5 months	0	5	19
5.25–6 months	0	1	9
Over 6 months	0	0	14

2.5 Discussion

2.5.1 Mastication

2.5.1.1 Opening

Opening deformation of the mandible has always been assumed to be dominated by the contraction of the lateral pterygoid muscles (DuBrul and Sicher, 1954). The expected strain pattern was medial transverse bending, manifested as rostral tension decreasing or changing to compression caudally (Fig. 2.1A), with some minor twisting resulting from asymmetrical contraction (Zhang, 2001). The results were only partially consistent with these expectations. Transverse tension was found in most subjects, but indications of a rostro-caudal gradient were weak. Both animals with gages in multiple locations, #368 (Table 2.2) and #362 (Table 2.3), showed greater tension in the more rostral gage. However, there was no caudal compression.

The fact that the predictions of medial transverse bending were not completely fulfilled indicates that some corrections to this model are necessary. First, the assumption of a strong rostro-caudal gradient is predicated on the center of bending being located in or near the symphysis. However, given the fact that the center of mass of the pig mandible is at the back of the tooth row (Zhang et al., 2001), it is likely that the center of bending is well caudal to the entire symphysis. This would explain the absence of a strong rostro-caudal gradient. A second consideration is that the lateral pterygoid muscles, which attach well above the center of mass, should produce not only medial transverse bending but also eversion of the rami (Fig. 2.1D). Eversion would be expected to produce transverse tension on the inferior labial surface with no particular gradient, as was observed. Eversion should also produce compression on the superior lingual surface, for which our data were equivocal, with one pig (#102) showing compression and one (#103) tension. One feature that was consistent with eversion but not with medial transverse bending was the finding of increasing distance between the mandibular rami, as measured by sonomicrometry. This increase in dimension is the reverse of the narrowing documented in the human mandible during opening (Chen et al., 2000). Unfortunately, our finding was based on only two animals, one of which had no strain gage information. It is also possible that the contraction of the overlying digastric muscles disturbed the position of the sonomicrography crystals during opening and that the finding of widening was spurious.

Somewhat surprisingly, given that the osteotomy separated a fragment with the attachment of the right lateral pterygoid from the rest of the mandible, the results from the osteotomized pigs were indistinguishable from those of the intact pigs (Tables 2.2 and 2.3). This suggests: (1) the fixation of the osteotomy was rigid and transmitted strains efficiently, and/or (2) forces from just one side are sufficient to produce characteristic symphyseal strain patterns, and/or (3) additional muscles, such as the digastric, mylohyoid, and geniohyoid, are at least as important as the lateral pterygoid in producing symphyseal strain.

In short, the symphysis is almost always tensed transversely on its labial surface during opening. However, neither can we distinguish between medial transverse bending and eversion as the primary cause of the tension nor can we eliminate a role for muscles other than the lateral pterygoid.

2.5.1.2 Closing

Closing is a more complicated situation than opening in that (1) muscle activity is always asymmetrical; and (2) in addition to the muscle forces acting on the mandible, there are reaction forces at the teeth and at the jaw joints. In pigs, molar occlusal forces and jaw joint reactions are usually similar on the working and balancing sides, but incisor occlusal forces are highly asymmetrical because they depend on the direction of jaw movement (Rafferty et al., 2003).

We had expected to see two distinct phases of closing. Early in the movement, as the teeth engaged the bolus, we expected dorso-ventral shear, manifested as 45° strains on the symphysis. Toward the end of the movement, as the jaw moved toward the balancing side, we expected lateral transverse bending, manifested as rostral compression and caudal tension. However, the comparison between early and late closing proved to be impossible, except in two pigs. These two subjects did seem to fit the model in that tensile strain was more sagittal for late closing, as expected for rostral areas during lateral transverse bending (Fig. 2.1B). However, a true sagittal orientation (7°) was only seen in the more rostral gage of #362 (Table 2.3).

In the remaining 15 pigs, although the time course of symphyseal strain was often complex, the phases of closing were so blended that we could not discern separate phases. Because the averages reported in Tables 2.2 and 2.3 are for the peak strains, they would represent either overlapping occlusal and grinding phases or the larger of the two. Because of the preponderance of tensile strain orientations near 45° (Table 2.3), dorso-ventral shear may be the largest force imposed on the symphysis, as was suggested by Zhang's kinetic model (Zhang, 2001). However, the frequency of 45° tension could have other origins as well, including torque from incisor or molar occlusal loads. In this context, it is interesting to note that #103, one of the few pigs that never showed the 45° orientation, was a Yucatan, a breed lacking incisor occlusion.

The most surprising aspect of the closing results is that except for the late closing of #362 and #387, the observed pattern is clearly not lateral transverse bending as depicted in Fig. 2.1B. Transverse strains in the symphysis were everywhere tensile on the labial side, and there was no pattern of increasing tension caudally in the sample as a whole. The two pigs with dual gages showed opposite patterns. Pig #362 did show higher caudal strain levels, a reversal of its opening pattern (Table 2.3). However, #368 had much lower transverse tension in the caudal than in the rostral gage, the same pattern it had for opening (Table 2.2). Admittedly, our failure to find a consistent rostro-caudal gradient could be due to the non-uniformity of the sample, with only two animals multiply instrumented. Nevertheless, the deformations observed are all consistent with an alternate explanation, that of eversion (Fig. 2.1D). Eversion accounts for the widening of the mandibular rami and explains the compression on the lingual surface of the symphysis as well as the tension on all locations along the labial surface. Eversion is also consistent with the torsional strains observed bilaterally on the mandibular corpus of chewing pigs (Herring et al., 2001). Asymmetrical eversion of the rami could also have had a role in producing the modal orientation of 45° seen on the labial surface. Asymmetrical eversion during closing could arise from masseter contraction. Late in the closing movement, the working-side masseter is highly active even though the balancing-side activity has ceased (Herring and Scapino, 1973). In addition, the morphology of the jaw joint may be important (Zhang, 2001). Because the articular surface slopes from dorso-lateral to ventro-medial (Herring et al., 2002), the vertical reaction loads on the condyles would tend to push them medially, thus causing eversion (Fig. 2.1D).

The comparison of pigs with intact mandibles and those with stabilized osteotomies is also helpful in considering the origin of the apparent eversion during closing. The overall level of strain was higher in the osteotomy sample, at least for the more thorough rosette data. This finding may be due to the younger average age of the osteotomy pigs (Table 2.1) and their less fused symphyseal sutures, as discussed further below. The general 45°/transverse orientation of peak tensile strain was the same in both groups. From previous work on the osteotomy sample, we know that during closing (as opposed to opening) the osteotomy site was compliant rather than stiff (Sun et al., 2006), that the animals were chewing primarily on the opposite side (Rafferty et al., 2006), and that the jaw joint loads on the osteotomy side were decreased and no longer vertically oriented, while those on the opposite side were normal in magnitude and orientation (Rafferty et al., 2006). By elimination, then, the most likely hypothesis to explain mandibular eversion is contraction of the working-side masseter, perhaps coupled with working-side incisor contact.

Summarizing, strain patterns in the pig symphysis suggest that dorso-ventral shear or twisting, possibly involving incisor contact, are important during closing. Transverse strains do occur, but they appear to arise primarily from eversion of the mandibular rami, not from lateral transverse bending.

2.5.2 Symphyseal Strain, Fusion, and Morphology

The range of shear strain (approximately 200–900 $\mu\varepsilon$, calculated from Table 2.3) is similar to or higher than the strain magnitudes reported for fused primate symphyses

(Hylander, 1984; Hylander, 1985). Indeed, the maximal values fall at the high end of our previous recordings from other areas of the pig skull. Symphyseal strain was in general greater than strain on the tooth-bearing mandibular body, maxillary and premaxillary, and comparable to strain on the mandibular condyle and squamosal bone (Herring et al., 2001; Rafferty et al., 2003; Rafferty et al., 2006). However, the higher values (400–900 $\mu\varepsilon$) belong to younger pigs with unfused symphyses, and the lowest values (200–350 $\mu\varepsilon$) belong to the older pigs with fused symphyses (#347, #362, #387). These latter values are quite similar to strains on the other tooth-bearing parts of the skull. There is a strong implication that fusion decreases symphyseal strain, a conclusion that also arose from our previous study of the sagittal suture (Sun et al., 2004). It is difficult to compare the rosette strain magnitudes to our previous studies on sutures, which used single-element gages, because single-element gages typically report higher strains from similar locations, probably because they are less stiff. The single-element data on the symphysis are skimpy, but the top values of 1600–3000 $\mu\varepsilon$ (Table 2.2) are as high as any we have recorded from open cranial sutures (Rafferty and Herring, 1999; Herring et al., 2001; Rafferty et al., 2003; Sun et al., 2004). In short, as hypothesized by Zhang (2001), the pig symphysis appears to be a very highly stressed region of the skull, and fusion serves to decrease strain to a level more comparable to that of the neighboring tooth-bearing bones.

The two animals that showed indications of lateral transverse bending during the power stroke were both older pigs with fused symphyses, suggesting an association. We therefore carefully examined the recordings from the third older pig, #347 (Table 2.1), for which separate early and late closing peaks could not be calculated. The closing strains from this animal were, in fact, complex (Fig. 2.3B). Our inability to identify separate peaks was due to a trimodal strain pattern plus side differences. It is quite possible that lateral transverse bending was one element of this blend. If so, then fusion of the symphysis may be a prerequisite for the detection of strain patterns characteristic of the different phases of closing. In younger animals with unfused symphyses, the large strains associated with separation of the suture may overwhelm pattern details.

Although the pig symphysis shows high strain levels and the pig oral apparatus has all of the features of muscle anatomy and contraction pattern that should promote transverse bending, this pattern of deformation was not dominant even in older individuals. Instead, most of the strains observed were more compatible with a rotation that everts the lower border of the mandible. In retrospect, symphyseal morphology itself may account for this. The rostro-caudally elongated symphysis is stiffened against transverse bending, but its comparatively small dorso-ventral dimension leaves it vulnerable to rotation and dorso-ventral shear. Thus the fact that we seldom observed transverse bending does not mean that it does not occur, just that the pig symphysis is fortified against this type of loading. Because much of the mechanical environment of the pig symphysis is related to its horizontal orientation, which helps it resist deformation in the transverse plane, it is interesting to note that symphyseal orientation can become more vertical under certain dietary or occlusal modifications (Ferrari and Herring, 1995; Ciochon et al., 1997). In Yucatan swine, which characteristically lack incisor occlusion, the symphysis is more vertical than in normally configured pigs.

The timing of fusion of the pig symphysis may not be coincidental. Under natural conditions, piglets are not weaned before 2 months of age (Pond and Houpt, 1978), the age at which symphyseal fusion begins (Table 2.4). Masticatory strains could conceivably have a role in inducing fusion. The first region of the symphysis to fuse, the rostral part of the lingual surface, is the only sampled part of the symphysis that is often under compressive loading. The labial surface shows no rostro-caudal order of fusion, which corresponds to the lack of any rostro-caudal gradient in strain. We have no data from the late-fusing caudal edge, but this is a complex area with many muscle attachments and large nutrient foramina, and is likely to be under a unique strain regime.

2.6 Conclusions

The pig symphysis undergoes two major peaks of deformation in each chewing cycle, a smaller one coinciding with opening and a larger one coinciding with contraction of the jaw closing musculature. Separate phases for early and late power stroke were a rarity, seen only in the oldest animals of the sample. Opening and closing deformations were more similar in pattern than was expected. Both opening and closing featured tensile strain on the labial surface of the symphysis and separation of the mandibular rami. Neither opening nor closing peaks showed rostro-caudal gradations in strain. However, tensile strain was more often transverse during opening and 45° during closing. Further, during closing the lingual surface showed transverse compression, whereas this surface was inconsistent during opening. We consider the opening pattern as consistent with a mild degree of medial transverse bending plus some eversion of the mandibular rami, possibly caused by the lateral pterygoid muscles. The closing pattern indicates dorso-ventral shear and strong eversion of the ramus, for which working-side masseter muscle contraction seems the most likely cause. Lateral transverse bending was rarely observed.

These patterns of deformation differ somewhat from those of anthropoid primates (Hylander, 1984; Hylander and Johnson, 1994), particularly in the absence of unequivocal transverse bending in either direction. Anatomical differences between suoids and anthropoids may account for the difference. The relative absence of transverse bending in pigs may reflect stiffening from the rostro-caudal elongation of the symphysis rather than the absence of load. In addition, the suoid symphysis is distant from the mandible's center of mass (Zhang et al., 2001), which would tend to obscure the rostro-caudal gradients for which we were looking. Finally, the occlusion of the procumbent incisors and the curvature of the symphyseal surfaces are clearly different in pigs from anthropoid primates, but as yet we do not know how these factors influence strain.

The status of the symphyseal junction was an important determinant of strain. In addition to the fact that lateral transverse bending was only discernible in pigs with fused symphyses, magnitudes of strain were higher in young pigs with unfused symphyses and decreased to more typical bone levels when fusion began. Nevertheless, it is clear that even the youngest pigs did not show the degree of movement seen in prosimians and artiodactyls with permanently unfused symphyses (Beecher, 1977; Lieberman and Crompton, 2000), and that the unfused pig symphysis is functionally more similar to a cranial suture (Sun et al., 2004) than to a mobile joint. The first part of the symphysis to fuse, the rostral section on the lingual surface, is the most compressed section. The onset of fusion at 2 months of age corresponds with weaning and suggests that masticatory forces have played an evolutionary or developmental role in fusing the symphysis.

Acknowledgments This paper is affectionately dedicated to Bill Hylander, and was undertaken in homage to his many contributions to understanding the primate skull and in particular to the obsession with the mandibular symphysis, which he has spread throughout the physical anthropology community. The experiments were supported by PHS awards P60 DE13061 (pilot study funding to Z.J. Liu), R03 DE14457 (to K.L. Rafferty), R01 DE14336, and R01 DE08513 (to S.W. Herring), all from NIDCR. Hannah Hook gathered the data on symphyseal fusion. We thank Emi Katzenberger, Frank Starr, and Patricia Emry for help with experiments.

References

Beecher, R. M. (1977). Function and fusion at the mandibular symphysis. *Am. J. Phys. Anthrop.* 47:325–336.
Beecher, R. M. (1979). Functional significance of the mandibular symphysis. *J. Morphol.* 159:117–130.
Bermejo, A., González, O., and González, J. M. (1993). The pig as an animal model for experimentation on the temporomandibular articular complex. *Oral Surg. Oral Med., Oral Pathol.* 75:18–23.
Chen, D. C., Lai, Y. L., Chi, L. Y., and Lee, S. Y. (2000). Contributing factors of mandibular deformation during mouth opening. *J. Dentistry* 28:583–588.
Ciochon, R. L., Nisbett, R. A., and Corruccini, R. S. (1997). Dietary consistency and craniofacial development related to masticatory function in minipigs. *J. Craniofac. Gen. Dev. Biol.* 17:96–102.
DuBrul, E. L., and Sicher, H. (1954). *The Adaptive Chin*. Charles C. Thomas, Springfield, IL.
Ferrari, C. S., and Herring, S. W. (1995). Use of a bite-opening appliance in the miniature pig: modification of craniofacial growth. *Acta Anat.* 154:205–215.
Greaves, W. S. (1988). A functional consequence of an ossified mandibular symphysis. *Am. J. Phys. Anthrop.* 77:53–56.
Herring, S. W. (1976). The dynamics of mastication in pigs. *Arch. Oral Biol.* 21:473–480.
Herring, S. W. (1998). How can animal models answer clinical questions? In Carels, C., and Willems, G. (eds.), *The Future of Orthodontics*. Leuven University Press, Leuven, Belgium, pp. 89–96.
Herring, S. W., and Scapino, R. P. (1973). Physiology of feeding in miniature pigs. *J. Morphol.* 141:427–460.
Herring, S. W., Rafferty, K. L., Liu, Z. J., and Marshall, C. D. (2001). Jaw muscles and the skull in mammals: the biomechanics of mastication. *Comp. Biochem. Physiol. A* 131:207–219.
Herring, S. W., Decker, J. D., Liu, Z. J., and Ma, T. (2002). The temporomandibular joint in miniature pigs: anatomy, cell replication, and relation to loading. *Anat. Rec.* 266:152–166.

Hogue, A. S., and Ravosa, M. J. (2001). Transverse masticatory movements, occlusal orientation, and symphyseal fusion in selenodont artiodactyls. *J. Morphol.* 249:221–241.

Hylander, W. L. (1984). Stress and strain in the mandibular symphysis of primates: a test of competing hypotheses. *Am. J. Phys. Anthrop.* 64:1–46.

Hylander, W. L. (1985). Mandibular function and biomechanical stress and scaling. *Am. Zool.* 25:315–330.

Hylander, W. L., and Johnson, K. R. (1994). Jaw muscle function and wishboning of the mandible during mastication in macaques and baboons. *Am. J. Phys. Anthrop.* 94:523–547.

Lieberman, D. E., and Crompton, A. W. (2000). Why fuse the mandibular symphysis? A comparative analysis. *Am. J. Phys. Anthrop.* 112:517–540.

Pond, W. G., and Houpt, K. A. (1978). *The Biology of the Pig.* Cornell University Press, Ithaca, NY.

Rafferty, K. L., and Herring, S. W. (1999). Craniofacial sutures: morphology, growth and *in vivo* masticatory strains. *J. Morphol.* 242:167–179.

Rafferty, K. L., Herring, S. W., and Artese, F. (2000). Three-dimensional loading and growth of the zygomatic arch. *J. Exptl. Biol.* 203:2093–3004.

Rafferty, K. L., Herring, S. W., and Marshall, C. D. (2003). The biomechanics of the rostrum and the role of facial sutures. *J. Morphol.* 257:33–44.

Rafferty, K. L., Sun, Z., Egbert, M. A., Baird, E. E., and Herring, S. W. (2006). Mandibular mechanics following osteotomy and appliance placement. II. Bone strain on the body and condylar neck. *J. Oral Maxillofac. Surg.* 64:620–627.

Ravosa, M. J., and Hylander, W. L. (1994). Function and fusion of the mandibular symphysis in primates: stiffness or strength? In: Fleagle, J., and Kay, R. (eds.), *Anthropoid Origins.* Plenum Press, New York, pp. 447–468.

Scapino, R. P. (1981). Morphological investigation into functions of the jaw symphysis in carnivorans. *J. Morphol.* 167:339–375.

Ström, D., Holm, S., Clemensson, E., Haraldson, T., and Carlsson, G. E. (1986). Gross anatomy of the mandibular joint and masticatory muscles in the domestic pig *(Sus scrofa). Arch. Oral Biol.* 31:763–768.

Sun, Z., Lee, E., and Herring, S. W. (2004). Cranial sutures and bones: growth and fusion in relation to masticatory strain. *Anat. Rec.* 276A:150–161.

Sun, Z., Rafferty, K. L., Egbert, M. A., and Herring, S. W. (2006). Mandibular mechanics following osteotomy and appliance placement. I. Postoperative mobility of the osteotomy site. *J. Oral Maxillofac. Surg.* 64:610–619.

Zhang, F. (2001). *Biomechanics of the Mammalian Jaw.* Doctoral dissertation, University of British Columbia.

Zhang, F., Langenbach, G. E. J., Hannam, A. G., and Herring, S. W. (2001). Mass properties of the pig mandible. *J. Dent. Res.* 80:327–335.

Chapter 3
Symphyseal Fusion in Selenodont Artiodactyls: New Insights from In Vivo and Comparative Data

Susan H. Williams, Christine E. Wall, Christopher J. Vinyard, and William L. Hylander

Contents

3.1 Introduction	39
3.2 Methods	43
3.2.1 In Vitro Stresses – Gauge Placement, Recording, and Analysis	43
3.2.2 In Vivo Stresses – Gauge Placement, Recording, and Analysis	44
3.3 Results	45
3.3.1 Symphyseal Strains During in vitro Transverse Bending	45
3.3.2 Symphyseal Strains During In Vivo Simulated Stresses and Mastication	46
3.4 Discussion	51
3.4.1 Symphyseal Strains During Simulated Transverse Bending In Vitro and In Vivo: Implications for Interpreting Masticatory Strains	51
3.4.2 Symphyseal Strains During Rhythmic Mastication	52
3.4.3 Symphyseal Strains and Jaw Morphology in Alpacas	55
3.5 Conclusions	59
References	60

3.1 Introduction

Comparative and in vivo studies in primates (e.g., Hylander, 1977, 1979a, b, 1984, 1985; Hylander and Johnson, 1994; Hylander et al., 1987, 2000; Luschei and Goodwin, 1974; Vinyard et al., 2001) have contributed significantly to our understanding of mammalian craniofacial biology and function. Hylander and colleagues, in particular, have amassed a unique data set, arguably unmatched in any clade of mammals, in their effort to elucidate form–function relationships in the

---

S.H. Williams
Department of Biomedical Sciences, Ohio University College of Osteopathic Medicine, Ohio University, Athens, OH 45701
e-mail: willias7@ohio.edu

These comparative findings are supported by recent EMG studies in two species of selenodont artiodactyls – goats and alpacas. These studies show that alpacas have a wishboning jaw-adductor motor pattern like anthropoids. That is, they recruit their balancing-side deep masseter relatively late in the power stroke (Williams, 2004; Williams et al., 2003, 2007). In contrast to alpacas, goats and other non-camelid selenodont artiodactyls are similar to strepsirrhines in having mobile symphyses united by ligaments, fibrocartilage, and interdigitating bony rugosities (Lieberman and Crompton, 2000). As in many strepsirrhines, goats also recruit their balancing-side deep masseter relatively early in the power stroke (Fig. 3.1) (Williams, 2004; Williams et al., 2003, 2007). Thus, there is also strong association in selenodont artiodactyls between symphyseal fusion, symphyseal shapes advantageous for resisting wishboning, and the wishboning jaw-muscle activity pattern.

The combined comparative and in vivo evidence outlined above suggests that wishboning would be a predominant loading regime along the alpaca symphysis. If true, symphyseal fusion in camelids may have a similar adaptive significance as has been proposed for anthropoids. Here, we summarize symphyseal strain data from a series of experiments on alpacas (*Lama pacos*) to determine whether alpacas wishbone their symphyses.

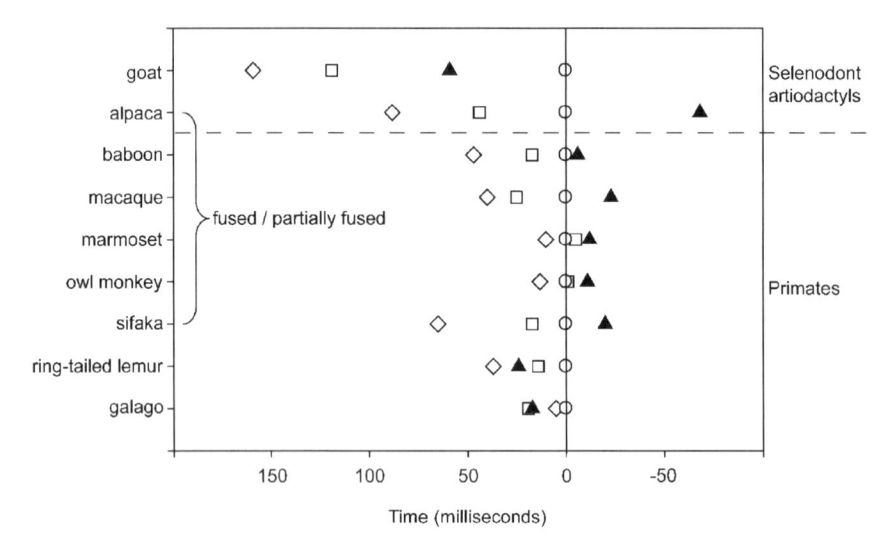

Fig. 3.1 Summary data of peak masseter activity in primates and selenodont artiodactyls. diamonds, working-side deep masseter; squares, balancing-side superficial masseter; triangles, balancing-side deep masseter; circles, working-side superficial masseter. The balancing-side deep masseter peaks later relative to other jaw muscles in species that fuse their symphyses early during ontogeny or have the tendency to develop partial fusion (e.g., sifakas). Primate data are from Hylander and Johnson (1994), Hylander et al. (1987, 2000, 2002, 2003, 2004), and Vinyard et al. (2001, 2006)

3 Symphyseal Fusion in Selenodont Artiodactyls

3.2 Methods

Three types of strain data from alpacas are presented in this paper: (1) in vitro symphyseal strains resulting from simulated stresses on a fresh cadaveric specimen; (2) in vivo symphyseal strains resulting from simulated stresses on anesthetized subjects; (3) in vivo symphyseal strains from alert individuals during normal mastication. The in vitro data set is useful for characterizing the expected strain patterns along the entire symphysis during simulated loading regimes, whereas the simulated stresses on live alpacas provide a baseline for interpreting the strains recorded from these same animals during mastication. Because these last two data sets are from the same gauges used for recording masticatory strains, methods for these data sets are discussed together.

3.2.1 In Vitro Stresses – Gauge Placement, Recording, and Analysis

Eight rectangular rosette strain gauges (WA-06-060WR-120, Micro-Measurements, Raleigh, NC) and single-element gauges (FRA-1-11-1L, Sokki Kenkyujo Co., Tokyo) were attached along the horizontally oriented labial and lingual surfaces of the symphysis of a fresh alpaca mandible harvested from a cadaveric specimen. Each rosette was placed along the midline of the symphysis with one element oriented perpendicular to the long axis of the symphysis. In the case of the rosettes, this was the B-element. A single-element gauge oriented similarly was also placed directly on the vertically oriented and curved lingual surface of the symphysis Fig. 3.2).

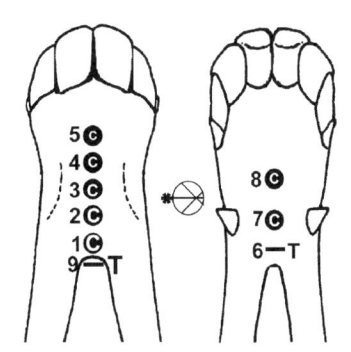

Fig. 3.2 Summary data of in vitro symphyseal strains along the labial (*left*) and lingual (*right*) surfaces of the alpaca symphysis during simulating lateral transverse bending, or wishboning. C, compression; T, tension. Numbers refer to gauges discussed in the text. Gauges 1–5 and 7–8 are rectangular rosettes and gauges 6 and 9 are single element gauges. The orientation of each rosette B-element and the single-element gauges is indicated by the *asterik*. Gauge 9 is attached to the symphysis directly on the curved, vertically oriented lingual surface. The strain gradient of tension or compression along the B- and single-elements is indicated by the letter size

Wishboning was simulated by pulling the two halves of the mandible apart at the level of the insertion of the deep masseter just below the mandibular notch. Reverse wishboning was simulated by pushing the two halves of the mandible together at this same level. The raw strain output for each element during simulated medial and lateral transverse bending – or reverse wishboning and wishboning, respectively – were recorded directly to paper via a Gould Brush 260 chart recorder. A calibration signal was also recorded to the chart recorder paper. The polarity and relative magnitude of the raw strains from each of the elements were determined from these chart recordings.

3.2.2 In Vivo Stresses – Gauge Placement, Recording, and Analysis

Symphyseal strain data were collected from four adult female alpacas. These animals were on loan from the Camelid Research Program of the College of Veterinary Medicine of Ohio State University. Prior to gauge placement, animals were fully anesthetized using an intramuscular injection of 5 mg/kg ketamine, 0.05–0.1 mg/kg butorphanol, 0.5 mg/kg xylazine (Mama, 2000). The skin over the symphysis was shaved and then infiltrated with a local anesthetic (lidocaine HCl) containing epinephrine (1:100,000). A small incision (< 2 cm) was made along the caudal border of the symphysis, and the skin and underlying periosteum reflected to expose the bone. The bone was then degreased and neutralized, and the gauge was attached to the bone using a cyanoacrylate adhesive. Delta rosettes (SA-06-030WY-120, Micro-Measurements, Raleigh, NC) were bonded in the midline on the labial surface of the symphysis, in the position of gauge 1 in Fig. 3.2. The A-element of the rosette was oriented parallel to the long axis of the symphysis. The incision was then sutured closed around the lead wires, which exited the incision and connected to the strain bridges.

Prior to the animal's recovery from anesthesia, the symphyses of two of the four alpacas (Alpacas 2 and 4) were gently reverse wishboned as described above. This was done to ensure that the gauges functioned properly and to facilitate the interpretation of the strain data recorded during mastication. Once fully alert and standing, the animals were fed hay. The resulting voltage output from each of the strain-gauge elements during the load simulations and mastication was conditioned and amplified (Vishay 2100 System, Vishay Instruments), and recorded at 15 in/s with a multiple-channel FM tape recorder (Honeywell 101e). Sequences were identified as left or right chews from an audio track on the tape recorder. Strain data were also monitored on a six-channel chart recorder (Gould Brush 260). In one experiment, (Alpaca 4 Experiment 2), a split screen of the subject and the strain tracings on the chart recorder was recorded to video using special effects generator so that bone strain could be qualitatively correlated with jaw movements. After sufficient data were collected, the animals were re-anesthetized and the gauges were removed. All recoveries were uneventful and the animals were administered prophylactic doses of antibiotics (Dual-Cillin) the day of and for several days following the experiment.

The raw strain data from the simulated stress sequences and from selected sequences of vigorous rhythmic chewing were digitized at a sampling rate of 500 Hz and filtered with a digital Butterworth low-pass filter set at 40 Hz. The magnitude of the maximum and minimum principal strains (ε_1 and ε_2, respectively), shear strain (γmax), and the angular value of ε_1 (the angle Φ) were calculated over 2-ms intervals. γmax is calculated as $\varepsilon_1 - \varepsilon_2$; because ε_1 is usually positive and ε_2 is usually negative, γmax is usually greater than both ε_1 and ε_2. The peak strains for each power stroke were identified as those coinciding with γmax. The angular value of ε_1 was measured relative to the long axis of the A-element of the delta-rosette strain gauge (and parallel to the long axis of the symphysis). Positive values were measured counterclockwise to the A-element and negative values were measured clockwise to the A-element. The angular value of ε_2 is always at $\pm 90°$ to the angular value of ε_1.

Descriptive statistics of ε_1, ε_2, $\varepsilon_1/\varepsilon_2$ and the direction of ε_1 were calculated at γmax and for the 25% levels of peak γmax along the mandibular symphysis during loading and unloading for each experiment. Data for each chewing side (left or right) are not combined in order to determine if there is a chewing side effect on the strain patterns, particularly with respect to the direction of ε_1. The grand means reported here for left and right chews are calculated from these experimental means. The largest single peak maximum and minimum principal strain at loading, peak, and unloading from all experiments are also reported.

3.3 Results

3.3.1 Symphyseal Strains During in vitro Transverse Bending

During simulated wishboning of the fresh alpaca jaw, all elements on gauges 1–5 (labial surface) and 7–8 (lingual surface) sensed compression (Fig. 3.2). Moreover, the B-element of each rosette, oriented perpendicular to the long axis of the symphysis, sensed greater levels of tension as compared to the A- and C-elements. On the other hand, the two single-elements, gauges 6 and 9, sensed tension. Gauge 9, located on the curved, vertically oriented surface of the symphysis and sensed greater tension than gauge 6, the most caudally placed lingual gauge. Thus, during wishboning, most of the symphysis is in compression with only the caudal portion of the symphysis sensing tension. During simulated reverse wishboning, the above pattern is reversed. That is, gauges 1 through 8 record tensile strains along the labial and lingual surfaces of the symphysis, whereas gauges 6 and 9 sensed compression. Based on these strain patterns, gauges 1 through 5, 7, and 8 are on one side of the neutral axis during transverse bending, and gauges 6 and 9 are on the other side (Fig. 3.3).

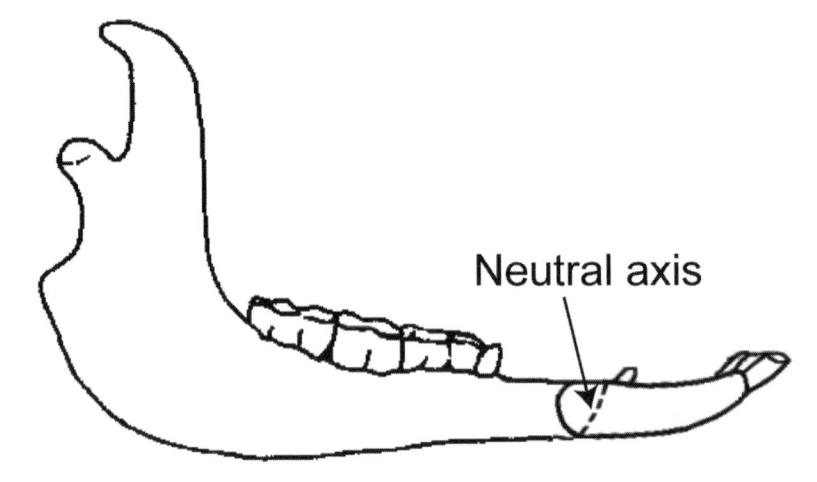

Fig. 3.3 Drawing of an alpaca mandible sectioned in the sagittal plane through the symphysis. The dashed line shows the approximate neutral axis during transverse bending as determined from the simulated loads and resulting strains presented in Fig. 3.2.

3.3.2 Symphyseal Strains During In Vivo Simulated Stresses and Mastication

3.3.2.1 Simulated Reverse Wishboning Strains

During simulated reverse wishboning, the maximum principal strain exceeds the minimum principal strain by a factor of 2–3.5 (Fig. 3.4). Moreover, the maximum principal strain is oriented between 90° and 105° relative to the long axis of the symphysis. The expected patterns of strain during medial and lateral transverse bending can be derived from the in vitro and in vivo strain data. During medial transverse bending, the maximum principal strain at the gauge site should be oriented at 90° relative to the long axis of the symphysis and about two to three times higher than the minimum principal strain (i.e., $\varepsilon_1/\varepsilon_2$ should range from about 2.0–3.0) (Fig. 3.5). The opposite should occur during lateral transverse bending. Maximum principal strain at the gauge site should be oriented at 0° relative to the long axis of the symphysis, and principal compression should exceed principal tension. Theoretically, shifts in chewing side should have no effect on transverse bending strain patterns.

3.3.2.2 Overview of Masticatory Strains

In vivo masticatory strain data were collected from a total of five experiments on four animals. Alpaca 4 was used in two experiments. The animals vigorously chewed hay unilaterally on both the left and right sides in all but one experiment (Alpaca 2), yielding 315 left and 297 right analyzed chews. Raw and principal strains during a vigorous chewing bout by Alpaca 4 are presented in Fig. 3.6, along with associated information on jaw movements. There are two notable bouts of

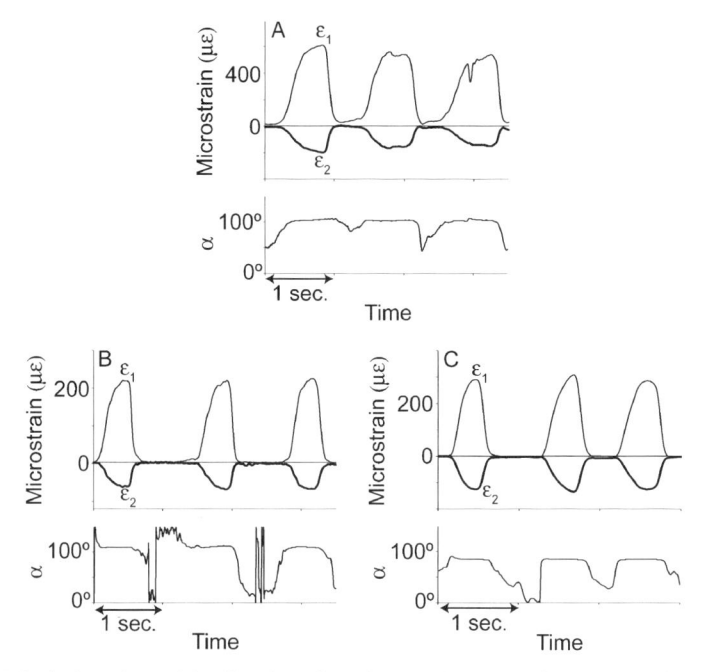

3 Symphyseal Fusion in Selenodont Artiodactyls

Fig. 3.4 Principal strains and the direction of maximum principal tension recorded from the symphyses of anesthetized alpacas during simulated manual reverse wishboning. (**A**) Alpaca 2 Experiment 1; (**B**) Alpaca 4 Experiment 1; (**C**) Alpaca 4 Experiment 2; ε_1, maximum principal strain (tension); ε_2, minimum principal strain (compression); α, direction of ε_1 relative to the A-element of the rosette and the long axis of the symphysis; sec., second

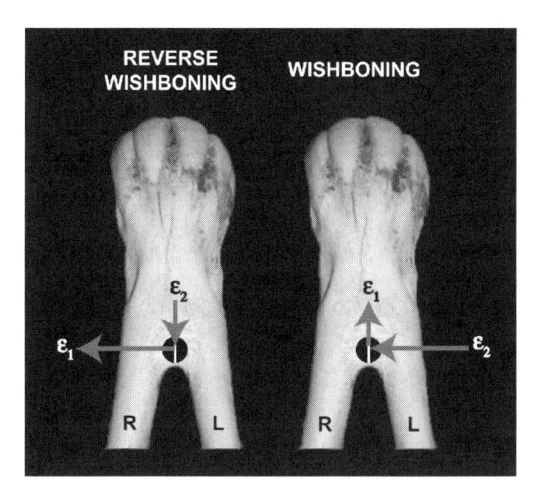

Fig. 3.5 Expected patterns of strain during reverse wishboning (i.e., medial transverse bending) and wishboning (i.e., lateral transverse bending) of the alpaca symphysis. The size of the arrow indicates the relative magnitude of the maximum and minimum principal strains (ε_1 and ε_2, respectively). The direction of ε_1 and ε_2 are determined relative to the A-element. All symbols as in previous figures

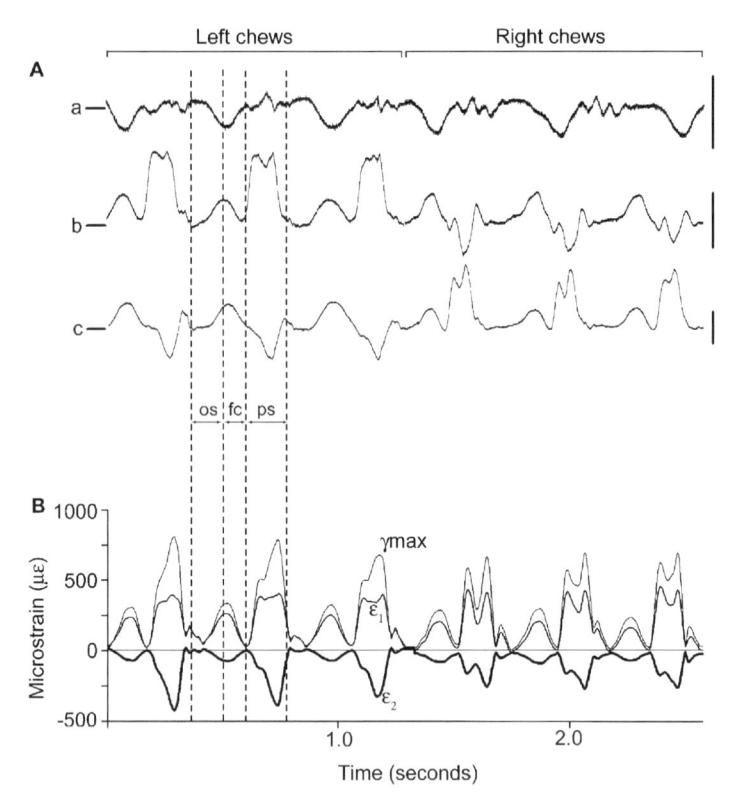

Fig. 3.6 Raw (**A**) and transformed (**B**) symphyseal strains during mastication by Alpaca 4 (Experiment 2). In **A**, raw strain traces from each of the gauge elements (a, b, and c) are shown. The 0 level of strain is shown by the horizontal bar next to each gauge-element label. Positive strains are tensile and negative strains are compressive. The scale bars to the right of each trace is 250 $\mu\varepsilon$. The opening stroke (os), fast-closing (fc), and the power stroke (ps) for one chewing cycle are indicated by the dashed vertical lines. In **B**, the principal (ε_1, ε_2) and shear (γmax) strains are shown

strains. During jaw opening, strains rise and peak at or near maximum gape. During fast-closing, strains decrease and subsequently increase again at the start of power stroke. Power stroke strains are complex, and there may be two to three minor peaks between which symphyseal strain does not fall to zero. Because the largest principal strains typically occur during the power stroke, we focus the remainder of this paper on strains during loading at 25% of peak γmax (25% loading), peak γmax, and unloading at 25% of peak γmax (25% unloading).

3.3.2.3 Summary Data of Principal Strains, $\varepsilon_1/\varepsilon_2$ and α at 25% Loading, Peak, and 25% Unloading

At 25% loading, the maximum and minimum principal strains average 101 $\mu\varepsilon$ (s.d. = 52) and $-25\,\mu\varepsilon$ (s.d. = 33), respectively, for left-side chews (Table 3.1; Fig. 3.7). During right-side chews, ε_1 averages 85 $\mu\varepsilon$ (s.d. = 48) and ε_2

3 Symphyseal Fusion in Selenodont Artiodactyls

Table 3.1 Summary data of symphyseal strains at 25% loading, peak, and 25% unloading of the power stroke of mastication

Chewing side	N	Total cycles	Phase	Maximum principal strain (ε_1)	Minimum principal strain (ε_2)	$\varepsilon_1/\varepsilon_2$	Direction of $\varepsilon_1(\alpha)$
Left	5	315	25% Load	101 ± 52	−25 ± 33	5.1 ± 5.0	131° ± 16
			Peak	308 ± 92	−155 ± 84	3.7 ± 4.2	132° ± 14
			25% Unload	96 ± 50	−28 ± 24	6.3 ± 6.7	124° ± 34
Right	4	297	25% Load	85 ± 48	−44 ± 9	2.1 ± 1.2	46° ± 14
			Peak	301 ± 138	−170 ± 64	1.9 ± 0.7	48° ± 9
			25% Unload	98 ± 41	−26 ± 25	4.8 ± 4.1	51° ± 14
Largest (side)			25% Load	195 (R)	−90 (L)	12.6 (L)	
			Peak	551 (L)	−439 (L)	11.2 (L)	
			25% Unload	212 (L)	−101 (L)	17.0 (L)	

N, total number of experiments across animals.

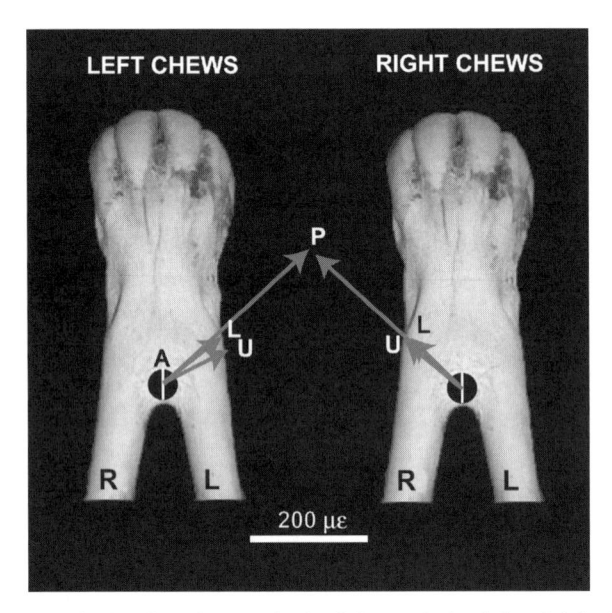

Fig. 3.7 Summary strain data from the symphysis of alpacas during left and right chews. Rt, right dentary; Lt, left dentary. The orientation of the A-element in each experiment is indicated by the white line on the gauge (*black circle*). Arrows indicate the magnitude and orientation of the maximum principal strains at 25% loading (L), peak (P), and 25% unloading (U)

averages $-44\,\mu\varepsilon$ (s.d. $= 9$). There is approximately a three- to fivefold increase in maximum principal strain magnitude from 25% loading to peak γmax, with ε_1 averaging 308 $\mu\varepsilon$ (s.d. $= 92$) and 301 $\mu\varepsilon$ (s.d. $= 138$) during left and right chews, respectively. The increase in minimum principal strain magnitude is slightly higher from 25% loading to peak, with left-side chews averaging $-155\,\mu\varepsilon$ (s.d. $= 84$) and right-side chews averaging $-170\,\mu\varepsilon$ (s.d. $= 64$). At 25% of peak γmax during unloading, ε_1 and ε_2 decrease to levels similar to those recorded at 25% loading. When chewing on the left, ε_1 averages 96 $\mu\varepsilon$ (s.d. $= 50$) and ε_2 averages $-28\,\mu\varepsilon$ (s.d. $= 24$). When chewing on the right, ε_1 averages 98 $\mu\varepsilon$ (s.d. $= 41$) and ε_2 averages $-26\,\mu\varepsilon$ (s.d. $= 25$).

The ratio of the principal strains (i.e., $\varepsilon_1/\varepsilon_2$) during all phases of the power stroke is greater than 1.0 for both left and right chews, indicating that the maximum principal strain exceeds the minimum principal strain (see Table 3.1). Principal strain magnitudes are most similar at peak γmax during both left and right chews as compared to at 25% loading and unloading. During left chews, $\varepsilon_1/\varepsilon_2$ averages 3.7 (s.d. $= 4.2$) whereas during right chews, $\varepsilon_1/\varepsilon_2$ averages 1.9 (s.d. $= 0.7$) at peak γmax. However, there is a significant amount of variation across experiments, with experimental means ranging from 1.0 to 12.6 at 25% loading, 1.1–11.2 at peak γmax, and 0.8–17.0 at 25% unloading.

The direction of ε_1 (i.e., α) varies as a function of chewing side in all experiments at each phase of jaw-closing (see Table 3.1; see Fig. 3.7). At 25% loading, α

averages 127° (s.d. = 15.3) when the animals chew on the left. When the animals chew on the right, α averages 53° (s.d. = 6.4). Thus, on average there is a 74° shift in the orientation of principal tension during loading. Mean angular data for ε_1 at peak γmax indicate that there is an 84° difference in the direction of tension along the symphysis between chewing sides. Maximum principal tension is oriented at 132° (s.d. = 14.0) and 48° (s.d. = 8.5) relative to the long axis of the symphysis during left- and right-side chews, respectively. At 25% unloading, this difference decreases to only 47°, with ε_1 oriented at 116° (s.d. = 26.7) during left-side chews and at 69° (s.d. = 27.2) during right-side chews.

Because α is determined relative to the A-element on the rosette, which is in line with the long-axis of the symphysis in the mid-sagittal plane, on average principal tension is oriented rostrally and to the left during left chews and rostrally and to the right during right chews (see Fig. 3.7). In the one experiment in which no right chews were recorded (Alpaca 2), the general orientation of principal tension during the left-side chews is similar to the other experiments at all phases of jaw-closing. Therefore, there probably would be a significant change in the orientation of principal tension if the animal had chewed on the right.

3.4 Discussion

3.4.1 Symphyseal Strains During Simulated Transverse Bending In Vitro and In Vivo: Implications for Interpreting Masticatory Strains

Based on the in vitro strains, expected strain patterns at the gauge sites associated with wishboning are straightforward: principal compression should exceed principal tension, and principal tension should be oriented at 0° relative to the long axis of the corpus. Chewing side should have no impact on the magnitude or orientation of principal strains during pure transverse bending.

The significance of the strain gradient along the lower border of the symphysis is less straightforward. Theoretically, the absolute value of the raw strains should increase at gauge sites farther from the neutral axis. However, according to the transverse bending simulation data on the alpaca jaw, strains from the B-element of each rosette on the upper and lower border of the symphysis tend to *decrease* rostrally. Therefore, if wishboning occurs during mastication, the recorded strains should be among the highest compressive strains along the labial aspect of the symphysis rather than the lowest. Of course, the largest strains along any portion of the symphysis will theoretically be tensile strains located along the caudal-most lingual margin during wishboning.

The above in vitro strain patterns associated with transverse bending could be the result of twisting moments introduced unintentionally during simulated transverse bending. An equally plausible explanation is that during transverse bending the alpaca symphysis does not behave as a curved beam being bent in its plane of curvature. In addition to the external loads placed on the mandible, the shape and internal architecture of the symphysis influence strains along its outer surface. If the alpaca mandible does not behave as a curved beam during transverse bending, the identification of anteroposterior shear or transverse twisting from in vivo masticatory strain data is still possible. However, the identification of wishboning and medial transverse bending from masticatory strain data will not be as obvious without corresponding strain data from the same gauge site during simulated loading. More importantly, comparative analyses of symphyseal size and shape may not be sufficient for assessing relative wishboning resistance in camelid versus non-camelid selenodont artiodactyls (cf. Hogue and Ravosa, 2001).

3.4.2 Symphyseal Strains During Rhythmic Mastication

In alpacas, there appears to be two distinct phases of significant symphyseal loading during rhythmic mastication, one during jaw opening and the other during the power stroke. Although the jaw-opening strains were not analyzed in this study, a cursory review of the magnitude and orientation of the principal strains suggests that ε_1 typically exceeds ε_2 and ε_1 is oriented at 90° relative to the long axis of the corpus regardless of chewing side. This strain pattern is consistent with medial transverse bending of the symphysis as described above for the simulated stresses. Hylander (1984) found a similar pattern for jaw-opening strains in macaques.

Chewing side consistently influences the orientation of the principal strains during the power stroke. Regardless of the magnitude of the principal strains or the $\varepsilon_1/\varepsilon_2$ values, α differed by approximately 90° between left- and right-side chews at peak γmax and somewhat less during loading and unloading. Because the orientation of principal strains during left and right chews would be the same if the symphysis was bent transversely, symphyseal strain patterns are not consistent with wishboning. This does not mean that transverse bending does not occur. However, it does mean that it is not the predominant strain pattern during mastication.

According to Hylander (1984), there are several strain patterns that would show a 90° shift in the orientation of the maximum principal strain associated with changes in chewing side. These are dorsoventral shear, anteroposterior shear, and/or twisting about a transverse axis (Fig. 3.8). Dorsoventral shear is due to the upward vertical components of the muscle force on the balancing-side and the oppositely directed vertical components of the bite force on the working side (Hylander, 1984). The presence of this loading regime cannot be verified in alpacas because the symphysis is horizontally inclined, placing the gauge out of plane of the applied load (see Fig. 3.8). Moreover, as pointed out by Hogue and Ravosa (2001), the posteriorly placed masticatory muscles and relatively long jaws of selenodont artiodactyls are a mechanically unfavorable system for producing vertical bite force at the incisors, resulting in significantly reduced vertical reaction forces at the symphysis. Thus, while we cannot definitely rule out the presence of dorsoventral shear, it is unlikely that it would be a significant loading regime in alpacas. This is in contrast to

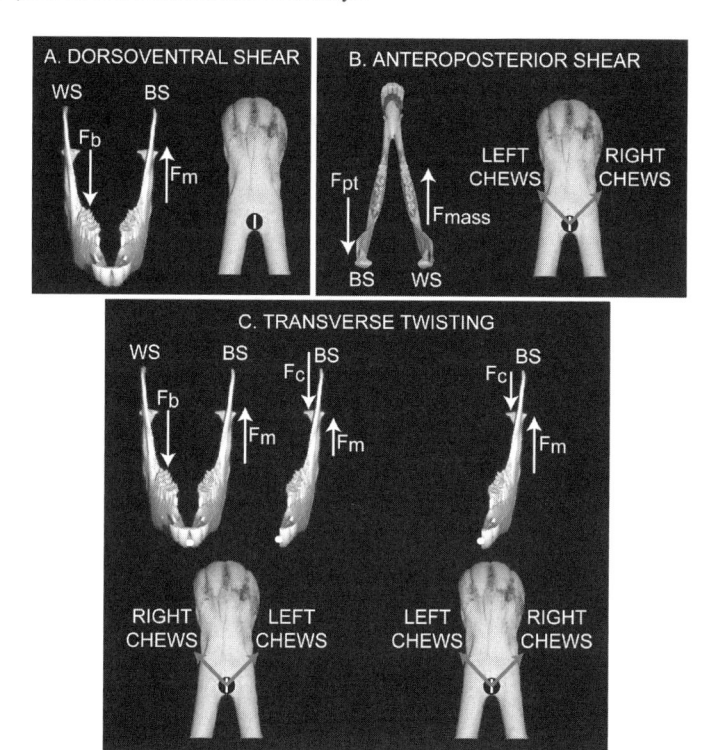

Fig. 3.8 Theoretical stresses and their expected patterns of strain during dorsoventral shear (**A**), anteroposterior shear (**B**), and transverse twisting (**C**) of the symphysis. WS, working side; BS balancing side; F_{pt}; F_{mass}; F_m, muscle force; F_b, bite force; F_c, condylar reaction force. White arrows indicate only the direction of the force. Gray arrows indicate the expected orientation of the maximum principal strain (ε_1) during left and right chews for each of the three loading regimes. Expected patterns of strain are indicated for the gauge site used in the in vivo experiments. During dorsoventral shear, there is no expected strain pattern because the gauge is out of plane of the applied load. See text and Hylander (1984) for additional details

macaques in which dorsoventral shear is a significant component of symphyseal loading during the power stroke (Hylander, 1984, 1985).

In contrast to dorsoventral shear, the gauges are aligned parallel to the shearing force due to anteroposterior shear. Anteroposterior shear of the symphysis was first described by Beecher (1977), who noted that when some mammals chew, the working-side dentary is displaced anteriorly relative to the balancing-side dentary during the power stroke. He hypothesized that this was due to the posteriorly directed pull of the balancing-side temporalis and the anteriorly directed pull of the working-side masticatory force. There are several reasons to conclude that the strain data do not reflect this loading regime either. If anteroposterior shear is the predominant loading regime, ε_1 would be directed at 45° during left-side chews and ε_1 and at 135° during right-side chews (see Fig. 3.8) (Hylander, 1984). The data presented above are directly opposite to this predicted pattern and, therefore, anteroposterior shear is likely not the predominant loading regime during the power stroke along the alpaca symphysis.

The strain data are most consistent with transverse twisting of the symphysis. Transverse twisting of the symphysis is thought to be the result of the oppositely directed vertical components of the muscle force, bite force, and/or condylar reaction forces from the two sides of the jaw (see Fig. 3.8). However, the expected patterns of strain associated with transverse twisting are dependent on the combination of forces involved. For example, if twisting is a result of the vertical components of the bite force directed downward on the working side and the muscle force directed upward on the balancing side, the working-side corpus should be depressed and the balancing-side corpus should be elevated. For left-sided chews, the pattern of strain along the lower border of the labial aspect of the symphysis should indicate ε_1 directed at 45°, and during chews on the right side, ε_1 should be directed at 135° (Hylander, 1984). On the other hand, if transverse twisting results from the oppositely directed vertical components of the muscle force and condylar reaction force on the same side of the jaw, then the magnitude of the moments associated with these forces need to be considered. If the moment associated with the balancing-side condylar reaction force is larger than that of the balancing-side muscle force, then the balancing-side corpus will rotate in a counter-clockwise direction about the twisting axis of neutrality through the symphysis. This will result in ε_1 directed at 135° for chews on the left side and 45° for chews on the right side. If the moment associated with the balancing-side condylar reaction force is smaller than that of the bite force, then the balancing-side corpus will be rotated clockwise about the twisting axis of neutrality and the opposite strain pattern should occur. That is, ε_1 should be directed at 45° for left-side chews and 135° for right-side chews (see Fig. 3.8) (Hylander, 1984).

According to the strain data, the mandibular symphysis of alpacas is twisted about a transverse axis as due to the vertical components of the bite force on the working side depressing the corpus and those from the balancing-side muscles elevating the balancing-side corpus or from the vertical components of the balancing-side muscle force and oppositely directed balancing-side condylar reaction force. If this latter explanation is the case, then the moment associated with the condylar reaction force is larger than the moment associated with the balancing-side muscle force (see Fig. 3.8). This type of transverse twisting is similar to what has been observed in macaques (Hylander, 1984, 1985).

Although there is strong evidence in favor of transverse twisting of the symphysis, symphyseal strains reflect a combination of loading regimes because the maximum principal strains are rarely exactly at 45° and 135° during the power stroke and because principal strain magnitudes are not always equal. Deviations from this predicted pattern are most notable at 25% loading and 25% unloading, but they are also evident at peak loading. During loading and unloading, the ratio of the principal strains is on average higher than at peak loading and the orientation of ε_1 tends to be more transversely oriented across the symphysis, i.e., $\alpha > 45$ for right chews and $\alpha < 135$ for left chews. These patterns are consistent with frontal bending of the symphysis in which the labial aspect of the symphysis is loaded in tension and the lingual aspect is loaded in compression. This may be due to the tendency of the mandibular corpora to twist about their long axes, inverting the alveolar process and everting the lower border (Hylander, 1984).

In summary, given the in vivo strain data, the alpaca symphysis is primarily twisted about a transverse axis upon which frontal bending may be superposed during portions of the closing phase of the chewing cycle.

3.4.3 *Symphyseal Strains and Jaw Morphology in Alpacas*

Based on the data presented above, there is little evidence for wishboning as the predominant masticatory loading regime of the alpaca symphysis. However, because strains were only quantified at peak γmax and at 25% of peak γmax during loading and unloading, it is unclear that these selected data points adequately characterize the complexity of symphyseal strains during the entire power stroke. Moreover, it is difficult to fully characterize what effect the late activity of the balancing-side deep masseter has on symphyseal strains when only these data points are considered. Therefore, a closer examination of the relationship between the late activity of the balancing-side deep masseter and symphyseal strains may shed some light on whether this jaw-muscle activity pattern can be directly tied to any component of masticatory strains along the symphysis.

Figure 3.9 provides an example of simultaneously recorded electromyographic and strain data from Alpaca 3 (Experiment 1). Of interest here is the second peak of balancing-side deep masseter activity, which appears to be correlated with a shift in the orientation and relative magnitude of the strains. During this time, principal compression slightly exceeds principal tension (i.e., $\varepsilon_1/\varepsilon_2 < 1.0$) and the orientation of ε_1 drops to around 0°, both of which are consistent with wishboning. In this particular chewing sequence, principal strains are typically negligible at less than 20 με. This second peak of balancing-side deep masseter activity is not always present across experiments or animals (Williams, 2004). However, when it does occur, it appears to affect symphyseal strains in the predicted manner, resulting in some noticeable but arguably negligible wishboning. The absence of wishboning associated with the first peak of balancing-side deep masseter activity (which is delayed relative to the other jaw adductors) is surprising, particularly given that these animals have strongly curved symphyses and a relatively long moment arm associated with wishboning. Perhaps, this laterally directed force exerted by the main burst of activity of the balancing-side deep masseter is resisted by a medially directed force from other muscles, such as the lateral pterygoid. Currently, we have no EMG data from alpacas to verify this hypothesis.

Given these results, can symphyseal strain patterns explain differences in symphyseal morphology between camelids and non-camelids? In order to answer this question, we undertook a broader consideration of symphyseal morphology in contrast to more traditional comparative biomechanical approaches that use beam theory to determine relative load resistance capabilities of the symphysis (e.g.,

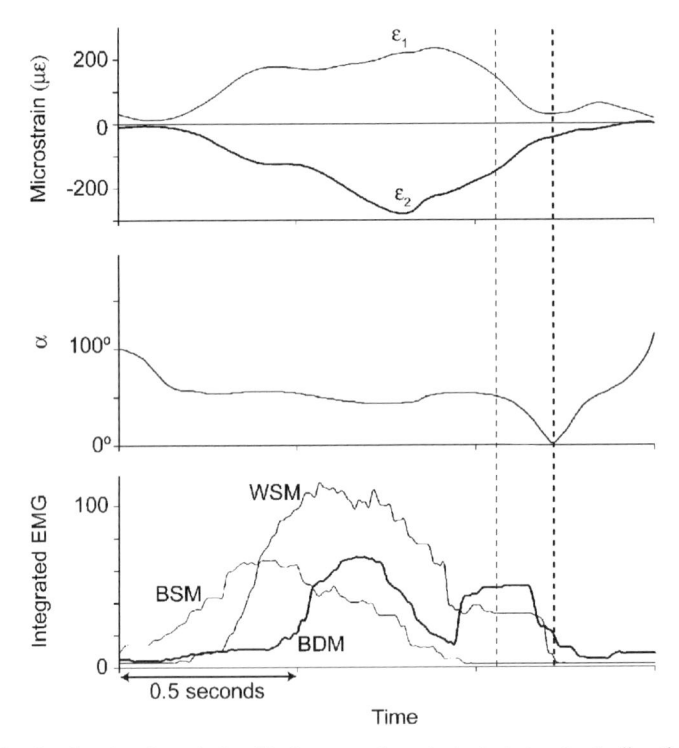

Fig. 3.9 Graphs showing the relationship between the principal strains (*top*), direction of $\varepsilon_1(\alpha)$ (*middle*) and integrated masseter EMG (*bottom*) from a single power stroke during chewing by an alpaca. BSM, balancing-side superficial masseter; BDM, balancing-side deep masseter; WSM, working-side superficial masseter. Dashed vertical lines through the graphs indicate the time at the end of the power stroke when strains indicate an increase in wishboning of the symphysis in association with a second peak of activity of the balancing-side deep masseter

Bouvier, 1986; Daegling, 1992, 2001; Daegling and Hylander, 1998; Hogue and Ravosa, 2001; Hylander, 1984, 1985; Ravosa, 1991, 1996a, b, 2000; Ravosa and Hogue, 2004; Vinyard and Ravosa, 1998; Vinyard et al., 2003; Williams et al., 2002). Following the classification system developed by Scapino (1981) for carnivorans, gross symphyseal morphology of unfused selenodont artiodactyls was examined using osteological specimens of two to five adult male and female individuals housed in the American Museum of Natural History. These observations highlighted the potential role that bony interdigitations in the symphysis can play in resisting the observed in vivo symphyseal loads. Bony interdigitations in the symphysis project from the surface of one dentary and fit into a corresponding depression in the symphyseal surface of the opposite dentary. Symphyses with relatively flat symphyseal surfaces are considered to be Class I symphyses whereas those that are fully fused are Class IV symphyses. Class II and Class III symphyses are both interdigitated, with Class III being the more heavily interdigitated.

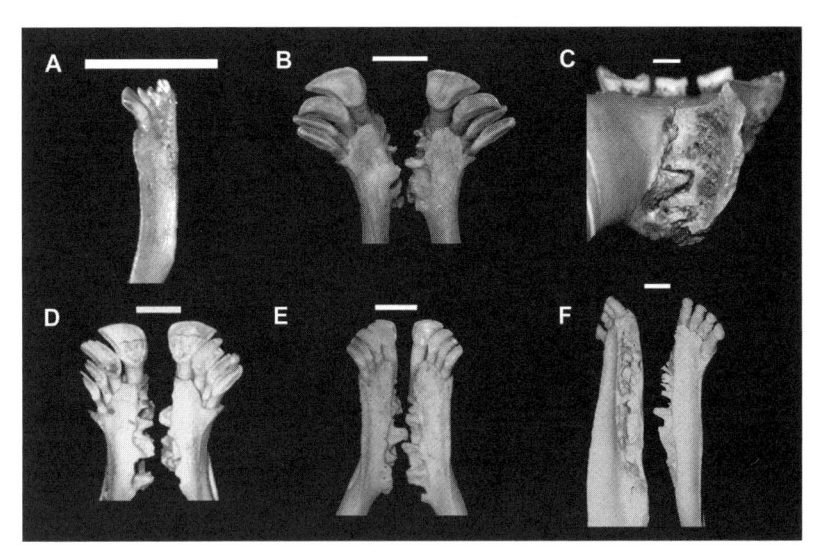

3 Symphyseal Fusion in Selenodont Artiodactyls

Fig. 3.10 Symphyses of four species of unfused selenodont artiodactyls: (**A**) *Tragulus javanicus*, Tragulidae; (**B**) *Procapra gutterosa*, Bovidae; (**C**) *Kobus ellipsiprymnus*, Bovidae; (**D**) *Dama dama*, Cervidae; (**E**) *Mazama mazama*, Cervidae; (**F**) *Okapi johnstoni*, Giraffidae. All scale bars are 1 cm. **A**, **B**, **D**, and **E** are lingual views. In **C**, the right dentary is broken through the symphyseal portion, exposing the interdigitating rugosities connecting the two hemimandibles. In **F**, the symphyseal plate of the left dentary is shown on the left and the lingual view of the right dentary is on the right. *Tragulus* is an example of a Class I symphysis. All other specimens are Class III symphyses

Of the 34 selenodont artiodactyl species examined, including representatives from all non-camelid extant families and 12 subfamilies, all but two species have Class III symphyses. The tragulids *Hyemoschus* and *Tragulus* have Class I symphyses (Fig. 3.10). In the species with Class III symphyses, the symphyseal plates are flatter in the region of the incisor roots where there are only minor interdigitations. Caudally, these interdigitations, which are offset both dorsoventrally and anteroposteriorly, typically increase in number and/or size throughout the symphysis. According to Lieberman and Crompton (2000), this is the region of the fibrocartilaginous pad in goats (see also Beecher, 1977, 1979; Scapino, 1981). Interestingly, examination of several infant and juvenile osteological specimens (e.g., *Ourebia ourebia, Antilocapra americana, Connochaetes gnou, Hippotragus equinus, Alcelaphus buselaphus*) indicates that they typically have relatively flat symphyseal surfaces (Class I), and the interdigitations become more pronounced with age in both number and size.

While these interdigitations increase the surface area for ligamentous attachment, they also help to resist forces loading the joint (Beecher, 1977; Lieberman and Crompton, 2000; Ravosa and Hylander, 1994; Rigler and Mlinsek, 1968; Scapino, 1981). Because they are interlocking and offset from one another in multiple vertical and anteroposterior planes, they can effectively resist anteroposterior shear, dorsoventral shear, and twisting of the symphysis about the transverse axis.

However, transverse twisting of the symphysis appears to be the dominant loading regime during the power stroke of mastication in alpacas. Thus, it is reasonable to assume that goats and other unfused selenodont artiodactyls also transversely twist their symphyses, for which symphyseal interdigitations in addition to ligaments and other connective tissues, likely provide some resistance.

These observations do not directly address the question of why camelids fuse their symphyses, particularly if all selenodont artiodactyls transversely twist their symphyses. A comparison on the mandibles of unfused and fused selenodont artiodactyls highlights one possible explanation linking symphyseal fusion to the strain data from alpacas, as well as taking into consideration the morphology of the unfused symphyses described above. In addition to fusion, camelids differ from unfused selenodont artiodactyls in having large, hypsodont, and deeply rooted incisors. Moreover, in contrast to unfused selenodont artiodactyls, the incisors of the South American camelids (alpaca, lama, vicuña, and guanaco) tend to radiate from the midline, leaving little room for a series of bony interdigitations. Finally, in young, pre-fused individuals, the incisors are very well developed at birth and it appears that they do not develop interdigitations prior to the initiation of fusion (Fig. 3.11).

Thus, based on the combined in vivo and morphological observations, we propose that selenodont artiodactyls will fuse if (1) they have to resist transverse twisting of the symphysis and (2) they have large, deeply rooted incisors precluding the accommodation of numerous deeply interdigitating rugosities. The link between incisor size and fusion has been proposed previously by other researchers including Greaves (1988) and Hiiemae and Kay (1972). However, in those hypotheses, symphyseal fusion is linked to incisor use during food procurement, for which there is very little behavioral data and no strain data. The hypothesis proposed here is the first to link relative incisor size, fusion, and symphyseal strains incurred during

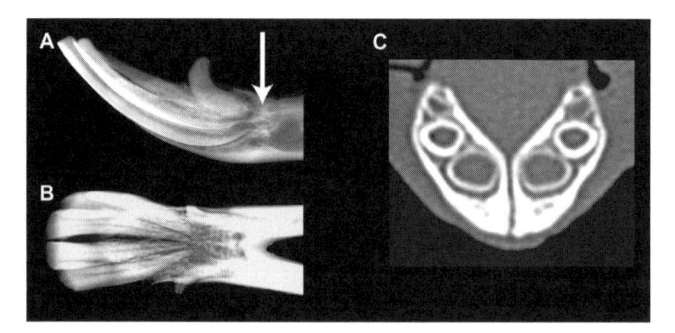

Fig. 3.11 The symphyseal region in South American camelids. (**A**) Lateral radiograph of the symphyses of an adult alpaca; (**B**) radiograph through the symphysis of a guanaco (*Lama guanicoe*); (**C**) computed tomography scan (sagittal plane) of the symphysis of a 4-month old alpaca. In **A**, the arrow indicates the caudalmost border of the symphysis in this specimen. Note the relatively large incisors and narrow symphyses in **A** and **B**. In the infant, the incisors are large and this region of the symphysis is unfused with flat symphyseal plates. The anterior 1/3 of the symphysis of this animal is already fused

mastication. Additional in vivo data as well as comparative data from extant and extinct camelids are currently being collected to evaluate this new hypothesis. Given the relatively robust experimental data set for cercopithecoid primates linking symphyseal fusion to wishboning, we stress that this hypothesis may only be applicable to selenodont artiodactyls.

3.5 Conclusions

Strain data from alpacas indicate that the symphysis is primarily twisted about a transverse axis during the power stroke of mastication. There is little evidence of wishboning at the levels of strain greater than 25% of peak strain. However, wishboning does occur toward the end of the power stroke at very minimal strain levels and is associated with the activity of the balancing-side deep masseter. While relative symphyseal dimensions in both primates and selenodont artiodactyls indicate that species with fused symphyses are better able to resist wishboning (Hogue and Ravosa, 2001; Hylander, 1985), the strain data from alpacas are not consistent with this biomechanical interpretation of mandibular form. Thus, whereas similar comparative studies in conjunction with in vivo data in primates coincide to offer a plausible explanation for fusion in anthropoids, biomechanical interpretations of symphyseal form and jaw-muscle activity patterns in selenodont artiodactyls are not necessarily indicative of symphyseal loading patterns (cf. Hogue and Ravosa, 2001; Williams et al., 2003).

Combined with the strain data, subsequent preliminary investigation into gross symphyseal morphology provides the foundation for a hypothesis linking fusion, incisor size, and the observed masticatory strains in selenodont artiodactyls. However, because additional data are required to more fully test this hypothesis, these observations should be treated as preliminary. Regardless, findings from this study suggest that while anthropoids and camelids may exhibit convergent and derived symphyseal morphologies as well as similarities in the firing patterns of their balancing-side deep masseters, there is likely more than one loading regime driving fusion of the symphysis in mammals. Moreover, unraveling its functional and adaptive significance will best be accomplished using multiple and complementary approaches including in vivo, in vitro, and comparative data.

Acknowledgments This chapter is based on a presentation given at the 74th Annual Meeting of the American Association of Physical Anthropologists in a symposium in honor of William L. Hylander. We would like to thank Dr. David Anderson, Head of Farm Animal Surgery and Director of the International Camelid Initiative, Department of Veterinary Clinical Sciences at The Ohio State University College of Veterinary Medicine, for the loan of the four alpacas used in this study. Many, many thanks go to the staff of the Duke University Department of Laboratory Animal Resources. We extend our thanks to Calvin Davis and Robert Parker of the Duke University Research Farm, who cared for the animals used in this study as well as assisted with animal handling during experiments. Kirk Johnson provided technical support during all phases of this project for which we are most grateful. Finally, we thank Eileen Westwig for facilitating the collection of data in the Division of Mammalogy in the American Museum of Natural History.

This research was supported by a National Science Foundation (NSF) Dissertation Improvement Grant (BCS-02-41652), grants from Sigma Xi and the Ford Foundation, as well as NSF research grants (BCS-01-38565 and IOS-05-020855).

References

Beecher, R.M. (1977). Function and fusion at the mandibular symphysis. *Am. J. Phys. Anthropol.* 47:325–336.
Beecher, R.M. (1979). Functional significance of the mandibular symphysis. *J. Morphol.* 159:117–130.
Bouvier, M. (1986). Biomechanical scaling of mandibular dimensions in New World Monkeys. *Int. J. Primatol.* 7:551–567.
Carter, D.R., Caler, W.E., Spengler, D.M., Frankel, V.H. (1981). Fatigue behavior of adult cortical bone: the influence of mean strain and strain range. *Acta Orthop. Scand.* 52:481–490.
Carter, D.R., Spengler, D., Frankel, V.H. (1977). Bone fatigue in uniaxial loading at physiologic strain rates. *IRCS Med. Sci.* 5:592.
Daegling, D.J. (1992). Mandibular morphology and diet in the genus *Cebus*. *Int. J. Primatol.* 13:545–570.
Daegling, D.J. (2001). Biomechanical scaling of the hominoid mandibular symphysis. *J. Morphol.* 250:12–23.
Daegling, D.J., Hylander, W.L. (1998). Biomechanics of torsion in the human mandible. *Am. J. Phys. Anthropol.* 105:73–87.
Greaves, W.S. (1988). A functional consequence of an ossified mandibular symphysis. *Am. J. Phys. Anthropol.* 77:53–56.
Hiiemae, K., Kay, R.F. (1972). Trends in the evolution of primate mastication. *Nature* 240:486–487.
Hogue, A.S., Ravosa, M.J. (2001). Transverse masticatory movements, occlusal orientation, and symphyseal fusion in selenodont artiodactyls. *J. Morphol.* 249:221–241.
Hylander, W.L. (1977). *In vivo* bone strain in the mandible of *Galago crassicaudatus*. *Am. J. Phys. Anthropol.* 46:309–326.
Hylander, W.L. (1979a). Functional significance of primate mandibular form. *J. Morphol.* 160:223–240.
Hylander, W.L. (1979b). Mandibular function in *Galago crassicaudatus* and *Macaca fascicularis*: an in-vivo approach to stress analysis of the mandible. *J. Morphol.* 159:253–296.
Hylander, W.L. (1984). Stress and strain in the mandibular symphysis of primates: a test of competing hypotheses. *Am. J. Phys. Anthropol.* 64:1–46.
Hylander, W.L. (1985). Mandibular function and biomechanical stress and scaling. *Am. Zool.* 25:315–330.
Hylander, W.L., Johnson, K.R. (1994). Jaw muscle function and wishboning of the mandible during mastication in macaques and baboons. *Am. J. Phys. Anthropol.* 94:523–547.
Hylander, W.L., Johnson, K.R., Crompton, A.W. (1987). Loading patterns and jaw movements during mastication in *Macaca fascicularis*: a bone-strain, electromyographic, and cineradiographic analysis. *Am. J. Phys. Anthropol.* 72:287–314.
Hylander, W.L., Ravosa, M.J., Ross, C.F., Wall, C.E., Johnson, K.R. (2000). Symphyseal fusion and jaw-adductor muscle force: an EMG study. *Am. J. Phys. Anthropol.* 112:469–492.
Hylander, W.L., Vinyard, C.J., Wall, C.E., Williams, S.H., Johnson, K.R. (2002). Recruitment and firing patterns of jaw muscles during mastication in ring-tailed lemurs. *Am. J. Phys. Anthropol. Suppl.* 34:88.
Hylander, W.L., Vinyard, C.J., Wall, C.E., Williams, S.H., Johnson, K.R. (2003). Convergence of the "wishboning" jaw-muscle activity pattern in anthropoids and strepsirrhines: the recruitment and firing of jaw muscles in *Propithecus verreauxi*. *Am. J. Phys. Anthropol. Suppl.* 36:120.

Hylander, W.L., Wall, C.E., Vinyard, C.E., Ross, C.F., Ravosa, M.J. (2004). Jaw adductor force and symphyseal fusion. In: Anapol, F., German, R.Z., Jablonski, N.G. (eds.), *Shaping Primate Evolution: Papers in Honor of Charles Oxnard*. Cambridge University Press, Cambridge, pp. 229–257.

Keaveny, T., Hayes, W. (1993). Mechanical properties of cortical and trabecular bone. In: Hall, B.K. (ed.), *Bone Growth*. CRC Press, Boca Raton, pp. 285–344.

Lieberman, D.E., Crompton, A.W. (2000). Why fuse the mandibular symphysis? A comparative analysis. *Am. J. Phys. Anthropol.* 112:517–540.

Luschei, E.S., Goodwin, G.M. (1974). Patterns of mandibular movement and jaw muscle activity during mastication in the monkey. *J. Neurophysiol.* 37:954–966.

Mama, K. (2000). Anesthetic management of camelids. In: Steffey, E.P. (ed.), *Recent Advances in Anesthetic Management of Large Domestic Animals*. International Veterinary Information Service, Ithaca (September 4, 2000; http://www.ivis.org/advances/Steffey_Anesthesia/mama_camelids/chapter_frm.asp?LA=1).

Ravosa, M.J. (1991). Structural allometry of the prosimian mandibular corpus and symphysis. *J. Hum. Evol.* 20:3–20.

Ravosa, M.J. (1996a). Jaw morphology and function in living and fossil Old World monkeys. *Int. J. Primatol.* 1996:909–932.

Ravosa, M.J. (1996b). Mandibular form and function in North American and European Adapidae and Omomyidae. *J. Morphol.* 229:171–190.

Ravosa, M.J. (1999). Anthropoid origins and the modern symphysis. *Folia Primatol.* 70:65–78.

Ravosa, M.J. (2000). Size and scaling in the mandible of living and extinct apes. *Folia Primatol.* 71:305–322.

Ravosa, M.J., Hogue, A.S. (2004). Function and fusion of the mandibular symphysis in mammals: a comparative and experimental perspective. In: Ross, C.F. Kay, R.K. (eds.), *Anthropoid Origins: New Visions*. Kluwer Press/Plenum Publishers, New York, pp. 399–488.

Ravosa, M.J., Hylander, W.L. (1994). Function and fusion of the mandibular symphysis in primates: stiffness or strength? In: Fleagle, J.G. Kay, R.F. (eds.), *Anthropoid Origins*. Plenum Press, New York, pp. 447–468.

Ravosa, M.J., Simons, E.L. (1994). Mandibular growth and function in Archaeolemur. *Am. J. Phys. Anthropol.* 95:63–76.

Ravosa, M.J., Vinyard, C.J., Gagnon, M., Islam, S.A. (2000). Evolution of anthropoid jaw loading and kinematic patterns. *Am. J. Phys. Anthropol.* 112:493–516.

Rigler, L., Mlinsek, B. (1968). Die Symphyse der Mandibula beim Rinde. Ein Beitrag zur Kenntnis ihrer Struktur und Funktion. *Anato. Anz.* 122:293–314.

Scapino, R.P. (1981). Morphological investigation into the function of the jaw symphysis in carnivorans. *J. Morphol.* 167:339–375.

Vinyard, C.J., Ravosa, M.J. (1998). Ontogeny, function, and scaling of the mandibular symphysis in papionin primates. *J. Morphol.* 235:157–75.

Vinyard, C.J., Wall, C.E., Williams, S.H., Hylander, W.L. (2003). A comparative functional analysis of skull morphology of tree-gouging primates. *Am. J. Phys. Anthropol.* 120:153–170.

Vinyard, C.J., Wall, C.E., Williams, S.H., Johnson, K.R., Hylander, W.L. (2006). Masseter electromyography during chewing in ring-tailed lemurs (*Lemur catta*). *Am. J. Phys. Anthropol.* 130:85–95.

Vinyard, C.J., Williams, S.H., Wall, C.E., Johnson, K.R., Hylander, W.L. (2001). Deep masseter recruitment patterns during chewing in callitrichids. *Am. J. Phys. Anthropol. Suppl.* 32:156.

Williams, S.H. (2004). Mastication in selenodont artiodactyls: an *in vivo* study of masticatory form and function in goats and alpacas. Ph.D. Thesis, Duke University, Durham.

Williams, S.H., Vinyard, C.J., Wall, C.E., Hylander, W.L. (2003). Symphyseal fusion in ungulates and anthropoids: a case of functional convergence? *Am. J. Phys. Anthropol. Suppl.* 36:226.

Williams, S.H., Vinyard, C.J., Wall, C.E., Hylander, W.L. (2007). Masticatory motor patterns in ungulates: a quantitative assessment of jaw-muscle coordination in goats, alpacas and horses. *J. Exp. Zool. Part A: Ecol. Genet. Physiol.*, 307:226–240.

Williams, S.H., Wall, C.E., Vinyard, C.J., Hylander, W.L. (2002). A biomechanical analysis of skull form in gum-harvesting galagids. *Folia Primatol.* 73:197–209.

Chapter 4
Does the Primate Face Torque?

Callum F. Ross

Contents

4.1 Introduction	63
4.2 Materials and Methods	68
4.3 Results	69
4.3.1 Postorbital Bar	71
4.3.2 Dorsal Skull Surface	72
4.4 Discussion	74
4.4.1 Skull Loading in *Eulemur*	74
4.4.2 Comparisons with Other Primates	78
4.5 Conclusions	79
References	80

4.1 Introduction

The nature and strength of the relationship between the morphology of the primate facial skeleton and feeding behavior are not well understood. This lack of understanding is perhaps to be expected, given that we have a poor understanding of both the external forces acting on the skeleton and the resulting patterns of stress and strain in the skeleton. What we do know about in vivo patterns of strain in the primate facial skeleton to date derives in large part from the work of W. L. Hylander, who pioneered the use of strain gages and electromyography in the study of primate feeding, and that of his students, and colleagues.

Hylander et al.'s 1991 paper on in vivo bone strain in the brow-ridges of macaques and baboons was a seminal contribution in two respects. First, it showed that in macaques and baboons strain magnitudes above the orbits are much lower than those recorded from the mandible, zygomatic arch, and anterior root of the zygoma. These data force the conclusion that the cercopithecine cranium is not optimized for resisting and dissipating feeding forces, where optimality is defined

---

C.F. Ross
Organismal Biology & Anatomy, University of Chicago, 1027 East 57th Street, Chicago, IL 60637
e-mail: rossc@uchicago.edu

as maximum strength with minimum material. Despite persistent disregard of these results by some workers (e.g., Prossinger et al., 2000; Wolpoff et al., 2002), it is clear that cercopithecine supraorbital structures are much larger than they need to be, given their shapes, to dissipate the forces experienced during feeding. Thus, their shape could be very different and/or their size could be radically reduced without compromising their ability to resist and dissipate feeding forces. Hylander et al. (1991) suggested that "enough bone must be present within the supraorbital and bridged regions to prevent structural failure due to relatively infrequent non-masticatory external forces associated with highly active primates (e.g., traumatic accidental forces applied to the orbits and neurocranium)" (Hylander et al., 1991). This hypothesis was later reiterated by Ravosa and colleagues to explain the relatively low bone strain magnitudes recorded from the galago postorbital bar during mastication (Ravosa et al., 2000).

Subsequent work has shown that extremely low bone strain magnitudes also characterize the thin bony plates forming the medial orbital walls of owl monkeys, macaques, and galagos (Ross, 2001). As these thin plates are unlikely to provide protection against "traumatic accidental forces," it seems reasonable to conclude that circumorbital bones might perform a range of functions, including not only protecting against trauma but also augmenting attachment areas for epithelia and muscles, and deflecting muscles away from the eye. Thus, the bone strain data gathered to date suggest that the facial skeleton performs many functions and, in so doing, becomes over-designed for dissipating feeding forces. This suggests that the morphology of the primate circumorbital region at least, and maybe other parts of the face as well, is not determined by the nature of the stresses to which it is subjected during feeding.

Second, Hylander et al.'s 1991 paper was also seminal in showing that patterns of loading in the primate face are complex and cannot be predicted and interpreted by modeling the face as a single simple structural member, such as a beam or a cylinder. Rather, the face needs to be conceptualized as simpler structures, subjected to simple loads. In this vein, Hylander et al. (1991) suggested that the supraorbital region is bent in a frontal plane, and Hylander and Johnson suggested that the zygomatic arch of macaques is bent in parasagittal and transverse planes, and twisted about an anteroposterior axis (Hylander and Johnson, 1997). The success of modeling the face as simpler structures does not mean that large-scale, global loading regimes do not act on the primate face; indeed, they probably do. However, it does suggest that these global loading regimes cannot be simplistically used to predict loading regimes in local regions of the face.

The most salient example of a global loading regime in the primate face is Greaves' hypothesis that the primate face twists on the braincase about an anteroposterior axis during mastication (Greaves, 1985). Consideration of the external forces acting on the face suggests that this hypothesis must be true of the pre-orbital facial skeleton (Fig. 4.1). The face must twist on the braincase when the torques (twisting moments) acting on the face in one direction exceed those acting in the other direction. During unilateral biting and mastication, the bite force is the only external force acting on the face anterior to the attachments of the chewing muscles

4 Does the Primate Face Torque? 65

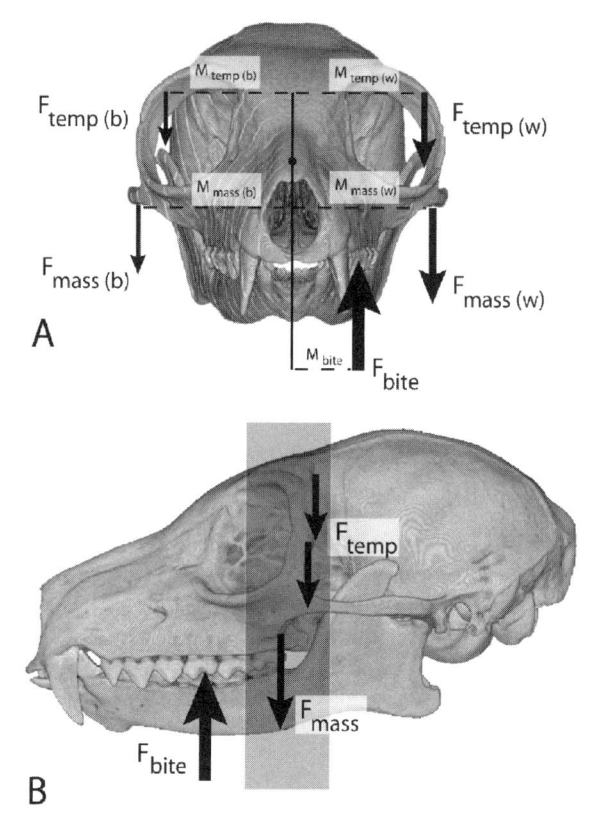

Fig. 4.1 Diagram of rostral and lateral views of *Eulemur fulvus* skull illustrating the principal external forces hypothesized to impose twisting moments (torques) on the facial skeleton. F_{bite} = bite force during unilateral biting or mastication; F_{mass} = inferiorly directed components of masseter muscle force acting bilaterally; F_{temp} = inferiorly directed components of temporalis muscle force acting bilaterally; M_{bite} = moment arm of bite force about section centroid; $M_{mass(w)}$ or $M_{mass(b)}$ = moment arm of working (w) or balancing (b) side masseter muscle force; $M_{temp(w)}$ or $M_{temp(b)}$ = moment arm of working side (w) or balancing side (b) temporalis muscle force. Inferiorly directed components of medial pterygoid muscle force acting bilaterally are omitted for clarity. The *Eulemur* face will twist in a counterclockwise direction when $(F_{bite} \times M_{bite} + F_{mass(b)} \times M_{mass(b)} + F_{temp(b)} \times M_{temp(b)}) > (F_{mass(w)} \times M_{mass(w)} + F_{temp(w)} \times M_{temp(w)})$. Shaded box in B illustrates vertical sections in which gages were placed in these experiments. The facial skeleton of *Eulemur* anterior to the rostrad insertion of masseter will be subjected to an external torque during unilateral biting or mastication, because there are no other external forces to resist the torque of the bite force (B)

(Fig. 4.1B). Consequently, the facial skeleton anterior to those muscle attachments must be twisting (counterclockwise in Fig. 4.1A). Moreover, twisting will also occur in progressively more posterior frontal sections, until a section is reached in which clockwise and counterclockwise torques become equal.

The sections of interest in this study (from which bone strain data were recorded) are indicated by the shaded box in Fig. 4.1B. As Hylander et al. (1991) pointed out, to determine whether twisting occurs in these sections, only forces acting to one side of them need be considered, and it is simplest to examine the forces acting anterior to them. Twisting will occur in these sections when the combined torques of the bite force and the balancing-side masseter and temporalis muscle forces ($F_{bite} \times M_{bite} + F_{mass(b)} M_{mass(b)} + F_{temp(b)} \times M_{temp(b)}$) exceed those exerted by the working-side masseter ($F_{mass(w)} \times M_{mass(w)}$) and temporalis ($F_{temp(w)} \times M_{temp(w)}$) muscle forces. In strepsirrhines, including *Eulemur* (unpublished data), the working-side masseter and temporalis muscles are more powerfully recruited than the balancing-side muscles (Hylander et al., 2000, 2002, 2003), so the working-side muscle torques must be larger than the balancing-side muscle torques. However, the superiorly directed components of bite force are probably larger than the inferiorly directed force components of the working-side masseter acting at the anterior root of the zygoma and the working-side anterior temporalis acting on the postorbital bar (i.e., those portions of masseter and temporalis acting within the shaded box in Fig. 4.1B). This is because the working-side masticatory muscles (masseters, temporales, and medial pterygoids) attaching to the cranium behind the sections of interest all contribute to the bite force torque, whereas only the most anterior fibers of these muscles contribute to the counteracting torque in the circumorbital region. Thus, although the torque due to bite force may be countered by equal muscle torques in more posterior sections, it seems likely that all of the sections examined in this study experience twisting due to bite force torques during mastication and unilateral biting.

In order to generate predictions with which to test his twisting hypothesis, Greaves (1985) assumed that the skull behaves under twisting like a simple cylinder. A cylinder subjected to an external torque experiences strain orientations oriented at 45° to the twisting axis of the cylinder, reversing by 90° with a change in torque direction (Fig. 4.2). The assumption that the skull behaves like a cylinder under twisting has received mixed support. The study of the supraorbital region in macaques and baboons by Hylander et al. (1991) revealed strain orientations that did not match those of a twisting cylinder. In their study of the circumorbital region of *Aotus*, Ross and Hylander (1996) found that strain orientations from the dorsal interorbital region in one experiment matched those predicted for a twisting cylinder, but those from the same region in another experiment, and the lateral and medial orbital walls in a number of others, did not (Ross and Hylander, 1996): Fig. 11; Ross, 2001). In contrast with these results from anthropoids, strain orientations recorded from the postorbital bar, dorsal interorbital region, and medial orbital wall of the strepsirrhine *Otolemur* consistently match those predicted for a twisting cylinder (Ravosa et al., 2000; Ross, 2001).

These results raise the possibility that the skulls of strepsirrhines behave like simple twisting cylinders during mastication while the skulls of anthropoids do not, either because they do not twist during mastication or because they do not behave like simple cylinders. Either of these possibilities would have significant implications for how biomechanical hypotheses regarding evolution of the primate facial skeleton are tested. In particular, an understanding of how strepsirrhine and anthropoid facial bones are stressed during feeding can be used to test hypotheses

Fig. 4.2 Diagrams of cylinders fixed at one end and subjected to opposite torques at the free end. The orientation of ε_1 is illustrated: 45° to the long axis of the cylinder, reversing by 90° with a change in torque direction (adapted from (Metzger et al., 2005). Large arrow indicates the direction of torque applied to cylinder

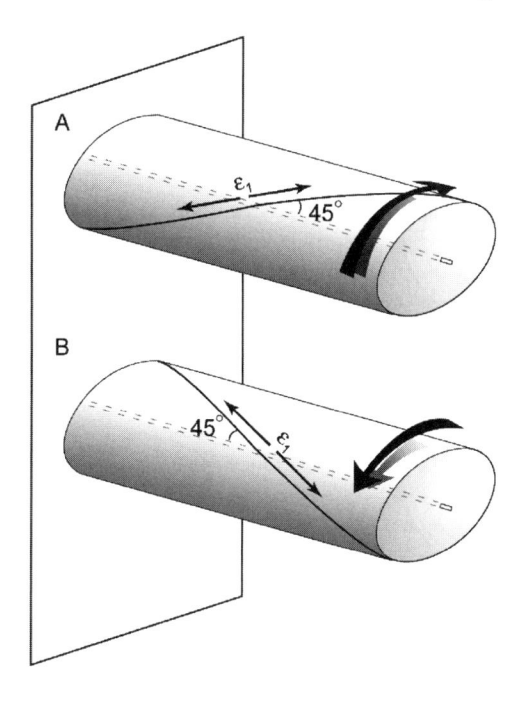

regarding the functional significance of the suite of features that evolved along the anthropoid stem lineage (Ravosa, 1991; Rosenberger, 1986; Ross and Hylander, 1996; Ross and Ravosa, 1993; Ross, 2000). If the *Eulemur* skull behaves like that of *Otolemur* during chewing, this would suggest a fundamental difference between anthropoids and strepsirrhines in patterns of loading in the facial skeleton. Distinct patterns of loading in the facial bones of strepsirrhines and anthropoids would support the idea that changes in circumorbital morphology at anthropoid origins were functionally related to dissipating or resisting the feeding forces associated with those loading regimes. In contrast, if strepsirrhines and anthropoids do not exhibit significant differences in patterns of facial loading, it is difficult to support hypotheses that the differences in morphology between the two groups are functionally related to different loading regimes.

To determine whether the skull of another strepsirrhine, *Eulemur fulvus*, twists like a simple cylinder during mastication, this paper presents the strain orientation data recorded from the circumorbital regions of three representatives of this species. Strain orientations in these animals are then compared with those in *Otolemur*, *Aotus*, *Macaca*, and *Papio* in order to determine whether there is a "strepsirrhine pattern" of facial loading, i.e., twisting, and whether the behavior of the skull of *Eulemur* can be modeled as that of a simple cylinder.

4.2 Materials and Methods

The three *Eulemur fulvus* (1 male, 2 females) that served as subjects were borrowed from the Duke Lemur Center for a period of 9 months, during which time they were housed at Stony Brook University. All procedures were approved by the Institutional Animal Care and Use Committees at Duke University and Stony Brook University, and by the Scientific Committee at the Duke Lemur Center. Data reported here come from five experiments on the three animals.

Strain data were collected using delta rosette strain gages wired in a three-wire quarter-bridge circuit. The wires from the gage were soldered onto nine pins of a twelve-pin connector (PMC 012, Cristek Interconnects Inc., Anaheim CA), and the gage-wire-connector assembly was gas sterilized. Strain gages were placed with the animal anesthetized with inhalant 2% isoflurane delivered in O_2 administered through a nose cone. Once the animal was sufficiently anesthetized, it was intubated and anesthesia maintained with isoflurane. The skin overlying the gage site was washed with betadine, a small (1–2 cm) incision was made, the tissues overlying the gage site were reflected, and the bone exposed. The bone was then degreased with chloroform and the strain gage bonded to the bone using a cyanoacrylate adhesive and pressure administered through a cotton-tipped applicator coated with sterile surgical lubricant. After checking that the gage elements balanced, the surgical wound was sutured closed around the wires from the strain gage exiting the wound. The wires were sutured to the skin overlying the posterior root of the zygomatic arch and/or the top of the head.

Following strain gage placement, the animal was placed in the restraining apparatus to wake up, while inhalant O_2 was administered through the intubation tube, then a nose cone. For data recording, the animals were restrained in a commercially available primate restraint chair (XPL-517-CM, PlasLabs, Lansing, MI) in a sitting position with their arms and legs restrained but their head and neck freely moving. After recovery from anesthesia (30–60 min.) the animals were presented with several kinds of food cut into pieces roughly 5–10 mm in maximum diameter: apple, raisin, grape, prune, and almond. The food was presented one piece at a time with a pair of forceps.

Each of the three elements of the rosette strain gages was connected to form one arm of a Wheatstone bridge. Bridge excitation was 2 V. Voltage changes were conditioned and amplified on a Vishay 2100 system, and then recorded on a PC at 1 kHz using MiDAS data acquisition software (Xcitex, Inc., Cambridge, MA, www.xcitex.com). The bone strain data were analyzed in IGOR Pro 4.0 (WaveMetrics, Inc., Lake Oswego, OR), using custom-written software. The strain data were converted to microstrain using the calibration files made during the recording sessions following Ross (2001). Strain (ε), a dimensionless unit equaling the change in length of an object divided by its original length, is measured in microstrain ($\mu\varepsilon$) units that are equal to 1×10^{-6} strain. Tensile strain is registered as a positive value, and compressive strain as a negative value. The maximum principal strain (ε_1) is usually the largest tensile strain value, while the minimum principal strain is usually the largest compressive strain value (ε_2).

4 Does the Primate Face Torque?

To visually display the bone strain data on the images of *Eulemur* skulls, vectors were computed from the magnitude and orientation of ε_1 in each power stroke. The orientations of the strain gages on the skulls were measured from the radiographs taken of the animals after gage placement. Orientations of gages on top of the skull were measured relative to a prosthion-inion line in superoinferior radiographs, and orientations of gages on the postorbital bars were measured relative to a prosthion-inion line in lateral radiographs. Images of *Eulemur fulvus* skulls were extracted from reconstructions of a *Eulemur fulvus* skull CT scanned by J. Rossie, and generously made available on the Digimorph website: www.digimorph.org. The strain vector plots were then rotated into the coordinate frame of the skull images using the prosthion-inion reference line in both superior and lateral views. It should be noted that, using this method, the orientations of the strain vectors relative to the A-element are exact, but the orientations relative to the skull images include errors associated with measuring gage orientation and position from radiographs, as well as intraspecific variation in skull form.

4.3 Results

Strain orientations and magnitudes recorded during the five experiments are illustrated in Figs. 4.3–4.7. Descriptive statistics for the strain data are given in Tables 4.1

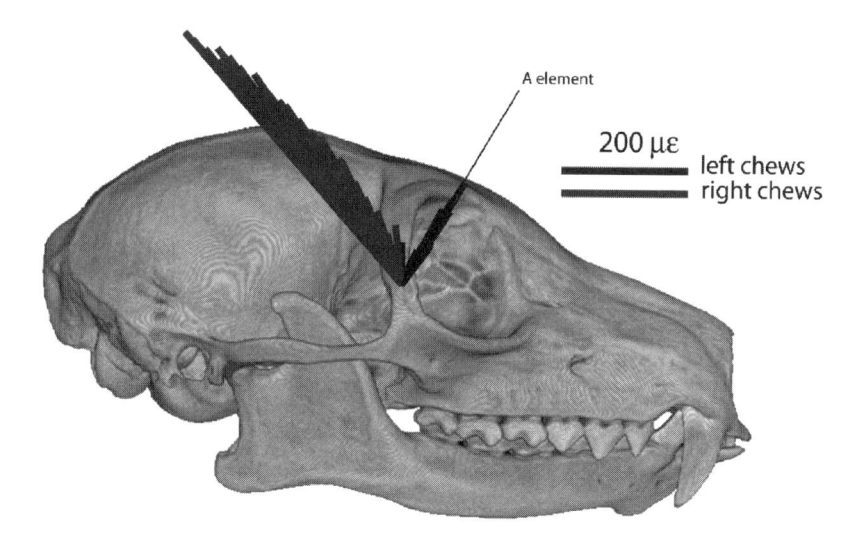

Experiment 72
(Bao)

Fig. 4.3 Lateral view of skull of *Eulemur fulvus* showing position of strain gage in experiment 72, and the orientation of the A-element of the gage (indicated by *black line*). The orientation of the maximum principal strain (ε_1) during mastication is illustrated by the vectors. Each line represents ε_1 orientation and magnitude recorded at the time of peak strain magnitude during one power stroke. Scale bars indicate strain magnitude

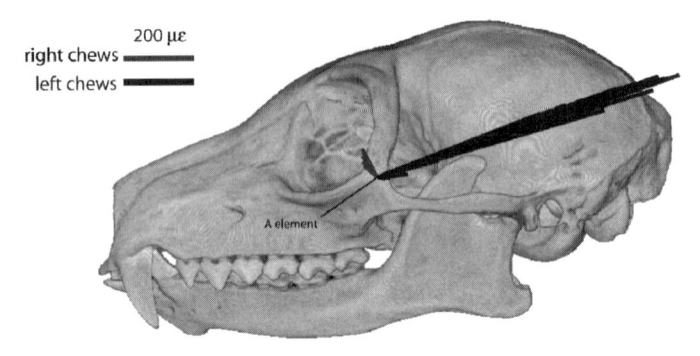

Fig. 4.4 Lateral views of skull of *Eulemur fulvus* showing position of strain gages in experiment 74, and the orientation of the A-elements of the gages (indicated by *black line*). The orientation of the maximum principal strain (ε_1) during mastication is illustrated by the vectors. Each line represents ε_1 orientation and magnitude recorded at the time of peak strain magnitude during one power stroke. Scale bars indicate strain magnitude

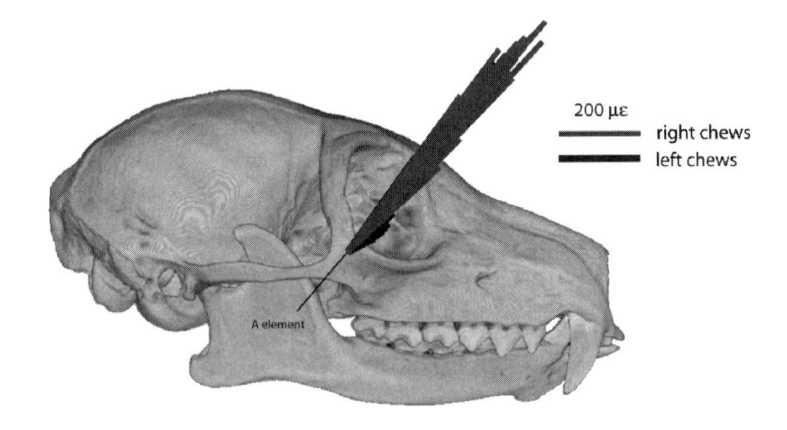

and 4.2. In the figures, strain orientations and magnitudes recorded at peak are shown for all power strokes during each experiment, separated by chewing side. Each line is a vector, where the length indicates strain magnitude and the orientation indicates the orientation of ε_1 relative to the A-element of the gage and the bony skeleton. In the tables, the strain orientation statistics are given relative to the A-element of the gage. The data from the postorbital bar are discussed first, followed by those from the dorsal surface of the skull.

4 Does the Primate Face Torque? 71

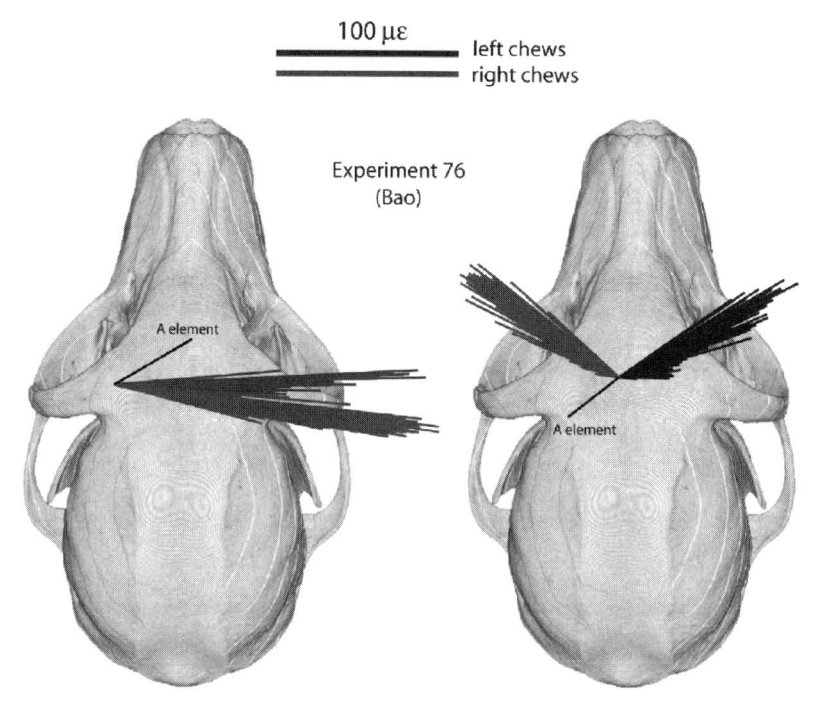

Fig. 4.5 Dorsal views of skull of *Eulemur fulvus* showing position of strain gages in experiment 76, and the orientation of the A-elements of the gages (indicated by *black line*). The orientation of the maximum principal strain (ε_1) during mastication is illustrated by the vectors. Each line represents ε_1 orientation and magnitude recorded at the time of peak strain magnitude during one power stroke. Scale bars indicate strain magnitude

4.3.1 Postorbital Bar

In experiment 72, strain data were recorded from the postorbital bar at mid-orbital height (Fig. 4.3). In this experiment, ε_1 is oriented upward and backward during ipsilateral (right) chewing and upward and forward during contralateral (left) chewing. These strain orientations are the reverse of those predicted by the torsion hypothesis. In experiment 74, strain gages were simultaneously bonded to the lower half of the left and right postorbital bars, near the bar's junction with the zygomatic arch (Fig. 4.4). On the left side, ε_1 orientations resemble those recorded during experiment 72: i.e., upward and backward during ipsilateral (left) chewing, and upward and forward during contralateral (right) chewing. These results are also the reverse of the predictions of the torsion hypothesis. Strains recorded simultaneously from the right-side strain gage reveal the strains oriented upward and forward during both ipsilateral and contralateral chewing, also providing no support for the torsion hypothesis.

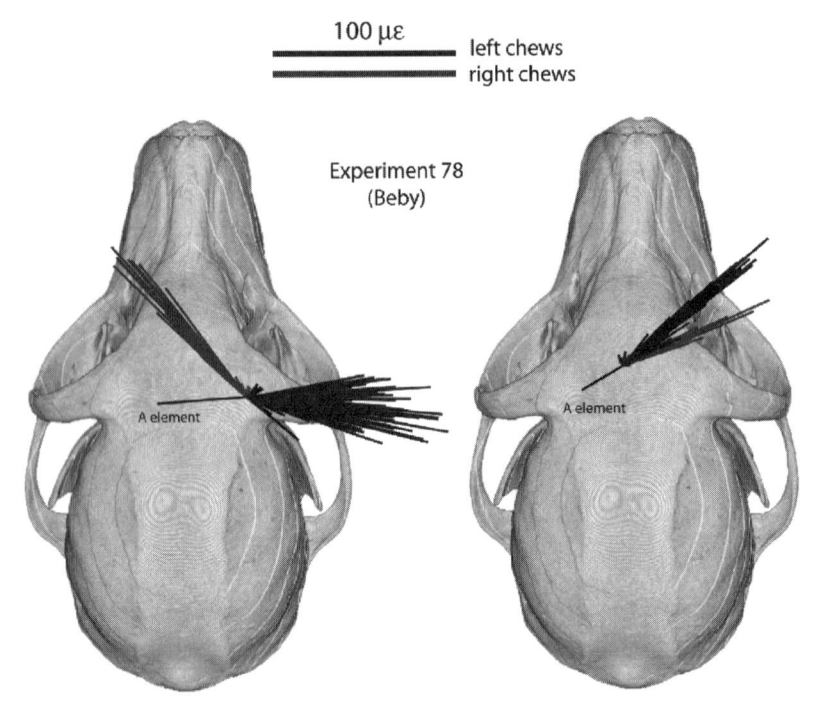

Fig. 4.6 Dorsal views of skull of *Eulemur fulvus* showing position of strain gages in experiment 78, and the orientation of the A-elements of the gages (indicated by *black line*). The orientation of the maximum principal strain (ε_1) during mastication is illustrated by the vectors. Each line represents ε_1 orientation and magnitude recorded at the time of peak strain magnitude during one power stroke. Scale bars indicate strain magnitude

In all postorbital bar experiments, strain magnitudes recorded during chewing ipsilateral to the strain gages are larger than those recorded during contralateral chewing (Table 4.1). These working balancing differences are contrary to the predictions of the torsion hypothesis, which predicts similar strain magnitudes at similar distances from the twisting axis (Metzger et al., 2005).

4.3.2 Dorsal Skull Surface

Strain was recorded from the dorsal interorbital region of three animals during experiments 76, 78, and 79 (Figs. 4.5–4.7). In experiments 76 and 79, strain orientations in the dorsal interorbital region approximate those predicted by the torsion model of skull loading: i.e., rostral to the left during right chews and rostral to the right during left chews (Figs. 4.5 and 4.7). In both experiments, the mean differences between the strain orientations recorded during left and right chews exceed the 90°

4 Does the Primate Face Torque? 73

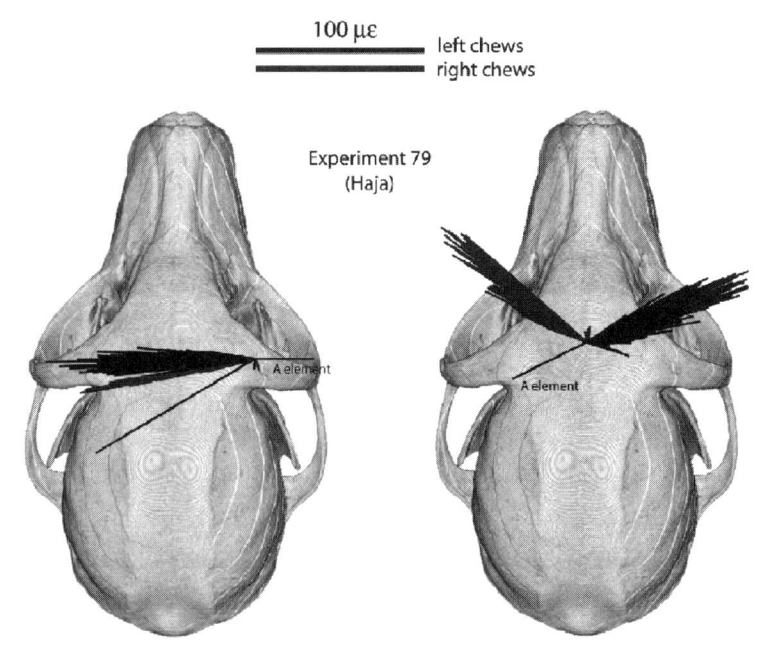

Fig. 4.7 Dorsal views of skull of *Eulemur fulvus* showing position of strain gages in experiment 79, and the orientation of the A-elements of the gages (indicated by *black line*). The orientation of the maximum principal strain (ε_1) during mastication is illustrated by the vectors. Each line represents ε_1 orientation and magnitude recorded at the time of peak strain magnitude during one power stroke. Scale bars indicate strain magnitude

predicted by the torsion model. In experiments 76, 78, and 79, strain magnitudes recorded during right and left chews are similar.

In experiment 78, maximum principal strain orientations recorded during left chews are those predicted by the torsion model, but those recorded during right chews are not, being oriented only slightly clockwise of those recorded during left chews (Fig. 4.6).

Strain was recorded from dorsal orbital sites during all three experiments in which interorbital strains were recorded. In all experiments, ε_1 orientations recorded during ipsilateral chews are significantly different from those recorded during contralateral chews. In experiments 76 and 79, ε_1 orientations are predominantly laterally directed in both cases, although rotated slightly clockwise of transverse during ipsilateral chews in experiment 76 (Fig. 4.5) and slightly counter clockwise of transverse during contralateral chews in experiment 79 (Fig. 4.7). In experiment 78, ε_1 orientations recorded from the right dorsal orbital region reveal larger differences during ipsilateral and contralateral chewing. Contralateral chewing is associated with laterally directed maximum principal strains, whereas ipsilateral chewing is associated with ε_1 orientations directed rostrally and to the left, as predicted by the torsion model (Fig. 4.6).

Table 4.1 Descriptive statistics for maximum principal strain magnitudes and directions recorded from the postorbital bars during experiments 72, 73, and 74

	Left chews		Right chews	
	ε_1 magnitude (µε)	ε_1 angle (degrees)	ε_1 magnitude (µε)	ε_1 angle (degrees)
Experiment 72				
Right bar				
N	172	172	79	79
Mean	65.1785	−0.36702	306.343	65.2761
Std dev	23.8227	6.17923	106.268	6.34651
Maximum	187.196	61.1979	523.629	72.934
Minimum	8.25833	−6.18927	71.8719	37.6693
Experiment 73				
Right bar				
N	34	34	12	12
Mean	33.0329	−24.8461	140.671	98.3596
Std dev	9.37954	19.3885	68.0765	64.5019
Maximum	56.7686	24.7175	229.841	134.976
Minimum	18.9326	−44.101	34.5249	−40.0129
Left bar				
N	152	152	71	71
Mean	100.871	−71.7817	62.6369	59.1382
Std dev	48.0295	13.7739	28.7024	26.2354
Maximum	187.774	85.1118	104.355	88.308
Minimum	4.73436	−89.362	7.23607	−89.0285
Experiment 74				
Right bar				
N	232	232	153	153
Mean	98.1365	−9.37696	463.684	0.398199
Std dev	104.543	7.90181	122.757	2.06507
Maximum	586.787	5.04501	745.064	4.81093
Minimum	8.75987	−22.7513	56.8501	−6.98161
Left bar				
N	231	231	154	154
Mean	308.166	−17.9153	40.633	89.0858
Std dev	261.921	5.42242	19.2392	11.944
Maximum	885.724	−10.6587	89.8145	133.995
Minimum	6.87972	−44.8496	5.8225	−18.4243

4.4 Discussion

4.4.1 Skull Loading in *Eulemur*

It was argued in the Introduction that during chewing the strepsirrhine face is subjected to torques about an anteroposterior axis due to muscle and bite forces. The data presented here were collected to evaluate whether the *Eulemur* facial skeleton resembles that of *Otolemur* in behaving like a simple twisting cylinder under this

4 Does the Primate Face Torque?

Table 4.2 Descriptive statistics for maximum principal strain magnitudes and directions recorded from the dorsal orbital and dorsal interorbital regions during experiments 76, 78, and 79

	Left chews		Right chews	
	ε_1 magnitude (µε)	ε_1 angle (degrees)	ε_1 magnitude (µε)	ε_1 angle (degrees)
Experiment 76				
dorsal interorbital				
n	132	132	160	160
Mean	59.3123	−14.3615	64.1074	111.396
Std dev	25.6719	8.01206	25.5442	5.9231
Maximum	110.498	−3.74682	102.274	134.292
Minimum	4.49833	−42.304	6.566	100.075
dorsal orbital				
n	132	132	160	160
Mean	77.8581	−31.2262	112.94	−39.0409
Std dev	38.5223	3.22521	41.9114	1.96287
Maximum	174.077	−25.1838	183.467	−29.3778
Minimum	4.02162	−42.7565	15.5649	−44.2856
Experiment 78				
dorsal interorbital				
n	274	274	50	50
Mean	36.4337	10.5307	22.9133	5.00056
Std dev	24.1668	12.7961	20.8195	26.7039
Maximum	104.093	93.0543	87.1213	109.029
Minimum	2.40216	−28.32	2.62694	−14.8994
dorsal orbital				
n	274	274	50	50
Mean	40.1949	−4.89311	39.6547	123.478
Std dev	25.7892	16.7573	35.3085	15.0308
Maximum	112.443	48.285	110.945	134.512
Minimum	2.59234	−44.2271	3.2958	68.8197
Experiment 79				
dorsal interorbital				
n	201	201	288	288
Mean	39.7356	122.585	46.2631	−9.81579
Std dev	26.7438	12.1698	24.7989	10.5469
Maximum	108.641	134.517	104.031	7.77136
Minimum	3.10753	33.8178	4.08605	−43.6497
dorsal orbital				
n	201	201	288	288
Mean	41.2939	12.8046	53.8626	3.98819
Std dev	29.8236	11.6194	26.3793	7.87611
Maximum	110.483	108.402	126.039	82.0444
Minimum	2.29545	3.3321	1.38455	−7.30827

loading regime. If the *Eulemur* skull does behave like that of *Otolemur* during chewing, it would suggest a strepsirrhine pattern of loading in the facial skeleton, distinct from that reported for anthropoids (Hylander et al., 1991; Ross and Hylander, 1996). The existence of distinct patterns of loading in the facial bones of strepsirrhines and anthropoids would lend support to the idea that the changes in circumorbital morphology at anthropoid origins were functionally related to dissipating or resisting the feeding forces responsible for generating those loads. In contrast, if there are not obvious differences between strepsirrhines and anthropoids in patterns of facial loading, then the differences in morphology between the two groups cannot be related to different loading regimes.

The in vivo bone strain data reported here reveal that the facial skeleton of *Eulemur* is not loaded the same way as that of *Otolemur*. Strains in the circumorbital region of *Eulemur* only match those predicted for a twisting cylinder, and resemble those of *Otolemur*, in the dorsal interorbital region. In two experiments (76 and 79), ε_1 orientations recorded from the dorsal interorbital region approximate those predicted by the torsion hypothesis, oriented rostrally and to the left during right chews, and rostrally and to the right during left chews. However, in both these experiments, ε_1 is oriented more laterally than predicted for a twisting cylinder, yielding differences between chewing sides of greater than 90° predicted for a twisting cylinder. Moreover, in experiment 78, ε_1 orientations are predominantly lateral during both left and right chewing.

In contrast with the strain data from the dorsal interorbital region, strain orientations from the dorsal orbital region and the lateral and posterior aspects of the postorbital bars are not those predicted for a twisting cylinder, and differ from those of *Otolemur*. Strains in the dorsal orbital region are laterally directed in experiments 76 and 79, and in experiment 78 laterally directed during left (contralateral) chews and rostral and to the left during right chews. Strain orientations recorded from the lateral aspect of the postorbital bar in experiment 72 and the left bar in experiment 74 are the reverse of those predicted for a twisting cylinder: i.e., upward and backward during ipsilateral chewing, and upward and forward during contralateral chewing.

These data reveal that the *Eulemur* skull is not acting like a twisting cylinder during mastication. Is the *Eulemur* skull twisting and not behaving like a simple cylinder? The data from the dorsal interorbital region in two experiments are close to those predicted for a twisting cylinder, albeit more laterally directed than predicted. The more laterally directed ε_1 orientations at these sites might reflect combinations of twisting about an anteroposterior axis and either upward bending in the sagittal plane, or bending in the frontal plane (cf. Hylander et al., 1991). Various authors have argued against the idea that upward bending in sagittal planes is an important loading regime in most primate skulls. Bending is only an important loading regime in long beams (i.e., beams longer than four times their diameter): in shorter beams, shear is a more important loading regime than bending (Demes, 1982; Hylander et al., 1991; Ross, 2001). The position of the bite point during mastication must lie somewhere between the upper canine and the M^3, and the distance to any of these locations is less than the depth of the skull in the circumorbital region. It therefore seems unlikely that upward bending in sagittal planes is a significant loading regime in *Eulemur* skulls (cf. Hylander et al., 1991).

Hylander et al. (1991) suggested that bending in the frontal plane might account for the laterally directed strain orientations observed in the dorsal and rostral interorbital regions in macaques and baboons, and these arguments can reasonably be applied to *Eulemur* also. This loading regime might result from inferiorly directed components of temporalis muscle forces acting at the temporal lines, or bending and deformation of the orbital margin due to inferiorly directed components of masseter muscle forces (Endo, 1966). Bending in the frontal plane also provides an explanation for the laterally directed strains observed in the dorsal orbital region in experiment 76 and 79. Thus, strain orientations in the dorsal orbital and dorsal interorbital regions can be hypothesized to be due to a combination of twisting of the facial skeleton about an anteroposterior axis and bending of the circumorbital region in a frontal plane. Can the strain orientations in the postorbital bar be fit to this model?

Strain data collected from the left bars in experiments 72 and 74 are the reverse of those predicted for a twisting cylinder: i.e., upward and backward during ipsilateral chewing and upward and forward during contralateral chewing. These strain orientations are certainly suggestive of twisting, but not of a global twisting regime induced in the facial skeleton by bite forces. Rather, these strain orientations suggest that the postorbital bar is twisted as a unit independent of any twisting occurring in the snout. How might such twisting be produced?

One possible source of twisting regimes in the postorbital bars is the external forces acting on the inferior end of the bars as a result of twisting of the face on the braincase or rocking of the palate about the interorbital region (Fig. 4.8), as suggested by Ross and Hylander (1996) for *Aotus*. Under this loading regime, the lower end of the postorbital bar is subjected to superiorly directed components of bite force on the working side and posteroinferiorly directed masseter muscle force on the balancing side. Importantly, however, the postorbital bar of *Eulemur* does not lie perpendicular to the frontal or coronal planes, but oblique to them. As a result, the superiorly and inferiorly directed forces acting on the inferior end of the bars act obliquely to the long axis of the bar, not only producing bending and "unbending" (i.e., bending such as to reduce the curvature) of the bar on working and balancing sides, respectively, but also twisting moments (Fig. 4.8).

Strain orientations recorded from the right bar in experiment 74 (Fig. 4.4) are always oriented upward and forward. The position of this gage is not obviously different from that on the left side in the same experiment, although it is clearly experiencing a different loading regime during ipsilateral chews. One possible explanation for this strain pattern is that the right gage was placed closer to the origin of the superficial masseter. The strain orientations might reflect local effects of masseter contraction, or shearing of the anterior root of the zygoma due to inferiorly directed components of masseter muscle force acting behind the gage and superiorly directed components of bite force acting anterior to it.

In sum, the bone strain data from the circumorbital region of *Eulemur* are interpreted here as resulting from twisting of the face on the braincase and bending of

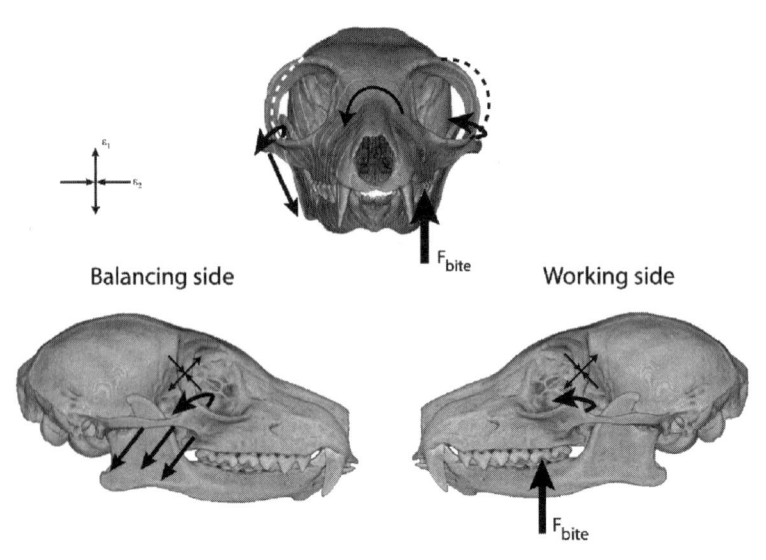

Fig. 4.8 Anterior and lateral views of skull of *Eulemur fulvus* illustrating loading regimes hypothesized for the skull of *Eulemur fulvus* during unilateral biting or mastication. In the anterior view, the upwardly directed components of bite force (F_{bite}) twist the rostrum counterclockwise and act on the inferior root on the obliquely oriented working-side bar, twisting it about a roughly superoinferior axis. On the balancing side, the posteroinferiorly oriented superficial masseter "unbends" the bar (i.e., bends the bar such as to reduce its curvature) and twists its lower end down and out. On the working side, superiorly oriented components of bite force transmitted to the bar by the maxilla cause the bar to bend and, because it is oriented obliquely, twist as well. White and black curved, dashed lines beside the postorbital bars represent the amplified changes in shape hypothesized for the postorbital bars. Converging and diverging arrows represent the strain orientations predicted on the lateral surface of lower end of postorbital bar under the suggested loading regimes

the circumorbital region in the frontal plane (Fig. 4.8). These global loading regimes are hypothesized to produce local patterns of deformation in the postorbital bars that cannot be predicted by modeling the facial skeleton as a simple cylinder or beam. The local loading regimes in the postorbital bars are produced by the overall twisting regime, but they differ from it because of the local geometry of the bars themselves.

4.4.2 Comparisons with Other Primates

In vivo strain data from *Otolemur* suggest that the facial skeleton twists on the braincase about an anteroposterior axis during mastication (Ravosa et al., 2000; Ravosa et al., 2000; Ross, 2001), producing the strain orientations predicted for a simple cylinder under torsion (Greaves, 1975). The loading regime hypothesized here for *Eulemur* resembles that of *Otolemur* in that there is torsion of the face on the braincase, but differs from it in two important ways. First, the patterns of strain associated with twisting are not simply those predicted for a cylinder under torsion.

Rather, in *Eulemur* other loading regimes predominate in the postorbital bars and dorsal orbital region. Second, in *Eulemur* bending of the circumorbital region in the frontal plane appears to be an important loading regime.

Comparisons with in vivo strain data available for other primates suggest that *Eulemur* resembles anthropoid primates in these two respects. Strain data from the anthropoids *Aotus*, *Macaca*, and *Papio* resemble those from *Eulemur* in that a global twisting regime does not predict strain orientations throughout their facial skeletons, and the circumorbital region appears to be bent in the frontal plane. In particular, the loading regime hypothesized here for *Eulemur* resembles that hypothesized for *Aotus*. Ross and Hylander (1996) suggested that the working-side lateral orbital wall of *Aotus* is bent in frontal planes ("buckled") by superoinferiorly oriented components of bite force transmitted through the anterior root of the zygoma to the septum. On the balancing side, a combination of bending and twisting due to masseter muscle force was hypothesized. Strain orientations in the dorsal interorbital region of *Aotus* were variable, with one experiment matching the predictions of a twisting cylinder and one matching the laterally directed strains recorded from the dorsal interorbital sites in macaques and baboons (Hylander et al., 1991), and the dorsal orbital sites in *Eulemur*. Whether the face of *Aotus* is twisting on the braincase or not, it is clear that bending and twisting of the lateral orbital walls in the frontal planes are important in both *Eulemur* and *Aotus*. Endo argued that the human and gorilla circumorbital regions and lateral orbital walls are bent in the frontal plane during mastication as a result of inferiorly directed components of masseter and temporalis muscle forces acting on the lateral ends of the supraorbital region (Endo, 1966; Endo, 1970; Endo, 1973). This argument was later extended to australopithecines (Rak, 1983), is probably also true of macaques (Ross, 2001; Strait et al., 2008), and is therefore the loading patterns most commonly measured in primates to date.

4.5 Conclusions

The data presented here suggest that strepsirrhines and anthropoids do not exhibit distinct patterns of loading in the circumorbital region. Rather, *Eulemur* resembles anthropoids and differs from *Otolemur* in patterns of deformation in the circumorbital skeleton. Thus, there are not distinct "strepsirrhine" and "anthropoid" patterns of skeletal loading in the circumorbital regions of extant primates. If these data from extant primates are informative about the loading regimes in the stem anthropoids, in which many of the morphological differences between anthropoids and strepsirrhines arose, then these morphological differences cannot be attributed solely to the differences in the nature of the loading regimes in the circumorbital region during feeding. This suggests that rearrangements of the circumorbital region at anthropoid origins are better explained by hypotheses regarding functional systems other than the feeding apparatus (Cartmill, 1980; Ravosa, 1991; Ross and Ravosa, 1993; Ross and Hylander, 1996; Ross, 2000).

Acknowledgments Bill Hylander is one of my most important professional and personal influences. His generosity, friendship, and unflagging support have been instrumental in many of my successes. He made the phone call that got me into biological anthropology in the first place, and he continues to stimulate me through his critical approach to science and his enjoyment of life. My life is richer and better for having known him.

W.L. Hylander and the staff at the Duke Lemur Center facilitated the loan of the *Eulemur* used in this research. Without that assistance, this research would not have been possible. This paper is Duke Lemur Center publication 1012. The Division of Laboratory Animal Resources at Stony Brook University provided expert care and excellent housing facilities for the animals while they were on loan. Brigitte Demes, Susan Larson, and Jack Stern provided intellectual and logistical support for the experiments discussed here, which is greatly appreciated. This research was funded by NSF Physical Anthropology (BCS-010913). Images of *Eulemur fulvus* used in the figures were extracted from reconstructions of *Eulemur fulvus* skull CT scanned by J. Rossie and available on the Digimorph website: www.digimorph.org. I wish to thank the editors of this volume and reviewers of this paper for their comments and insights.

References

Cartmill, M. (1980). Morphology, function and evolution of the anthropoid postorbital septum. In: Ciochon, R. L. and Chiarelli, A. B. (eds.), *Evolutionary Biology of the New World Monkeys and Continental Drift*. Plenum, New York, pp. 243–274.

Demes, B. (1982). The resistance of primate skulls against mechanical stresses. *Journal of Human Evolution* 11: 687–691.

Endo, B. (1966). Experimental studies on the mechanical significance of the form of the human facial skeleton. *Journal of the Faculty of Sciences, University of Tokyo, Section V* III: 1–106.

Endo, B. (1970). Analysis of stresses around the orbit due to masseter and temporalis muscles respectively. *Journal of the Anthropological Society of Nippon* 78: 251–266.

Endo, B. (1973). Stress analysis on the facial skeleton of gorilla by means of the wire strain gauge method. *Primates* 14: 37–45.

Greaves, W. S. (1975). The mammalian postorbital bar as a torsion-resisting helical strut. *Journal of the Zoological Society (London)* 207: 125–136.

Greaves, W. S. (1985). The mammalian postorbital bar as a torsion-resisting helical strut. *Journal of Zoology, London* 207: 125–136.

Hylander, W. L. and Johnson, K. R. (1997). *In vivo* bone strain patterns in the zygomatic arch of macaques and the significance of these patterns for functional interpretations of craniofacial form. *American Journal of Physical Anthropology* 102: 203–232.

Hylander, W. L., Picq, P. G. and Johnson, K. R. (1991). Masticatory-stress hypotheses and the supraorbital region of primates. *American Journal of Physical Anthropology* 86: 1–36.

Hylander, W. L., Ravosa, M. J., Ross, C. F., Wall, C. E. and Johnson, K. R. (2000). Symphyseal fusion and jaw-adductor muscle force: An EMG study. *American Journal of Physical Anthropology* 112: 469–492.

Hylander, W. L., Vinyard, C. J., Wall, C. E., Williams, S. H. and Johnson, K. R. (2002). Recruitment and firing patterns of jaw muscles during mastication in ring-tailed lemurs. *American Journal of Physical Anthropology* Supplement 34: 88.

Hylander, W. L., Vinyard, C. J., Wall, C. E., Williams, S. H. and Johnson, K. R. (2003). Convergence of the wishboning jaw-muscle activity pattern in anthropoids and strepsirrhines: Recruitment and firing patterns of jaw muscles in *Propithecus verreauxi*. *American Journal of Physical Anthropology* Supplement 35: 120.

Metzger, K. A., Daniel, W. and Ross, C. F. (2005). Comparison of beam theory and finite-element analysis to in vivo bone strain in the alligator cranium. *Anatomical Record A* 283A: 331–348.

Prossinger, H., Bookstein, F., Schafer, K. and Seidler, H. (2000). Reemerging stress: supraorbital torus morphology in the mid-sagittal plane. *Anatomical Record* 261: 170–172.

Rak, Y. (1983). *The Australopithecine Face*. Academic Press, New York.

Ravosa, M. J. (1991). Interspecific perspective on mechanical and nonmechanical models of primate circumorbital morphology. *American Journal of Physical Anthropology* 86: 369–396.

Ravosa, M. J., Johnson, K. R. and Hylander, W. L. (2000). Strain in the galago facial skull. *Journal of Morphology* 245: 51–66.

Ravosa, M. J., Noble, V. E., Johnson, K. R., Kowalski, E. M. and Hylander, W. L. (2000). Masticatory stress, orbital orientation, and the evolution of the primate postorbital bar. *Journal of Human Evolution* 38: 667–693.

Rosenberger, A. L. (1986). Platyrrhines, catarrhines and the anthropoid transition. In: Wood, B., Martin, L. and Andrews, P. (eds.), *Major Topics in Primate and Human Evolution*. Cambridge University Press, Cambridge, pp. 66–88.

Ross, C. F. (2000). Into the light: The origin of Anthropoidea. *Annual Reviews of Anthropology* 29: 147–194.

Ross, C. F. (2001). In vivo function of the craniofacial haft: The interorbital "pillar". *American Journal of Physical Anthropology* 116: 108–139.

Ross, C. F. (2001). In vivo intraorbital bone strain from the lateral orbital wall of *Macaca* and the functioning of the craniofacial haft. *American Journal of Physical Anthropology Suppl.* 32: 128.

Ross, C. and Hylander, W. L. (1996). *In vivo* and *in vitro* bone strain in owl monkey circumorbital region and the function of the postorbital septum. *American Journal of Physical Anthropology* 101: 183–215.

Ross, C. F. and Hylander, W. L. (1996). In vivo and in vitro bone strain in the owl monkey circumorbital region and the function of the postorbital septum. *American Journal of Physical Anthropology* 101: 183–215.

Ross, C. F. and Ravosa, M. J. (1993). Basicranial flexion, relative brain size and facial kyphosis in nonhuman primates. *American Journal of Physical Anthropology* 91: 305–324.

Strait, D. S., Wright, B., Richmond, B. G., Ross, C. F., Dechow, P. C., Spencer, M. and Wang, Q. (2008). Craniofacial strain patterns during premolar loading: implications for human evolution. In: Vinyard, C. J., Ravosa, M. J. and Wall, C. E. (eds.), *Primate Craniofacial Function and Biology*. Springer, New York, pp. 173–198.

Wolpoff, M. H., Senut, B., Pickford, M. and Hawks, J. (2002). *Sahelanthropus* or '*Sahelpithecus*'? *Nature* 419: 581.

Chapter 5
Motor Control of Masticatory Movements in the Southern Hairy-Nosed Wombat
(Lasiorhinus latifrons)

Alfred W. Crompton, Daniel E. Lieberman, Tomasz Owerkowicz, Russell V. Baudinette*, and Jayne Skinner

Contents

5.1 Introduction	84
5.2 Anatomy of the Wombat Masticatory System	88
5.2.1 Dentition	88
5.2.2 Skull	90
5.2.3 Muscles	93
5.3 Hypotheses to be Tested	94
5.3.1 General Hypothesis	94
5.3.2 Alternate Hypothesis	94
5.4 Material and Methods	95
5.4.1 Subjects	95
5.4.2 Surgery	96
5.4.3 Recording	97
5.4.4 EMG and Strain Gauge Analysis	97
5.5 Results	98
5.5.1 Muscle Activity	98
5.5.2 Strain	101
5.6 Discussion	105
References	109

Abbreviations **AT** and **a tem** anterior temporalis, **as ram** ascending ramus, **bdm** balancing-side deep masseter, **bmpt** balancing-side medial pterygoid, **btem** balancing-side temporalis, **bsm** balancing-side superficial masseter, **bu** buccinator, **ca** transverse orientation of condyle, **cran cav** cranial cavity, **DM** and **dm** deep

---

A.W. Crompton
Museum of Comparative Zoology, Harvard University, 26 Oxford St., Cambridge MA, 02138, USA
e-mail: acrompton@oeb.harvard.edu

*We regret to note that in the time period between completing this work and publication that Dr. Baudinette died on March 17, 2004.

masseter, **d mas** deep masseter, **DMPt** and **d mpt** deep part of the medial pterygoid, **ex s mas** external part of the superficial masseter, **fc** fibrous cartilage disc, **in rt** incisor tooth, **in s mas** internal part of the superficial masseter, **inf ang** inflected mandibular angle, **lpt** lateral pterygoid, **mas r** masseteric ridge, **MPt** medial pterygoid, **NC** nasal cavity, **o smas** origin area of superficial masseter, **o mpt** origin of medial pterygoid, **or** orbit, **ps** direction of jaw movement during the power stroke, **PT** and **p tem** posterior temporalis, **s** sinus, **SM** and **sm** superficial masseter, **SMPt** and **s mpt** superficial part of the medial pterygoid, **t** tendinous sheet, **tem** temporalis, **wdm** working-side deep masseter, **wmpt** working-side medial pterygoid, **wsm** working-side superficial masseter, **wtem** working-side temporalis **z** zygoma.

5.1 Introduction

Placental and marsupial herbivores have independently developed similar masticatory mechanisms to break down plant material. In both groups, chewing is accomplished by drawing the lower molars of the working side medially across the upper molars. The patterns of adductor muscle activity that controls jaw movement and generates strain in the mandibular corpus and symphysis are well documented in many placental mammals (Hogue and Ravosa, 2001; Hylander et al., 1998, 2000, 2004, 2005; Hylander and Johnson, 1994; Ravosa and Hylander, 1994; Vinyard et al., 2006; Weijs and Dantuma, 1981; Williams et al., 2003a, b, c, 2004). However, with the exception of the American opossum (*Didelphis virginiana*) (Crompton and Hylander, 1986; Crompton, 1995; Lieberman and Crompton, 2000), this information is not available for marsupials, including herbivorous taxa. The purpose of this paper is to describe the control of jaw movements in a marsupial herbivore, the Southern hairy-nosed wombat (*Lasiorhinus latifrons*), a large (22–24 kg), burrowing mammal from semi-arid South Australia, and to test the extent to which these patterns resemble those of placental herbivores.

In her extensive review of oral activity during feeding, Hiiemae (1978) stated: "the pattern of EMG activity was broadly similar in all mammals studied so far (1976) despite the differences in their profile of jaw movement." Hiiemae (1978, 2000) recognized numerous themes common to all mammals in jaw and tongue movements and activity patterns of the adductor, supra- and infrahyoid muscles, and suggested that a basic mammalian pattern was established early on in their evolution. The review of the mammalian masticatory motor patterns by Weijs (1994), based upon the timing of the activity patterns of the jaw adductors during rhythmic chewing, expanded on these generalizations and proposed that the variations evident in jaw adductor patterns within extant mammals were modifications of an ancestral pattern.

Weijs divided the adductor muscles into three functional groups:

1. Symmetric closers (zygomaticomandibularis, and vertical fibers of the anterior temporalis) that fire early during the fast close phase (FC);

2. Triplet I (working-side temporalis and balancing-side masseter and medial pterygoid) that moves the working side of the jaw laterally during FC;
3. Triplet II (balancing-side temporalis and working-side masseter and medial pterygoid) that moves the working side of the jaw medially during the power stroke (PC). For convenience, we will continue to use the terms Triplet I and II despite the fact that more than three muscles are involved in controlling transverse jaw movements.

While details of how the adductor triplets function have yet to be worked out, it is important to note that several adductor timing patterns are evident among mammals. In the "primitive" motor pattern exemplified by the American opossum (Weijs, 1994; Lieberman and Crompton, 2000) and the treeshrew (Vinyard et al., 2005), the time differential between the onset and the offset of the muscle activity in Triplets I and II is relatively short, and consequently their respective periods of activity overlap considerably (Fig. 5.1). In the opossum, peak activity in the first muscle to fire (WDM) is only separated from that of the last (WSM) by 20–30 ms.

Primitive mammals have a highly mobile symphysis. In the opossum (Lieberman and Crompton, 2000) and the tenrec (Oron and Crompton, 1985), the hemi-mandibles rotate around their longitudinal axes. During the power stroke, the ventral border of the working-side hemi-mandible is everted (Fig. 5.1, bold arrow marked PS). As the molars lie above the axis of rotation (symphysis to jaw joint), medially directed transverse movement of the molars is effected by a combination of rotation and transverse movement of the whole hemi-mandible. Weijs (1994) demonstrated how in more specialized mammals the "primitive" motor pattern was altered by changing the timing of the onset, offset, and duration of the activity periods of the two triplets. For example, in carnivorous mammals that accentuate vertical jaw movements during the power stroke, Triplets I and II fire symmetrically. In ungulates like the goat (Fig. 5.1) that accentuate transverse movement of the working-side jaw, the time between peak activity in the first Triplet I muscle to fire (WDM) and the last Triplet II muscle to fire (WSM) is considerably longer than in primitive mammals. Depending on the food type, this may last between 150 and 250 msec. In the goat (Williams et al., 2003b, c), galagos (Hylander et al., 2005), ring-tailed lemurs (Hylander et al., 2003, 2004, Vinyard et al., 2006), and opossums (Lieberman and Crompton, 2000), the balancing-side deep masseter reaches peak activity before that of the working-side superficial masseter. In the goat, Triplet II muscles draw the working side hemi-mandible dorso-medially (bold arrow marked PS in Fig. 5.1) during the power stroke (Becht, 1953). The symphysis in the goat is slightly mobile. Some hemi-mandibular rotation can occur during opening and the beginning of closing in the goat, but not during the power stroke (Lieberman and Crompton, 2000). A mobile symphysis is capable of transferring vertically directed forces from one hemi-mandible to the other (Crompton, 1995), but appears to be poorly designed to transfer transversely directed forces.

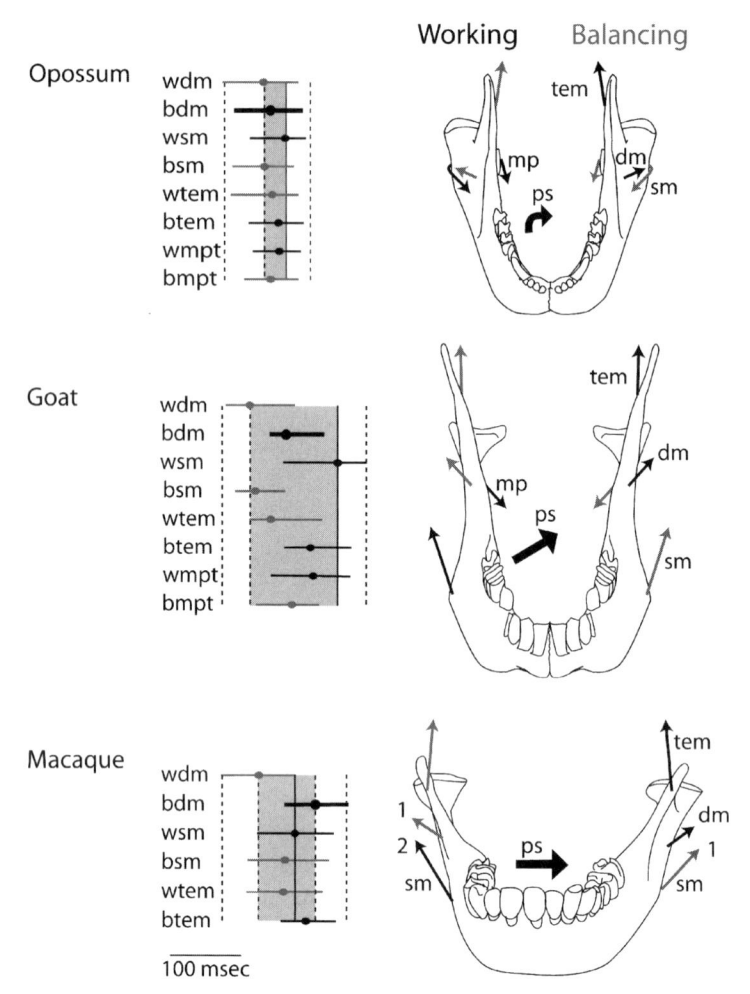

Fig. 5.1 Control of jaw movements in an opossum, goat, and macaque. On the left, the duration (indicated by horizontal lines) and time of peak activity (indicated by dots in the horizontal lines) in working- and balancing-side adductor muscles. On the right, the vectors of these muscles are shown on an oblique dorsal view of their lower jaws. Triplet I muscles are shown in gray and Triplet II muscles in black. The solid vertical line on the left is drawn through peak activity of the working-side superficial masseter. The shaded areas indicate the time between peak activity in the first and last muscles to fire. In the macaque there is a shift in the timing of activity in the balancing-side deep masseter

In animals with a fused symphysis (anthropoids, Hylander et al., 2000; Ravosa et al., 2000; sifakas, Hylander et al., 2003; and alpacas and horses, Williams et al., 2003b, c), the balancing-side deep masseter peaks after the working-side superficial masseter (Fig. 5.1). Hylander and colleagues suggest that the working- to balancing-side (W/B) EMG ratios and the muscle firing pattern of anthropoids support the hypothesis that symphyseal fusion and transversely directed muscle force are functionally linked (Hylander, 1984; Hylander and Johnson, 1994; Hylander et al., 2000, 2003, 2005).

In fused symphyses of placental mammals, the late firing of the balancing-side deep masseter generates lateral transverse bending, or "wishboning," of the mandibular symphysis in anthropoids (Hylander and Johnson, 1994). The fusion of the symphysis permits the complete transfer of a horizontal force from the balancing to the working side in order to draw the working side medially on a horizontal plane. Transverse movement early in the power stroke is generated by Triplet II muscles such as the working-side superficial masseter, working-side medial pterygoid, and balancing-side temporalis, but continued transverse movement toward the end of the power stroke is generated by the balancing-side deep masseter.

In placentals, a fused symphysis appears directly related to a shift in the timing of activity in the balancing-side deep masseter. In another placental, the alpaca, however, Williams et al. (2008) have shown that despite the late firing of the balancing-side deep masseter, the symphysis is fused and strengthened in order to resist twisting as opposed to any transversely directed forces. Strains associated with transverse masticatory forces in alpacas are small, it turns out. Williams et al. (2008) hypothesize that the fused symphysis functions to support the deep roots of the mandibular incisors.

While the general organization of the adductor muscles in marsupials and placentals is broadly similar (Turnbull, 1970; Abbie, 1939), the most striking difference between these two major groups is the presence of an inflected mandibular angle in marsupials and a deep non-inflected angle in placental herbivores. Sánchez-Villagra and Smith (1997) have shown that the inflected angle represents a synapomorphy of marsupials, but they could point to no consistent differences in mastication between marsupial and placentals that could be attributed to its presence.

Murray (1998) discussed the anatomy of the masticatory apparatus in the common wombat. He compared the mastication of wombats, rodents, and placental ungulates, and pointed out that the three groups exhibit different patterns of jaw movement: rodents emphasize propalinal jaw movements; ungulates emphasize wide translational movements of the mandible, slung between large medial pterygoids and superficial masseters; and wombats emphasize a more limited translational, but a more powerful compressive stroke during mastication. He concluded that the key attribute of the wombat's masticatory complex is an ability to exert extremely high occlusal forces as the lower molars are drawn medially in the horizontal plane, enabling the animal to grind down highly abrasive foods such as the fibrous perennial grasses and sedges typical of its diet (Finlayson et al., 2005). The anatomy of the masticatory apparatus in wombats differs from placental herbivores in many other respects, as well. Therefore, before presenting data on the kinematics and motor control of jaw movement during wombat mastication, we will begin with a review of the anatomy of wombat's teeth, mandible, cranium, and adductor muscles.

Transverse jaw movements and high occlusal forces are the hallmark of mammalian herbivores, but the anatomical features and neuro-muscular patterns commonly associated with transverse jaw movements are based entirely on placentals.

Wombats are ideal animals to test whether the models proposed for placentals are also true for marsupials.

5.2 Anatomy of the Wombat Masticatory System

5.2.1 Dentition

One of the most distinguishing features of the wombat is the postcanine dentition, characterized by tall, continuously growing, and semi-lophodont molars, well described by Murray (1998) and Ferreira et al. (1985, 1989). The height of the molar roots is illustrated in the transverse sections through the skull (Fig. 5.2, Sections 1[M^1], 2, 3). The upper and lower molars, as seen in these sections, form a sigmoid curve so that the exposed crowns of the upper molars are directed slightly laterally, and those of the lowers slightly medially. In occlusal view (Fig. 5.3B), the molars are bilobate with a deep embayment in the lingual aspect of the uppers and buccal aspect of the lowers, and smaller embayments on the opposite side. The enamel is thick on the buccal aspect of the lowers and lingual aspect of the uppers, and absent or extremely thin on the opposite sides. Most of the occlusal surface consists of dentine with the enamel forming a low ridge on the edge of the molar (Murray, 1998; Ferreira et al., 1985, 1989). The lower molars are positioned slightly anterior to the uppers (Fig. 5.3C).

In contrast to other herbivorous Diprodontia, such as the Phascolarctidae (koalas: Davison and Young, 1990) and macropodidae (kangaroos and wallabies: Sanson,

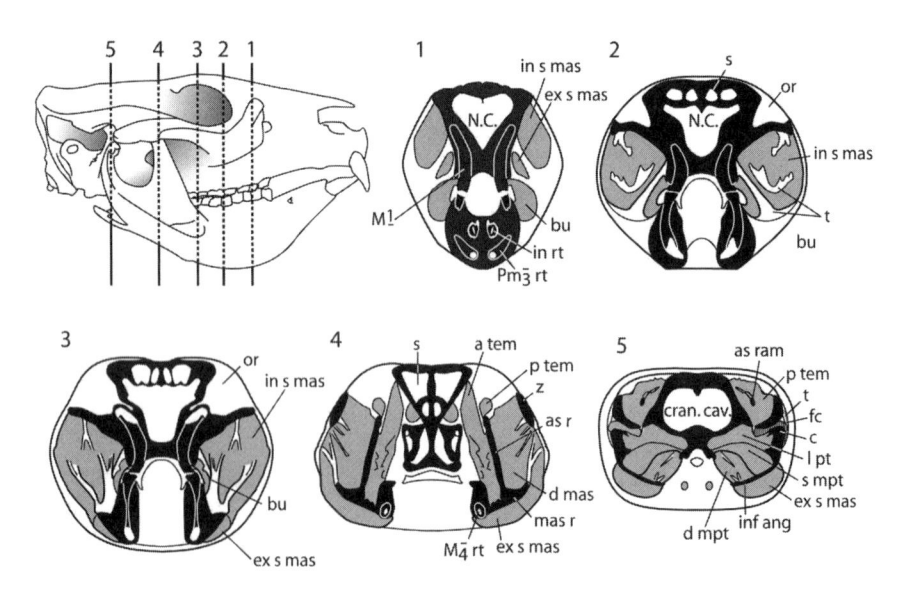

Fig. 5.2 *Lasiorhinus latifrons*. Five sections through the adductor musculature. The position of the sections is shown on a lateral view of the skull

5 Motor Control of Masticatory Movements

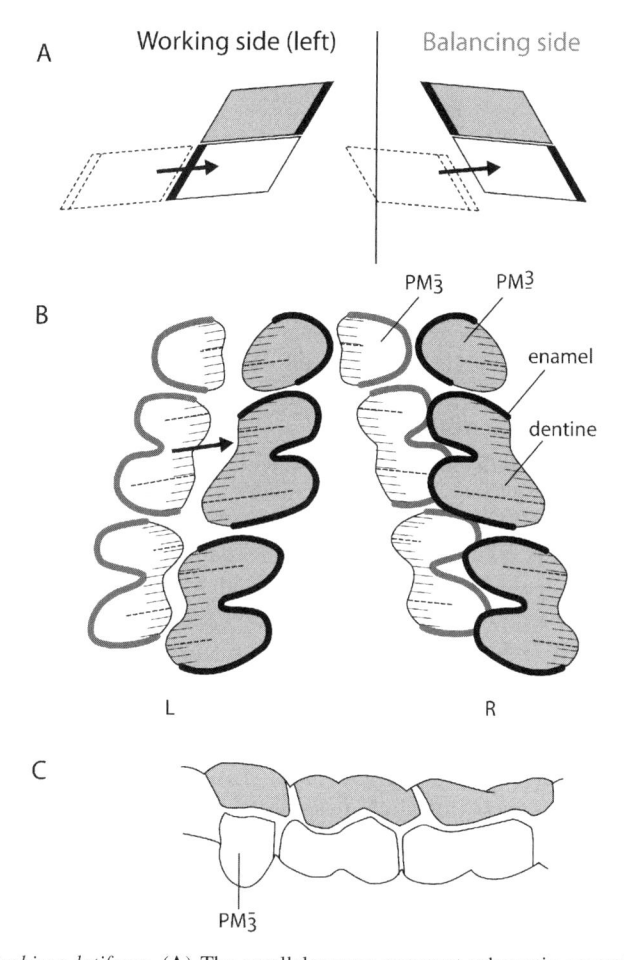

Fig. 5.3 *Lasiorhinus latifrons*. (**A**) The parallelograms represent schematic coronal (transverse) sections through the right- and left-side upper (*shaded*) and lower (*unshaded*) molars on either side of the midline (vertical line). The working side is on the left. At the beginning of occlusion, the positions of the lower molars are represented by parallelograms with dashed outlines and the end of occlusion with solid outlines. A thick black line represents the thick enamel on the lingual side of the uppers and buccal side of the lowers. (**B**) View of the occlusal surfaces of upper (*shaded*) and lower (*unshaded*) molars at the beginning of occlusion on the left side. Lingual enamel is indicated on the uppers by a thick black line and on the lowers by a thick gray line. (**C**) Lateral view of upper and lower PM3 and M1 and M2 to illustrate the position of the lower molars relative to the uppers

1980, 1989), who retain shearing crests on the sides of the main cusps of the molars, wombats rapidly wear down the occlusal surfaces of the molars leaving no remnant traces of the cusp pattern. In particular, the occlusal surfaces of the molars wear down heavily in the horizontal plane to produce low, rounded transverse ridges that fit into shallow valleys of the occluding tooth (Fig. 5.3C). The worn surfaces of the molars dip slightly in a ventral direction from lingual to buccal (Fig. 5.3A) so that the lingual edges of the molars are higher than the buccal edges. Judging from the observed lateral position of the lower incisors at the beginning of medially directed movement of the working-side mandible, the lingual edge of the posterior lower molar row meets or lies just lateral to the buccal aspect of the occluding uppers at the beginning of the power stroke. During transverse chewing movements, the exposed edges of dentine on the buccal edges of the upper molars and lingual edges of the lower molars rapidly wear down to form sharp cutting edges. Shallow transverse grooves are gouged out of the dentine on the lingual aspect of the lowers and buccal aspect of the uppers. Because of the orientation of the occlusal plane of the molars (Fig. 5.3A), the balancing-side molars do not contact each other at the beginning of the power stroke (dotted outline) and probably only come into full contact when centric occlusion is reached at the end of the power stroke. As Murray (1998) pointed out, food is trapped and broken down between the more resistant enamel ridges on the lingual side of the uppers and buccal side of the lowers. The wear pattern on the molar surfaces indicates that occlusion is unilateral.

Transverse jaw movement during the power stroke in wombats is probably achieved through the rotation of the jaw around the working-side condyle and the movement or translation of the balancing-side condyle in a posterior direction. However, as the molar rows are isognathic (the distance separating upper and lower molar rows are identical) and because transverse movement of the lower jaw as a whole does not exceed the transverse width of the upper molars, condylar translation during the power stroke is minimal. In wombats, the lower molars lie directly below the occluding uppers at the end of the power stroke. This limited medial movement contrasts with artiodactyls in which the narrower lower molars shear across the wider upper molars. At the end of the occlusal stroke in artiodactyls, both the active and the balancing-side lower molars lie internal to the matching upper molars (Crompton et al., 2005).

Another distinctive feature of the wombat dentition is the single pair of procumbent lower incisors (Figs. 5.2, 5.4). In the Southern hairy-nosed wombat (but not in the common wombat), the lower incisors shear past the posterior surfaces of the slightly larger upper incisor pair to produce a sharp apical cutting edge and beveled posterior surface on the uppers (Scott and Richardson, 1987). In order for the apices of the upper and lower incisors to meet, the lower jaw must be drawn slightly forward during opening and retracted as the jaw closes.

5.2.2 Skull

Several derived aspects of the wombat's skull are probably related to mastication. In general, wombat crania are broad and low with zygomatic arches that extend far laterally from the dentition (Figs. 5.2, 5.5), thereby extending the adductor musculature far laterally to the tooth row and increasing the leverage around the tooth row (Murray, 1998). The mandible is also large and broad. The mandibular symphysis, which houses the long roots of the lower incisors, is immobile because of a tightly interdigitating or partially fused suture that completely fuses in older individuals.

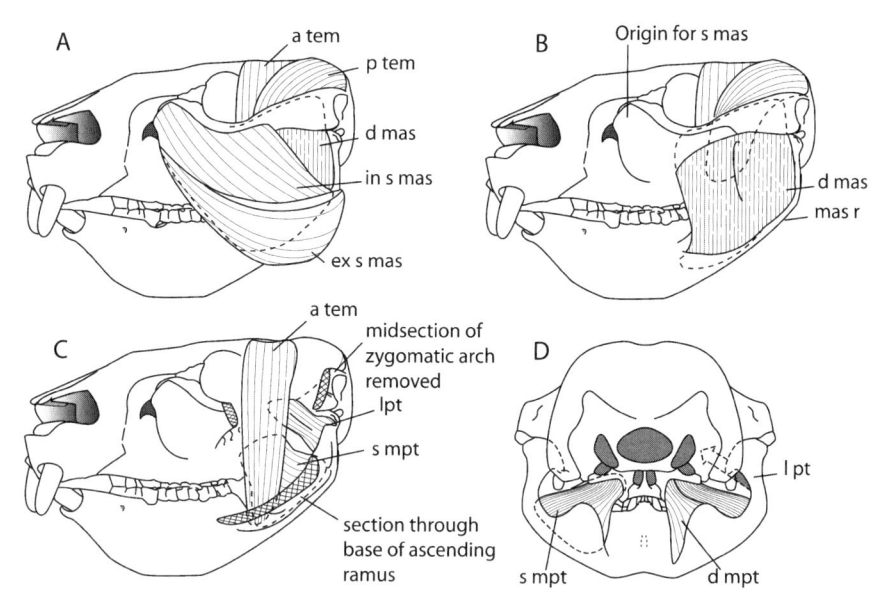

Fig. 5.4 *Lasiorhinus latifrons*. Lateral and posterior views of the adductor musculature. (**A**) shows the lateral view of the superficial masseter. In (**B**) this muscle is removed to expose the lateral surface of the deep masseter. In (**C**) the zygomatic arch and ascending ramus of the lower jaw has been removed to illustrate some of the deeper adductors. (**D**) is a posterior view.

Independent movement of the hemi-mandibles is therefore not possible. A large inflected mandibular angle extends internally far beyond the medial edge of the condyles (Fig. 5.5B, inf ang). The ventral surface of the angle forms a broad, thin, horizontal sheet (Fig. 5.2, Section 5.5). The massenteric ridge (Fig. 5.2, Section 5.4 and Fig. 5.5A, C, D: mas r) extends laterally so that a shallow depression separates it from the external surface of the ascending ramus.

The wombat temporomandibular joint (TMJ) is narrow antero-posteriorly, wide medio-laterally, and transversely oriented (Fig. 5.5A, C, D) in contrast to the long diarthroidal glenoid of the TMJ joint of placental ungulates. The axis of the narrow condylar head is oblique so that the medial edge lies posterior to the lateral edge (Fig. 5.5B, ca). In anterior and posterior views, the articular face of the mandibular condyle is convex and is separated from the concave and narrow glenoid (formed by both the squamosal and jugal) by a thick, tough fibrous disc (Fig. 5.2, Section 5.5, fc). A massive set of ligaments binds the temporal, the articular disc, and the condylar process of the mandible, restricting extensive antero-posterior movement of the mandibular condyle. Medially, a tendon of the lateral pterygoid muscle inserts on the medial aspect of the condyle and disc. In the common wombat, the medial aspect of the glenoid is buttressed by a prominent entoglenoid eminence. Murray (1998) suggests that the medial aspect of the condyle is forced against this structure, which acts as the point of rotation for the jaw as it moves in the horizontal plane.

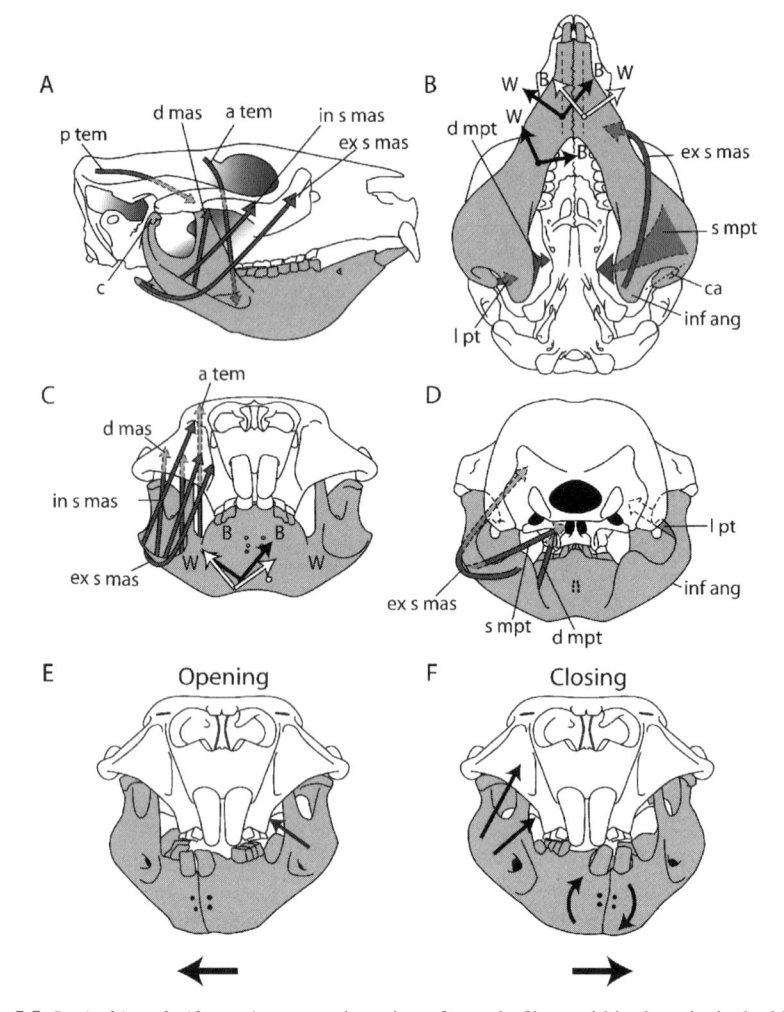

Fig. 5.5 *Lasiorhinus latifrons*. Average orientation of muscle fibers within the principal adductor muscles as seen in lateral (**A**), ventral (**B**), anterior (**C**), and posterior (**D**) views. The orientation of the principle strains (ε_1) on the ventral surface of the symphysis and ventral surface of the right hemi-mandible are indicated in **B**. The bold black arrows indicate strain orientation when chewing on the right side and the open arrows when chewing on the left side. Arrows labeled W indicates orientation of strain on the working side and those labeled B orientation of strain on the balancing side. Note that the orientation of working- and balancing-side strain orientations on each side are parallel to one another and shift through about 90° when the chewing side changes. In **B** and **C**, the orientation of principal strains is shown in ventral and anterior views, respectively. (**E**) At the end of opening and beginning of closing, the working side of the jaw (on the right side of this figure) is moved laterally by unilateral activity in the balancing (*left*) superficial medial pterygoid. (**F**) The working-side mandible is shifted in a medial direction by the synchronous activity of the working-side adductors. The working side of the jaw (*right*) is rotated clockwise (when viewed from the front) around its long axis (condyle to symphysis). Because of the tightly sutured mandibular symphysis, the balancing (*left*) side of the jaw is also rotated clockwise

5.2.3 Muscles

The jaw adductor musculature in wombats was described briefly by Murray (1998), and is illustrated in Fig. 5.4. The internal architecture of the muscles is complex and includes numerous intramuscular tendinous sheets, as shown in Fig. 5.2 (t). Figure 5.5 illustrates the principle directions of the muscle fibers within each muscle. The fibers of different muscle groups tend to interdigitate, and clear borders between muscles cannot be easily recognized. Weights and percentages of the total mass for the Southern hairy-nosed wombat (based on dissection of one specimen) appear in Table 5.1. These muscle ratios, specifically the relative dominance of the masseter, are similar to those in the common wombat (masseter, 65%; temporalis, 20%, pterygoid, 15%) (Murray, 1998). The masseter has a complex architecture. The superficial masseter is comprised of an external (ex s mas, Figs. 5.2, 5.4, 5.5) and an internal part (in s mas). The external part of the superficial masseter originates antero-ventrally on the root of the zygoma and inserts on and covers the ventral surface of the broad inflected mandibular angle. The internal part (in s mas) has a wide area of origin that extends posteriorly along the ventral edge of the anterior part of the zygoma, and it inserts on the massenteric ridge. The deep masseter (d mas) originates from the medial aspect of the zygoma and inserts in the basin formed between the massenteric ridge and the ascending ramus. The fibers of the two parts of the superficial masseter are oriented obliquely, while those of the deep, vertically. The anterior temporalis (a tem) originates on the exposed surfaces of the temporal fossa and inserts at the base and medial face of the ascending ramus; and the posterior temporalis inserts on both sides of the tip of the ascending ramus.

Although the pterygoid muscles are short and comprise a small percentage of the total adductor mass (Table 5.1), they have a large cross-sectional area. The wombat's medial pterygoid is divided into two: a deep part (d mpt) originates on the ventral edge of the pterygoid bone and inserts on the medial edge of the inflected angle of the mandible. A larger superficial part (s mpt) originates in a deep, concave pocket on the lateral surface of the alisphenoid above the pterygoid hamulus, and inserts on an extensive area that is bounded ventrally by the inflected angle and laterally by

Table 5.1 Weight and percentage of the total weight of the adductor muscle, in *Lasiorhinus latifrons*

Muscle	Weight	Percentage of total Weight
External Superficial Masseter	25.6 g	19.03
Internal Superficial Masseter	34.0 g	25.3
Deep Masseter	10.1 g	7.5
Total Masseter		51.8
Temporalis	42.1 g	31.3
Medial Pterygoid	16.0 g	11.9
Lateral Pterygoid	6.7 g	5.0
Total	134.5 g	

the medial surface of the ascending ramus and articular process (Fig. 5.4C, D and Fig. 5.2, Section 5.5). The fibers of the superficial medial pterygoid vary between horizontal and oblique, whereas those of the deep medial pterygoid are more vertical in orientation. The lateral pterygoid (l pt) has an insertion area on the medial aspect of the condyle and articular disc, and originates from the lateral aspect of the squamosal medially to the glenoid and above the origin of the superficial medial pterygoid.

5.3 Hypotheses to be Tested

5.3.1 General Hypothesis

The general hypothesis tested here is that the wombat, a grazer with molars designed to break down food by transverse movement of the lower teeth relative to the uppers, will, as Weijs (1994), Hiiemae (2000), and Kalvas (1999) all implied but did not explicitly state, resemble placental grazers in terms of the neuromuscular control of jaw movements. This hypothesis is tested via two more specific hypotheses:

3.1.1. The masticatory motor pattern for the control of jaw movements in wombats is predicted to resemble that of placental grazers such as goats and horses, wherein Triplet I muscles move the working-side jaw laterally and, during the power stroke, Triplet II muscles move the jaw medially. Both Triplet I and Triplet II muscles will also have a superior component of force. A substantial time difference is predicted to separate the activity periods of the two muscle groups.

3.1.2. The tightly sutured mandibular symphysis in wombats is predicted to transfer transversely directed forces from the balancing to the working side, as it does in placental herbivores with a fused symphysis. It is further predicted that these forces are generated early in the power stroke by Triplet II muscles and late in the power stroke by the balancing-side deep masseter muscle.

5.3.2 Alternate Hypothesis

The first two hypotheses do not take into consideration the fact that herbivory in Australian mammals developed independently and in isolation from the evolution of herbivory in placental mammals. The masticatory apparatus of wombats, despite superficial similarities such as a fused symphysis and transverse jaw movements, differs fundamentally from that of placental herbivores. Examples of these differences include the wombat's large inflected mandibular angle, the enlarged superficial part of its medial pterygoid muscle, the narrow transversely oriented mandibular condyle, and the position of enamel on its molars. Based upon these differences, it is predicted that the masticatory motor control pattern is also fundamentally unlike that of placental herbivores.

5.4 Material and Methods

5.4.1 Subjects

Data were collected from six mature hairy-nosed wombats (see Table 5.2) housed at the animal facility on the Waite campus of the University of Adelaide.[1] Data were collected from nine experiments: initially on six animals between September and November 2001 and then on three of the same animals in June 2004. The animals were housed in pens, 3/4 m in size, in which a stainless steel mesh was buried 1 m below the ground surface to prevent the animals digging deep tunnels and escaping. Each animal was provided with an artificial underground burrow that was covered with a removable roof.

The animals were given a daily diet of lucerne (alfalfa), oats, pellets, and freshly cut grass. Fresh water was provided daily. For the collection of data, the animals were housed temporarily in the animal quarters of the University of Adelaide.

Table 5.2 Data collected on nine hairy-nosed wombats

Animal	Electrode placement	Rosette strain gauge position
2001		
Wombat 1	r & l superficial masseter and medial pterygoid	r & l symphysis and ventral mandibular surface
Wombat 2	r & l superficial masseter and medial pterygoid	r & l symphysis and ventral mandibular surface
Wombat 3	r & l superficial masseter and medial pterygoid	r & l symphysis and ventral mandibular surface
Wombat 4	r & l anterior & posterior temporalis, deep masseter, superficial masseter and medial pterygoid	none
Wombat 5	r & l anterior & posterior temporalis, deep masseter, superficial masseter, and medial pterygoid	none
Wombat 6	r & l anterior & posterior temporalis, deep masseter, superficial masseter and medial pterygoid	none
2004		
Wombat 7	r & l temporalis, superficial masseter and deep & superficial medial pterygoid	r & l symphysis
Wombat 8	r & l temporalis, deep masseter, superficial masseter, and medial pterygoid	r & l symphysis
Wombat 9	r & l deep masseter, superficial masseter and deep & superficial medial pterygoid	r & l symphysis

---

[1] These animals were part of a wombat colony being studied by Glen Shimmin (Shimmin et al., 2001).

5.4.2 Surgery

In Wombats 1, 2, and 3 (Table 5.1), strain gauges were placed on either side of the suture on the ventral surface of the mandibular symphysis and onto the ventral surface of the right hemi-mandible below M_1. Six EMG electrodes were placed in the medial pterygoid and superficial masseter. In Wombats 4 through 6, 12 electrodes were implanted in the medial pterygoid, masseter, and temporalis muscles, but data could only be collected from six electrodes at any one time. In Wombats 7, 8, and 9, two strain gauges were positioned on the ventral surface of the symphysis. Twelve electrodes were implanted in the medial pterygoid, masseter, and temporalis muscles, but data could only be collected from ten electrodes at a time.

Prior to surgery, all the animals were sedated with Zoletil (1 mg/kg) and then maintained on a surgical plane of anesthesia with isoflurane administered through a facemask. All incision sites (skin as well as periosteum) were infused with a local analgesic Bupivacaine (diluted 1:10 v/v). In Wombats 1, 2, and 3, the ventral surface of the jaws and neck of all the animals were shaved and sterilized. Under sterile surgical conditions, an incision was made ventral to the mandibular symphysis and in the midline below the posterior region of the lower jaw. A plastic tube with an external diameter of 6 mm was inserted below the skin between these two incisions. Three insulated FRA-1-11 rosette strain gauges (Sokki Kenkyujo, Tokyo, Japan) of $120 \pm 0.05\,\Omega$ resistance and their leads, together with six electrode wires, were led forward through the tube that was subsequently removed, leaving the leads and wires below the skin. The surface of the bone at each gauge site was prepared by cutting a small window in the periosteum, cauterizing any vessels in the bone and degreasing with chloroform. Gauges were bonded to the bone using methyl-2-cyno-acrylate glue and the orientation of the A-element of each gauge relative to the longitudinal axis of the symphysis was measured. Enameled copper wire (0.125 mm) was used to prepare the EMG electrodes. The tips of the electrodes were inserted in the ends of either a 2- or 3-in, 20-gauge hypodermic needle. The needle was inserted through the posterior portion of the medial incision into a selected muscle and then withdrawn, leaving the hooked tips of the electrodes in place. Electrodes were inserted in the superficial masseter on both sides just below the inflected mandibular angle. An attempt was made to insert one electrode in the deep and one electrode in the superficial portions of the medial pterygoids on both sides of the skull. The incisions were closed and the leads from the strain gauges and the electrodes led to two 25-pin connectors (one for strain gauges and one for EMG electrodes) on the animal's back.

In Wombats 4 through 6, a similar procedure was followed to place six electrodes in the medial pterygoid and superficial masseter. In addition, electrodes were placed in the temporalis muscle. The skin over the temporal and neck region was shaved and sterilized. A medial incision was made in both regions and a plastic tube was led subcutaneously between the two incisions. Wires for six electrodes were passed through the tube and inserted in the anterior and posterior temporalis. An attempt was made to reach the deep masseter with the aid of a 3-in, 20-gauge needle inserted vertically, lateral to the ascending ramus of the dentary. EMGs were collected first from the six ventrally placed electrodes, then from the six dorsally placed electrodes, and finally from all six electrodes on the right side.

For Wombats 7 through 9, the same procedure was followed to place two rosette gauges on the symphysis and twelve electrodes in the masseter, temporalis, and medial pterygoid complexes. Leads from the electrodes and strain gauges were led to a 37-pin connector on the animal's back. The connectors were positioned between the shoulder blades and together with the superficial wires held in place by flexible bandages loosely wrapped around the animal's neck and chest. Strain and EMG data were first collected approximately 6 h after surgery and again 24 h after surgery.

5.4.3 Recording

During recording sessions, the strain gauges were connected via insulated wires to Vishay 2120A amplifiers (MicroMeasurements Inc., Raleigh, NC, USA) to form one arm of a Wheatstone bridge in quarter bridge mode; bridge excitation was 1 V. Voltage outputs were recorded on a TEAC^{TM} RD-145T DAT tape recorder (TEAC^{TM} Corp, Tokyo, Japan). Gauges were periodically balanced to adjust for zero offsets during the experiment and calibrated when the animal was stationary. EMG electrodes were connected to P511J amplifiers (Grass Inc., Quincy, MA, USA) and amplified (X1000–X10000) with a 300 Hz–3 kHz bandpass filter. All data were recorded digitally on the TEAC tape recorder. Amplification of each EMG electrode was held constant during the course of each feeding sequence.

Feeding behavior was recorded with a digital video recorder (DCR-TRV30 Sony). Because wombats eat with their heads directed downward, they were fed on a 3/4 in glass plate and filmed from below in order to document the chewing side. In order to synchronize the video, strain, and EMG data, a small diode was placed within the edge of the video field. Manually triggered short pulses, of varying duration and number, illuminated the video and generated a 5-V signal that was recorded synchronously with the EMG and strain data on the tape recorder.

5.4.4 EMG and Strain Gauge Analysis

Sections of EMG data that corresponded to video recordings of rhythmic molar chewing were sub-sampled at 6 kHz using TEAC's QuikVu II program. A program written by D. Hertweck processed the EMG data. This program eliminated any offset, fully wave-rectified the raw EMG signals, applied constant time (10 ms) reset integration, rejected randomly timed activity following Thexton (1996), and graphed the resultant EMG signals together with any other associated mechanical measure (strain, synchronization pulse, etc.). EMG records chosen for analysis were limited to feeding sequences in which the chewing side shifted several times. Final analysis of all data was carried out using Igor Pro™ 2.01 (WaveMetrics, Inc., Lake Oswego, OR). In order to determine working-to-balancing-side ratios, peak EMG activity for the recording of each muscle was scaled to 100 units and total activity calculated by totaling the activity levels at 10-msec intervals, that is, the area under the curve.

Selected sequences of strain data were sampled from the tape recorder on a Macintosh G4 computer using an Ionet™ A-D board (GW Instruments, Somerville, MA, USA) at 250 Hz. A Superscope 3.0™ (GW Instruments, Somerville, MA, USA) virtual instrument, written by D.E.L., was used to determine the zero offset and calculate strains (in microstrain, $\mu\varepsilon$) from raw voltage data using shunt calibration signals recorded during the experiment. For each gauge, principal tension (ε_1), compression (ε_2), and the orientation (ϕ) of the principal tensile strain relative to the longitudinal axis of the symphysis were calculated following equations in Biewener (1992). Processed EMG and strain data were then correlated with the synchronization signal.

5.5 Results

5.5.1 Muscle Activity

Figure 5.6 illustrates EMGs of muscles from which successful recordings were obtained in the nine wombat experiments. Sequences chosen for analysis include frequent shifts in the chewing side with right-side muscles in black and left-side muscles in gray; bars beneath the graphs indicate the working side (black for right, gray for left). Vertical lines (labeled in Fig. 5.6, W1) divide each cycle into two periods. Period 1 is the power stroke; during Period 2 the jaw opens and begins to close. The lengths of the masticatory cycle (measured from the time of onset of adductor activity of one cycle to the time of onset of activity of the next cycle) vary between 395 ± 12 msec and 600 ± 83 msec (Table 5.3). The cycle lengths are governed by food type: longer cycles are associated with lucerne, medium lengths with pellets, and the shortest with rolled oats. Adductor activity occupies about 50% of the cycle length (Table 5.3).

The most striking feature of wombat mastication is that, with the exception of some parts of the medial pterygoid, muscle activity during the power stroke is virtually restricted to the working side. Unilateral adductor activity in the wombat contrasts with all other placental herbivorous mammals in which both the working- and the balancing-side muscles are active during the power stroke. In several recordings, it is possible to compare the ratio of the level of activity of muscles on the working side with those on the balancing side. We only compared ratios of sequences that included a side shift. The maximum level activity in each muscle was scaled to 100 units, and the total activity during the power stroke of a single cycle was based on the area below the curve rather than peak values. We divided the value associated with each working-side muscle by that of a matching balancing-side muscle value to

5 Motor Control of Masticatory Movements

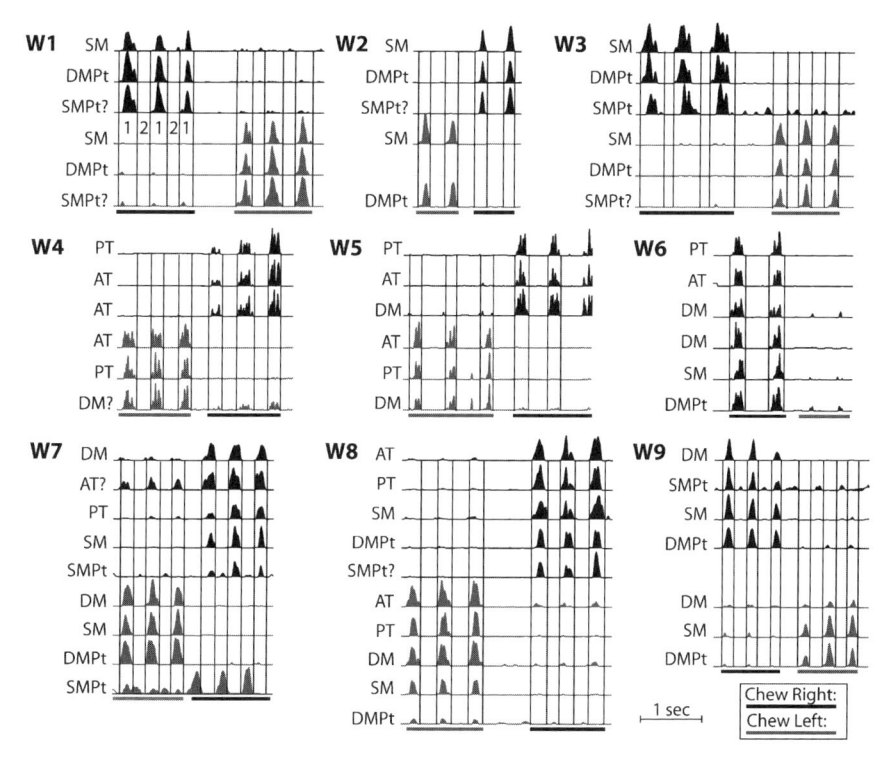

Fig. 5.6 *Lasiorhinus latifrons*. EMGs of a selection of adductor muscles during shifts in the chewing side in nine different wombat experiments. EMGs of muscles on the *right* are shaded black and those on the *left*, gray. The black bar *below* the EMGs indicates that the animal was chewing on the *right*, the gray bar, chewing on the *left*. Chewing cycles are divided into two periods by vertical lines. Period 1: adductor activity is on the working side when the working side of the jaw is drawn medially. Period 2: there is no activity in the working-side adductors

Table 5.3 Summary of cycle length and percentage of cycle with active adductors in six wombats feeding on different foods

	Animal	n	Median cycle length in ms	SD	Duration power stroke as % cycle length	SD
Oats	W9	5	395	12.9	49.4	4.4
	W7	10	402	19.3	54.5	3.4
	W2	10	432	38.2	49.9	7.5
Pellets	W1	10	458	45.6	50.9	4.3
	W4	10	467	23.6	49.5	5.1
	W3	10	497	52.7	49.6	5.7
	W8	10	507	52.5	46.9	6.4
Lucerne	W5	10	591	86.5	45.1	6.7
	W6	8	600	83.1	44.7	6.1

Table 5.4 Working-to-balancing side ratios of adductor activity during the power stroke in different adductors of *Lasiorhinus latifrons*

	AT	PT	DM	SM	MPt
W1				>50	>50
W2				>50	>50
W3				>50	>50
W4	>50	>50			
W5	>50		>50		
W7			>50	>50	34
W8	14	22	12		>50
W9			19	>50	>50

determine the working-to-balancing (W/B) side ratio (Table 5.4). In those instances where it was possible to record simultaneously from the same muscle on both the working and balancing sides, the lowest ratio was 12:1; however, the W/B ratio was highly variable and usually greater than 50:1 because the levels of most balancing-side muscles were either zero or extremely small. Muscles that tended to register the smallest ratios included the temporalis and deep masseter. In some cases, simultaneous recordings of working- and balancing-side muscles were not obtained. In these cases, we compared the levels of activity in the same muscle during a side shift when it was first on the balancing side and then on the working side. In two such cases, the deep masseter in Wombat 4 and the anterior temporalis in Wombat 7 (Fig. 5.6), the W/B ratio varied between 2.5 and 5.7. For these recordings, the electrodes had been placed deep in the muscles and it could not be determined whether they were in the intended locations.

In Wombats 1, 2, 3, 7, 8, and 9, the medial pterygoid was accessed from the ventral surface of the jaw. An attempt was made to place electrodes in both the superficial and deep parts of this muscle, but the actual position within the muscle could not be confirmed because the animals were not sacrificed at the end of the experiment. In Wombats 1, 2, 3, 7, 8, and 9, the superficial part of the medial pterygoid is active during the power stroke when it is on the working side. However, in Wombats 3, 7, and 9 (Fig. 5.6), the right-side superficial medial pterygoid (SMPt) is also active during the opening stroke when it is the balancing-side muscle and active during closing when it is a working-side muscle. In Wombat 7, major activity in the left superficial medial pterygoid occurs during opening when it is a balancing-side muscle. In Wombats 7 and 9, the working-side superficial medial pterygoid is active both during opening and closing, whereas the balancing-side superficial medial pterygoids in these two wombats are active during opening and silent during closing. In most instances (Wombat 1, 2, 3, 6, 7, and 9), the deep medial pterygoid (DMPt) is only active during the power stroke when it is on the working side and silent when it is on the balancing side. A plausible explanation for this variability, which needs further testing, is that in some experiments the electrodes that were originally intended to be placed in the superficial part moved or partially withdrew, so that they ended up in the deep part or between the deep and superficial parts.

5.5.2 Strain

Figure 5.7 shows the recordings of strain on either side of the symphyseal suture, together with five EMGs on the right and five on the left. Figures 5.8, 5.9, and 5.10 show strains both at the symphysis and on the ventral surface of the right mandible below M_1, along with the EMGs of selected muscles during a chewing-side shift. As in the case of adductor muscle activity, there is a clear change in the strain pattern of the jaw as the animal shifts chewing sides. Peak strain (Figs. 5.9, 5.10) occurs at the end of activity in the superficial masseter and medial pterygoid.

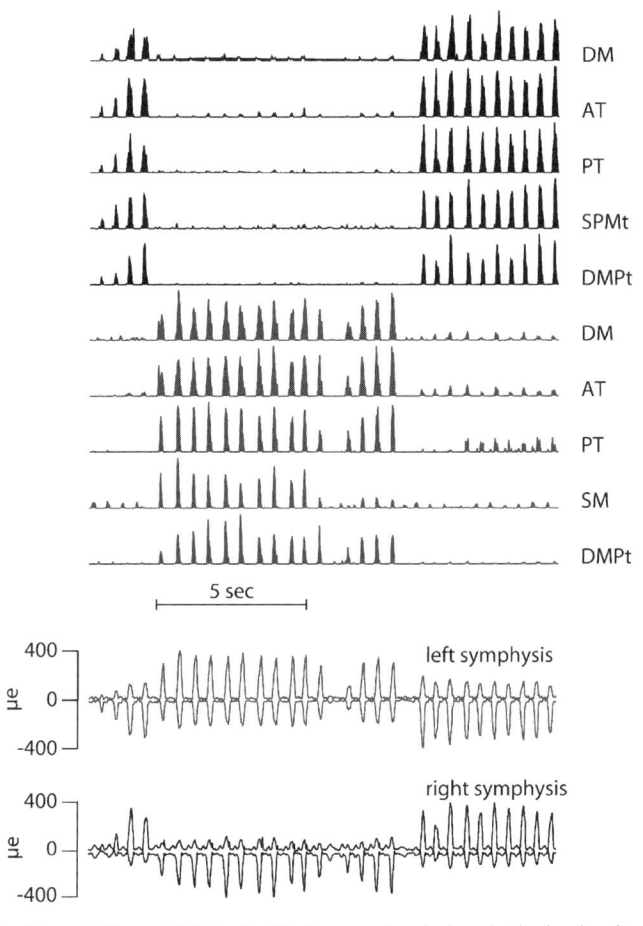

Fig. 5.7 *Lasiorhinus latifrons.* EMGs of adductor muscles during rhythmic chewing on lucerne, first on the right, then on the left, and finally on the right together with synchronous strain on the right and left sides of the ventral surface of the mandibular symphysis. EMGs and strain on the right are shown in black and those on the left in gray. There is a marked change in both the EMG and the strain pattern when the chewing side changes

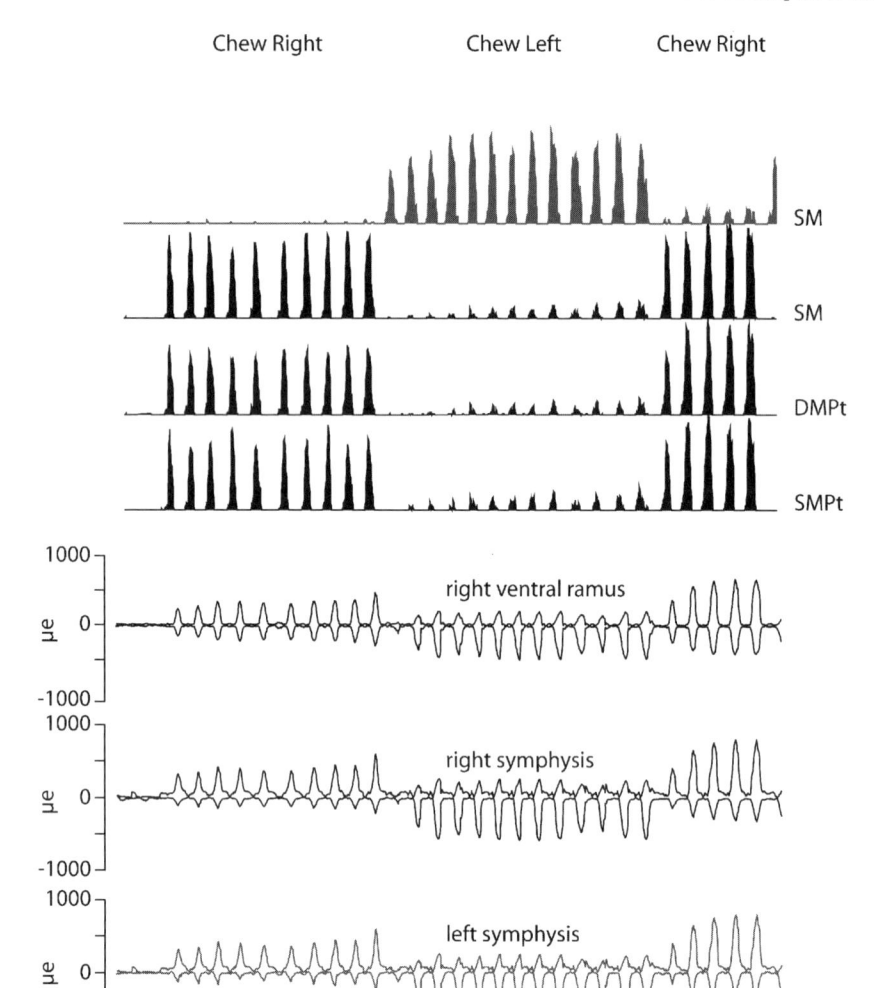

Fig. 5.8 *Lasiorhinus latifrons.* Synchronous EMGs of four adductors and strain on the ventral surface of the mandible on both sides of the symphyseal suture and ventral surface of the right mandibular ramus to illustrate the change in the muscle activity and strain as the animal shifts the chewing side, first from right to left and then back to right

Table 5.5 shows average tension (ε_1), compression (ε_2), and the angle of principle strain ($\varepsilon_1\phi$) relative to the sagittal plane, for 10–15 consecutive right- and left-side chewing cycles.

Figure 5.5B illustrates the orientation of principle strains on the working and balancing sides.

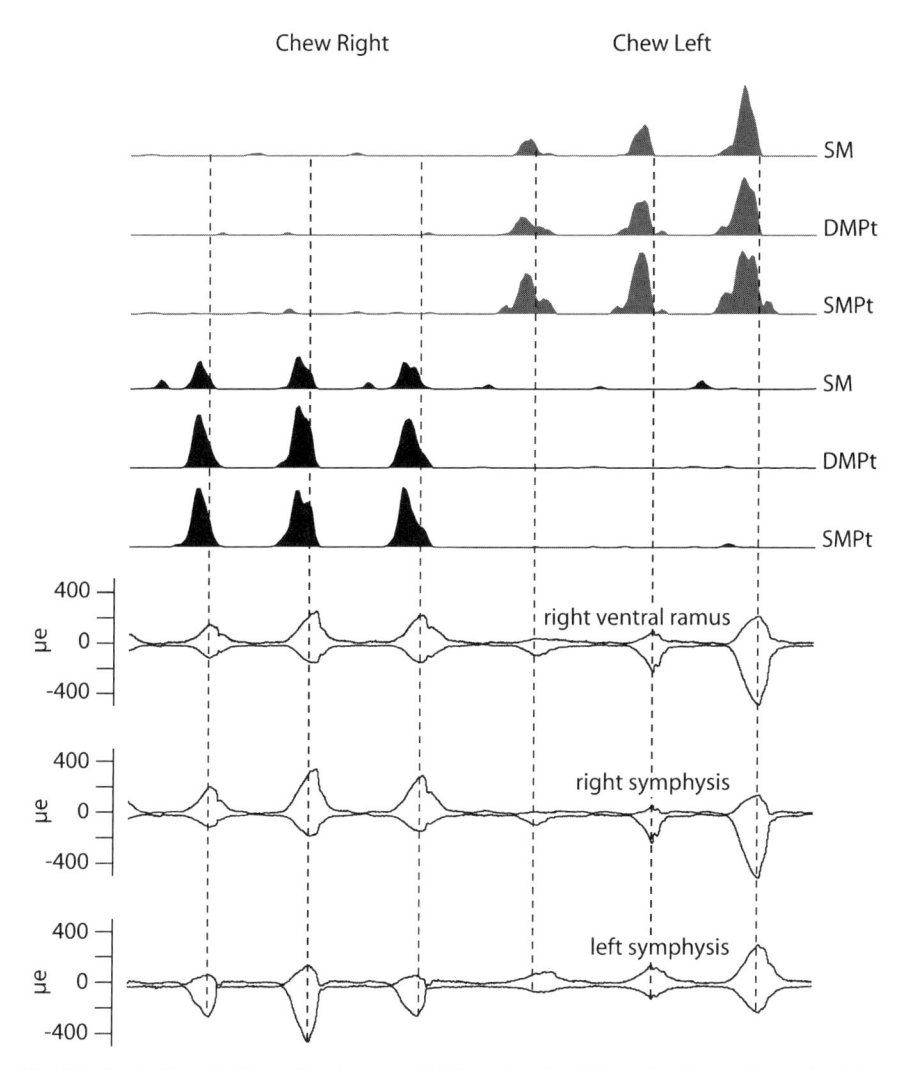

Fig. 5.9 *Lasiorhinus latifrons.* Synchronous EMGs and strain of three chewing cycles on the right, followed by three on the left, to illustrate the relation between muscle activity and mandibular strain. Vertical dotted lines indicate relationships between peak strain and the timing of adductor EMGs

With each change of chewing side, the angle of principle strain ($\varepsilon_1\phi$) on the ventral surface of the symphysis shifts approximately 90°. The $\varepsilon_1\phi$ on working and balancing sides are parallel to one another. For the gauge on the ventral surface of the right dentary, the shift is slightly greater (106–116°) (Fig. 5.5B). When an animal chews on the right side, $\varepsilon_1\phi$ on the right side of the symphyseal region (working side) varies between $-32°$ and $-55°$ relative to the long axis of the symphysis, while that on the left (balancing side) varies between $-41°$ and $-60°$. When the

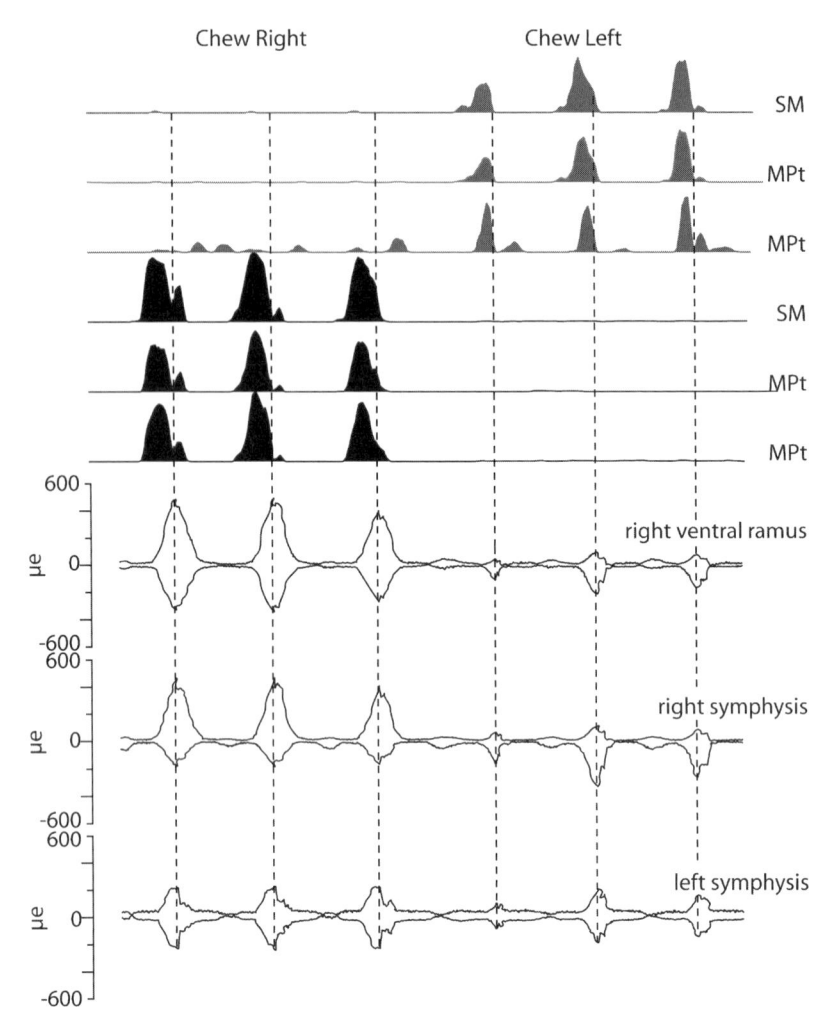

Fig. 5.10 *Lasiorhinus latifrons.* Synchronous EMGs and strain during three chewing cycles on the right, followed by three on the left

animal chews on the left, the angle on the left (working) side varies between 33° and 53° while that on the right (balancing) varies between 25° and 35°. The principle strain angle of a gauge on the ventral surface of the right mandibular ramus varies between −25° and −37° when it is on the working side, and 79° and 91° when it is on the balancing side. The parallel orientation of the angles of principal strain on the working and balancing sides suggests that the whole symphyseal region is twisted in the same direction. It acts as a single unit, and only one side generates all the torque force at any one time.

5 Motor Control of Masticatory Movements

Table 5.5 Descriptive statistics for peak principal strains (ε_1 and ε_2) and angle of peak tensile strain ($\varepsilon_1\phi$) (standard deviation) on the ventral margin of the right hemi-mandible and ventral surface of the mandibular symphysis on either side of the midline during left- and right-side chews. $\varepsilon_1\phi$ is relative to the mid-sagittal plane

	n	Chew right			Chew left		
		ε_1	ε_2	$\varepsilon_1\phi$	ε_1	ε_2	$\varepsilon_1\phi$
Ventral Margin:							
W3	15	409.1 ± 46.9	−282 ± 42.6	−25.6 ± 1.1	91.9 ± 18.3	−259.7 ± 46.4	80.4 ± 4.9
W2	15	320 ± 52.7	−214 ± 37	−24.8 ± 1.3	184.4 ± 46.9	−460 ± 35.1	91.2 ± 2.1
W1	15	231.4 ± 27.9	−127.2 ± 18.9	−36.9 ± 2.5	140.2 ± 32.5	−314.6 ± 68.6	79.4 ± 1.5
Right Symphysis:							
W3	15	395 ± 56	−160 ± 18.8	−55.0 ± 0.63	109.3 ± 34	−394.6 ± 85	35.7 ± 0.5
W2	15	403 ± 75.3	−137.7 ± 30	−55.5 ± 1.0	225.2 ± 31.4	−536.6 ± 43.2	33.9 ± 0.3
W8	10	331 ± 50	−81 ± 9.4	−32 ± 1.4	112.5 ± 14.3	−222 ± 39.4	25.1 ± 0.35
Left Symphysis:							
W3	15	236.6 ± 44.1	−249.9 ± 37.4	−55.9 ± 0.8	284.2 ± 77.3	−292.9 ± 77.3	39.4 ± 1.7
W2	15	412.6 ± 76.3	−155.9 ± 47.2	−55.3 ± 1.3	226.5 ± 34.4	−532.1 ± 43.1	41.0 ± 1.3
W1	10	106.3 ± 20.2	−274.6 ± 33.2	−54.6 ± 0.5	254.6 ± 38.5	−148.5 ± 29.9	32.7 ± 1.3
W8	10	143.5 ± 18.4	−337 ± 37.2	−41.5 ± 0.96	363.5 ± 36.5	−224 ± 23.3	53.2 ± 0.98

5.6 Discussion

All herbivores (marsupial and placental) generate a transverse force to drag the lower molars in a medial direction across the uppers during the power stroke of mastication. In placentals, this movement involves the transfer of some force from the balancing to the working side. Placental herbivores with fused symphyses and those with unfused symphyses accomplish transverse movements in slightly different ways. For example, in an artiodactyl with an unfused symphysis, the working side is drawn medially by Triplet II muscles in which the activity in the balancing-side deep masseter precedes that of the working-side superficial masseter (Williams et al., 2003c). In anthropoids and ungulates with fused symphyses (Hylander et al., 2000, Williams et al., 2008), the timing of the firing of Triplet II muscles is changed so that the balancing-side deep masseter is the last muscle to fire. In the Southern hairy-nosed wombat, transverse jaw movements during the power stroke are generated almost entirely by the working-side adductor muscles. Forces are transferred from the working to the balancing and not, as in placentals, from the balancing to the working side. During the initial stages of jaw closure, transverse movements to draw the working side of the jaw laterally are generated by the balancing-side superficial medial pterygoid. It is not possible to divide the adductor muscles in wombats into Triplet I and II muscles. On rare occasions, low levels of activity were noted in the balancing-side temporalis and masseter during the power stroke. Judging from video recordings recorded synchronously with EMGs, this activity occurs when the incisors manipulate food. At such moments, the working- to balancing-side ratio

(W/B) of adductor muscles may drop to 12:1. In all other chewing cycles, the ratio is greater than 50:1.

In placental herbivores, bite force along the molar row can be increased by raising the level of activity in the working-side muscles, while at the same time recruiting increasing amounts of activity in the balancing-side muscles and W/B ratio may approach 1:1 (opossum: Crompton, 1995; anthropoids: Hylander et al., 2000; goats: Lieberman and Crompton, 2000). In wombats increased bite force is accomplished by recruiting increasing amounts of working-side musculature. In wombats the masseter muscle complex lies far lateral to the molar row, and because it is so wide medio-laterally, it has a large cross-sectional area presumably to generate high occlusal forces without the involvement of balancing-side muscles.

Another distinctive feature of wombats is the greatly enlarged superficial medial pterygoid that inserts on the wide and horizontally oriented inflected angle of the mandible. In contrast to placental herbivores that have a deep mandibular angle, the insertion of both parts of the medial pterygoid muscle in wombats is close to the condyle. As a result, the fibers of the superficial part of the medial pterygoid muscle are nearly horizontal. The superficial medial pterygoid has a large cross-sectional area and, together with the working-side superficial masseter, it plays an important role in drawing the working side of the mandible medially during the power stroke.

The activity pattern of the medial pterygoid muscle is complex in wombats. The deep medial pterygoid on the working side appears to be active during the power stroke and silent on the balancing side. This is the typical pattern reported for the medial pterygoid in placental herbivores (Herring and Scapino, 1973; Hylander and Johnson, 1994; Lieberman and Crompton, 2000). The firing pattern of the superficial part is more variable. During opening and the beginning of closing it is active on the balancing side; whereas on the working side it can be active during both closing and opening. In placental herbivores such as artiodactyls and perissodactyls (Nickel et al., 1986), and primates (Gray, 1858 [35th edition published 1973]), the medial pterygoid is also divided into two parts: a larger rostromedial part and a smaller superficial slip or caudolateral part. These divisions may be homologous to the two parts of this muscle in wombats, but this possibility still needs to be established. The presence of two parts of the medial pterygoid in placental herbivores may account for the atypical firing pattern reported for the medial pterygoid in primates (macaque, Hiiemae and Crompton, 1985) and artiodactyls (goat, Lieberman and Crompton, 2000). In both of these species, activity in the medial pterygoid is biphasic with activity during both the opening and the closing phases, and sometimes only during opening or closing. In goats, synchronous activity in the medial pterygoid, digastric, and geniohyoid during opening helps pull the lower jaw forward; during closing, synchronous activity of the medial pterygoid and the other adductors adduct the jaw vertically (Crompton et al., 2005). It must be noted that this variable pattern may simply reflect the positions of the electrodes within the medial pterygoid muscle complex. Nevertheless, it appears that in all marsupial herbivores, with the exception of koalas (Davison and Young, 1990), the superficial part of the medial pterygoid has been enlarged to become the dominant part of the medial pterygoid. The inflected angle provides an area of insertion for this muscle (Crompton and Lieberman, 2004a, b).

Recordings from strain gauges placed on the ventral surface of the massive, tightly sutured symphysis of wombats indicate that the whole symphyseal region is twisted clockwise (when viewed from the front) while chewing on the right. When the skull is viewed from the front, the symphysis is twisted clockwise when chewing on the right (Fig. 5.5F) and counter clockwise when chewing on the left. Torque is probably generated primarily by the superficial masseter, because the force vector of this muscle lies lateral to the working-side condyle. Because the mandibular symphysis is stiff and immobile, the torque force generated by the working side will be transferred to the balancing side and will tend to rotate the whole lower jaw around the working-side condyle (which rotates in the coronal plane), placing the balancing-side TMJ under tension. It was not possible to determine the magnitude of the twisting force, but the shear strains on the ventral surface of the symphysis are seldom in excess of 500 microstrain. Balancing-side condylar distraction is apparently resisted by the massive set of ligaments around the TMJ. Manipulation of the wombat skull shows that only a very small amount of rotation is necessary to prevent the balancing-side molars from coming into contact when the lower working-side molars are dragged across the uppers. Because wombat molars are isognathic, even low levels of activity in the balancing-side adductors would result in balancing-side occlusion. The wear pattern of the molars in wombats confirms that occlusion is unilateral.

There appear to be multiple reasons for developing a fused or immobile mandibular symphysis. In anthropoids the fused symphysis resists lateral transverse bending; but this is not the case in alpacas, where it is twisted rather than bent (Williams et al., 2008). Williams et al. (2008) suggest that the alpaca symphysis is fused in order to support long incisor roots. This could also be true for the wombat, since it has long incisor roots that extend the entire length of the symphysis. On the other hand, as kangaroos also have long incisor roots that extend the length of a highly mobile symphysis, there may be an alternative reason for the development of an immobile symphysis in wombats.

In contrast to wombat molars, those of the kangaroo molars are adapted for shearing rather than crushing and grinding. As Sanson (1989) pointed out, they have not evolved common adaptations to resist heavy wear of the molars such as hypsodonty and molar crown patterns that are not changed significantly by wear. Medially directed transverse movement of the whole working-side hemi-mandible in kangaroos is severely limited by the narrow upper incisal arcade. Kangaroos are only mildly anisognathic, and consequently there is not sufficient space between the upper molar rows to permit extensive transverse movements as in the case of most placental ungulates. However, because the occlusal surfaces of kangaroo lower molars lie above the axis of rotation of the hemi-mandible, rotation alone contributes to the transverse movements of the lower molars relative to the uppers (Lentle et al., 1998, 2003; Ride, 1959). Our as yet unpublished data on kangaroo mastication indicate that, as in wombats, only the working-side muscles move the hemi-mandible medially while simultaneously rotating it during the power stroke.

This suggests that the common ancestor of kangaroos and wombats possessed a similar masticatory motor pattern that has been modified in different ways in wombats and macropods. Wombats evolved a masticatory system designed to deal with tough, abrasive foods that require high occlusal forces to be broken down. However, the masticatory motor control pattern of wombats appears to rule out the possibility of recruiting balancing-side musculature to generate large bite forces. Wombats have increased the mass of the adductor muscles by moving the zygoma laterally, and further increasing the space for adductors by moving the molar rows closer to one another.

Figure 5.11 compares the areas of origin for the superficial masseter and medial pterygoid in a kangaroo and a wombat. This comparison is not to suggest that wombats arose from kangaroos, but to show the adaptations required to increase the bite force across the molars. Moving the origin of the adductors laterally increased the torque force acting on the working-side mandible. To deal with a tougher and more abrasive diet, wombats evolved ever-growing hypsodont molars. A tightly sutured symphysis prevents the balancing-side molars from coming into contact, because the torque force generated by the working side separates the molars on the balancing side. The whole mandible rotates around the long axis of the working side.

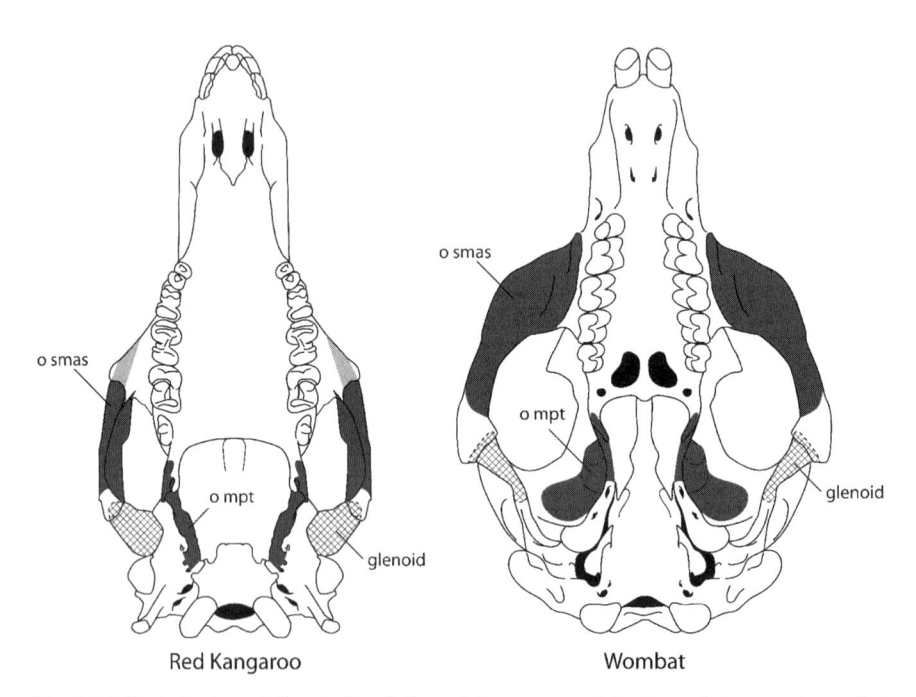

Fig. 5.11 Ventral view of the skulls of the red kangaroo and Southern hairy-nosed wombat, comparing the areas of origin (*shaded in gray*) of the superficial masseter (o smas) and the medial pterygoid (o mpt) muscles. (Light gray shading indicated a part of the origin of the superficial masseter that is not visible in ventral view). The hatched areas indicate the glenoid of the TMJ

This hypothesis is based upon the neuromuscular control of jaw movements of a single species, and with reference to unpublished data on the red kangaroo: as such it is still tentative and speculative. It does, however, establish that the masticatory motor pattern of wombats and kangaroos are fundamentally different from that of placental herbivores, and that the masticatory motor patterns in mammals are far more varied than the current literature suggests (Weijs, 1994; Hiiemae, 2000; Langenbach and van Eijden, 2001). This study stresses the fact that the ubiquitous feature of mammalian herbivores, viz. transverse jaw movements, can be generated and controlled in several different ways and accomplished with fundamentally different biomechanical systems. In order to understand the function, origin and, evolution of the diverse feeding mechanisms of Australian marsupials, it will be necessary to broaden the scope of this study to include, at a minimum, representatives of the major groups of Australian mammals, and then to integrate these findings with the known fossil record.

Acknowledgments This work was carried out with the aid of a grant from the Putman Fund of the Museum of Comparative Zoology, Harvard University, and a grant from the University of Adelaide.

References

Abbie, A. (1939) A masticatory adaptation peculiar to some diprotodont marsupials. *Proc Zool Soc Lond* 109: 261–279.

Becht, G. (1953) Comparative biologic-anatomical researches on mastication in some mammals. *Proc K Ned Akad Wet Ser C Biol Med Sci* C. 56: 508–527.

Biewener, A. (1992) In vivo measurement of bone strain and tendon force. In: Biewener, A. (ed.) *Biomechanics – tructure and systems: A practical approach*. Oxford University Press, Oxford, England pp 123–147.

Crompton, A. W. (1995) Masticatory function in nonmammalian cynodonts and early mammals. In: Thomason, J. (ed.) *Functional morphology in vertebrate paleontology*. Cambridge University Press, Cambridge pp 55–75.

Crompton, A. W., Hylander, W. L. (1986) Changes in mandibular function following the acquisition of a dentary-squamosal jaw articulation. In: Hotton III, N., Mchean, P., Roth, J., Roth, E. (eds.) *The ecology and biology of mammal-like reptiles*. Smithsonian Institute, Washington pp 263–282.

Crompton, A.W., Lieberman, D.E. (2004a) Motor control of mastication in mammals: Marsupials and placentals compared. *Int Comp Biol* 44: 541.

Crompton, A. W., Lieberman, D. E. (2004b) Functional significance of the inflected mandibular angle. *J Vert Paleo* 24: 49A.

Crompton, A. W., Lieberman, D. E., Aboelela, S. (2005). Tooth orientation during occlusion and the functional significance of condylar translation in primates and herbivores. In: Carrano, M., Blob, R., Gaudin, T., Wible, J. (eds.) *Amniote paleobiology* pp 367–388.

Davison, C., Young, G. (1990) The muscles of mastication of *phascolarctos cinereus* (phascolactidae: Marsupiala). *Aust J Zool* 38: 227–240.

Ferreira, J. M., Phakey, P. P., Palamara, J., Rachinger, W. A., Orams, H. J. (1989) Electron microscopic investigation relating the occlusal morphology to the underlying enamel structure of molar teeth of the wombat vombatus-ursinus. *J Morphol* 200: 141–150.

Ferreira, J. M., Phakey, P. P., Rachinger, W. A., Palamara, J., Orams, H. J. (1985) A microscopic investigation of enamel in wombat*(vombatus ursinus)*. *Cell Tissue Res* 242: 349–355.

Finlayson, G. R., Shimmin, G. A., Temple-Smith, P. D., Handasyde, K. A., Taggart, D. A. (2005) Burrow use and ranging behaviour of the southern hairy-nosed wombat (lasiorhinus latifrons) in the murraylands, south australia. *J Zool* 265: 189–200.

Gray, H. (1858) *Gray's anatomy*. W. B. Saunders, Philadelphia. (1471).

Herring, S. W., Scapino, R. P. (1973) Physiology of feeding in miniature pigs. *J Morphol* 141: 427–460.

Hiiemae, K. M. (1978) Mammalian mastication: A review of the activity of the jaw muscles and the movements they produce in chewing. In: Butler, P., Joysey, K. A. (eds.) *Development, function and evolution of teeth*. Academic Press, London pp 359–398.

Hiiemae, K. M. (2000) Feeding in mammals. In: Schwenk, K. (ed.) *Form, function and evolution in tetrapod vertebrates*. Academic Press, San Diego pp 399–436.

Hiiemae, K. M., Crompton, A. W. (1985) Mastication, food transport and swallowing. In: Hildebrand, M., Bramble, D., Liem, K., Wake, D. (eds.) *Functional vertebrate morphology*. Belknapp Press, Harvard University Press, Cambridge pp 262–290.

Hogue, A. S., Ravosa, M. J. (2001) Transverse masticatory movements, occlusal orientation, and symphyseal fusion in selenodont artiodactyls. *J Morphol* 249: 221–241.

Hylander, W. L. (1984) Stress and strain in the mandibular symphysis of primates a test of competing hypotheses. *Am J Phys Anthropol* 64: 1–46.

Hylander, W. L., Johnson, K. R. (1994) Jaw muscle function and wishboning of the mandible during mastication in macaques and baboons. *Am J Phys Anthropol* 94: 523–547.

Hylander, W. L., Ravosa, M. J., Ross, C. F., Johnson, K. R. (1998) Mandibular corpus strain in primates: Further evidence for a functional link between symphyseal fusion and jaw-adductor muscle force. *Am J Phys Anthropol* 107: 257–271.

Hylander, W. L., Ravosa, M. J., Ross, C. F., Wall, C. E., Johnson, K. R. (2000) Symphyseal fusion and jaw-adductor muscle force: An emg study. *Am J Phys Anthropol* 112: 469–492.

Hylander, W. L., Vinyard, C. J., Ravosa, M. J., Ross, C. F., Wall, C. E., Johnson, K. R. (2004) Jaw adductor force and symphyseal fusion. In: Anapol, F., German, R., Jablonski, N. (eds.) *Shaping primate evolution: Papers in honor of charles oxnard*. Cambridge University Press, Cambridge pp 229–257.

Hylander, W. L., Vinyard, C. J., Wall, C. E., Williams, S. H., Johnson, K. R. (2003) Convergence of the "wishboning" jaw-muscle activity pattern in anthropoids and strepsirrhines: The recruitment and firing of jaw muscles in propithecus verreauxi. *Am J Phys Anthropol Suppl* 36: 120.

Hylander, W. L., Wall, C. E., Vinyard, C. J., Ross, C., Ravosa, M. R., Williams, S. H., Johnson, K. R. (2005) Temporalis function in anthropoids and strepsirrhines: An emg study. *Am J Phys Anthropol* 128: 35–56.

Kalvas, J. (1999) The masticatory musculature and temporomandibular joint of a grazing macropodid (macropus giganteus); its morphology and function as compared with that of a grazing bovid (ovis aries). Department of Anatomical Sciences. University of Adelaide, Adelaide p 52.

Langenbach, G. E., van Eijden, T. M. G. J. (2001) Mammalian feeding motor patterns. *Am Zool* 41: 1338–1351.

Lentle, R., Hume, I., Stafford, K., Kennedy, M., Haslett, S., Springett, B. (2003) Comparison of tooth morphology and wear patterns in four species of wallabies. *Aust J Zool* 51.

Lentle, R., Stafford, K., Potter, M., Springett, B., Haslett, S. (1998) Incisor and molar wear in the tammar wallaby. *Aust J Zool* 46: 509–527.

Lieberman, D., Crompton, A. (2000) Why fuse the mandibular symphysis? A comparative analysis. *Am J Phys Anthropol* 112: 517–540.

Murray, P. F. (1998) Palaeontology and palaeobiology of wombats. In: Wells, R. T., Pridmore, P. A. (eds.) *Wombats*. Surrey Beatty and Sons Pty Ltd., Adelaide pp 1–33.

Nickel, R., Schummer, A., Seiferle, E., Frewein, J., Wilkens, H., Wille, K.-H. (1986) *The locomotor system of the domestic animals*. Springer-Verlag, New York

Oron, U., Crompton, A. (1985) A cineradiographic and electromyographic study of mastication in *Tenrec ecaudatus*. *J Morphol* 185:155–182.

Ravosa, M. J., Hylander, W. L. (1994) Function and fusion of the mandibular symphysis in primates: Stiffness or strength? In: Fleagle, J. G., Kay, R. (eds.) *Advances in primatology; Anthropoid origins*. Plenum, New York pp 447–468.
Ravosa, M. J., Vinyard, C. J., Gagnon, M., Islam, S. A. (2000) Evolution of anthropoid jaw loading and kinematic patterns. *Am J Phys Anthropol* 112: 493–516.
Ride, W. (1959) Mastication and taxonomy in the macropodine skull. In: Cain, A. (ed.) *Function and taxonomic importance*. Oxford University Press, Oxford Systematics association publication, Vol 3, pp 33–59.
Sánchez-Villagra, M. R., Smith, K. (1997) Diversity and evolution of the marsupial mandibular angular process. *J Mammal Evol* 4: 119–144.
Sanson, G. (1980) The morphology and occlusion of the molariform cheek teeth in some macropodinae (marsupialia: Macropodidae). *Aust J Zool* 28: 341–365.
Sanson, G. (1989) Morphological adaptations of teeth to diets and feeding in the macropodoidea. In: Grigg, G., Jarman, P., Hume, I. (eds.) *Kangaroos, wallabies and rat-kangaroos*. Surrey Beatty & Sons Pty Limited, New South Wales pp 151–168.
Scott, G., Richardson, K. (1987) Osteological differences of the axial skeleton between lasiorhinus latifrons (owen, 1845) and vombatus ursinus (shaw, 1800) (marsupialia: Vombatidae). *Rec S Afr Mus* 22: 29–39.
Shimmin, G., Skinner, J., Baudinette, R. (2001) Life in a wombat burrow. *Nature Aust* 27:62–69.
Thexton, A. (1966) A randomisation method for discriminating between signal and noise recordings of rhythmic electromyographic activity. *J Neurosci Methods*. 66:93–98.
Turnbull, W. (1970) Mammalian masticatory apparatus. *Fieldiana: Geol* 18: 153–356.
Vinyard, C. J., Wall, C. E., Williams, S. H., Johnson, K. R., Hylander, W. L. (2006) Masseter electomyography during chewing in ring-tailed lemurs (lemur catta). *Am J Phys Anthropol* 130:85–95.
Vinyard, C. J., Williams, S. H., Wall, C. D., Johnson, K. R., Hylander, W. L. (2005) Jaw-muscle electromyography during chewing in belanger's treeshrews (tupaia belangeri). *J Am Phys Anthropol* 127:26–45.
Weijs, W. (1994) Evolutionary approach of masticatory motor patterns. In: Bels, V. L., Chardon, M., Vandewalle, P. (eds.) *Biomechanics of feeding in vertebrates*. Springer-Verlag, Belgium Advances in comparative and environmental physiology, Vol 18, pp 281–320.
Weijs, W., Dantuma, R. (1981) Functional anatomy of the masticatory apparatus in the rabbit *(orytolagus cuniculus l.)*. *Netherlands J Zool* 31: 99–147.
Williams S. H., Vinyard, C. J., Wall, C. E., Hylander, W. L. (2004) Masticatory strains in the mandibular corpus of selenodont artiodactyls. *Int Comp Biol* 282A
Williams, S. H., Wall, C. E., Vinyard, C. J., Hylander, W. L. (2003a) Strain in the mandibular symphysis of alpacas and the evolution of symphyseal fusion in camelids. *J Vert Paleo* Sup 23: 110A.
Williams, S. H., Wall, C. E., Vinyard, C. J., Hylander, W. L. (2003b) Jaw-muscle motor patterns in ungulates: Is there a transverse pattern?*SICB Annual Meeting & Exhibition Final Program & Abstracts*: 342.
Williams, S. H., Vinyard, C. J., Wall, C. E., Hylander, W. L. (2003c) Symphyseal fusion in anthropoids and ungulates: A case of functional convergence?*Am J Phys Anthropol Suppl* 36: 226.
Williams, S. H., Wall, C. E., Vinyard, C. J., Hylander, W. L. (2008) Symphyseal fusion in selenodont artiodactyls: New insights from *in vivo* and comparative data. In: Vinyard, C. J., Ravosa, M. J., Wall, C. E. (eds.) *Primate Craniofacial Function and Biology*. Springer. New York. pp 39–62.

Chapter 6
Specialization of the Superficial Anterior Temporalis in Baboons for Mastication of Hard Foods

Christine E. Wall, Christopher J. Vinyard, Susan H. Williams, Kirk R. Johnson, William L. Hylander

Contents

6.1 Introduction 113
6.2 Background 115
6.3 Materials and Methods 117
6.4 Results 118
6.5 Discussion 120
    6.5.1 Functional and Structural Heterogeneity 120
    6.5.2 The Temporalis Muscle of Humans 122
    References 122

6.1 Introduction

The temporalis muscle is the largest of the jaw adductors in primates, typically comprising greater than 50% of the total jaw adductor muscle mass (Schumacher, 1961; Turnbull, 1970; Cachel, 1979, 1984). It is also a three-dimensionally complex muscle, divided along its anteroposterior length into a superficial part and a deep part by a central tendon that thins superiorly. In coronal view, the central tendon is easily identified (Fig. 6.1). In baboons, the deep part of the anterior temporalis is thinner than the superficial part (Fig. 6.1) and weighs less than the superficial part (Wall et al., 2007).

    The deep fibers exert a medial and superior pull on the mandible (black arrow, Fig. 6.1). The superficial fibers exert a lateral and superior pull on the mandible (white arrow, Fig. 6.1). The superficial part is divisible into anterior, middle, and posterior parts on the basis of vector orientation (Fig. 6.2). As can be seen in Figs. 6.1 and 6.2, the superior component of force is relatively large in all parts

---

C.E. Wall
Department of Evolutionary Anthropology, Duke University, Box 90383, Durham, NC 27708
e-mail: christine_wall@baa.mc.duke.edu

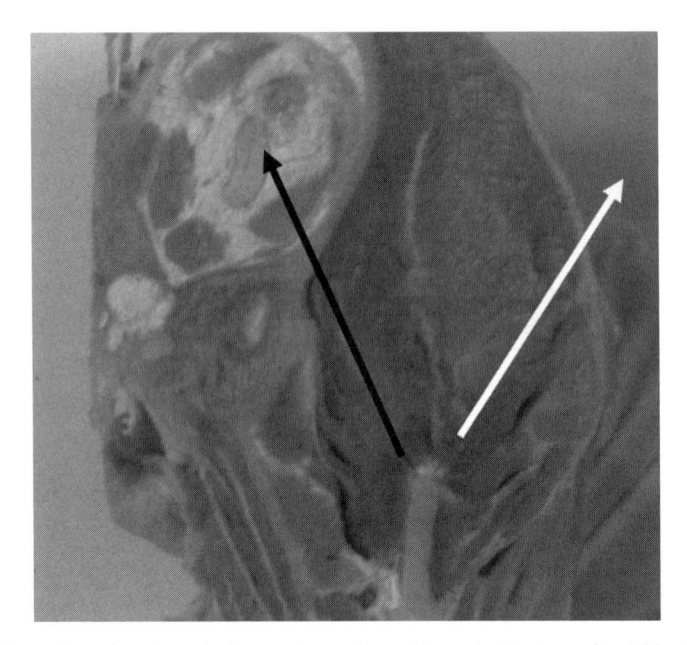

Fig. 6.1 Coronal section through the cranium of an adult male *Papio anubis*. This photograph shows all but the superior-most part of anterior temporalis. The midline is toward the left, lateral is toward the right, and superior is toward the top. The direction of pull is simplified by the black arrow showing the medial and superior pull of the deep part, and the white arrow showing the lateral and superior pull of the superficial part

of the muscle (except possibly the most posteriorly directed fibers of the posterior part). Some of the anterior-most fibers have a small anterior component, and the middle fibers have a small posterior component (Fig. 6.1). These vector orientations are similar in the deep part.

Variation in fiber orientation means that the temporalis muscle can move the mandible in a variety of directions depending on which fibers are activated. Previous studies of muscle function in primates and other mammals indicate that differences in fiber orientation and the presence of fascial planes are correlated with differences in EMG recruitment during feeding (e.g., Hylander et al., 2005; Weijs, 1994). One reason this occurs is because during the chewing cycle the requirements for bite force and jaw movement change. For example, the strong posterior vector of the balancing-side posterior temporalis rotates the jaw toward the balancing side. Hylander et al. (2005) show that in baboons there is a late peak in activity level of the balancing-side posterior temporalis in comparison to the deep anterior temporalis and the working-side posterior temporalis. This late peak of the balancing-side posterior temporalis moves the working-side mandible medially toward the balancing side during the late part of the power stroke.

This paper reports on a portion of a larger study we are undertaking to document the links between EMG recruitment during mastication as well as fiber architecture, fiber type, and myosin expression for the jaw adductor muscles in baboons and macaques. Identifying these links is important for understanding the evolution of

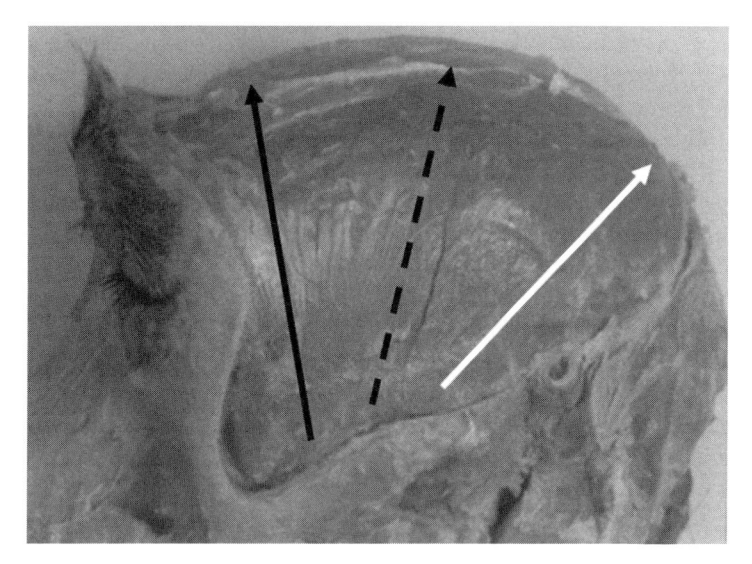

Fig. 6.2 Lateral view of the temporalis muscle in an adult male *Papio anubis*. The solid black arrow indicates the anterior and superior pull of the superficial anterior temporalis, the dashed black arrow shows the posterior and superior pull of the superficial middle temporalis, and the white arrow shows the posterior and superior pull of the superficial posterior temporalis

craniofacial structure and function in primates. Here we focus on a comparison of EMG recruitment in the deep and superficial parts of the anterior temporalis. The central hypothesis evaluated by this study is that a high proportion of type II fibers is directly related to the rapid production of high bite forces during mastication (Gibbs et al., 1984; Herring et al., 1979; Nielsen and Miller, 1988). One prediction of this hypothesis is that increased EMG activity during feeding on hard objects will be greater in the superficial anterior temporalis if, as a predominately "type II muscle," it is specialized for activity when high force is required. Here, as a partial test of this prediction, we provide data on the magnitude of peak activity in the deep and superficial anterior temporalis during chewing of hard objects relative to soft objects. We use a ratio of EMG activity during hard object feeding divided by EMG activity during soft object feeding and look at the magnitude of that ratio in four muscles: the working- and balancing-side deep anterior temporalis and the working- and balancing-side superficial anterior temporalis. If the prediction is supported, the ratios for the superficial anterior temporalis will be significantly higher than those for the deep anterior temporalis.

6.2 Background

The muscle fibers in a whole muscle can be characterized according to histochemical and contractile properties (Pette and Staron, 1990). These properties are conveyed in part by the different proteins that make up a muscle cell, the most well studied being myosin. The jaw adductor muscles of primates are composed of a variety of histochemically defined types of fibers (Korfage et al., 2005a; Maxwell et al., 1979; Rowlerson et al., 1983). In adult muscle, most fibers can be grouped into one of two categories, Type I and Type II, based on the type of myosin present in the fiber. Although a single fiber can contain two or more types of myosin (Korfage et al. 2005b), fibers usually express one type of myosin. In the jaw adductors of non-human primates as well as some other mammal species, the IIM isoform of the myosin heavy chain (MHC) is commonly expressed in the Type II fibers (Hoh, 2002; Rowlerson et al., 1981, 1983).

Fiber type composition has implications for the functional properties of a muscle fiber, motor unit, part of a muscle, or whole muscle. This includes contractile properties such as time to peak tension, the resistance to fatigue, and both the twitch and the tetanic tension that the motor unit produces (Henneman and Olson, 1965; Barany, 1967; Edström and Kugelberg, 1968; Burke et al., 1971; Hoh, 2002). Although numerous sub-types of type II and type I fibers have been identified, the behavior of type II fibers can be generalized in contrast to the behavior of type I fibers. Type II fibers can be either fatigue-resistant (as are type I fibers) or fatigueable, but type II fibers have faster contraction times, larger cross-sectional areas (typically two to three times larger), develop higher twitch and tetanic tension (i.e., from 4 to 12 times higher) than type I fibers, and occur in larger motor units.

Little is known of the relation between histochemical properties of muscle fibers and whole muscle function for the muscles of mastication. Herring et al. (1979) and Anapol and Herring (2000) found regional variation in the duration of activity in the masseter muscle that correlated well with fiber type and fiber length. In these

Table 6.1 Fiber type proportions in the deep and superficial anterior temporalis based on myosin ATPase reactivity.[1] %II = percentage of type II fibers in the fascicle

	Deep anterior temporalis	Superficial anterior temporalis
Male		
I	64	6
II	65	174
Total	129	180
%II	50.4	96.6
Female		
I	52	7
II	62	125
Total	114	132
%II	54.4	94.7

[1] The sample was "fresh-frozen" and stored in a conventional freezer at $-10°$F. After thawing, 1-cm^3 blocks were removed and immediately snap re-frozen in isopentane chilled in liquid nitrogen and transferred to a $-70°$C freezer. A technician in the Muscle Pathology Laboratory at the Duke University Medical Center then prepared and stained 10-μm thin sections. Myosin ATPase reactivity was evaluated following pre-incubation in an alkaline buffer at pH 9.4 and in an acid buffer at pH 4.35, following the method of Brooke and Kaiser (1970).

studies on miniature swine, high-amplitude EMGs during feeding were observed in regions that contained high proportions of type II fibers. The link between histochemistry and function should be greatest in regions, such as the superficial head of the temporalis muscle of baboons, that are dominated by a single fiber type (Maxwell et al., 1979; Wall et al., 2007; Table 6.1).

Prior to the EMG study, we collected preliminary data on ATPase reactivity for one adult male and one adult female baboon. These data suggest that approximately 50% of the fibers in the deep anterior temporalis are type I, whereas fewer than 6% of the fibers in the superficial anterior temporalis are type I (Table 6.1). Work in progress on the types of myosin heavy chains (MHCs) present in the temporalis muscle for most of the EMG subjects (three male and two female baboons) confirm this finding by showing that the type I MHC is common in the deep temporalis and that the type IIM MHC dominates in the superficial temporalis (Wall et al., 2006, 2007).

6.3 Materials and Methods

The experimental subjects are three adult female *Papio anubis* and three adult male *P. anubis*. Indwelling, fine wire electrodes were placed in the deep anterior temporalis and the superficial anterior temporalis. Conventions for naming the parts of the temporalis muscle vary widely. For example, some authors consider the muscle fibers that originate on the most medial margin of the inferior surface of the zygomatic arch and insert on the lateral surface of the mandibular ramus (i.e., the zygomaticomandibularis muscle) to be part of the temporalis muscle, whereas others consider it to be the deepest part of the masseter muscle (Gaspard, 1972; Gaspard et al., 1973a,b; Turnbull, 1970; Cordell, 1991; Perry and Wall, 2004). We did not record EMGs from this region.

The male and female baboons have thick temporalis muscles, and electrode depth could be reliably controlled for by using long (1½") needles for the deep anterior temporalis, and short (1/2") needles for the superficial anterior temporalis.

For the deep anterior temporalis, the goal is to insert the electrode into the thickest and the deepest part in each subject. An electrode-bearing needle is inserted into the muscle posterior to the superolateral margin of the orbit superior to the zygomatic arch at the level where the superior and inferior tarsal plates meet. The needle is driven into the superficial anterior temporalis, through the central tendon, and into the deep anterior temporalis until it hits the bone.

For the superficial anterior temporalis, the electrode-bearing needle pierces the skin superior to the zygomatic arch and is inserted at an angle nearly parallel to the side of the head. This insures stable electrode insertion at a superficial location.

The methods used to collect and analyze the EMG interference patterns follow those detailed in Hylander et al. (2005) and Vinyard et al. (2005). We simultaneously record the EMG interference patterns from up to ten fine-wire, indwelling electrodes onto a 14-channel analog tape recorder. These analog data are then digitally converted to yield the raw-EMGs (Fig. 6.3). The raw-EMGs are then rectified and

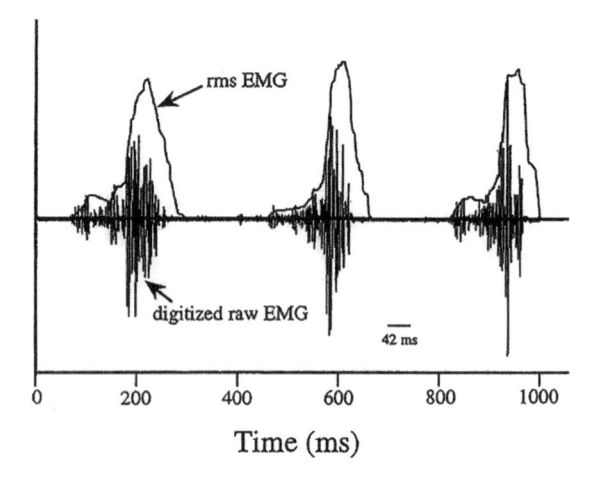

Fig. 6.3 Digitized raw-EMGs and the corresponding root-mean-square (rms-) EMGs. Time is in milliseconds (msec). The methods used to derive the rms-EMGs are detailed in Hylander and Johnson (1993)

root-mean-squared to yield an integrated curve of the amplitude of the EMG interference pattern (Fig. 6.3).

Hard foods included in calculation of the hard/soft ratio are unpopped popcorn kernels, cherry pits, and pieces of monkey chow. Soft foods included in the calculation of this ratio are small pieces of apple with skin and pear with skin (approximately $2\,cm^3$), short and thin strips of banana peel (approximately 1" × 0.5"), and thin carrot slices (approximately 5–10 mm thick). The hard foods are both hard and tough in comparison to the soft foods (Williams et al., 2005).

To create the hard/soft ratio, the mean peak EMG amplitude during hard-object cycles was divided by the mean peak EMG amplitude during soft-object cycles for each electrode. For each individual, per-electrode means were used to calculate a grand mean. The balancing and working sides were calculated separately, as were data for males and females. The Mann-Whitney U-test (Zar, 1999) was used to test for significant differences between the deep anterior temporalis and the superficial anterior temporalis.

6.4 Results

The number of chews, mean, and standard deviation of the data that are used to create the hard/soft ratio are provided for each subject in Table 6.2. These values do not provide information about absolute muscle force since raw amplitudes cannot be meaningfully compared across electrodes. They are provided only to document intra-individual variation in EMG amplitude during a chewing sequence. One thing to note is the small standard deviations for peak EMG amplitude in the balancing-side superficial anterior temporalis.

6 Specialization of the Superficial Anterior Temporalis in Baboons

Table 6.2 EMG amplitude during hard- and soft-object feeding in the deep anterior temporalis (DAT) and the superficial anterior temporalis (SAT) in adult male and adult female *Papio*. N = number of chews (in italics); sd = standard deviation

	Working-side DAT		Working-side SAT	
	Hard chews	Soft chews	Hard chews	Soft chews
Sample	*N*, mean (sd)	*N*, mean (sd)	*N*, mean (sd)	*N*, mean (sd)
Female A	*15*, 2.37(0.2)	*10*, 1.99(0.4)	*15*, 1.25(0.2)	*9*, 0.73(0.4)
Female B	*14*, 2.35(0.3)	*12*, 1.53(0.3)	*15*, 0.99(0.5)	*11*, 1.51(0.6)
Female C	*18*, 2.34(0.3)	*11*, 2.01(0.2)	*18*, 2.97(0.6)	*11*, 1.29(0.1)
All Females	*47*, 2.35(0.3)	*33*, 1.83(0.3)	*48*, 2.44(0.4)	*31*, 1.21(0.4)
Male A	*57*, 0.69(0.1)	*50*, 0.75(0.4)	*57*, 0.64(0.2)	*50*, 0.26(0.2)
Male B	*272*, 0.60(0.1)	*165*, 0.47(0.1)	*272*, 0.28(0.1)	*165*, 0.20(0.1)
Male C	*128*, 0.67(0.1)	*77*, 0.38(0.1)	*128*, 0.47(0.2)	*77*, 0.14(0.1)
All Males	*457*, 0.60(0.1)	*292*, 0.50(0.2)	*457*, 0.38(0.2)	*292*, 0.19(0.1)
	Balancing-side DAT		Balancing-side SAT	
	Hard chews	Soft chews	Hard chews	Soft chews
Sample	*N*, mean (sd)	*N*, mean (sd)	*N*, mean (sd)	*N*, mean (sd)
Female A	*14*, 2.02(0.3)	*7*, 0.87(0.3)	*13*, 1.05(0.1)	*7*, 0.37(0.3)
Female B	*14*, 1.99(0.5)	*7*, 0.79(0.5)	*13*, 3.55(0.7)	*8*, 0.13(0.1)
Female C	*8*, 1.28(0.3)	*8*, 0.32(0.1)	*8*, 1.06(0.3)	*8*, 0.11(0.03)
All Females	*36*, 1.84(0.3)	*22*, 0.64(0.3)	*34*, 2.01(0.4)	*23*, 0.19(0.1)
Male A	*57*, 0.78(0.1)	*50*, 0.20(0.1)	*57*, 0.48(0.2)	*50*, 0.02(0.01)
Male B	*272*, 0.52(0.1)	*165*, 0.18(0.1)	*272*, 0.26(0.1)	*165*, 0.05(0.02)
Male C	*128*, 0.66(0.1)	*77*, 0.15(0.04)	*128*, 0.55(0.2)	*77*, 0.11(0.1)
All Males	*457*, 0.58(0.1)	*292*, 0.17(0.1)	*457*, 0.37(0.2)	*292*, 0.06(0.05)

Table 6.3 provides the hard/soft ratio for each muscle part that is derived from the data in Table 6.2. The prediction is that increased activity during feeding on hard objects will be greater in the superficial anterior temporalis if it is specialized for activity when high force is required. The Mann-Whitney U-test is significant for comparison of both the working-side muscles and the balancing-side muscles. As shown graphically in Fig. 6.4, all muscles increase their activity during hard-object feeding. However, the high values for the superficial anterior temporalis show that this muscle shows a greater increase in activity than that seen in the deep anterior temporalis. The ratio is high in the superficial anterior temporalis because it is recruited at very low activity levels, particularly on the balancing side, during the majority of cycles in which soft foods are chewed.

The males and females show similar values for the deep anterior temporalis and the superficial anterior temporalis on the working side, and for the deep anterior temporalis on the balancing side. There is substantially more intra-individual variation in the balancing-side superficial anterior temporalis. This reflects the fact that Females B and C and Male A showed almost no activity in the balancing-side superficial anterior temporalis when chewing the soft foods, whereas the other subjects showed slightly higher levels of activity.

Table 6.3 Hard/soft ratio in the working- and balancing-side deep anterior temporalis (WDAT and BDAT) and superficial anterior temporalis (WSAT and BSAT) in adult male and female *Papio*. The grand mean for each sex for hard-object feeding is divided by the grand mean for each sex for soft-object feeding to derive the hard/soft ratio for each sex. Refer to Table 6.2 for information about the standard deviation associated with the numerator and the denominator

	WDAT	WSAT
Female A	1.2	1.7
Female B	1.5	2.0
Female C	1.2	2.8
All Females	1.3	2.0
Male A	0.9	2.5
Male B	1.3	1.4
Male C	1.8	3.4
All Males	1.2	2.0
	BDAT	BSAT
Female A	2.3	2.9
Female B	2.5	28.4
Female C	4.0	10.0
All Females	2.9	10.4
Male A	4.0	32.0
Male B	3.0	5.3
Male C	4.4	5.0
All Males	3.4	6.2

Fig. 6.4 The hard/soft ratio in the deep anterior temporalis and superficial anterior temporalis on both the working side and the balancing side in the male and female baboons. The Mann Whitney U-test is significant for both the working- and the balancing-side comparisons

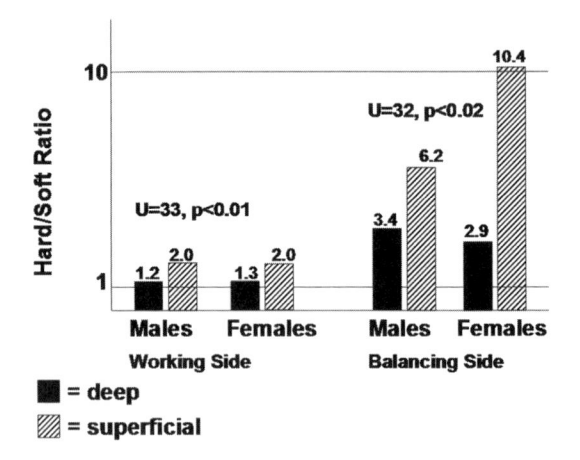

6.5 Discussion

6.5.1 Functional and Structural Heterogeneity

These data document a remarkable amount of functional heterogeneity in the anterior temporalis muscle of both female and male baboons. Furthermore, this functional heterogeneity is strongly correlated with structural heterogeneity with respect to whole muscle architecture, intramuscular fiber type distribution, and MHC composition.

The peak EMG comparisons indicate that these four muscle parts function during chewing as three functional units. The working-side muscles (working-side superficial anterior temporalis and working-side deep anterior temporalis) form a group that shows the least amount of variation in peak EMG amplitude related to food consistency (Table 6.3, Fig. 6.4). However, we also want to stress that EMG amplitudes are relatively reduced in the working-side superficial anterior temporalis as compared to the working-side deep anterior temporalis (approximately 50% as active during soft object cycles as compared with approximately 80% as active during soft object cycles). The balancing-side deep anterior temporalis forms the second functional unit and is about 30% as active during soft object cycles as it is during hard object cycles. The balancing-side superficial anterior temporalis forms a third functional unit showing negligible activity during feeding on soft objects. The superficial temporalis is larger than the deep temporalis (Fig. 6.1). We plan further work on leverage, pinnation, fiber length, and cross-sectional area in the functional units identified electromyographically to estimate how much force the superficial temporalis can produce compared to the deep temporalis.

In sum, these data suggest that one important way that primates generate the high bite forces required by hard foods is to recruit the superficial anterior temporalis. The data presented here suggest that the temporalis muscle of baboons is remarkable in having a large superficial, anterior part that is recruited at low levels during chewing of soft foods. This demonstrates that a large part of the largest jaw adductor muscle is not strongly recruited during mastication of soft foods. Further study of the fiber architecture, fiber type, and EMG activity in the temporalis muscle of prosimians and anthropoids will allow an evaluation of how common this solution is to the problem of generating high bite forces.

The data also provide evidence in support of the hypothesis that the presence of high proportions of type II fibers is directly related to the production of high bite forces during chewing (Gibbs et al., 1984; Herring et al., 1979; Nielsen and Miller, 1988). Type II fibers are specialized for *rapid* force production. Predictions related to the importance of rapid force production in the superficial part of temporalis are based on the idea that the transition between the closing stroke and the power stroke is a relatively short period of time (Wall et al., 2006).

For a number of years, it has been known that the temporalis muscle of macaques shows heterogeneity in the size and distribution of fiber types (Maxwell et al., 1979; Miller and Farias, 1988). As noted earlier, type II fibers tend to be relatively more numerous in the superficial head of temporalis compared to the deep head. Work to characterize the fiber population of the temporalis muscle and to determine if differences in fiber type and EMG recruitment exist between males and females is underway in these subjects (Wall et al., 2006, 2007). Our data indicate that baboons are similar to macaques with more type II fibers superficially and with smaller type II fibers in females compared to males (Wall et al., 2006, 2007). These data also suggest that baboons are remarkable in having an overwhelming predominance of type II fibers and IIM MHC in the anterior, middle, and posterior parts of the superficial head of the temporalis muscle. If this is borne out by further work, it will support the idea that the superficial head of temporalis is a highly specialized, force-producing muscle in baboons.

6.5.2 The Temporalis Muscle of Humans

Along those lines, it is very interesting that humans appear to have lost the superficial head of the temporalis muscle during evolution (Oxnard, 1984; Oxnard and Wealthall, 2003). Loss of this muscle is part of a group of features that together contribute to the gracilization of the feeding apparatus that characterizes humans. Some other features include the reduction in the dimensions of the postcanine dentition, the reduction of the maxillary canine and loss of the canine/P3 honing complex, the absence of sagittal and occipital crests for attachment of the temporalis muscle, overall reduction in the "robusticity" of the jaws, the small diameter of type II fibers in the chewing muscles, and the loss of expression of IIM MHC in the chewing muscles (Bosley and Rowlerson, 1980; Brace, 1977; Daegling, 1993; Ringqvist, 1974a,b; Rowlerson et al., 1983; Sciote et al., 1994; Stedman et al., 2004). There is a strong temptation to link these features. However, it is unclear when, how, and why these features evolved during human evolution. Rowlerson (1990) has suggested that if the small diameter of type II fibers in human chewing muscles is related to disuse because of a soft diet (Ringqvist, 1974a,b), then we should expect to find large-diameter type II fibers in people who have a more "primitive" diet. Oxnard and Wealthall, (2003) suggest that the presence of a well-developed superficial temporalis can possibly be inferred from muscle markings on the side of the braincase. The high diversity of MHCs found in modern human jaw muscles in comparison to those of other mammals may be related functionally to a soft diet and speech, but probably also reflects sampling bias caused by differences in age and gender (Korfage et al., 2005b). Further study of the distribution of fiber types in the chewing muscles of primates as well as the bony anatomy of fossil hominins will shed light on the pattern of evolution of the modern human masticatory apparatus.

Acknowledgments The authors wish to thank Margaret Briggs, Fred Schachat, Kathleen Smith, Fred Anapol, and Dan Schmitt for valuable advice and help on this project. This research was supported by the NSF.

References

Anapol, F., and Herring, S. W., 2000, Ontogeny of histochemical fiber types and muscle function in the masseter muscle of miniature swine. *Am. J. Phys. Anth.* 112:595–613.
Barany, M., 1967, ATPase activity correlated with speed of muscle shortening. *J. Gen. Physiol.* 220:410–414.
Bosley, M., and Rowlerson, A., 1980, A new fiber type associated with the fast fibre type in the cat jaw-closer muscles? *J. Physiol.* 306:18P–19P.

Brace, C.L., 1977, Occlusion to the anthropological eye. In (McNamara, JA, Jr., ed.): *Monograph Number 7, Craniofacial Growth Series*. Ann Arbor, Center for Human Growth and Development, University of Michigan, pp. 179–209.
Brooke, M. H., and Kaiser, K. K., 1970, Muscle fibers. How many and what kind? *Archs. Neurol.* 23:369–379.
Burke, R. E., Levine, D. N., Zajac, F. E., III, Tsairis, P., and Engel, W. K., 1971, Mammalian motor units: Physiological-histochemical correlation in three types in cat gastrocnemius. *Science* 174:709–712.
Cachel, S. M., 1979, A functional analysis of the primate masticatory system and the origin of the anthropoid post-orbital septum. *Am. J. Phys. Anthl.* 50:1–18.
Cachel, S. M., 1984, Growth and allometry in primate masticatory muscles. *Archs. Oral Biol.* 29:287–293.
Cordell, N. N., 1991, *Craniofacial Anatomy and Dietary Specialization in the Galagidae*. Ph.D. Dissertation, University of Washington, Seattle.
Daegling, D. J., 1993, Functional morphology of the human chin. *Evol. Anth.* 1:170–176.
Edström, L., and Kugelberg, E., 1968, Histochemical composition, distribution of fibres and fatiguability of single motor units. *J. Neurol. Neurosurg. Psychiat.* 31:424–433.
Gaspard, M., 1972, *Les muscles masticateurs superficiels des singes à l'homme. Anatomie comparée et anatomie-physiologie*. Librairie Maloine S. A. Éditeur, Paris.
Gaspard, M., Laison, F., and Mailland, M., 1973a, Organisation architecturale du muscle temporal et des faisceaux de transition du complexe temporo-massétérin chez les Primates et l'Homme. *J. Biol. Buccale* 1:171–196.
Gaspard, M., Laison, F., and Mailland, M., 1973b, Organisation architeturale et texture du muscle masseter chez les Primates et l'Homme. *J. Biol. Buccale* 1:7–20.
Gibbs, C. H., Mahan, P. E., Wilkinson, T. M., and Mauderli, A., 1984, EMG activity of the superior belly of the lateral pterygoid muscle in relation to other jaw muscles. *J. Prosthet. Dent.* 51:691–701.
Henneman, E., and Olson, C. B., 1965, Relations between structure and function in the design of skeletal muscles. *J. Neurophysiol.* 28:581–598.
Herring, S. W., Grimm, A. F., and Grimm, B. R., 1979, Functional heterogeneity of a multipinnate muscle. *Am. J. Anatom* 154:563–578.
Hoh, J. Y. H., 2002, "Superfast" or masticatory myosin and the evolution of jaw-closing muscles of vertebrates. *J. Exp. Biol.* 205:2203–2210.
Hylander, W. L., and Johnson, K. R., 1993, Modelling relative masseter force from surface electromyograms during mastication in nonhuman primates. *Archs. Oral Biol.* 38:233–240.
Hylander, W. L., Wall, C. E., Vinyard, C. J., Ross, C., Ravosa, M. J., Williams, S. H., and Johnson, K. R., 2005, Temporalis function in anthropoids and strepsirrhines: An EMG study. *Am. J. Phys. Anth.* 128:35–56.
Korfage, J. A. M., Koolstra, J. H., Langenbach, G. E. J., and van Eijden, T. M. G. J., 2005a, Fiber-type composition of the human jaw muscles – (part 1) origin and functional significance of fiber-type diversity. *J. Dent. Res.* 84:774–783.
Korfage, J. A. M., Koolstra, J. H., Langenbach, G. E. J., and van Eijden, T. M. G. J., 2005b, Fiber-type coposition of the human jaw muscles – (part 2) role of hybrid fibers and factors responsible for inter-individual variation. *J. Dent. Res.* 84:784–793.
Maxwell, L. C., Carlson, D. S., McNamara, J. A., Jr., and Faulkner, J. A., 1979, Histochemical characteristics of the masseter and temporalis muscles of the rhesus monkey (*Macaca mulatta*). *Anat. Rec.* 193:389–402.
Miller, A. J., and Farias, M., 1988, Histochemical and electromyographic analysis of craniomandibular muscles in the rhesus monkey. *J. Oral Maxillofac. Surg.* 46:767–776.
Nielsen, I. L., and Miller, A. J., 1988, Response patterns of craniomandibular muscles with and without alterations in sensory feedback. *J. Prosthet. Dent.* 59:352–361.
Oxnard, C. E., 1984, *The Order of Man: A Biomathematical Anatomy of the Primates*. New Haven, Yale University Press. 366pp.
Oxnard, C. E., and Wealthall, R., 2003, Ghosts of the past: Temporal muscles, fasciae and bones in some primates. *Am. J. Phys. Anth.* 120(S36):163.

Perry, J. M. G., and Wall, C. E., (2004) Theoretical expectations and empirical features of prosimian chewing muscles. *J. Vert. Paleontol.* 24(3, Suppl.):101A.

Pette, D., and Staron, R. S., 1990, Cellular and molecular diversities of mammalian skeletal muscle fibers. *Rev. Physiol. Biochem. Pharmacol.* 116:2–47.

Ringqvist, M., 1974a, A histochemical study of temporal muscle fibers in denture wearers and subjects with normal dentition. *Scand. J. Dent. Res.* 82:28–39.

Ringqvist, M., 1974b, Fiber types in human masticatory muscles. Relation to function. *Scand. J. Dent. Res.* 82:333–355.

Rowlerson, A. M., 1990, Specialization of mammalian jaw muscles: Fibre type composition and the distribution of muscle spindles. In (Taylor, A, ed.): *Neurophysiology of the Jaws and Teeth.* Macmillan Press, London. pp. 1–51.

Rowlerson, A. M., Pope, B., Murray, J., Whalen, R. G., and Weeds, A. G., 1981, A novel myosin present in cat jaw-closing muscles. *J. Muscle Res. Cell Motil.* 2:415–438.

Rowlerson, A. M., Mascarello, F., Veggetti, A., Carpene, E., 1983, The fibre-type composition of the first branchial arch muscles in Carnivora and Primates. *J. Muscle Res. Cell Motil.* 4:443–472.

Schumacher, G. H., 1961, *Funktionelle Morphologie der Kaumuskulatur.* Gustav Fischer Verlag, Jena. 262pp.

Sciote, J. J., Rowlerson, A. M., Hopper, C., and Hunt, N. P., 1994, Fiber type classification and myosin isoforms in the human masseter muscle. *J. Neurol. Sci.* 126:15–24.

Stedman, H. H., Kozyak, B. W., Nelson, A., Thesler, D. M., Su, L. T., Low, D. W., Bridges, C. R., Shrager, J. B., Minugh-Purvis, N., and Mitchell, M. A., 2004, Myosin gene mutation correlates with anatomical changes in the human lineage. *Nature* 428:415–418.

Turnbull, W. D., 1970, Mammalian masticatory apparatus. *Fieldiana: Geol.* 18:148–355.

Vinyard, C. J., Williams, S. H., Wall, C. E., Johnson, K. R., and Hylander, W. L., 2005, Jaw-muscle electromyography during chewing in Belanger's treeshrews (*Tupaia belangeri*). *Am. J. Phys. Anth.* 127:26–45.

Wall, C. E., Briggs, M., Schachat, F., Vinyard, C., Williams, S., and Hylander, W., 2006, Anatomical and functional specializations of the anterior temporalis muscle of baboons. *Comp. Biochem. Physiol.* 134A(Number 4):S70.

Wall, C. E., Perry, J. M. G., Briggs, M., and Schachat, F., 2007, Mechanical correlates of sexual dimorphism in the jaw muscles and bones of baboons. *Am. J. Phys. Anth.* 132, S44:242.

Weijs, W. A., 1994, Evolutionary approach of masticatory motor patterns in mammals. *Adv. Comp. Env. Physiol.* 18:281–320.

Williams, S. H., Wright, B. W., van Den, T., Daubert, C. R., Vinyard, C. J., 2005, Mechanical properties of foods used in experimental studies of primate masticatory function. *Am. J. Prim.* 67:329–346.

Zar, J. H., 1999, Biostatistical Analysis, Fourth Edition. Prentice Hall, New Jersey. 663 pp.

Part III
Modeling Masticatory Apparatus Function

In vitro and finite element studies modeling bone behavior during feeding are a rapidly growing area of research in primatology. Typically, these studies apply in vivo data, in many cases collected by Hylander and colleagues, to set parameters in their models. Part III includes three chapters that illustrate the broad range of potential applications for using modeling approaches in studying primate craniofacial function.

Daegling et al. consider the mechanical consequences of the tooth sockets in the mandibular corpus during masticatory loading. Most analyses of primate mandibular form tend to model the corpus as an elliptical beam focusing on external linear dimensions, thereby disregarding the intrusion of the teeth into the corpus. This chapter considers several important questions related to the mechanics of this dentoalveolar interface, including whether stresses are raised at alveoli, how tooth roots influence these stress patterns, and whether we should be modeling the jaw as an open versus closed section. The results presented here suggest that alveoli may not act as traditional stress raisers, despite their tendency to be stiffer than surrounding corporal bone. In the end, Daegling et al. use their results to identify new questions and directions that will help us understand how the teeth influence the mechanical behavior of the mandibular corpus.

Wang et al. use an in vitro approach to examine the patterns of strains on and around craniofacial sutures in the macaque skull. Ultimately, they are interested in applying their in vitro results to improve finite element models by determining how sutural patency potentially influences a finite element model. As seen in vivo, they observe significant differences in strain patterns between sutures and adjacent bone, as well as between bones on the two sides of patent sutures. Their results indicate the potential importance of detailing the state of sutural fusion in finite element modeling of craniofacial loading.

In the final chapter of this part, Strait et al. use a finite element approach to compare the patterns of facial strains during premolar and molar loading in macaques. These comparisons are built around exploring the question of how robust australopithecines loaded their faces during chewing. Strait et al. observed the highest strains when bite locations were restricted to the premolars, demonstrating the potential for differential adaptation to localized bite points. Based on their results, they hypothesize that certain features of the robust australopithecines faces, such as anterior pillars or an anteriorly positioned root of the zygoma, might be adapted for feeding on large, resistant food objects with their premolars.

Chapter 7
Effects of Dental Alveoli on the Biomechanical Behavior of the Mandibular Corpus

David J. Daegling, Jennifer L. Hotzman, and Andrew J. Rapoff

Contents

7.1 The Problem Under Investigation... 127
    7.1.1 The Morphology of Alveolar Bone....................................... 128
    7.1.2 The Weakening Effects of Dental Alveoli 129
    7.1.3 Skeletal Strategies for Minimizing Stress Concentrations 130
    7.1.4 The Mechanical Consequences of Alveoli 131
7.2 Mechanical Behavior of Dental Alveoli 134
    7.2.1 Is the Mandible an Open or Closed Section?............................. 134
    7.2.2 The Role of Tooth Roots in Mitigating Stress Concentrations 136
    7.2.3 Material Property Variation in Alveolar Bone 139
7.3 Modeling Issues and Recommendations 142
    References 146

7.1 The Problem Under Investigation

The in vivo strain environment of the mandible is probably the most thoroughly understood of the varied elements of the primate skeleton, due in large part to Hylander's experimental work (1977, 1979a, b, 1981, 1984; Hylander and Crompton, 1986; Hylander et al., 1987, 1998). Interestingly, a precise stress analysis of the primate mandible has remained elusive, as several issues of modeling the structure and composition of the mandibular corpus remain unresolved (Daegling and Hylander, 1997, 1998, 2000; Dechow and Hylander, 2000; Schwartz-Dabney and Dechow, 2003). Among these issues was the recognition that the presence of teeth within alveoli precluded application of simple, homogeneous geometric models to predict stress and strain gradients in the corpus. Hylander (1979b, 1981) suggested that consideration of periodontal anatomy might lead to the conclusion that the mandible behaved as a member with open sections; Smith (1983) later suggested that failure to account for such effects could produce sizable errors, perhaps

---

D.J. Daegling
Department of Anthropology, University of Florida, Box 117305, Gainesville, FL 32611-7305
e-mail: daegling@anthro.ufl.edu

approaching an order of magnitude. Despite the articulation of these concerns, most comparative analyses of the mandible have since relied on the linear measures of corpus contours to estimate mechanical competence (reviewed in Taylor, 2002). While tooth root and alveolar morphology have been the subject of functional investigation, (Ward and Molnar, 1980; Spencer, 2003), the mechanical costs of alveoli as structural defects are largely unexplored, with the exception of clinical concerns relating to dental implant performance (Gambrell and Allen, 1976; Mason et al., 1990; Ishigaki et al., 2003).

7.1.1 The Morphology of Alveolar Bone

Alveolar bone is continuous with the basal bone of the jaw, yet it is structurally distinct in several ways. Superficial surfaces comprise cortical plates that consist of compact, lamellar bone. The bone surrounding the tooth roots is another layer of relatively compact bone known as the cribriform plate (Moss-Salentijin and Hendricks-Klyvert, 1990). This layer is named for the numerous small openings present, which allow the passage of blood vessels and nerves. The term lamina dura is also used to describe the structure, as radiographically the cribriform plate presents as a light line that reveals its relative density relative to adjacent structures. Depending on the location of the alveoli in the jaw, there may be a layer of trabecular bone present between the two cortical layers. For example, in the human mandible there is trabecular bone present opposite only to the apical third of the root in a bicuspid tooth, but trabecular bone is found surrounding much of the molar tooth socket (Saffar et al., 1997). The presence of trabecular bone not only varies between locations but can also vary among species. The comparative anatomy of alveolar bone in primates has not been systematically described functionally or structurally in the primary literature. Non-human primates, such as macaques, exhibit alveolar bone morphology similar to humans (Schou et al., 1993). The periodontal ligament provides a fibrous connection between alveolar bone and the cementum that comprises the outer layer of the tooth roots.

There are two conceptually distinct sources of mechanical stress that must be accommodated by alveolar bone. First, there are local effects of occlusal forces that are transmitted from the tooth roots to the alveolar bone via the periodontal ligament. Second, there are remotely arising muscular and reaction forces that produce bending and twisting moments – as well as shearing forces – all of which produce stresses in alveolar bone. The relative contributions of each of these sources to overall stress levels is not well understood, but occlusal forces do not occur in isolation from muscular and reaction forces. In any case, analytical approaches to modeling these distinct types of load require different theoretical perspectives.

The in vivo data of Hylander (1979a, b; 1981) suggest that alveolar bone experiences significant stress even in the absence of locally occurring occlusal forces. An interesting question from the standpoint of the function of alveolar bone is whether it is optimized for resisting occlusal forces specifically, or whether this bone is mechanically "tuned" to resist the bending and twisting moments that occur in mastication. While we do not yet have the data needed to inform speculation on this question, it is nevertheless obvious that both of these sources of stress must be successfully resisted by alveolar bone.

7.1.2 The Weakening Effects of Dental Alveoli

Ignoring the intrusion of alveoli into the bony corpus fails to account for two sources of weakening. First, the alveoli can be thought of as a determinable amount of missing material in terms of how the corpus is usually modeled. That is, the weakening effect of alveoli might simply be accounted for by computationally removing the bone mass that corresponds to alveolar size and geometry. Analytically, this does not pose significant problems as a region containing an alveolus of known dimensions can be modeled as a net section, with the estimates of stress revised accordingly.

The second means by which alveoli pose mechanical risk is their potential to function as sites of stress concentration. The effect of holes and notches in structures may be more insidious than a simple accounting of missing material would indicate. That is, structural defects can weaken severely because they cause local stresses to pile up in the vicinity of the defect; a body of research in engineering focuses on this very problem (see Savin, 1961; Peterson, 1974). The textbook example is the thin plate with a hole in its center subjected to uniaxial tension (Fig. 7.1). In

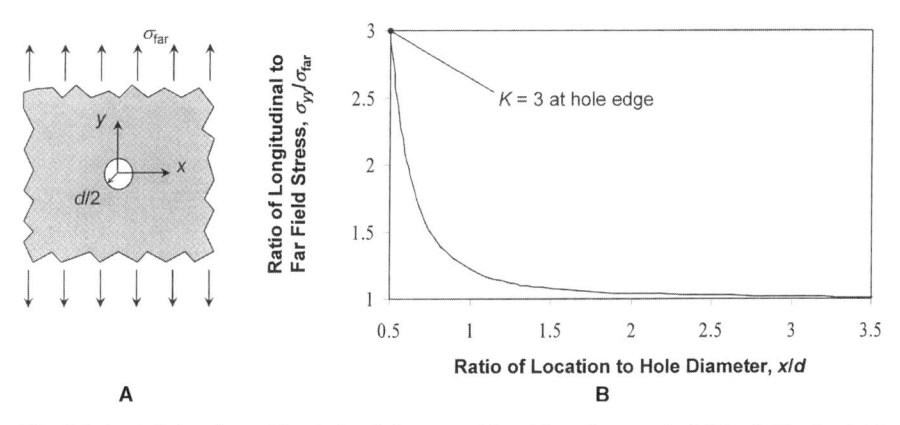

Fig. 7.1 An infinite plate with a hole of diameter d is subjected to a uniaxial load. The far-field stress σ_{far} is equivalent to the nominal stress found in an infinite plate without a hole. Because of the hole, the stress is not uniform throughout the plate; the maximum stress σ_{max} is found at the hole margins coincident with the x-axis ($x = d/2$). The relationship of maximum to far-field stress is defined as the stress concentration factor K, which in this example is 3.0 The longitudinal stress σ_{yy} equals σ_{max} at the hole edge, but decays rapidly as one progresses away from the hole edge. At 1.3 hole diameters from the edge, the longitudinal stress is within 10% of the far-field value, and it is within 1% of this value at 3.6 diameters removed from the edge. For circular holes in plates of finite width K_t exceeds 3.0, and larger hole diameter to plate width ratios produce larger values of K_t. The form of the stress decay curve is sensitive to this ratio as well. These values assume material isotropy and structural homogeneity. Anisotropic materials generally exacerbate stress concentrations, as in this example if the stiffer directions were oriented in the direction of tension

the general case of a circular hole in a very large (i.e., infinite) plate, the stress concentration factor K is 3, where the hole edge tangent is parallel to the applied load. That is, the stress at the hole edge is three times what the far-field stress is; this far-field stress can also be conceived as a nominal stress, which in the plate example will be uniform throughout the structure, except at the defect borders parallel to the applied load. In this and all other cases, the magnitude of K is dependent on the size and shape of the defect, as well as the overall dimensions of the structure in which the defect is embedded. As defects represent a sudden change in cross-sectional geometry of specimen, such local geometric changes can be thought of as contributing to stress concentration severity. Sharper defects generally produce more severe stress concentrations (cracks representing an extreme case). Theoretically speaking, in an otherwise solid homogeneous structure, a cube-shaped defect would be expected to have a larger local stress-raising effect than a spherical one, and the effect would be exacerbated if the cubic defect was larger than the spherical one.

7.1.3 Skeletal Strategies for Minimizing Stress Concentrations

For obvious functional reasons, the skeleton has holes in the form of various foramina, dental alveoli, and orbital and nasal apertures. The presence of such defects does not require any special mechanical explanation if physiological loads are routinely small in their vicinity; that is, the stress-concentrating effects of the defects are probably unimportant (from a strength criterion) if the peak loads never result in local deformations beyond a few hundred microstrain. For example, in vivo strains in the circumorbital region of primates are relatively small in comparison to those in the mandible or near the zygomatic root (Hylander and Johnson, 1991, 1998; Ross and Hylander, 1996; Ross, 2001); thus, it is unlikely that the stress concentrations at the orbital margins (if they arise) would ever put the circumorbital bone in danger of failure from cyclical physiological loads.

The nutrient foramina of long bones, however, pose an interesting biomechanical paradox. Locomotor forces produce large functional strains in mammalian long bones, yet the nutrient foramina rarely present as sites of fracture in clinical or veterinary contexts. A series of studies using the equine metacarpal as a model have revealed how the stress-concentrating effect of the foramen is mitigated. Microindentation of bone surrounding the nutrient foramen reveals that the bone at the foramen edge is relatively compliant (low modulus), while there exists a nearby region of bone, roughly annular in extent but removed from the foramen edge, that is relatively stiff (Rapoff et al., 2003). The effect of this material distribution was evaluated by Götzen et al. (2003) through comparison of two finite element models (FEM) of the cortical bone in the foramen vicinity. They compared a hypothetical FEM in which the entire model was populated with far-field material properties (i.e., not local to the foramen) to an FEM using properties specific to locations, as determined through an elastic constants algorithm that utilized composition and architecture variables as inputs. Götzen et al. (2003) determined that the compliant bone adjacent to the foramen edge effectively redirects stress to the stiffer region remote to the foramen. The effect of this structural heterogeneity is that the bone adjacent to the foramen is prone to large strain concentrations, but the expected stress concentration (based on an assumption of homogeneity) is mitigated. Local variations in mineralization and density are implicated in the spatial variation in elastic modulus, and the mechanical parameter being optimized appears to be the ratio of strength to stress, which is minimized around the foramen (Venkataraman et al., 2003; Huang et al., 2003). Variations in material stiffness axes in the immediate vicinity of the foramen further reduce the stress concentration at the foramen edge (Rapoff et al., 2003; Götzen et al., 2003). Examination of osteonal trajectories in the immediate vicinity of the foramen indicates that the microstructure of the bone here confers an increased toughness at the defect boundary as well, since osteon cement lines are arranged so that they do not become exposed at the foramen edge (Garita and Rapoff, 2003). Whether any of these tissue strategies are employed in the alveolar process is currently unknown, although it has been assumed that alveolar bone is less stiff than basal mandibular bone (Daegling and Hylander, 1997; Chen and Chen, 1998).

7.1.4 The Mechanical Consequences of Alveoli

In this paper, we offer experimental and theoretical evidence for assessing the mechanical consequences of dental alveoli. We use these data to address the following questions. (1) Do alveoli function as stress-raisers as conventionally conceived in engineering theory? (2) Do alveoli cause the mandibular corpus to function as an open section?, and (3) Do tooth roots bound to the alveolar bone by periodontal ligaments mitigate the potential stress-concentrating and open section effects posed by alveoli?

7.1.4.1 Theoretical Approaches

Unfortunately, the straightforward – if involved – formulas for determining K in simply loaded utilitarian structures may only provide a general approximation for estimating what K should be in a dental alveolus. The problems are many: the alveoli lack the symmetry that permits simple calculation; corpus geometry is nonuniform at adjacent sections; the in vivo loads on the corpus are superpositions of shearing, bending, and torsional forces; and the corpus itself is inhomogeneous. Contributing to this inhomogeneity are the tooth roots bound to the alveoli by the periodontal ligament. The question naturally arises as to whether the teeth act as mechanical plugs that mitigate any potential stress-concentrating effects. This latter consideration is insoluble by simple geometric formulae, but we present experimental evidence below that addresses this particular problem.

Ignoring the above issues for the moment, one can arrive at a general sense of what kind of stress concentration factors could potentially be present in the mandible by applying the standard calculations for determining K in homogeneous members

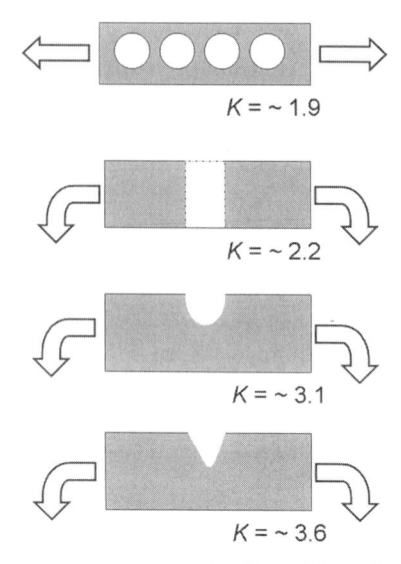

Fig. 7.2 Theoretical stress concentration factors for four different "virtual mandibles," based on the dimensions taken from an adult human mandible at a section through the P_4 alveolus. From top to bottom: (**A**) several circular alveoli in series under tension, (**B**) circular hole, centrally located, under bending, (**C**) semicircular notch under bending, (**D**) hyperbolic notch under bending. All K values are rough approximations as the formulae assume invariant and regular geometry of the member as well as the defect (Young, 1989). Drawings are for illustrative purposes only and are not to scale with the mandibular specimen used to calculate stress concentration factors

of regular geometry. Utilizing dimensions from a human mandibular corpus to define the necessary variables (dimensions of a premolar alveolus, corpus width and height), we examined several configurations (Fig. 7.2). For a member under uniaxial tension having a number of holes in series, the value of K is 1.9. Modeling the corpus as a bent bar with a circular hole embedded centrally yields a value of 2.2. Alveoli can also be modeled as notches rather than as complete holes: under bending in such "virtual mandibles," K assumes values between 3.1 and 3.5 if we consider the effects of a single alveolus; additional alveoli will obviously add points of stress concentration, but theoretically speaking (Peterson, 1974; Young, 1989), the extra defects will reduce the initial value of K in the single-defect case (under the dubious assumption of equivalent defect shape).

Torsion is also a major loading regime occurring during mastication and biting, and the effects of holes in the bones subject to this load have been evaluated through strain gage and photoelastic experimentation as well as finite element analysis (Hipp et al., 1990; Mahinfalah and Harms, 1994). These studies suggest that a single alveolar "hole" in a mandible could yield a K value above 7 under a twisting load.

None of the above idealized cases provides a completely accurate picture of the stress-concentrating effects of alveoli, especially considering that, in normal function, the mandible is not simply loaded in tension or pure bending, but is subjected to bending, twisting, and direct shearing loads simultaneously. But in the cases considered above, we can postulate that the effects of alveolar defects – ideally conceived – are such that in their vicinity local strains may rise from 2 to 7 times the nominal strain. The next question is whether these numbers are reasonable in light of empirical data.

7.1.4.2 Insights from In Vivo Experiments

There are ample strain data from in vivo investigations to allow informed speculation as to the accuracy of the various stress concentration estimates discussed above. Some experiments on macaques have sampled strains directly from the alveolar process and the basal corpus in the same experimental subjects (Hylander, 1979b; Hylander and Crompton, 1986), and a few have sampled strains from these two regions simultaneously (Hylander and Johnson, 1997). These data show that during mastication and biting the basal and alveolar strains are broadly similar in macaques. The high alveolar strains predicted by certain finite element models (Knoell, 1977) are not present during mastication in primates.

We should not immediately conclude from this finding that alveoli do not function as stress-raisers. First, it needs to be established what the nominal strains are expected to be at the sampled locations (i.e., if nominal alveolar strains are expected to be a fraction of basal strains, then sub-equal strain magnitudes might well indicate the presence of a stress concentration). Second, strains sampled from periosteal surfaces (a technical limitation of using strain gauges) are probably removed from the site of the maximal stress concentration. What this means is that the strains sampled – if they are in fact in the vicinity of a stress-raiser – are some fraction of the maximum value experienced at the defect margin. What the strain should be at the periosteal surface requires knowledge of how quickly the stress concentration decays as a function of distance from the alveolar edge.

During powerful biting, primates can induce strains of over 2000 $\mu\varepsilon$ in the corpus, and the available in vivo data suggest this is as true of the basal mandibular bone as it is for the bone in the alveolar process (Hylander and Johnson, 1997). If large stress concentrations are present deep to the periosteal surface of the alveolar process, then the failure to detect these might simply be attributed to a rapid decay of the stress concentration to far-field values at periosteal surfaces. The likelihood of this being the case for a defect the size of an alveolus is questionable, but we do not have to struggle through this calculation to conclude that the stress-concentration values cited above are inaccurate. Instead, we can simply ask what the theoretical values of K cited above would translate to given the reasonable assumption that periosteal strains represent nominal (i.e., far-field) values. These values at the site of stress concentration would result in local strains from 4,000 to 14,000 $\mu\varepsilon$. The latter figure is well beyond the yield strain and in the neighborhood of ultimate failure, while the former figure would place the alveolar bone in jeopardy of fatigue failure (assuming a limit of 3000 $\mu\varepsilon$ following 10^6 cycles, Hylander and Johnson, 2002). Thus, if the alveoli are acting as stress-raisers, the value of K could not be much more than 1.5, despite the various theoretical predictions. This conclusion, however, is only valid if we assume the alveolar bone deep to the perisoteal surface is similar in its material configuration (modulus, density, mineralization). This question is explored further below.

7.2 Mechanical Behavior of Dental Alveoli

7.2.1 Is the Mandible an Open or Closed Section?

The inference drawn above, that alveoli do not present as stress-raisers, is a separate mechanical question from whether they cause the mandible to behave as an open or a closed section. How critical the distinction is between the two depends on the load case under consideration. Under bending the difference between a mandible functioning as an open or closed section can be accounted for by recalculation of area moments of inertia. In the case of torsion, however, if the mandible functions as an open section, the conventional measures of torsional strength are meaningless (Daegling et al., 1992). Torsional strength and rigidity are sharply reduced by converting a member into an open section. For example, a hollow cylindrical rod that is transformed into an open section by an infinitesimal longitudinal slit (thus involving a trivial loss of material) may have less than 5% of the rigidity of the original section, and the differences in maximum stress can be an order of magnitude (Young, 1989).

Because the mechanical behavior of open versus closed sections is distinct, once the appropriate data are collected the choice of models is straightforward. In hollow closed sections, the shear flow (shearing force per unit length) increases as the wall thickness decreases; in a skeletal structure, this means that surface strains will be highest where the cortical thickness is least. By contrast, in open sections of any geometry, the highest stresses accumulate in the thickest portion of the member. Daegling and Hylander (1998) collected strain data from human mandibles twisted in vitro, sampling strain from four locations on corpus sections under the molars. These data showed the highest strains were situated medially (lingually) about halfway between alveolar and basal margins, and this location corresponded to the thinnest cortical bone in each of the coronal sections. As strains were uniformly low adjacent to the thickest areas of cortical bone, Daegling and Hylander (1998) concluded that the mandible behaves as a closed section despite the presence of multiple alveolar intrusions.

For the case of parasagittal bending, one does not expect fundamentally different mechanical behavior of open versus closed sections of members with similar geometry. In the case of the mandible, the alveolar intrusions are potentially involved in converting it into an open section, at least intermittently. One predicted difference between the alternative models is that as a closed section the mandible would be expected to have a neutral axis situated at or above the midpoint between basal and alveolar margins (depending on the extent of load-bearing by the tooth roots). By contrast, if the mandible is instead functioning as an open section, then the neutral axis will be situated below this midpoint to assure the equivalent distribution of

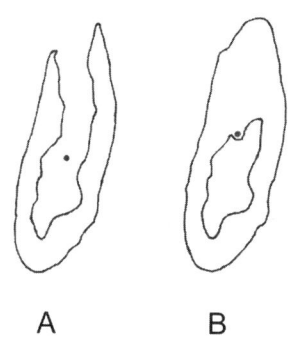

Fig. 7.3 Alternative models of corpus sections as open (**A**) or closed (**B**). An open section assumes no mechanical role for tooth roots, periodontal ligaments, or the lamina dura of the alveolus itself. The closed section as depicted here assumes these structures essentially fill the alveolus and can effect 100% load transfer. The different location of the centroid (pictured) in the two models suggests that the distribution of normal strains in parasagittal bending will differ. While these models are idealized, available evidence suggests that bending strains are more consistent with the open section model. The closed section model is more consistent for predicting the location of the highest shear strains under twisting loads. Drawings based on the minimum section at the P_4

material about this axis (Fig. 7.3). These alternatives are not readily tested in vivo since bending loads are not acting in isolation. There are, however, finite element and in vitro strain data that support the idea that the mandible functions as an open section in bending. The implications for functional interpretations are at present unclear, but given the goal of most comparative studies is the evaluation of relative stiffness and strength (as opposed to absolute values for these variables), the implicit assumption of closed versus open sections is probably not critically important. Most such studies are unconcerned with the effects of alveolar intrusions. If, however, there are significant systematic differences in alveolar size – in the absolute or allometric sense – then the consequences of not considering the distinction of the type of section is more serious.

Marniescu et al. (2005) conducted a finite element study of a *Macaca fascicularis* mandible to explore issues of model validation. The test of the finite element model consisted of an in vitro load case in which strain was recorded under the molars as a point load was applied to the ipsilateral central incisor. This approximated a cantilever in bending, although some superposition of torsion probably occurred as well. The initial model had tooth roots embedded within the alveoli with no interface to approximate the periodontal ligament, in effect permitting the tooth roots to be fully capable of load-bearing with or without the presence of local occlusal forces. The predicted strains at the site sampled were well below the empirical strains (nearly 500 µε difference in shear strain), and the predicted versus observed ratios of maximum to minimum principal strains were quite disparate (model $\varepsilon_1/\varepsilon_2 = 1.28$; observed $\varepsilon_1/\varepsilon_2 = 3.41$). When the teeth were removed entirely from the model, such that the virtual mandible contained unfilled alveoli, the agreement between the experimental and the mathematical models was greatly improved; the edentulous model principal strain ratio corresponding to the gage site was 3.03, and the difference in shear strains was nearly halved. This finding is consistent with the inference that removal of the teeth effectively lowered the neutral axis in bending to a more inferior position, such that the tensile normal strains at the virtual gage site increased relative to compressive strains. Marinescu et al. (2005) found that with further modification of boundary conditions, the edentulous model always fared better than the dentate model in recreating the experimental findings. They concluded that the roots of unloaded teeth do not contribute significantly to corpus stiffness and strength in bending. Thus, in contrast to the behavior in torsion, corpus behavior in bending is best modeled as an open section.

7.2.2 The Role of Tooth Roots in Mitigating Stress Concentrations

In vitro strain data on a sample of human mandibles have evaluated the stress-carrying capacity of intact teeth by examining surface strains before and after tooth removal (Daegling and Hylander, 1994a, b). Unpublished data from these studies suggest that tooth root play a relatively minor role in load-resistance when the corpus is primarily bent or twisted.

The sample involved ten fixed, wet human mandibles drawn primarily from a geriatric population. Most specimens were partially edentulous, although in each jaw strain gages were attached to medial and lateral alveolar processes and basal aspects in the same coronal plane inferior to a functional postcanine tooth with intact periodontal support. The alveolar gages were bonded as far superior on the periosteal surfaces as morphology would permit; thus, these gages were positioned opposite a functional alveolus in every case. Basal gages were affixed to the corpus inferior to midcorpus. Radiographic images confirmed that these gages were positioned below the inferior extent of the alveolus in question.

Each mandible was loaded in bending (with a small torsional component) and in torsion (with a small bending component). After these initial load cases, the tooth above the strain gages was removed and the specimens were again loaded with identical bending and twisting loads.

Regardless of the status of alveoli as true stress-raisers, it is expected that tooth removal will impact strain levels throughout the section if the roots are carrying any significant fraction of the load. In the present discussion, what is of interest is whether the alveolar strains display relative increases that are very much larger than the corresponding change in basal strains on a given side (medial or lateral) of the section. Basal strains, situated well below the inferior extent of the alveoli, can be considered far-field or "control" strains in such comparisons, in that any large change in strains in these regions are not attributable to the stress-concentrating effects of the alveoli, but presumably relate simply to the amount of lost material and the concomitant change in section properties. Of course, alveolar strains will also be subject to these effects, but additional strain adjacent to the alveolar defects may indicate the presence of a stress-concentration deep to the strain gage.

Results from the experiments are reported in Table 7.1. The results are not uniform for either load case; that is, removal of a tooth can cause strain values to

7 Effects of Dental Alveoli on the Biomechanical Behavior of the Mandibular Corpus

Table 7.1 Effects of tooth removal on mandibular bone strain

Bending loads

Specimen	Medial alveolar process Shear strain (γ) Tooth intact	Shear strain (γ) Tooth removed	$\Delta\gamma$(%)	Medial basal corpus Shear strain (γ) Tooth intact	Shear strain (γ) Tooth removed	$\Delta\gamma$(%)
1	47	48	+2.13	159	150	−5.66
2	77	81	+5.19	328	336	+2.44
3	84	84	0	135	135	0
4	185	197	+6.49	55	55	0
5	73	77	+5.48	137	134	−2.19
6	330	288	−12.73	1268	1318	+9.92
7	329	807	+145.3	738	752	+1.90
8	339	297	−12.39	238	252	+5.88
9	743	726	−2.29	601	600	−0.17
10	186	179	−3.76	153	165	+7.84

Bending loads

	Lateral alveolar process			Lateral basal corpus		
1	222	237	+6.76	200	188	−6.00
2	40	No data	—	278	282	+1.43
3	89	92	+3.37	98	98	0
4	81	79	−2.47	106	110	+3.77
5	381	329	−13.65	214	196	−8.41
6	No data	No data	—	569	602	+5.80
7	509	438	−13.95	284	312	+9.86
8	390	255	−34.65	323	333	+3.10
9	725	739	+1.93	471	466	−1.06
10	209	238	+13.88	253	254	+0.40

Twisting loads

	Medial alveolar process			Medial basal corpus		
1	135	140	+3.70	240	233	−2.92
2	453	471	+5.19	842	855	+2.44

Table 7.1 (continued)

Twisting loads

Specimen	Medial alveolar process			Medial basal corpus		
	Shear strain (γ) Tooth intact	Shear strain (γ) Tooth removed	$\Delta\gamma(\%)$	Shear strain (γ) Tooth intact	Shear strain (γ) Tooth removed	$\Delta\gamma(\%)$
3	284	290	+2.11	256	261	+1.95
4	197	210	+6.60	178	186	+4.49
5	316	307	−2.85	370	388	+4.86
6	1048	1152	+9.92	1693	1724	+1.83
7	668	1249	+86.98	1119	1172	+4.74
8	730	883	+20.96	604	643	+6.46
9	1794	2309	+28.71	1458	1521	+4.32
10	579	627	+8.29	418	431	+3.11

Twisting loads

Specimen	Lateral alveolar process			Lateral basal corpus		
	Shear strain (γ) Tooth intact	Shear strain (γ) Tooth removed	$\Delta\gamma(\%)$	Shear strain (γ) Tooth intact	Shear strain (γ) Tooth removed	$\Delta\gamma(\%)$
1	140	122	−12.86	103	111	+7.77
2	872	No data	—	367	383	+4.36
3	196	202	+3.06	175	182	+4.00
4	147	157	+6.80	128	130	+1.56
5	381	548	+43.83	482	527	+9.34
6	No data	No data	—	1472	1391	−5.50
7	477	399	−16.35	603	636	+5.47
8	477	779	+63.31	526	574	+9.13
9	1836	2209	+20.32	1285	1378	+7.24
10	779	847	+8.73	297	311	+4.71

Shear units in microstrain ($\mu\epsilon$). The rosette on the lateral alveolar process in specimen 6 malfunctioned throughout the experiments and data recorded were consequently unreliable. The rosette at this location in specimen 2 failed after tooth extraction. The change in shear strain ($\Delta\gamma$) is calculated as the difference between the shear strain following tooth extraction and the shear strain recorded with the tooth intact, divided by the shear strain with the tooth intact. A positive difference reflects an increase in shear strain following tooth extraction, while a negative value indicates a shear strain decrease.

increase, decrease or change trivially (if at all) at either alveolar or basal (control) sites. These results are attributable in part to differences in periodontal status (degree of alveolar resorption and overall integrity of the periodontal ligament) among specimens. In one specimen (#7), tooth removal had a dramatic influence on medial alveolar strains, with shear strains increasing 87% (torsion) and 145% (bending) while the basal strains were virtually unaffected; in most cases, however, changes in alveolar strains were much more modest and of the same order of magnitude seen in the nearby basal bone. Theoretically, the changes seen in alveolar strains might be tied to regional decay of potential stress concentrations, such that in cases of more pronounced effects of tooth removal one would expect to see that the gages were located nearer alveolar margins. Inspection of the specimens following experiments, however, reveal that the distance of the strain gages from the alveolar edge is uncorrelated with the magnitude of change following tooth removal. This absence of association may – to some degree – reflect technical limitations: strain gages are probably only capable of discerning the more general effects of stress concentration (Arjyal et al., 2000). Another possibility to consider is from early investigation into the decay of stress concentrations. The calculations of Howland (1930) suggest that the stress concentration reduces quickly a short distance from the defect, even in cases of fairly large defects (i.e., those having similar proportions as alveoli, which may occupy over half the width of the overall member), such that sampling strains along the edge may indicate strains very similar to far-field values. Whether this applies to the particular case of mandibles is unknown.

Strain gage technology is a limiting factor in interpretation for other reasons as well. If there exists a strain gradient beneath the functional area of the gage, the investigator is none the wiser because the voltage changes sensed by the gage represent the average effects of strain over that area. Thus, it is possible that the scale over which a stress concentration decays cannot be detected because the gages lack the appropriate spatial resolution. In any case, we are using periosteal surface strains to guess what might be going on at the lamina dura; our discrimination is obviously imprecise.

A simple conclusion regarding the mechanical role of tooth roots cannot be drawn from these data. Euphemistically, one might conclude that the mechanical behavior of the periodontium is exceedingly complex. In jargon-free terms, we do not know exactly what is going on in the alveolar bone given the variability of mechanical responses to tooth extraction. The strain data suggest that the teeth are not shielding the stress-raising potential of alveoli (at least in most cases), but beyond this much remains to be learned.

7.2.3 Material Property Variation in Alveolar Bone

The material property data from the investigations of bone surrounding nutrient foramina prompt the question of similar material organization in alveolar bone. If there is relatively compliant bone at the alveolar margins and stiffer bone away from the defect edges, the possibility emerges that alveolar bone could experience high strain concentrations, yet avoid catastrophic stress concentrations.

We investigated this possibility by examining cortical bone hardness via microindentation in the anterior symphysis in three sections corresponding to cardinal anatomical planes. We used the same specimen used in a prior investigation (Rapoff et al., 2003b), in which no significant differences were noted between alveolar and basal bone stiffness in the postcanine corpus. Here we examined material variation on a more local scale, with the null hypothesis being no regional dependence of elastic modulus. Areas sampled by microindentation for the assessment of alveolar bone stiffness are depicted in Fig. 7.4.

We used a Knoop indenter to determine hardness, rotating the indenter through $0°$, $45°$, and $90°$ (defined relative to a labiolingual axis for sagittal and transverse sections, and a mesiodistal axis for the coronal section) to test the assumption of transverse isotropy. If the assumption holds, then one can derive a single in-plane estimate of elastic modulus E through two regressions, converting Knoop hardness to Vickers hardness and Vickers hardness to modulus (Rapoff et al., 2003a). If modulus is directionally dependent, then comparative analyses must be restricted to common indenter orientations. In this study, we used a 100 g indenter mass at 10 s dwell time; microhardness is figured in units of mass/area (kg/mm^2).

A single-classification analysis of variance (ANOVA) indicates that the directionality of indentation has no significant effect on hardness values ($P = 0.70$ in transverse section; $P = 0.33$ in sagittal section; $P = 0.06$ in coronal section; these data are summarized in Table 7.2). Large amounts of variation at each orientation (as opposed to uniform, invariant hardness at all orientations) explain the nonsignificant results. This provides a weak case for arguing that the bone sampled is transversely

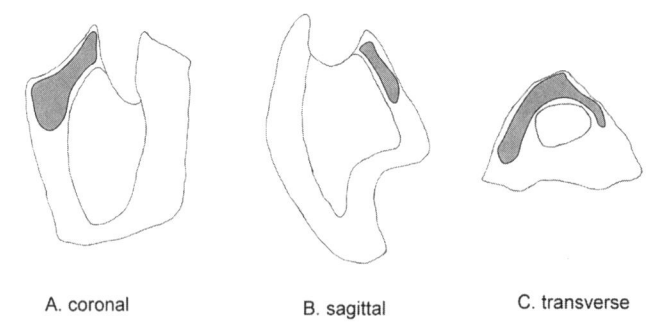

Fig. 7.4 Mandibular sections from the left anterior corpus of *Macaca fascicularis*. Shaded areas indicate the areas sampled by microindentation for assessing stiffness of alveolar bone. For testing the assumption of transverse isotropy, bone throughout each section was sampled. Lingual is to the left in **A** (viewed from anterior perspective), to the right in **B** (viewed in lateral perspective), and is to the right and bottom in **C**. The transverse section (**C**) is incomplete; the lingual and distal (bottom) borders are missing. The scale of Section **A** is approximately 16.5 mm (from top to bottom); for Section **B** this dimension is approximately 20 mm; for section **C** it is approximately 13 mm

Table 7.2 Effect of indenter orientation on hardness

Sagittal Section

Orientation	N	Mean	SD	95% CI	CV (%)
0 (labiolingual)	25	51.2	14.11	45.3–57.0	27.6
45	25	59.2	23.87	49.3–69.0	40.3
90 (superoinferior)	19	53.1	19.85	43.5–62.7	37.4

Differences among orientations are not significant by ANOVA ($P = 0.33$).

Transverse Section

Orientation	N	Mean	SD	95% CI	CV (%)
0 (labiolingual)	27	47.6	9.12	44.0–51.2	19.2
45	18	50.2	13.50	43.5–57.0	26.9
90 (mesiodistal)	16	49.2	8.34	44.8–53.7	17.0

Differences among orientations are not significant by ANOVA ($P = 0.70$).

Coronal Section

Orientation	N	Mean	SD	95% CI	CV (%)
0 (mesiodistal)	38	61.7	31.95	51.2–72.2	51.8
45	32	48.9	17.76	42.4–55.3	36.4
90 (superoinferior)	30	53.0	12.45	48.3–57.6	23.5

Differences among orientations are not significant by ANOVA ($P = 0.06$).
Hardness units in kg/mm^2. Orientation refers to the alignment of the long axis of the Knoop indenter; values reflect hardness perpendicular to this long axis.

isotropic. However, by a statistical criterion we have no basis for treating the different orientations as separate samples. Consequently, we calculated elastic modulus without respect to indenter orientation. The low P value in the coronal section is due to the relatively low mean hardness with the indenter rotated 45° (48.9 kg/mm^2) compared to mean values determined at 0° and 90° (61.7 and 53.0 kg/mm^2, respectively).

We conducted two tests of the null hypothesis that elastic modulus values are independent of proximity to alveolar margins. The first test sampled 12 indentations (four of each orientation) adjacent to an alveolar margin and compared these to 12 far-field "control" indentations remotely situated from an alveolar edge. The second test consisted of assessing the covariation of modulus with distance from an alveolar edge. Both tests were conducted on each section.

In each test, the far-field was defined as a set of indentations that was not in the vicinity of any alveolar margin. Because alveolar intrusions and porosity features presented distinctly in each section, we were not able to collect far-field sets in the same way among the sections. In the coronal section, this set was confined to the periosteal half of the lingual cortex on the alveolar process superior to midcorpus. In the sagittal section, the set was sampled from the periosteal half of the labial cortex superior to midcorpus, as well as from lingual and labial cortices inferior to midcorpus (these samples are not depicted in Fig. 7.4). In the superior transverse section, because the alveolus was the dominant feature, the far-field set was restricted to the labial periosteal margins of the cortex.

Comparison of alveolar versus far-field bone modulus yielded significant results for each section. Only in the case of the sagittal section, however, was the alveolar bone found to be less stiff than control indentations ($P = 0.01$ by ANOVA). For coronal and transverse sections, the bone local to the alveolus was found to be significantly stiffer ($P = 0.005$ and 0.04, respectively). Thus, alveolar bone does not exhibit the type of increased material compliance seen in the vicinity of nutrient foramina, except when inferred from indentations in a sagittal plane. Our results in this particular case may be attributable to the finding of extremely stiff bone in the vicinity of the superior transverse torus.

For evaluating the effect of proximity to the alveolar margin on modulus, we assumed that the basal mandibular bone is likely to vary randomly with respect to its distance to the nearest alveolar edge, as this is situated remotely from areas of potential stress concentration. For these analyses, sampling was restricted to bone in the alveolar process. Results are summarized in Fig. 7.5. No predictable relationship characterizes modulus variation with distance from the alveolus edge for sagittal or transverse sections (95% confidence limits for least squares slope and product–moment correlation include zero). In coronal section, the two variables are negatively correlated ($r = -0.49$, $P = 0.0002$), such that one can expect to find stiffer bone adjacent to the alveolus. This contrasts dramatically from the material properties of the bone surrounding nutrient foramina, where the bone adjacent to the hole edge is relatively compliant.

Schwartz-Dabney and Dechow (2003) collected data on regional variation in material properties from a sample of 10 human mandibles using ultrasound. This study did not find consistent evidence that alveolar bone is less stiff than the inferiorly situated basal bone at a given locality. In addition, they established that there is substantial intra- and inter-individual variation in elastic modulus. While our microindentation data are restricted to a single individual and sampled hardness from a small portion of the anterior corpus, we can offer as a working hypothesis that there are comparable levels of variation in modulus on scales below those sampled by ultrasonic methods. For example, a range of 10–30 GPa encompasses the mean values for modulus along three material axes for the modern human samples (Schwartz-Dabney and Dechow, 2003), while within the single macaque – on a scale below that for obtaining single values via ultrasound ($\sim$4 mm) – hardness variation suggests that values of E are on the order of this 10–30 GPa range (Fig. 7.5).

7.3 Modeling Issues and Recommendations

Disparate lines of evidence suggest that alveoli do not behave as significant stress-raisers, although their presence in the mandible still impacts mechanical behavior, much as Hylander (1979b, 1981) originally supposed. In the specific context of

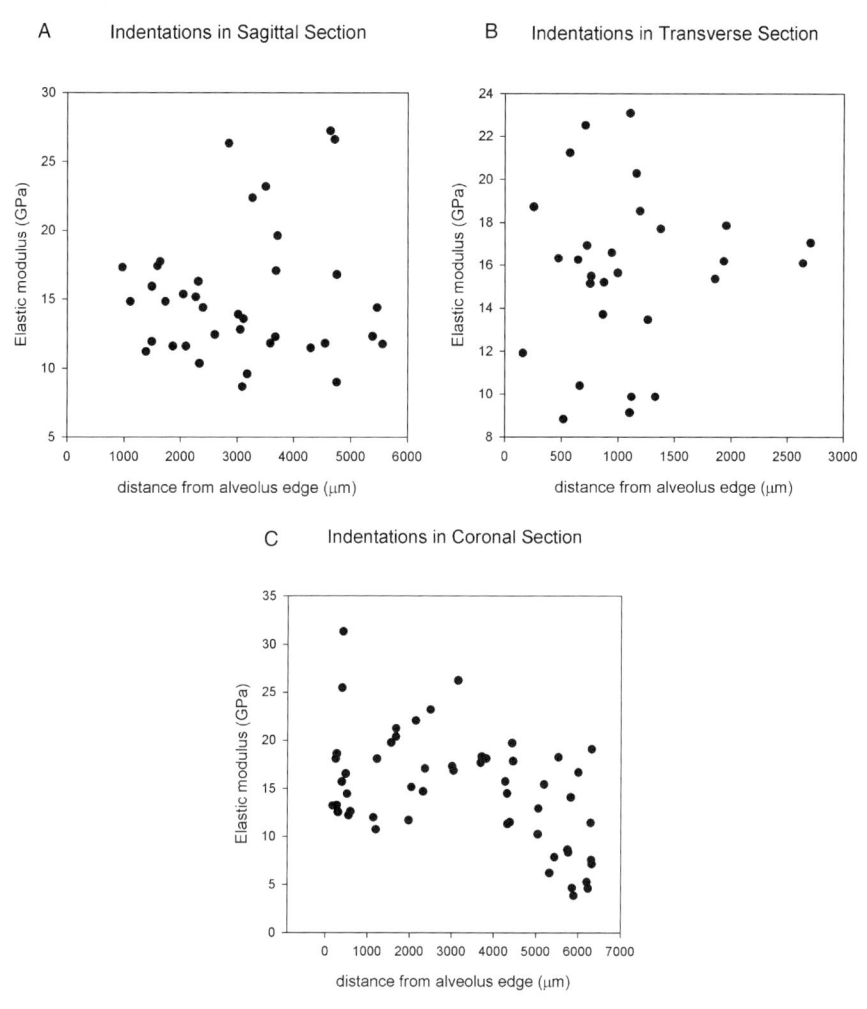

Fig. 7.5 Relationship of elastic modulus of cortical bone with proximity to alveolar margins as determined through microindentation. No predictable relationship exists between the variables in sagittal (**A**) or transverse (**B**) section. In coronal section (**C**), modulus decreases as a function of distance from the alveolus ($r=-0.49$)

bending loads, failure to consider the presence of alveoli (as an absence of material) will lead to significant overestimates of mechanical strength – irrespective of their role in creating local stress concentrations. The impact of alveoli on torsional rigidity and strength appears to be more subtle, despite theoretical predictions that they should have more severe weakening and stress-raising effects under twisting loads. Whether tooth roots can carry masticatory loads other than those arising directly from occlusal forces remains an open question. It seems likely that this capacity is a function of periodontal status, but evidence that the roots mitigate the weakening effects of alveoli is equivocal.

Whether these findings pose problems for comparative analysis depends on whether alveoli occupy a common fraction of mandibular corpus volume across taxa. Since comparative surveys of mandibular dimensions typically ignore alveolar morphology, allometric scaling of alveolar dimensions will introduce unacceptable errors in the estimates of comparative mechanical performance. That is, if alveolar structures are isometric across taxa, then we need not worry (for comparative purposes) that our stress and strength estimates are wrong, because the contrasts of relative stress and strength will remain valid.

Finite element analysis is theoretically capable of providing a comprehensive picture of the strain field in the periodontium, but there are large obstacles in terms of defining appropriate boundary conditions and material properties assignment before a valid model can be constructed. The raw data for populating such a model, where local variations in material behavior are critical, have yet to be collected.

How alveoli are mechanically invisible as stress-raisers is probably related to their presence as a set of defects in series. While this will have a large weakening effect on the mandibular corpus as a whole, the presence of multiple defects essentially prevents locally high stresses from accumulating, in marked contrast to the classical models of stress concentrations around single holes (Peterson, 1974). This does not mean the stress field is not influenced by the defects, in that a proper accounting of the "missing" material as a net section is required for realistic estimates of stresses to be calculated. The material distribution that promotes relocation of stress away from nutrient foramina is not in evidence in the alveolar bone we examined. This suggests that individual alveoli are not prone to excessive stress concentrations at their margins, as indicated above through multiple lines of evidence. The fact that we probably need not worry about highly localized stress gradients in mandibular bone is reassuring, but modeling problems remain. The most significant of these for the comparative context is that the jaw is best modeled as an open section for bending loads, while a closed section model is necessary for accurate assessment of the torsional strain field.

The indentation data are at odds with previous suppositions that the bone in the alveolar process is less stiff than the cortical bone elsewhere in the mandible (cf. Daegling and Hylander, 1997; Chen and Chen, 1998). Greater stiffness in alveolar bone predisposes it to higher stress concentrations. Structurally speaking, alveolar bone is prone to high stress concentrations arising from occlusal loads (Asuni and Kishen, 2000). A material solution to this involves making the bone at the alveolar margins relatively compliant, but this does not seem to be how the alveolar bone is organized based on the microindentation data. Certainly, the periodontal ligament is implicated in distributing bite forces in such a manner that dangerously high stresses do not arise, but forces other than occlusal loads are implicated in the generation of large strains during mastication (if this were not the case, we would expect to see very large differences between working- and balancing-side strains during mastication).

One possibility is that although materially stiff, the alveolar process is – structurally speaking – not particularly rigid. An appropriate analogy might be a thick steel mesh; very strong pound for pound but not terribly stiff because there is not

Fig. 7.6 MicroCT image of a transverse section through alveolar process of the anterior corpus of an adult male *Procolobus badius*. The large vacuity is the canine alveolus; the lingual margin is at the bottom of the image. Although materially speaking alveolar bone appears to be comparably stiff to basal mandibular bone, the porosity visualized here suggests that in terms of structural behavior the alveolar process may be relatively compliant

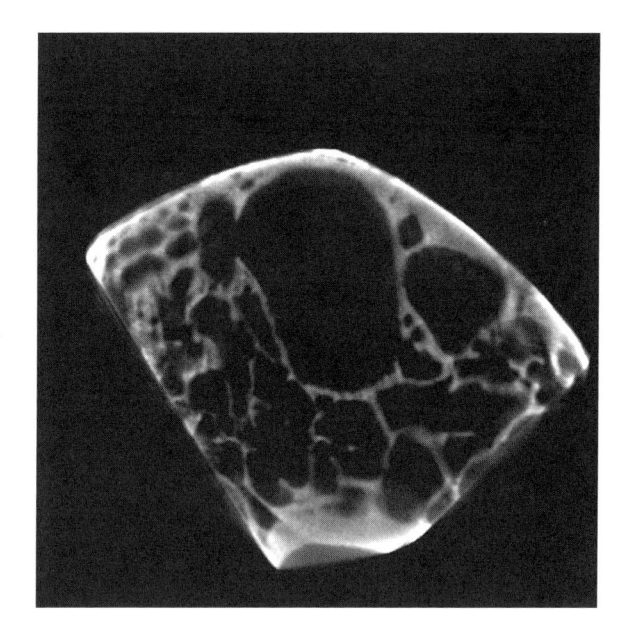

much of it per unit volume (Fig. 7.6). Whether this analogy is appropriate could be assessed through an investigation into density and volume fraction variation in mandibular bone. It is important to recognize that microindentation is sampling areas where cortical bone is present; that is, there is no accounting for the porosity that may exist on a less localized scale. Thus, our finding of very stiff alveolar bone is necessarily a highly localized phenomenon that may or may not extrapolate to an equivalent tissue property on a larger scale. The mesh analogy is particularly appropriate here, since if there is greater porosity in the vicinity of an alveolus, the effective modulus will be substantially lower than that implied by the microindentation data.

The expectation that alveolar bone must experience higher factors of strain than the bone elsewhere in the corpus is not empirically tenable (cf. Daegling and Hylander, 1997; Knoell, 1977), despite evidence that biting forces can precipitate steep strain gradients in the alveolar process (Daegling and Hylander, 2000; Gei et al., 2002). The reason these are not contradictory statements is related to the fact that occlusal forces never operate in isolation from muscular and reaction forces in vivo. The precise mechanisms by which alveolar bone avoids dangerously high stresses remains unknown, although we have presented and reviewed data that should serve to facilitate the articulation and testing of more decisive hypotheses.

Acknowledgments Bill Hylander inspired the senior author's dissertation topic as well as this contribution, and most of his research in the interim. Even though Bill's work has been internationally recognized for some time, the impact of this body of research will not be fully appreciated for many years to come. Bill's work has provided many answers to the questions of skull form and function, but equally important is the fact that his work inspires so many questions and has ignited so much research. We thank Matt Ravosa, Chris Vinyard, and Christine Wall for organizing a memorable symposium in his honor. We are all richer as scientists and as people for having been so lucky to work with Bill over the years. This work was supported by NSF (SBR-9307969, SBR-9514213, BCS 0096037 to DJD).

References

Arjyal, B. P., Katerelos, D. G., Filiou, C., and Galiotis, C. (2000). Measurement and modeling of stress concentration around a circular notch. *Exp. Mech.* 40:248–255.

Asundi, A., and Kishen, A. (2000). Stress distribution in the dento-alveolar system using digital photoelasticity. *Proc. Inst. Mech. Eng. [H].* 214:659–67.

Chen X., and Chen H. (1998). The influence of alveolar structures on the torsional strain field in a gorilla corporeal cross-section. *J. Hum. Evol.* 35:611–633.

Daegling, D. J., and Hylander, W. L. (1994a). Strain distribution in the human mandible. *Am. J. Phys. Anthropol.* Suppl. 18:75.

Daegling, D. J., and Hylander, W. L. (1994b). Profiles of strain in the human mandible. *J. Dent. Res.* 73 (Special Issue: IADR Abstracts):195.

Daegling, D. J., and Hylander, W. L. (1997). Occlusal forces and mandibular bone strain: Is the primate jaw "overdesigned?" *J. Human Evol.* 33:705–717.

Daegling, D. J., and Hylander, W. L. (1998). Biomechanics of torsion in the human mandible. *Am. J. Phys. Anthropol.* 105:73–87.

Daegling, D. J., and Hylander, W. L. (2000). Experimental observation, theoretical models and biomechanical inference in the study of mandibular form. *Am. J. Phys. Anthropol.* 112:541–551.

Daegling, D. J., Ravosa, M. J., Johnson, K. R., and Hylander, W. L. (1992). Influence of teeth and periodontal ligaments on torsional rigidity in human mandible. *Am. J. Phys. Anthropol.* 89:59–72.

Dechow, P. C., and Hylander, W. L. (2000). Elastic properties and masticatory bone stress in the macaque mandible. *Am. J. Phys. Anthropol.* 112:553–574.

Gambrell, S. C. Jr., and Allen, J. M. Jr. (1976). Stress concentrations at the apex of pinned, implanted teeth. *J. Dent. Res.* 55:59–65.

Garita, B., and Rapoff, A. J. (2003). Osteon trajectories near the equine metacarpus nutrient foramen. *Proc. ASME Summer Bioeng. Conf.*

Gei, M., Genna, F., and Bigoni, D. (2002). An interface model for the periodontal ligament. *Trans. ASME* 124:538–546.

Götzen, N., Cross, A. R., Ifju, P. G., and Rapoff, A. J. (2003). Understanding stress concentration about a nutrient foramen. *J. Biomech* 36:1511–1521.

Hipp, J. A., Edgerton, B. C., An, K.-N., and Hayes, W. C. (1990). Structural consequences of transcortical holes in long bones loaded in torsion. *J. Biomech* 23:1261–1268.

Howland, R. C. J. (1930). On the stresses in the neighborhood of a circular hole in a strip under tension. *Phil. Trans. Royal Soc. Lond. Ser A.* 226:49–86.

Huang, J., Venkataraman, S., Rapoff, A. J., and Haftka, R. T. (2003). Optimization of axisymmetric elastic modulus distributions around a hole for increased strength. *Struct. Multidisc. Optim.* 25:225–236.

Hylander, W. L. (1977). In vivo bone strain in the mandible of *Galago crassicaudatus*. *Am. J. Phys. Anthropol.* 46:309–326.

Hylander, W. L. (1979a). Mandibular function in *Galago crassicaudatus* and *Macaca fascicularis*: an in vivo approach to stress analysis of the mandible. *J. Morphol.* 159:253–296.

Hylander, W. L. (1979b). The functional significance of primate mandibular form. *J. Morphol.* 160:223–40.

Hylander, W. L. (1981). Patterns of stress and strain in the macaque mandible. In: Carlson, D. S. (ed.), *Craniofacial Biology*. Monograph 10, Craniofacial Growth Series. Center for Human Growth and Development, University of Michigan, Ann Arbor, pp. 1–37.

Hylander, W. L. (1984). Stress and strain in the mandibular symphysis of primates: a test of competing hypotheses. *Am. J. Phys. Anthropol.* 64:1–46.

Hylander, W. L., and Crompton, A. W. (1986). Jaw movements and patterns of mandibular bone strain during mastication in the monkey *Macaca fascicularis*. *Archs. Oral Biol.* 31:841–8.

Hylander, W. L., and Johnson, K. R. (1991). Strain gradients in the craniofacial region of primates. In: Davidovitch, Z., (ed.), *The Biological Mechanisms of Tooth Movement and Craniofacial Adaptation.* The Ohio State University Press, Columbus, pp. 559–569.

Hylander, W. L., and Johnson, K. R. (1997). In vivo alveolar bone strain and mandibular functional morphology. *Am. J. Phys. Anthropol.* Suppl. 24:134.

Hylander, W. L., and Johnson, K .R. (1998). In vivo bone strain patterns in the zygomatic arch of macaques and the significance of these patterns for functional interpretations of craniofacial form. *Am. J. Phys. Anthropol.* 102:203–232.

Hylander, W. L., and Johnson, K. R. (2002). Functional morphology and in vivo bone strain patterns in the craniofacial region of primates: Beware of biomechanical stories about fossil bones. In: Plavcan, J. M., Kay, R. F., Jungers, W. L., and van Schaik, C. P. (eds.),*Reconstructing Behavior in the Primate Fossil Record.* Kluwer Academic, New York, pp. 43–72.

Hylander, W. L., Crompton, A. W., and Johnson, K. R. (1987). Loading patterns and jaw movements during mastication in *Macaca fascicularis*: a bone-strain, electromyographic, and cineradiographic analysis. *Am. J. Phys. Anthropol.* 72:287–314.

Hylander, W. L., Ravosa, M. J., Ross, C. F., and Johnson, K. R. (1998). Mandibular corpus strain in primates: Further evidence for a functional link between symphyseal fusion and jaw-adductor muscle force. *Am. J. Phys. Anthropol.* 107:257–271.

Ishigaki, S., Nakano, T., Yamada, S., Nakamura, T., and Takashima, F. (2003). Biomechanical stress in bone surrounding an implant under simulated chewing. *Clin. Oral Implants Res.* 14:97–102.

Knoell, A. C. (1977). A mathematical model of an in vitro human mandible. *J. Biomech.* 10: 159–166.

Mahinfalah, M., and Harms, M. (1994). Stress concentrations associated with circular holes in cylinders and bone in torsion. *Exp.Mech.* 34:224–229.

Marinescu, R., Daegling, D. J., and Rapoff, A. J. (2005). Finite element modeling of the anthropoid mandible: the effects of altered boundary conditions. *Anat. Rec.* 283A: 300–309.

Mason, M. E., Triplett, R. G., Van Sickels, J. E., and Parel S. M. (1990). Mandibular fractures through endosseous cylinder implants: report of cases and review. *J. Oral Maxillofac. Surg.* 48:311–317.

Moss-Salentijin L., and Hendricks-Klyvert M. (1990). *Dental and Oral Tissues: An Introduction.* Lea and Febiger, Philadelphia.

Peterson, R. E. (1974). *Stress Concentration Factors.* John Wiley & Sons, New York.

Rapoff, A. J., Fontanel, O., and Venkataraman, S. (2003a). Heterogeneous orthotropic elasticity about a nutrient foramen via microindentation. *Proc. ASME Summer Bioeng Conf.*

Rapoff, A. J., Rinaldi, R. G., Johnson, W. M., Venkataraman, S., and Daegling, D. J. (2003b). Heterogeneous anisotropic elastic properties in a *Macaca fascicularis* mandible. *Am. J. Phys. Anthropol.* Suppl. 36:174–175.

Ross, C. F. (2001). In vivo function of the craniofacial haft: the interorbital "pillar". *Am. J. Phys. Anthropol.* 116:108–39.

Ross, C. F., and Hylander, W. L. (1996). In vivo and in vitro bone strain in the owl monkey circumorbital region and the function of the postorbital septum. *Am. J. Phys. Anthropol.* 101:183–215.

Saffar, J. L., Lasfargues, J. J., and Cherruau, M. (1997). Alveolar bone and the alveolar process: the socket that is never stable. *Periodontology 2000.* 13:76–90.

Savin, G. N. (1961). *Stress Concentration Around Holes.* Pergamon, Oxford.

Schou, S., Holmstrup, P., and Kornman, K. S. (1993). Non-human primates used in studies of periodontal disease pathogenesis: a review of the literature. *J Periodontol.* 64(6):497–508.

Schwartz-Dabney, C. L., and Dechow, P. C. (2003). Variations in cortical material properties throughout the human dentate mandible. *Am. J. Phys. Anthropol.* 120:252–277.

Smith, R. J. (1983). The mandibular corpus of female primates. taxonomic, dietary, and allometric correlates of interspecific variations in size and shape. *Am. J. Phys. Anthropol.* 61:315–30.

Spencer, M. A. (2003). Tooth root form and function in platyrrhine seed-eaters. *Am. J. Phys. Anthropol.* 122:325–335.

Taylor, A. B. (2002). Masticatory form and function in the African apes. *Am. J. Phys. Anthropol.* 117:133–156.

Venkataraman, S., Haftka, R. T., and Rapoff, A. J. (2003). Structural optimization using biological variables to help understand how bones design holes. *Struct. Multidisc. Optim.* 25:19–34.

Ward, S. C., and Molnar, S. (1980). Experimental stress analysis of topographic diversity in early hominid gnathic morphology. *Am. J. Phys. Anthropol.* 53:383–392.

Young, M. C. (1989). *Roark's Formulas for Stress and Strain*, 6th ed. McGraw-Hill, New York.

Chapter 8
Surface Strain on Bone and Sutures in a Monkey Facial Skeleton: An In Vitro Approach and its Relevance to Finite Element Analysis

Qian Wang, Paul C. Dechow, Barth W. Wright, Callum F. Ross, David S. Strait, Brian G. Richmond, Mark A. Spencer, and Craig D. Byron

Contents

8.1 Introduction and Background	149
8.2 Materials and Methods	151
8.2.1 Materials and Specimen Handling	151
8.2.2 Stabilization of the Head	152
8.2.3 Loading Forces and Loading Rate	152
8.2.4 Strain Gage Measurements	153
8.2.5 Data Analysis	154
8.3 Results	156
8.3.1 Reliability of In Vitro Strain Measurements	156
8.3.2 Strain Patterns	156
8.3.3 Global In Vitro Facial Strain Field	160
8.3.4 Effect of Patent and Fused Sutures	160
8.4 Discussion	166
8.4.1 Relevance of In Vitro Approaches to the Study of Functional Morphology	166
8.4.2 Sutural Morphology and Functional Analysis of Craniofacial Skeletons	168
8.5 Conclusions	169
References	169

8.1 Introduction and Background

There have been a number of studies on strain across craniofacial sutures in the past two decades, especially the work on pigs in Herring's lab (Herring and Mucci, 1991; Herring, 1993; Rafferty and Herring, 1999; Herring and Rafferty, 2000; Herring and Teng, 2000; Rafferty et al., 2003), which have identified important mechanical

---

Q. Wang
Division of Basic Medical Sciences, Mercer University School of Medicine, 1550 College St., Macon, GA 31207
e-mail: WANG_Q2@Mercer.edu

features of patent sutures. However, little research has been done on the structure, biomechanics, or ontogenetic changes of facial sutures in primates. Species-specific bone cell dynamics might produce different patterns of suture and bone biomechanics in different species (Carmody et al., 2006), so an investigation of primates is warranted. This study investigates the strain distribution across sutures in a primate skull.

In vivo experiments have demonstrated that sutures are strain dampeners, where strain magnitudes are significantly greater across sutures than on the surrounding bony surfaces. Many studies have amply born this out since strains over sutures were first measured by Behrents and colleagues (Behrents et al., 1978; Oudhof and van Doorenmaalen, 1983; Smith and Hylander, 1985; Herring and Mucci, 1991; Herring, 1993; Jaslow, 1990; Jaslow and Biewener, 1995; Rafferty and Herring, 1999; Herring and Rafferty, 2000; Herring and Teng, 2000; Liu and Herring, 2000; Rafferty et al., 2003; Lieberman et al., 2004).

Furthermore, there are many variations in sutural morphology (e.g., short, long, straight, "zigzag", direct edge-to-edge contact, overlapping) (Herring, 1972), configuration patterns (e.g., at the nasal and pterion areas) (Wang et al., 2006a, b), and the biomechanical properties of sutures can differ even within a single individual. For example, three facial sutures in rabbits, the pre-maxillo-maxillary suture, the naso-frontal suture, and the zygomatico-temporal suture and adjacent sutural mineralization fronts in rabbits have different elastic properties and different capacities for mechanical deformation (Mao, 2002; Mao et al., 2003; Radhakrishnan and Mao, 2004). In addition, sutural tissues and structures have been shown to change over time in humans and laboratory animals. For example, connective tissue cells and fibers in sutures progressively decrease in concentration, and collagen increases in tensile strength while decreasing in extensibility (Gross, 1961; Milch, 1966). In human faces, the overall structure of sutures becomes increasingly irregular, if not fused, with advancing age. This occurs due to the formation of bony projections or interdigitations at the sutural bony surface (Masseler and Schour, 1951; Kokick, 1976; Miroue and Rosenberg, 1975).

In vivo work using a mouse model demonstrates osteoid bone formation along these convex interdigitations (i.e., the convex bone front; Byron et al., 2004), and osteoclast resorption along concave interdigitations (i.e. the concave bone front; Byron, 2006). The iteration of these processes is responsible for cranial suture waveform patterning. Increases in the complexity of this pattern in mice are accompanied by increases in masticatory muscle function and suture extensibility (Byron et al., 2004; Byron et al., 2006a). Among the primates known to differentially exploit materially tough food items such as *Cebus apella*, increased cranial suture complexity is observed when compared to other congenerics that do not exploit such obdurate foods (Byron et al., 2006b). Thus it is proposed that mechanical information concerning mastication manifests itself, in part, through suture morphology.

Sutures clearly have an important influence on the strain distribution throughout the skull, but our understanding of the impact of sutures is incomplete. Some of the difficulties in attempting to address strain in and around craniofacial sutures in the studies of living primates can be resolved using an in vitro approach. In vitro methods have been used in biomechanical research for assessing deformation patterns of mandibles (Daegling and Hylander, 1998; Daegling and Hotzman, 2003), and in clinical and biomedical research for assessing the effects that prosthetic instruments have on bone surface strains (Yamashita et al., 2006; Dechow and Wang, 2006). In vitro experiments on cadavers allow loads and boundary conditions to be controlled (Marinescu et al., 2005); sites normally difficult to access can be measured and the sutural morphology (patent, fused, or degree of interdigitation) can be better evaluated; measurements at various points under identical loading and boundary conditions can provide a synchronous global strain pattern; and comparisons of the results of finite element analyses (FEA) to in vitro studies (in conjunction with comparisons to in vivo studies) permit validation of finite element (FE) models (Richmond et al., 2005).

Given the benefits of such an approach, we developed techniques for careful measurement and analysis of in vitro strain in primate skulls on bony surfaces and across sutures. In this paper, we introduce our design for in vitro experiments, and the reliability of our methods. We explore the strain patterns on the macaque facial skeleton during in vitro loading, and compare the results on bone with that found across several facial sutures.

The influence of sutures on strain patterns also holds important implications for the attempts to investigate strain distributions throughout the craniofacial skeleton, including FEA of craniofacial biomechanics (e.g., Strait et al., 2005; Richmond et al., 2005). Finite element analysis, which enables the examination of how objects of complex design deform and resist loads using advanced computational and engineering techniques, has become one of the most promising tools in the study of functional morphology, especially of the craniofacial skeleton in human and non-human primates (Richmond et al., 2005).

This study uses an in vitro model to explore the distribution of strain throughout the macaque craniofacial skeleton, with special focus on the influence of sutures.

8.2 Materials and Methods

8.2.1 Materials and Specimen Handling

One fresh head of a male long-tailed or crab-eating macaque (*Macaca fascicularis*), age 9.5 years, was used in this test. All permanent teeth including the third molars were in situ. Cause of death was not related to primary bone diseases. Animal's tissue use conformed to all NIH, state, and federal standards.

In order to prevent changes in bone material properties, special considerations were made concerning dehydration and temperature. Dehydration will increase bone stiffness but decrease bone strength (Evans and Lebow, 1951). In order to prevent dehydration, all soft tissues were kept on the head, except in the nuchal area where all muscles were removed. The posterior temporalis muscles on both the right and left sides of the skull were lifted with a periosteal elevator, in order to expose the underlying bony surface for stabilization. The head was wrapped with paper towels soaked in isotonic saline during the mounting process and during tests. Mechanical tests of bone material properties are also influenced by surrounding temperatures. For example, experiments at room temperature (23°C) produced 2–3% percent higher elastic moduli than the bone tested at 37°C (Bonfield and Li, 1966). The specimen was normally tested at a room temperature of 19°C.

8.2.2 Stabilization of the Head

The back of the head was fixed to orthodontic stone, and the latter was tightly fixed to an apparatus consisting of cross-slides and a rotary table (Sherline Products, Vista, CA), which permitted rotational and three-dimensional linear adjustments for proper positioning (Fig. 8.1).

8.2.3 Loading Forces and Loading Rate

Loads were applied by using a screw-driven DDL RT200 loading machine (Test Resources, Inc., Shakopee, MN, USA). The loading was normal to the functional occlusal plane (FOP), defined by the cusps of the second molar (M^2) and the third premolar (PM^3) (Thayers, 1990). Previous studies have measured voluntary bite forces in three adult female *M. fascicularis* on the second and third molars (Hylander, 1979). The average bite forces are less than 10 kg or 98 Newtons (N),

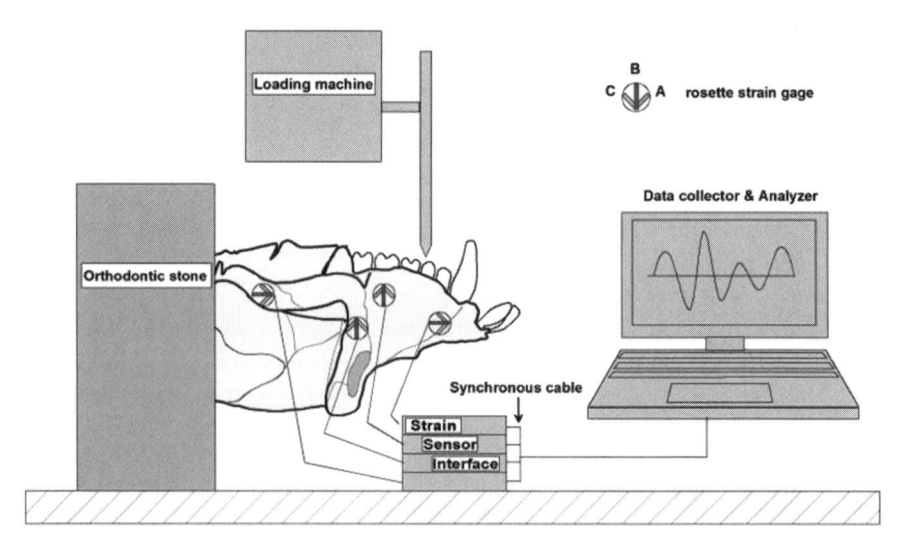

Fig. 8.1 Experiment setup. The head was fixed posterior to the external acoustic meatus using orthodontic stone, which was further fixed to an apparatus consisting of cross-slides and a rotary table

and the largest bite force is approximately 37 kg or 363 N (Hylander, 1979). The loads of 130 N were applied to the central incisor. Higher loads of 260 N were placed on the central fossa of the third molar, because in monkeys, maximum molar biting force is 2–2.5 times higher than maximum incisor biting force, based on the estimates derived from jaw geometry (Dechow and Carlson, 1990). Loads of 195 N (average of the former two) were placed on the buccal cusp of the fourth premolar. The resulting loads and displacements were recorded on computer directly via the software program MTestWR Windows (MTWR) (50 Hz). The loading speed was constant in all loading events. The loading rate was 10 N/s, and the highest strain rate on bone surfaces was 95 $\mu\varepsilon$/s, but most experiments exhibited strain rates lower than 20 $\mu\varepsilon$/s. The loading was immediately cancelled when the desired "bite force" was reached.

8.2.4 Strain Gage Measurements

Surface bone and suture strains were measured during artificial loadings by using small (Diameter 4 mm) 3-element rosette (45°) strain gages (UFRA-1-11-3LT; Tokyo Sokki Kenkyujo, Tokyo, Japan). After reflecting the overlying tissues, the periosteum at the gage site was incised and reflected. The exposed bone was cleaned with 100% acetone. The gage was affixed with Loctite cyanoacrylate. Gages were put on the right side of the head. Locations and orientations of the B-element of 26 gages are detailed in Table 8.1. Twenty-six strain gages were placed across the external surface of the facial skeleton, including hard palate, zygomatic arch, internal orbital wall, circumorbital area, alveolar area, and midface (Table 8.1; Fig. 8.2). Seven gages were put on six patent or fused sutures, including zygomatico-temporal, zygomatico-frontal, maxillo-zygomatic suture, premaxillo-maxillary, anterior midpalatal, and transpalatal. The mid-and transpalatal sutures and the inferior half of the premaxillo-maxillary suture were fused, all others were patent or open. On the bone surface, the B-element was placed along the long axis of bone, which was defined locally (see details in Table 8.1). When a gage was placed on a suture, the B-element was aligned perpendicular to the primary orientation of the suture. Periosteum was lifted regionally for applying strain gages. After the attachment of strain gages, the periosteum was flapped back and the skull was wrapped with paper towels soaked in isotonic saline.

At least five cycles of loading were applied on each loading location on the right side (i.e., gage side) and left side (i.e., non-gage side), respectively. During loading, strains were measured by four synchronized PCD-300A Sensor Interface (Kyowa Electronic Instruments, Tokyo, Japan), and strain measurements were directly stored in a computer (Sampling rate: 100 Hz). Each PCD-300A has four channels (16 channels in total), allowing simultaneous data collection from five rosette gages (i.e., 15 channels). The remaining channel was occupied by a single-element gage placed on the infraorbital area. This gage was used to monitor and calculate the variation in strains for the same loading regime when collecting data from different

Table 8.1 Strain gage sites, orientations, and descriptions

Gage number	Gage site code	Gage site	Orientation of the B-element
1	Parch	posterior zygomatic arch	along the axis of arch
2	Zts	zygomatico-temporal suture	normal to the suture
3	Marh	anterior zygomatico arch	along the axis of arch
4	aarch	zygomatic bone anterior to zygomatic arch	sup. – inf.
5	zfp	frontal process of zygomatic bone	along axis of process
6	zfs	zygomatico-frontal suture	normal to the suture
7	msot	median supraorbital torus	sup. – inf.
8	rio	Rostral interorbital area	sup. – inf.
9	mfp	lower part of the frontal process of the maxillary bone	sup. – inf.
10	pmmss	premaxillo-maxillary suture, superior part	normal to the suture
11	mlp	maxillary bone lateral to the piriform	sup. – inf.
12	prem	premaxillary bone, above the area between I^1 and I^2	sup. – inf.
13	pmmsi	premaxillo-maxillary suture, inferior part	normal to the suture
14	alc	alveolus above canine	sup. – inf.
15	alp	alveolus above PM^4	sup. – inf.
16	alm	alveolus above M^2	sup. – inf.
17	mbc	Center of maxillary bone body	sup. – inf.
18	mzs	maxillo-zygomatical suture	normal to the suture
19	zbi	inferior zygomatic bone	med. – lat.
20	orbwm	medial orbital wall	ant. – post.
21	orbws	superior orbital wall	ant. – post.
22	orbwl	Lateral orbital wall	ant. – post.
23	mps	midpalatal suture anterior to the transpalatal suture	normal to the suture
24	apm	anterior palatine process of maxilla, between C^1 and PM^3	ant. – post.
25	tps	transpalatal suture	normal to the suture
26	phpp	oral surface of horizontal plate of palatine bone	ant. – post.

Abbreviations: I^1 – central incisor; I^2 – lateral incisor; C^1 – canine; PM^3 – third premolar; PM^4 – fourth premolar; M^2 – second molar

gage groupings. The error in positioning for same loading regime was less than 6.3%. Thus global bone and suture behavior during in vitro tests can be summed with confidence from the experiments of different gage groupings using the same loading regime.

8.2.5 Data Analysis

Tensile (ε_1, positive in definition) and compressive (ε_2, negative in definition) strains and the orientation of tensile strain (ε_1°) were calculated by using the strain data analyzer program software DAS-100A (Kyowa Electronic Instruments, Tokyo,

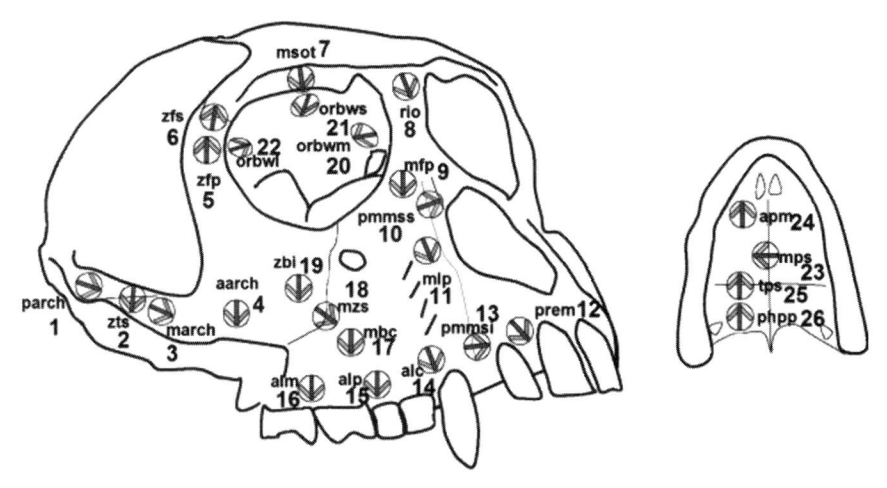

Fig. 8.2 Twenty-six gage sites. The orientation of B-element, the darkened central sensor of the gage, was assigned based on local skeletal morphology. Gages No. 1–3 were on the zygomatic arch. Gage No. 4 was in the lower part of the body of the zygomatic bone. Gages No. 5–7 were in the circumorbital area formed by the supraorbital torus, and the anterior surface of the lateral wall formed by the frontal process of the zygomatic bone and the zygomatic process of the frontal bones. Gage No. 8 was on the glabella area. Gages No. 9–11 was on the anterior middle face. Gages No. 12–16 were on the alveolar area, the alveolar profess of the maxillary and premaxillary bones. Gages No. 17–19 were on the middle part of middle face. Gages No. 20–22 inside the orbit were on the medial, lateral, and superior orbital walls. Gages No. 23–24 were on the oral roof. There were six patent sutural sites, including Site 2 on the zygomatico-temporal suture, Site 6 on the zygomatico-frontal suture, Site 10 on the superior premaxillo-maxillary suture, Site 18 on the maxillo-zygomatic suture, Site 23 on the intermaxillary suture, and Site 25 on the maxillio-palatine suture in the palatal area. Besides, Site 13 was put on the fused inferior section of the premaxillo-maxillary suture

Japan). The orientations of ε_1 were ultimately calculated in degrees relative to the B-element of each gage to indicate their relationships with the anatomical features of the skull (Table 8.1). Descriptive statistics, including means and standard deviation, were calculated for all measurements using Minitab statistical analysis program (MINITAB 14; Minitab, State College, PA). Circular descriptive statistics including angular means, circular standard deviation, and a Rayleigh's test of uniformity (Zar, 1999) were calculated with the Oriana Circular Statistical Analysis Program 2.02 (Kovach Computing Services, Wales, UK). Paired sample *t*-tests and paired circular sample tests (Hotelling test) (Zar, 1999) were used to assess the differences in strain patterns (tensile, compressive and shear strains, strains modes, and strain orientations) between repeated loadings, between gage-side and non-gage side loadings, and between sites at two sides of sutures ($\alpha = 0.05$). Paired data here refer to the analysis of tabulated pairs. For example, strain measurements on a site of two different experiments, or same strain parameters (i.e., strain magnitudes, modes, or orientations) of two different sites during the same loading experiments could be viewed as a paired data.

8.3 Results

8.3.1 Reliability of In Vitro Strain Measurements

The reliability of strain measurements was assessed by repeated loadings. Paired sample tests showed that there were no significant differences at all gage sites between paired mean (1) shear strains, and (2) strain modes (absolute value of the ratio between tensile and compressive strains). Nor were there any significant differences in the angle of principal strains according to Hotelling tests. For example, between two tests with a loading of 130 N placed on the right central incisor, the difference in 26 paired angles was ± 6.7°, which was not significant (Hotelling test: $F = 0.25$, $0.975 < P < 0.999$). Theoretically, the variance of surface suture and bone strain would be due solely to the methodological error since only a single skull was loaded under invariant conditions. The grand mean of the coefficient of variation of maximum tensile and compressive strains of 26 gage sites in all six experiments was 7.8%, suggesting an error of ±7.8%, or an overall precision level of 92.2% (Here the precision was defined as 1 minus CV).

8.3.2 Strain Patterns

Over all, at both bone surface and sutural sites, with increasing loading forces, the strain magnitudes increased gradually. However, the strain mode (ratio between tensile strain and compressive strain) and the orientation of the tensile strain remained constant (Fig. 8.3). For example, during the application of a load of 130 N was placed on the gage-side central incisor, both tensile and compressive strains at the posterior zygomatic arch (Site 1) were increasing in magnitude. The ratio between the tensile and compressive strains was constantly around 3.4. The orientation of the tensile strain was approximately about 80.9° in relation to Element B (Table 8.2). At the neighboring suture site, Site 2 on the zygomatico-temporal suture, the same phenomenon was observed (Fig. 8.3b). This site had strain about six times higher than the adjacent bone surface. During the loading process, similar to the bone surface site, the strain pattern remained constant except for the change in strain magnitude, the orientation of tensile strain was consistently around 21° to Element B. Curiously, the increase of strain magnitude across the sutures was more consistent than that on the bone surface, which might be related to different material structures and properties as well as bone and sutural tissues (Fig. 8.3).

Differences in strain pattern between the bone surfaces and the sutures were found during the period of post-loading. On the suture, the strains did not disappear totally: there were always residual strains in the sutures, indicating a viscoelastic response, if not plastic deformation. The residual tensile strain was 60 µɛ at an angle of −26.6° to Element A, and was 40 µɛ in compression (Fig. 8.3b).

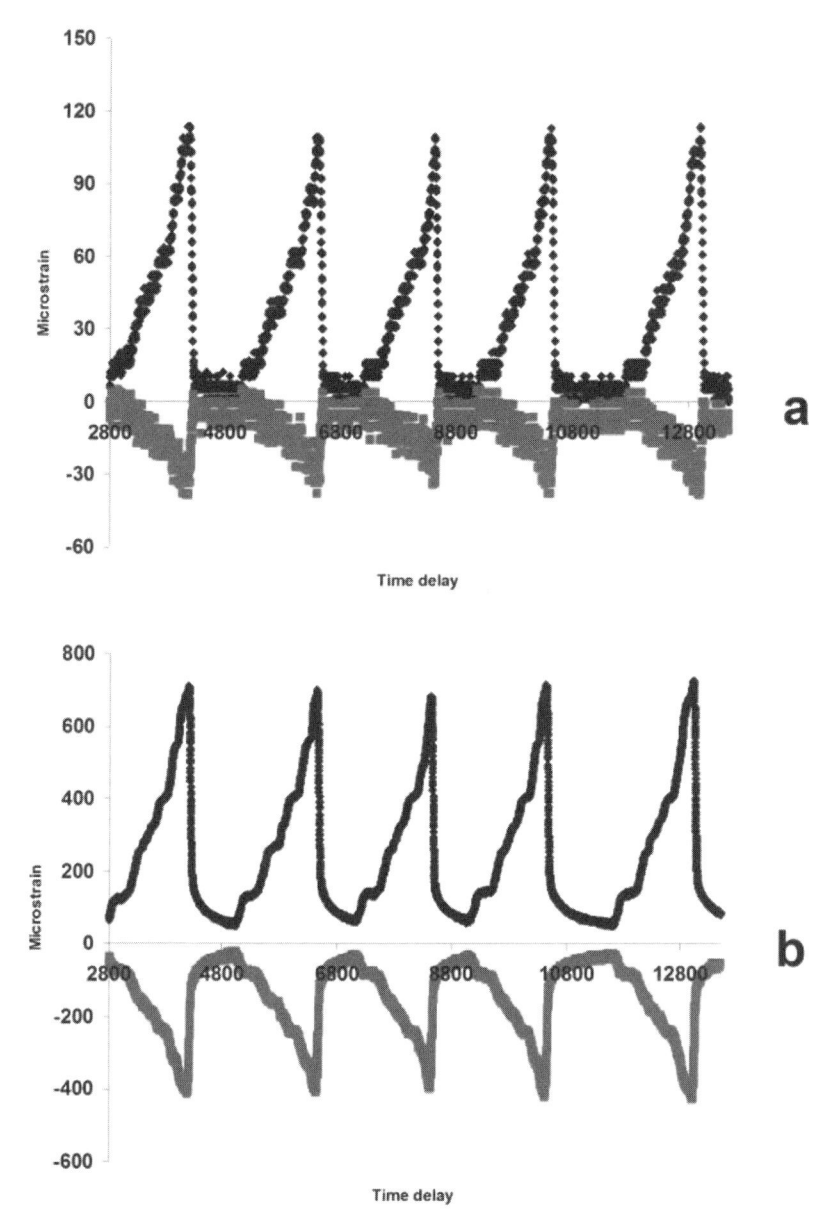

8 Surface Strain on Bone and Sutures in a Monkey Facial Skeleton 157

Fig. 8.3 Real-time strain records at a bone surface site, the posterior zygomatic arch (Gage Site No. 1) (**a**), and at a sutural site, the zygomatico-temporal suture (Gage Site No. 2) (**b**), during 130 N load on the gage-side central incisor. The positive values were tensile strains; the negative values were compressive strains. Notice the strain magnitudes were remarkably higher on the sutures than on the adjacent bone surface. With increasing loading forces, the strain magnitudes increased gradually. After the canceling of loads, there were always residual strains in the sutures (**a**), indicating a viscoelastic response. Conceivably, there should be residual strains on the bone surface, but this situation was not clearly demonstrated here (**b**)

Table 8.2 In vitro strain measurements when a load of 130 N was placed on right and left central incisors ($N = 5$)

130 N on right central incisor (gage side)

Gage site	Tension (ε_1) Mean	S.D.	Compression (ε_2) Mean	S.D.	Shear ($\gamma = \varepsilon_1 + |\varepsilon_2|$) Mean	S.D.	Mode ($\varepsilon_1/|\varepsilon_2|$) Mean	S.D.	Orientation(ε_1°) Mean	S.D.
1	108.5	3.0	−32.5	3.8	141.1	6.5	3.4	0.3	80.9	0.5
2	669.3	29.0	−386.3	21.9	1055.6	50.8	1.7	0.0	20.7	0.3
3	14.3	2.4	−45.3	1.7	59.6	3.6	0.3	0.0	20.5	1.6
4	200.2	3.7	−106.2	14.8	306.4	15.6	1.9	0.3	98.2	0.7
5	61.1	2.3	−71.1	2.3	132.2	4.7	0.9	0.0	98.8	0.3
6	268.4	6.5	−829.4	26.8	1097.9	33.2	0.3	0.0	28.9	0.2
7	88.9	2.4	−66.9	3.2	155.7	5.2	1.3	0.0	55.1	0.9
8	12.8	10.7	−175.8	8.4	188.7	7.2	0.1	0.1	69.7	0.7
9	61.7	7.4	−246.7	12.9	308.4	19.7	0.3	0.0	4.1	0.1
10	583.0	6.3	−973.0	8.2	1555.9	14.2	0.6	0.0	75.2	0.1
11	50.8	7.0	−180.2	8.3	231.0	14.3	0.3	0.0	21.6	5.0
12	925.4	8.8	−1211.4	13.4	2136.8	10.2	0.8	0.0	10.5	0.1
13	487.6	7.6	−315.6	1.9	803.2	8.7	1.5	0.0	9.4	0.1
14	237.9	4.0	−177.9	2.0	415.8	5.4	1.3	0.0	50.5	0.2
15	195.4	1.2	−76.4	0.8	271.8	0.4	2.6	0.0	71.8	0.3
16	121.6	2.0	−41.6	2.0	163.1	4.1	2.9	0.1	16.8	0.5
17	180.0	6.9	−86.0	4.8	266.0	10.3	2.1	0.1	85.0	0.4
18	354.4	2.6	−871.4	8.4	1225.7	7.7	0.4	0.0	88.1	0.2
19	147.7	3.6	−102.7	2.6	250.4	5.3	1.4	0.0	70.6	0.6
20	119.7	2.4	−504.7	5.3	624.4	7.6	0.2	0.0	12.2	0.2
21	190.3	4.4	−256.3	3.4	446.6	6.9	0.7	0.0	75.7	0.4
22	132.3	3.6	−162.3	3.8	294.6	3.7	0.8	0.0	25.7	0.3
23	346.6	2.7	−348.6	3.0	695.2	2.8	1.0	0.0	31.2	0.4
24	263.0	5.4	−176.0	5.4	439.0	10.0	1.5	0.0	14.1	0.7
25	253.3	2.9	−74.3	4.8	327.5	2.8	3.4	0.3	24.5	0.6
26	216.4	4.0	−42.4	2.8	258.8	6.7	5.1	0.2	14.6	0.2

8 Surface Strain on Bone and Sutures in a Monkey Facial Skeleton 159

Table 8.2 (continued)

130 N on left central incisor (non-gage side)

Gage site	Tension (ε_1) Mean	S.D.	Compression (ε_2) Mean	S.D.	Shear ($\gamma = \varepsilon_1 + |\varepsilon_2|$) Mean	S.D.	Mode ($\varepsilon_1/|\varepsilon_2|$) Mean	S.D.	Orientation(ε_1°) Mean	S.D.
1	158.1	2.0	−49.1	3.1	207.3	3.7	3.2	0.2	77.9	0.5
2	875.5	11.4	−395.5	12.6	1270.9	23.6	2.2	0.0	20.3	0.5
3	27.2	2.1	−103.2	2.0	130.4	3.6	0.3	0.0	10.3	0.6
4	201.0	3.7	−97.0	8.5	298.0	11.0	2.1	0.2	101.0	0.7
5	67.1	2.5	−61.1	2.0	128.2	4.0	1.1	0.0	178.7	0.4
6	187.6	3.7	−526.6	5.3	714.2	8.3	0.4	0.0	31.7	0.2
7	68.4	0.6	−38.4	3.0	106.8	2.9	1.8	0.1	56.6	1.8
8	7.7	5.3	−141.7	6.7	149.4	7.6	0.1	0.0	63.9	1.3
9	81.5	7.3	−266.5	16.3	348.0	19.6	0.3	0.0	2.9	0.1
10	436.1	11.0	−810.1	14.3	1246.2	24.9	0.5	0.0	74.2	0.1
11	62.4	23.2	−229.4	17.2	291.9	9.3	0.3	0.1	56.4	0.7
12	298.8	16.2	−202.8	13.8	501.6	29.2	1.5	0.0	80.8	0.3
13	9.3	3.5	−99.3	5.0	108.5	4.8	0.1	0.0	30.1	1.3
14	169.5	10.2	−77.5	3.9	247.1	13.8	2.2	0.1	47.6	0.8
15	93.4	8.5	−26.4	3.6	119.9	12.0	3.6	0.2	63.7	0.8
16	62.1	4.0	−6.1	2.0	68.2	5.9	10.8	1.9	88.2	0.9
17	132.1	2.9	−72.1	2.9	204.2	5.7	1.8	0.0	5.7	1.3
18	205.4	4.9	−110.4	43.1	315.8	47.1	2.3	1.1	76.7	2.7
19	120.2	3.3	−63.2	4.6	183.3	7.7	1.9	0.1	70.2	1.0
20	155.5	2.8	−530.5	8.4	686.0	10.8	0.3	0.0	11.5	0.1
21	141.6	4.3	−232.6	5.7	374.3	8.2	0.6	0.0	74.8	0.3
22	41.1	18.3	−189.1	15.9	230.2	5.4	0.2	0.1	25.3	0.4
23	263.9	2.8	−216.9	0.4	480.8	3.2	1.2	0.0	10.8	0.1
24	209.6	3.9	−372.6	3.7	582.2	6.5	0.6	0.0	87.0	0.1
25	253.2	3.7	−81.2	1.9	334.4	4.4	3.1	0.1	4.6	0.2
26	195.0	2.1	−45.0	2.1	240.0	2.9	4.3	0.2	20.0	0.5

8.3.3 Global In Vitro Facial Strain Field

Tables 8.2–8.4 summarize means and standard deviations for in vitro strain variables, including tensile strains (ε_1), compressive strains (ε_2), shear strains (γ), strain modes ($\varepsilon_1/|\varepsilon_2|$), and orientation of tensile strains (ε_1°). All mean angles of the tensile strains were significantly uniform in orientation (Rayleigh's test: $P < 0.001$), which means the orientation of strains at any specific sites were constant during identical loading regimes. Figures 8.4–8.6 illustrate the shear strain for each location under different loading conditions.

The loadings ipsilateral to the gages produced higher strains than the loads contralateral to the gages, especially when the loadings were placed on the posterior teeth, indicating various combinations of bending and twisting moments in the facial skeleton. When the central incisors and premolars were loaded, the principal strains suggested bending in the sagittal plane. The orientations of the maximum principal strains on the facial bone surfaces at 21 sites were comparable between the left and the right loadings on the incisors (Hotelling test: Incisor loadings, $F = 0.12$, $P = 0.99$; premolar loadings, ($F = 0.18$, $P = 0.97$). The differences in orientation between loadings on left and right molars were greater ($F = 2.184$, $0.10 < P < 0.25$), suggesting the presence of a strong torsional component during molar loading.

Theoretically, sites located in the median sagittal plane of the skull should have similar strain magnitudes when identical loadings are applied symmetrically. However, marked asymmetry was recorded. For example, when loadings of 130 N were put on the right and left central incisors respectively, the difference between the two resulting maximum shear strains was 21% at the glabella area (Site 8), and 31% at the midpalatal suture (Site 23), two sites located in the sagittal midline (Tables 8.3–8.4, Fig. 8.5). This might be explained by the asymmetry of the facial skeleton in the skull that was studied, or in the experimental rig.

Strain magnitudes demonstrate strain gradients related to the distance to the loading positions as seen in in vivo tests (i.e., Hylander and Johnson, 1997; Ross and Metzger, 2004). For example, when a loading of 130 N was placed on the gage-side central incisor, along the alveolar process, the mean shear values decreased significantly from 2136.8 µε at Site 12 (prem), to 803.2 µε at Site 13 (pmmsi), to 415.8 µε at Site 14 (alc), to 271.8 µε at Site 15 (alp), and to 163.1 µε at Site 16 (alm) (Table 8.2, Fig. 8.3).

8.3.4 Effect of Patent and Fused Sutures

Results showed that average strain values in patent sutures were remarkably higher than in adjacent regions of cortical bone (Tables 8.2–8.4, Figs. 8.4–8.6). For example, during the application of a load of 130 N on the gage-side central incisor, on five patent sutural sites on the face (Sites 2, 6, 12, and 18), the strain magnitude was on average about 7.2 times higher than that on the adjacent bone surface sites. Strain modes differed as well. Strain orientations were often considerably different on the cortical bone found on adjacent sides of patent sutures. For example, in all

Table 8.3 In vitro strains measurements when a load of 195 N was placed on right and left fourth premolars ($N = 5$)

195N on right fourth premolar (gage side)

Gage site	Tension (ε_1) Mean	S.D.	Compression (ε_2) Mean	S.D.	Shear ($\gamma = \varepsilon_1 + |\varepsilon_2|$) Mean	S.D.	Mode ($\varepsilon_1/|\varepsilon_2|$) Mean	S.D.	Orientation (ε_1°) Mean	S.D.
1	101.8	2.8	−29.8	5.0	131.6	5.4	3.5	0.6	14.1	0.7
2	772.0	15.3	−789.0	16.1	1561.1	29.7	1.0	0.0	29.6	0.3
3	58.0	1.3	−17.0	2.9	75.0	2.4	3.5	0.6	59.8	2.1
4	249.9	42.6	−321.9	29.7	571.7	63.3	0.8	0.1	172.9	1.8
5	122.6	3.1	−173.6	5.5	296.1	8.1	0.7	0.0	144.8	0.6
6	723.3	57.5	−2692.3	165.8	3415.7	223.2	0.3	0.0	13.1	1.5
7	108.5	5.3	−111.5	5.8	220.0	10.9	1.0	0.0	52.3	0.5
8	148.2	12.7	−185.2	19.2	333.5	27.1	0.8	0.1	77.4	1.0
9	65.1	8.6	−225.1	21.3	291.2	26.8	0.3	0.0	89.8	0.1
10	370.0	11.5	−391.0	12.0	760.9	22.6	0.9	0.0	79.6	0.2
11	207.2	6.0	−167.2	8.0	374.5	10.9	1.2	0.1	41.9	1.2
12	95.0	5.9	−14.0	1.6	109.0	5.7	6.9	1.1	21.8	0.9
13	99.3	3.6	−80.3	4.4	179.7	7.8	1.2	0.0	80.1	0.4
14	477.5	20.8	−177.5	5.7	655.0	26.4	2.7	0.0	65.5	0.3
15	405.5	8.0	−1541.5	39.1	1947.0	46.7	0.3	0.0	33.1	0.5
16	451.8	18.6	−410.8	15.0	862.7	33.6	1.1	0.0	35.8	0.1
17	259.8	28.0	−912.8	44.9	1172.5	59.0	0.3	0.0	86	0.4
18	716.0	20.5	−1332.0	67.7	2048.0	84.6	0.5	0.0	88.3	0.5
19	178.9	16.7	−136.9	14.2	315.8	28.2	1.3	0.1	78.8	0.7
20	132.3	2.4	−352.3	5.1	484.6	7.3	0.4	0.0	1.4	0.2
21	366.2	6.5	−368.2	8.1	734.3	14.5	1.0	0.0	78.7	0.2
22	206.5	8.5	−403.5	7.0	610.1	9.8	0.5	0.0	36.8	0.2
23	280.5	21.7	−242.5	9.7	522.9	28.6	1.2	0.1	50.8	0.5
24	30.5	2.7	−132.6	7.4	163.1	7.6	0.2	0.0	50.3	1.8
25	230.7	16.1	−221.7	10.3	452.4	25.8	1.0	0.0	52	0.3
26	79.0	4.7	−34.0	5.7	113.0	7.7	2.4	0.4	7.4	1.8

Table 8.3 (continued)

	195 N on left fourth premolar (non-gage side)									
	Tension (ε_1)		Compression (ε_2)		Shear ($\gamma = \varepsilon_1 + |\varepsilon_2|$)		Mode ($\varepsilon_1/|\varepsilon_2|$)		Orientation (ε_1°)	
Gage site	Mean	S.D.	Mean	S.D.	Mean	S.D.	Mean	S.D.	Mean	S.D.
1	113.3	4.0	−25.3	2.0	138.5	5.9	4.5	0.3	84.6	0.7
2	699.1	27.3	−402.1	14.8	1101.2	40.6	1.7	0.0	26.5	0.5
3	23.9	3.8	−69.9	5.3	93.7	8.5	0.3	0.0	8.1	1.1
4	117.0	14.3	−55.0	7.9	172.0	22.1	2.1	0.1	169.5	2.0
5	43.5	7.0	−27.5	3.5	71.1	10.4	1.6	0.1	159.5	1.8
6	447.9	29.0	−192.9	16.7	640.7	45.5	2.3	0.1	81.9	0.7
7	47.5	3.5	−9.5	2.5	57.0	3.4	5.5	1.9	54.3	1.5
8	4.1	2.7	−158.7	20.4	162.7	20.2	0.0	0.0	10.8	1.4
9	58.2	7.3	−222.2	13.2	280.4	20.0	0.3	0.0	68.6	1.1
10	27.1	2.8	−341.1	17.3	368.2	20.1	0.1	0.0	50.1	0.9
11	30.9	5.6	−47.9	3.6	78.8	3.6	0.7	0.2	59	3.3
12	3.0	3.7	−22.3	4.0	25.4	6.1	0.1	0.1	49.1	6.8
13	5.4	3.2	−39.8	6.6	45.1	6.5	0.1	0.1	11.5	4.4
14	33.2	7.5	−7.2	2.5	40.4	8.5	5.1	1.9	46.6	2.4
15	21.9	4.6	−20.9	4.7	42.7	4.9	1.2	0.6	59.7	2.8
16	35.7	3.2	−44.7	10.1	80.3	12.6	0.8	0.2	40.5	2.3
17	58.7	6.4	−10.6	9.4	69.4	5.6	14.8	12.1	12.8	4.7
18	333.6	48.7	−27.4	19.4	361.0	35.7	44.9	54.4	12.3	2.5
19	65.2	8.0	−11.2	2.3	76.5	8.6	6.0	1.4	74.8	2.6
20	205.8	9.0	−488.8	21.3	694.6	30.2	0.4	0.0	15.9	0.2
21	113.6	6.0	−194.6	3.6	308.2	7.4	0.6	0.0	2.5	0.4
22	127.9	3.5	−77.9	8.1	205.9	10.8	1.7	0.2	41.3	1.4
23	133.0	10.2	−29.0	3.3	161.9	8.6	4.7	0.8	36	0.6
24	119.1	4.9	−28.9	8.0	148.0	12.5	4.4	1.1	88.7	0.6
25	210.4	8.2	−16.4	2.2	226.7	7.7	13.1	1.7	43	1.1
26	108.5	6.7	−43.5	8.7	152.0	3.7	2.6	0.7	86.8	0.5

8 Surface Strain on Bone and Sutures in a Monkey Facial Skeleton

Table 8.4 In vitro strains measurements when a load of 260 N was placed on right and left third molars ($N=5$)

Gage site	Tension (ε_1) Mean	S.D.	Compression (ε_2) Mean	S.D.	Shear ($\gamma=\varepsilon_1+|\varepsilon_2|$) Mean	S.D.	Mode ($\varepsilon_1/|\varepsilon_2|$) Mean	S.D.	Orientation (ε_1°) Mean	S.D.
1	88.5	3.4	−185.5	16.7	274.0	17.5	0.5	0.0	62.8	0.9
2	29.3	10.7	−1329.5	85.8	1358.8	92.1	0.0	0.0	5.1	1.3
3	101.3	6.8	−31.3	6.2	132.6	12.2	3.3	0.5	53.4	6.0
4	111.1	13.8	−294.1	16.8	405.3	22.3	0.4	0.1	149.2	2.2
5	88.0	5.3	−61.0	3.9	149.0	9.0	1.4	0.0	144.2	1.9
6	634.7	45.4	−2717.7	119.8	3352.4	164.9	0.2	0.0	9.7	0.5
7	55.5	4.1	−113.5	8.5	169.0	6.5	0.5	0.1	63.2	1.3
8	192.3	2.5	−36.3	6.0	228.6	7.1	5.4	0.8	88	0.7
9	69.6	4.5	−261.6	12.6	331.1	16.4	0.3	0.0	68	0.9
10	19.4	2.1	−27.4	5.4	46.8	7.2	0.7	0.1	46.8	4.7
11	25.8	5.8	−29.8	6.1	55.7	6.2	0.9	0.3	22	13.1
12	39.1	4.8	−16.1	0.5	55.2	4.5	2.4	0.4	82	2.0
13	48.6	2.8	−47.6	2.8	96.2	4.3	1.0	0.1	58.9	0.4
14	30.6	3.5	−93.6	2.4	124.2	3.3	0.3	0.0	53.2	1.3
15	90.8	3.1	−96.8	3.7	187.6	6.5	0.9	0.0	41.3	0.4
16	294.8	18.7	−184.8	9.4	479.5	27.9	1.6	0.0	35.3	0.2
17	48.2	13.3	−131.8	10.7	180.0	15.8	0.4	0.1	49.3	4.8
18	267.6	36.7	−332.6	54.4	600.1	90.5	0.8	0.0	83.1	1.5
19	110.3	5.6	−56.3	8.9	166.5	13.7	2.0	0.3	14.5	4.2
20	137.0	5.0	−212.0	9.8	349.0	14.5	0.6	0.0	83.1	0.3
21	399.3	15.6	−337.3	13.2	736.7	28.8	1.2	0.0	79.7	0.3
22	225.1	5.4	−405.1	12.2	630.1	17.5	0.6	0.0	44.4	0.2
23	15.4	13.8	−173.1	12.3	188.5	9.6	0.1	0.1	54.4	1.4
24	53.7	22.5	−193.3	8.7	247.0	30.1	0.3	0.1	61.5	3.9
25	33.1	10.9	−225.9	3.3	259.0	12.8	0.1	0.0	72.3	2.2
26	36.4	20.6	−215.2	27.6	251.6	12.9	0.2	0.1	49.4	3.6

Table 8.4 (continued)

260 N on left third molar (non-gage side)

Gage site	Tension (ε_1) Mean	S.D.	Compression (ε_2) Mean	S.D.	Shear ($\gamma = \varepsilon_1 + |\varepsilon_2|$) Mean	S.D.	Mode ($\varepsilon_1/|\varepsilon_2|$) Mean	S.D.	Orientation (ε_1°) Mean	S.D.
1	15.7	2.7	−24.7	4.2	40.4	4.0	0.7	0.2	51.4	4.0
2	34.3	14.5	−220.7	3.7	255.0	18.2	0.2	0.1	66.5	5.7
3	15.9	3.0	−2.4	1.2	18.2	4.1	0.0	0.0	36	18.0
4	55.6	2.3	−52.6	3.1	108.2	4.8	1.1	0.0	157	3.3
5	48.3	4.0	−25.3	2.2	73.5	5.1	1.9	0.2	172.5	0.6
6	291.0	15.4	−354.0	6.8	645.0	22.0	0.8	0.0	55.9	0.3
7	44.6	5.2	−16.6	3.8	61.3	2.8	2.9	1.0	58.8	3.4
8	110.3	9.1	−33.7	3.2	144.0	8.6	3.3	0.5	33.8	3.1
9	111.1	1.1	−125.1	16.8	236.2	17.1	0.9	0.1	50.2	3.7
10	9.9	3.9	−239.9	25.4	249.9	26.9	0.0	0.0	42.4	2.5
11	39.9	2.7	−12.9	3.1	52.7	2.9	3.4	1.1	36.9	11.5
12	8.8	2.7	−22.8	7.0	31.7	8.8	0.4	0.1	80.7	1.4
13	15.4	4.4	−38.4	4.1	53.8	6.1	0.4	0.1	4.8	0.9
14	32.2	1.7	−11.2	0.9	43.4	1.8	2.9	0.3	53.8	4.0
15	3.3	2.1	−20.5	4.3	23.8	5.4	0.2	0.1	22.9	5.2
16	41.3	1.7	−68.3	3.4	109.5	3.7	0.6	0.0	25.4	0.9
17	91.1	3.1	−49.9	4.5	141.0	3.7	1.8	0.2	57.2	4.9
18	241.9	13.1	−408.9	29.8	650.7	41.6	0.6	0.0	86.9	0.4
19	76.5	6.2	−40.5	2.3	116.9	7.3	1.9	0.2	71	1.1
20	101.2	2.1	−281.2	2.7	382.4	3.6	0.4	0.0	10.3	0.8
21	211.4	2.3	−212.4	4.1	423.9	5.5	1.0	0.0	7.6	0.6
22	62.3	7.2	−114.3	5.8	176.5	2.4	0.5	0.1	36.1	0.6
23	129.7	8.6	−286.7	17.5	416.4	12.1	0.5	0.1	47.3	0.6
24	121.9	10.8	−45.9	8.9	167.8	4.0	2.8	0.8	54.8	1.1
25	126.2	3.0	−30.2	9.6	156.4	12.2	4.6	1.5	81.8	0.7
26	15.8	12.0	−41.8	25.7	57.5	13.8	1.3	1.8	83.9	4.4

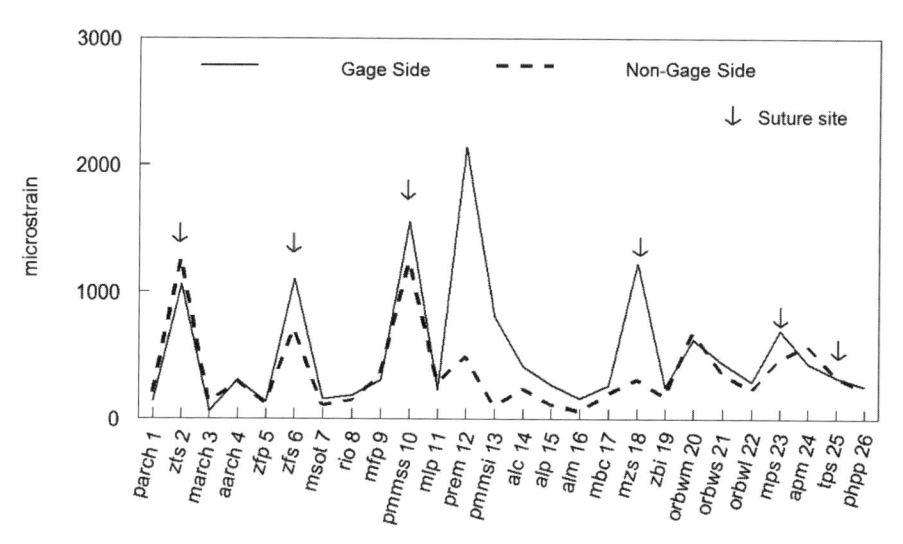

8 Surface Strain on Bone and Sutures in a Monkey Facial Skeleton

Fig. 8.4 Shear strain during 130 N load on central incisor. The solid line represents the result of gage-side loading experiments; the dashed line represents the result of non-gage side loading experiments. Gage sites were arranged from left to right following the sequence used in tables. Points of strain values were connected to form lines and peaks on the line corresponded to high strain values at sutural sites. Note the peaks at patent sutural sites including Site 2 (zygomatico-temporal suture), Site 6 (zygomatico-frontal suture), Site 10 (upper part of the premaxillo-maxillary suture), and Site 18 (maxillo-zygomatic suture). Note the strain gradients from Site 12 to Site 16. Site 13 was on fused part of the premaxillo-maxillary suture

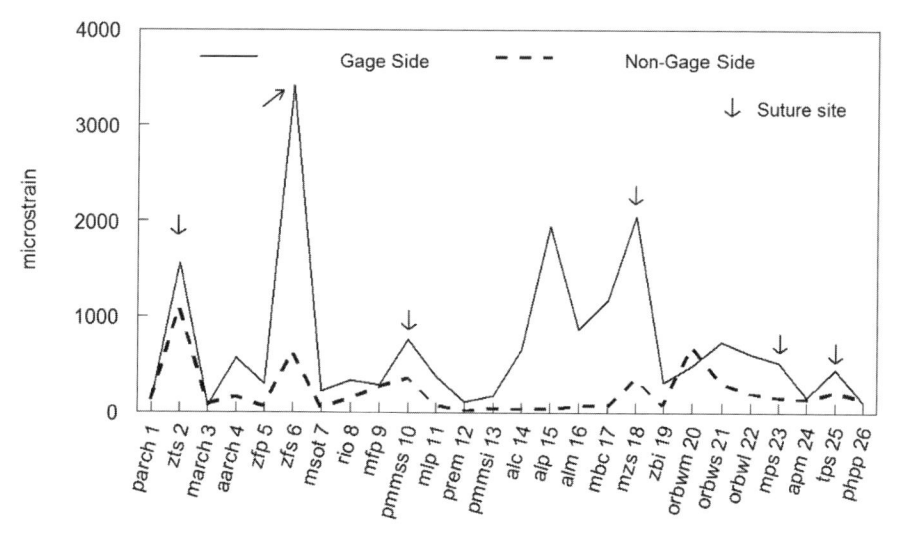

Fig. 8.5 Shear strain during 195 N load on fourth premolar. The solid line represents the result of gage-side loading experiments; the dashed line represents the result of non-gage side loading experiments. Note high strains on the facial sutural sites and on the alveolar process, Sites 14–16

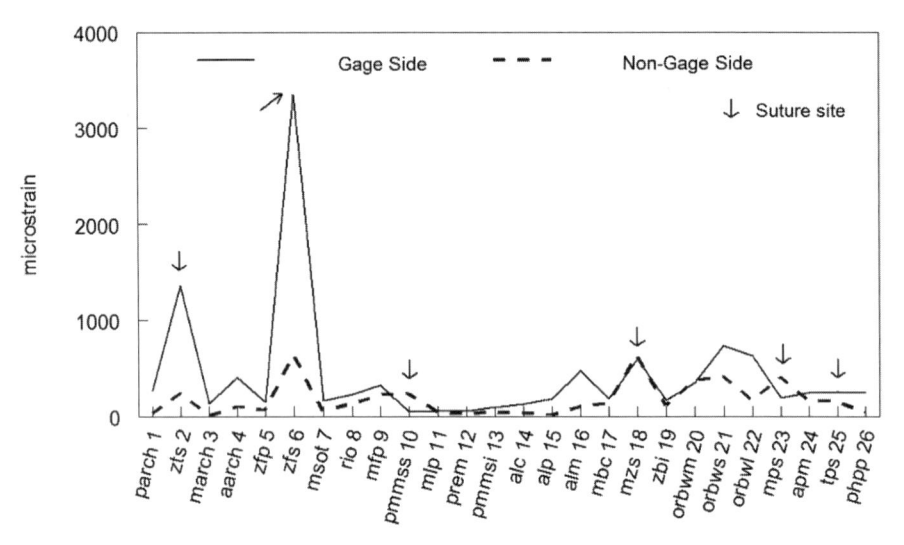

Fig. 8.6 Shear strain during 260 N load on third molar. The solid line represents the result of gage-side loading experiments; the dashed line represents the result of non-gage side loading experiments. Note high strains on two sutures at the posterior part of the face, Sites 2 and 6, but relatively low strains on other part of the facial skeleton

experiments, across the zygomatico-temporal suture and maxillo-zygomatic suture, the orientations of maximum principal strain changed significantly (Hotelling test: $F = 8.33, 0.025 < P < 0.050$). Across fused sutural sections, such as the inferior part of the premaxillo-maxillary suture and the transverse palatal suture, the shear strains were comparable to the surrounding bone strains or compatible with a surrounding strain gradient. The fused sutures or sutural sections behaved mechanically like the bones around them.

The various sutures exhibit a range of mechanical behaviors during different loading regimes. The zygomatico-temporal sutures on the gage-side exhibited similar strain magnitudes during all incisor, premolar, and molar loading experiments. The zygomatico-frontal sutures differed under posterior tooth loading compared to other loading regimes. Other sutures also showed remarkable changes in strain magnitude with shifts in the point of loading. The facial sutures on the balancing side had smaller strains than on the loading side, similar to strains on the bone surface.

8.4 Discussion

8.4.1 Relevance of In Vitro Approaches to the Study of Functional Morphology

Global surface bone and sutural strains were measured using an in vitro method. The method was found to be reliable and provides the data necessary for the validation of finite element models. Loading results demonstrate that the bending and twisting of the macaque facial skeleton, and the strain gradients on the bony surfaces, are generally comparable to the findings of in vivo experiments in pigs, although differing in detail (Rafferty et al., 2003). These results, in conjunction with in vivo studies (i.e., Hylander, 1986; Hylander et al., 1991; Daegling, 1993; Hylander and Johnson, 1997; Dechow and Hylander, 2000; Herring and Teng, 2000; Ravosa et al., 2000; Ross, 2001; Ross et al., 2002), can provide a baseline for validating FE models. For example, in vitro validation of FE models provides an assessment of how accurately geometry and material properties have been modeled under conditions in which loads and constraints can be tightly controlled. Once it has been established that geometry and material properties have been modeled well, then in vivo validation studies can assess how well loads and constraints have been modeled under conditions that are more physiologically realistic. Thus, a validation procedure that employs both in vitro and in vivo data allows an assessment of model validity that is more precise than one that employs in vivo data alone.

Physiological deformation of the craniofacial skeleton during mastication takes place due to a combination of forces at the teeth and the TMJ generated by the contraction of the masticatory muscles at their respective regions of attachment. In the in vitro tests presented here, the external loads and reaction forces are different. External loadings were put on a single tooth, and the posterior part of the skull was stabilized. The lack of loadings at muscle attachment sites would call into question any meaningful comparison between in vivo and in vitro deformations near muscle insertions. Such differences were obvious, especially in the zygomatic arch. Very high strains are generally exhibited in this region, both during in vivo experiments (Hylander et al., 1992; Hylander and Johnson, 1997) and in simulations with in vivo loading condition using FEA (Strait et al., 2005; Wright et al., in review). However, this lack of congruence near muscle insertion sites does not cast into doubt the fact that patterns of in vitro facial strains are essentially similar to in vivo facial strains.

It should be noted that while using in vitro data, the loading rates should be considered. The rate at which loading is applied during biomechanical experiments has an influence on the apparent stiffness of bone, as bone, like nearly all biological materials, is viscoelastic in its natural state (Lakes, 2001). When the loading speed is high, the strain will be increased by an order of magnitude, and measured bone strength will increase by about 15% (Carter and Hayes, 1977). In this analysis, as the loading speed is significantly slower than that in physiological conditions, it is reasonable to conclude that the in vitro bone strain is lower than in in vivo conditions if the loading regimes are identical.

The residual strains on the sutural sites are likely due to higher loading forces and longer loading durations in vitro. The increase in strain magnitude measured on the sutures was more consistent than on the bone surfaces when loads were increasing, which may relate to differences in the response of these tissues at low loading rates.

It is necessary here to remind readers that the in vitro tests are not equivalent to in vivo tests, and will never replace in vivo tests where the latter are feasible. There are some problems associated with in vitro experiments, such as changes in elastic properties of bone and sutural tissues postmortem, and using non-physiological loading forces in terms of source and speed of loads. However, the magnitudes of these problems can be determined experimentally and adjustments can be made. An examination of the correspondence between in vitro and in vivo experiments with FE analyses that model crania with sutures will provide valuable information about how to interpret analysis of in vivo strains when experimental subjects have patent sutures. In vivo, in vitro, and FEA methods have different yet mutual-supporting merits for the common goal of discerning patterns of stain during biting and chewing.

8.4.2 Sutural Morphology and Functional Analysis of Craniofacial Skeletons

The in vitro experiments demonstrate the role of patent sutures in dampening cortical bone strain, a finding that is in agreement with previous in vivo studies. The orientations of the maximum principal strains often differed on either side of a suture, suggesting a redistribution of strain along with strain dampening. If sutures are considered fused when they are not, interpretation of the results of strain gage studies in a global scale and the use of these results for FE models might be biased. On the other hand, patterns of sutural closure could be of great importance for understanding craniofacial form and adaptation in primates, and the inclusion of sutures might enhance the precision of FE models and the accuracy of the study of functional morphology.

The effect of sutural fusion on patterns of stress and strain in the face needs further investigation. Ironically, the greatest chance of successfully measuring the impact of sutures lies with FEA, in which the patterns of sutural closure and the properties of sutures can be modeled. With accurate assignment of bone material properties, there is great improvement in the accuracy and precision of FE models (Strait et al., 2005). The inclusion of sutural morphology and sutural material properties, especially in the facial skeleton, where sutures often remain patent in adults, will further increase the accuracy of these models. As demonstrated here, this 9.5-year-old monkey, with a fully occluding dentition, still has a majority of patent sutures.

There is limited information in the literature on individual or global patterns of sutural closure in primates (Krogman, 1930; Chopra, 1957; Mooney and Siegel, 1991; Leigh and Shea, 1995; Falk et al., 1989; Hershkovitz et al., 1997; Brag, 1998; Wang et al., 2006b). This information is of great importance for modeling craniofacial biomechanics, and a systematic and applicable dataset has yet been established. Many questions still remain unanswered, such as: What are the patterns of fusion of all primary individual sutures? Are there species-specific patterns of sutural fusion? How does the fusion of sutures affect craniofacial growth and biomechanics? Our observations of various monkey skulls demonstrate different fusion patterns among functional areas, among different ontogenetic stages, and between the sexes (Wang et al., 2006b). We further postulated that sutural fusion patterns could be species-specific. Given these findings and postulations, specific sutural morphology must be considered, along with species-specific skeletal elastic properties (e.g., Strait et al., 2005; Wang and Dechow, 2006; Wang et al., 2006c), when attempting to construct accurate FE models of particular species.

8.5 Conclusions

Global surface bone and sutural strains were measured using an in vitro method. This in vitro method was found to be a reliable approach for gleaning important, though non-physiological data, and can be easily modeled using Finite Element Analysis. The results of this experiment can also provide the data to validate FE models, besides in vivo or other strain measure experiments. Strain magnitudes and orientations were considerably different on the cortical bone adjacent to opposite sides of patent sutures. This pattern was generally not observed on either side of fused sutures. Considering sutures as fused when they are not may lead to biased interpretations of strain gage results and their extension to FEA. These findings demonstrate that in vitro experimental data, and information on sutural patency, may help to refine FE models, and increase our ability to understand anatomical function.

Acknowledgments This paper is dedicated to Dr. William L. Hylander, whose work on biomechanics and the function of primate skulls has provided inspiration for many researchers and has influenced a range of biomechanical and evolutionary projects, including this study. This research was supported by NSF-HOMINID grants to Q. Wang and C.D. Byron (BCS-0725183), P.C. Dechow (BCS-0725141), B.W. Wright (BCS-0725136), C.F. Ross (BCS-0725147), D.S. Strait (BCS-0725126), B.G. Richmond (BCS-0725122), and M.A. Spencer (BCS0725219). We thank Mr. Stanley Richardson for helping mount the monkey head for the loading tests. We also thank the editors Drs. Chris Vinyard, Matthew Ravosa, and Christine Wall, as well as two anonymous reviewers, for providing valuable advice for improving this manuscript.

References

Behrents, R.G., Carlson, D.S., and Abdelnour, T. (1978) . *In vivo* analysis of bone strain about the sagittal suture in *Macaca mulatta* during masticatory movements. *J. Dent. Res.* 57:904–908.
Bonfield, W., and Li, C.H. (1966). Deformation and fracture of bone. *J. Appl. Phys.* 37:869–875.
Brag, J. (1998). Chimpanzee variation facilitates the interpretation of the incisive suture closure in South African Plio-Pleistocene hominids. *Am. J. Phys. Anthopol.* 105:121–135.
Byron, C.D. (2006). Role of the osteoclast in cranial suture waveform patterning. *Anat. Rec.* 288A:552–563.
Byron, C.D., Borke, J., Yu, J., Pashley, D., Wingard, C.J., and Hamrick, M. (2004). Effects of increased muscle mass on mouse sagittal suture morphology and mechanics. *Anat. Rec.* 279A:676–684.
Byron, C.D., Hamrick, M.W., and Wingard, C.J. (2006a). Alterations of temporalis muscle contractile force and histological content from the myostatin and Mdx deficient mouse. *Arch. Oral Biol.* 51(5):396–405.

Byron, C.D., Hamrick, M., and Yu, J. (2006b). Cranial suture morphology: understanding how dietary strategy and brain size influence primate craniofacial bone growth (abstract). *Am. J Phys. Anthropol.* Suppl. 42:71.

Carmody, K.A., Mooney, M.P., Cooper, G.M., Bonar, C.J., Ciegel, M.I., and Smith, T.D. (2006). Bone cell dynamics of the premaxillary bone and its sutures (abstract). *Am. J. Phys. Anthropol.* Suppl. 42:73.

Carter, D.R., and Hayes, W.C. (1977). Compact bone fatigue damage—I. Residual strength and stiffness. *J. Biomech.* 10:325–337.

Chopra, S.R.K. (1957). The cranial suture closure in monkeys. *Pro. Zool. Soc.* London. 128:67–112.

Daegling, D.J. (1993). The relationship of *in vivo* bone strain to mandibular corpus morphology in *Macaca fascicularis*. *J. Hum. Evol.* 25:247–269.

Daegling, D., and Hotzman, J. (2003). Functional significance of cortical bone distribution in anthropoid mandibles: An *in vitro* assessment of bone strain under combined loads. *Am. J. Phys. Anthropol.* 122:38–50.

Daegling, D.J., and Hylander, W.L. (1998). Biomechanics of torsion in the human mandible. *Am. J. Phys. Anthropol.* 105:73–87.

Dechow, P.C., and Carlson, D.S. (1990). Occlusal force and craniofacial biomechanics during growth in rhesus monkeys. *Am. J. Phys. Anthropol.* 83:219–237.

Dechow, P.C., and Hylander, W.L. (2000). Elastic properties and masticatory bone stress in the macaque mandible. *Am. J. Phys. Anthropol.* 112:553–574.

Dechow, P.C., and Wang, Q. (2006). Strain gages. In:Webster, J.G.(ed.), *Encyclopedia of Medical Devices and Instrumentation*, 2nd edition, Vol. 6. Wiley, New Jersy, pp. 282–290.

Evans, F.G., and Lebow, M. (1951). Regional differences in some of the physical properties of the human femur. *J. Appl. Physiol.* 3:563–572.

Falk, D., Konigsberg, L., Helmkamp, R.C., Cheverud, J., Vannier, M., and Hildebolt, C. (1989). Endocranial suture closure in rhesus macaques (*Macaca mulatta*). *Am. J. Phys. Anthropol.* 80:417–428.

Gross, J. (1961). Aging of connective tissue, the extracellular components. In: Bourne, G.H., (ed.), *Structural Aspects of Aging*. Hafner Publishing Co. Inc., New York, pp:179–192.

Herring, S.W. (1972). Sutures – a tool in functional cranial analysis. *Acta Anatomica* 83: 222–247.

Herring, S.W. (1993). Epigenetic and functional influence on skull growth. In: Hanken, J. and Hall, B.K. (eds.), *The Vertebrate Skull*, Vol 1. University of Chicago Press, Chicago, pp 153–206.

Herring, S. W., and Mucci, R. J. (1991). *In vivo* strain in cranial sutures: the zygomatic arch. *J. Morphol.* 207:225–239.

Herring, S. W., and Rafferty, K. L. (2000). Cranial and facial sutures: functional loading in relation to growth and morphology. In: Davidovitch, Z. and Mah, J. (eds.), *Biological Mechanisms of Tooth Eruption, Resorption and Replacement by Implants*. Harvard Society for Advanced Orthodontics, Boston, pp 269–276.

Herring, S.W., and Teng, S. (2000). Strain in the braincase and its sutures during function. *Am. J. Phys. Anthropol.* 112:575–593.

Hershkovitz, I., Latimer, B., Dutour, O., Jellema, L.M., Wish-Baratz, S., Rothschild, C., and Rothschild, B.M. (1997). Why do we fail in aging the skull from the sagittal suture? *Am. J. Phys. Anthropol.* 103:393–399.

Hylander, W.L. (1979). Mandibular function in *Galago crassicaudatus* and *Macaca fascicularis*: an *in vivo* approach to stress analysis of the mandible. *J. Morphol.* 159:253–296.

Hylander, W.L. 1986. In vivo bone strain as an indicator of masticatory bite force in *Macaca fascicularis*. *Archs. Oral. Biol.* 31:149–157.

Hylander, W.L., and Johnson, K.R. (1997). *In vivo* bone strain patterns in the zygomatic arch of macaques and the significance of these patterns for functional interpretations of craniodental form. *Am. J. Phys. Anthropol.* 102:203–232.

Hylander, W.L., Johnson, K.R., and Crompton, A.W. (1992). Muscle force recruitment and biomechanical modeling: an analysis of masseter muscle function during mastication in *Macaca fascicularis*. *Am. J. Phys. Anthropol.* 88:365–387.

Hylander, W.L., Picq, P.G., and Johnson, K.R. (1991). Masticatory-stress hypotheses and the supraorbital region of primates. *Am. J. Phys. Anthropol.* 86:1–36.

Jaslow, C.R. (1990). Mechanical properties of cranial sutures. *J. Biomech.* 23:313–321.

Jaslow, C.R., and Biewener, A.A. (1995). Strain patterns in the horncores, cranial bones and sutures of goats (*Capra hircus*) during impact loading. *J. Zool.* 235:193–210.

Kokich, V.G. (1976). Age changes in the human frontozygomatic suture from 20 to 95 years. *Am. J. Orthod.* 69:411–430.

Krogman, W.M. (1930). Studies in growth changes in the skull and face of anthropoids: Ectocranial and endocranial suture closure in anthropoids and old world apes. *Am. J. Phys. Anthropol.* 46:315–353.

Lakes, R. (2001). Viscoelastic properties of cortical bone. In Cowin SC (ed.), Bone *Mechanical Handbook*, 2nd edition, Chapter 11. CRC, Bioca Raton, pp 1–15.

Leigh, S.R., and Shea, B.T. (1995). Ontogeny and the evolution of adult body size dimorphism in apes. *Am. J. Primatol.* 36:37–60.

Lieberman, D.E., Krovitz, G.E., Yates, F.W., Devlin, M., and Clairem M. St. (2004). Effects of food processing on masticatory strain and craniofacial growth in a retrognathic face. *J. Hum. Evol.* 46:655–677.

Liu, Z.J., and Herring, S.E. (2000). Bone surface strains and internal bony pressures at the jaw joint of the miniature pig during masticatory muscle contraction. *Archs. Oral. Biol.* 45:95–112.

Mao, J.J. (2002). Mechanobiology of craniofacial sutures. J. Dent. Res. 81:810–816.

Mao, J.J., Wang, X., and Kopher, R.A. (2003). Biomechanics of craniofacial sutures: orthopedic implications. *Angle Orthod.* 732:128–135.

Marinescu, R., Daegling, D.J., and Rapoff, A.J. (2005). Finite-element modeling of the anthropoid mandible: The effects of altered boundary conditions. *Anat. Rec.* 283A:300–309.

Massler, M., and Schour, I. (1951). The growth pattern of the cranial vault in the albino rat as measured by vital staining with alizarine red "S". *Anat. Rec.* 110:83–101.

Milch, R.A. (1966). Aging of connective tissues. In: Schock, N.W. (ed.), *Perspectives in Experimental Gerontology*. Charles C Thomas, Springfield, pp 109–124.

Miroue, M., and Rosenberg, L. (1975). *The Human Facial Sutures: A Morphological and Histological Study of Age Changes from 20 to 95 Years*. M.S.D. thesis, University of Washington.

Mooney, M.P., and Siegel, M.I. (1991). Premaxillary-maxillary suture fusion and anterior nasal tubercle morphology in the chimpanzee. *Am. J. Phys. Anthropol.* 85:451–456.

Oudhof, H.A., and van Doorenmaalen, W.J. (1983). Skull morphogenesis and growth: hemodynamic influence. *Acta. Anat.* (Basel) 117:181–186.

Radhakrishnan, P., and Mao, J.J. (2004). Nanomechanical properties of facial sutures and sutural mineralization front. *J. Dent. Res.* 83:470–475.

Rafferty, K.L., and Herring, S.W. (1999). Craniofacial sutures: morphology, growth, and *in vivo* masticatory strains. *J. Morph.* 242:167–179.

Rafferty, K.L., Herring, S.W., and Marshall, C. (2003). The biomechanics of the rostrum and the role of facial sutures. *J. Morph.* 257:33–44.

Ravosa, M.J., Johnson, K.R., and Hylander, W.L. (2000). Strain in the galago facial skull. *J. Morphol.* 245:51–66.

Richmond, B.G., Wright, B., Grosse, I., Dechow, P.C., Ross, C.F., Spencer, M.A., and Strait, D.S. (2005). Finite element analysis in functional morphology. *Anat. Rec.* 283A:259–274.

Ross, C.F. (2001). *In vivo* function of the craniofacial haft: the interorbital pillar. *Am. J. Phys. Anthropol.* 116:108–139.

Ross, C.F., and Hylander, W.L. (1996.) *In vivo* and *in vitro* bone strain in the owl monkey circumorbital region and the function of the postorbital septum. *Am. J. Phys. Anthropol.* 101:183–215.

Ross, C.F., and Metzger, K.A. (2004). Bone strain gradients and optimization in vertebrate skulls. *Ann. Anat.* 186:387–396.

Ross, C.F., Strait, D.S., Richmond, B.G., and Spencer, M.A. (2002). *In vivo* bone strain and finite-element modeling of the anterior root of the zygoma in *Macaca* (abstract). *Am. J. Phys. Anthropol.* Suppl. 34:133.

Smith, K.K., and Hylander, W.L. (1985). Strain Gage measurement of mesokinetic movement in the lizard *Varanus exanthematicus*. *J. Exp. Biol.* 114:53–70.

Strait, D.S., Wang, Q., Dechow, P.C., Ross, C.F., Richmond, B.G., Spencer, M.A., and Patel, B.A. (2005). Modeling elastic properties in finite element analysis: How much precision is needed to produce an accurate model? *Anat. Rec.* 283A:275–287.

Thayers, T.A. (1990). Effects of functional versus bisected occlusal planes on the "Wits" appraisal. *Am. J. Orthod. Dentofacial Orthop.* 97:422–426.

Wang, Q., and Dechow, P.C. (2006). Elastic properties of external cortical bones in the craniofacial skeletons of the rhesus monkey. *Am. J. Phys. Anthropol.* 131:402–415.

Wang, Q., Opperman, L.A., Havill, L.M., Carlson, D.S., and Dechow, P.C. (2006a). Inheritance of sutural patterns at pterion in rhesus monkeys skulls. *Anat. Rec.* 288A:1042–1049.

Wang, Q., Strait, D.S., and Dechow, P.C. (2006b). Fusion patterns of craniofacial sutures in rhesus monkey skulls of known age and sex from Cayo Santiago. *Am. J. Phys. Anthropol.* 131:469–485.

Wang, Q., Strait, D.S., and Dechow, P.C. (2006c). A comparison of cortical elastic properties in the craniofacial skeletons of three primate species and its relevance to human evolution. *J. Hum. Evol.* 51:375–382.

Wright, B., Richmond, B.G., Strait, D.S., Ross C.F., Spencer, M.A., Wang Q., and Dechow P.C., (n.d.). Finite element analysis of bite point location and craniofacial strain using a macaque model. *Am. J. Phys. Anthropol.* (In review).

Yamashita, J., Wang, Q., and Dechow, P.C. (2006). Biomechanical effects of fixed partial denture therapy on strain patterns of the mandible. *J. Prosthet. Dent.* 95:55–62.

Zar, J.H. (1999). *Biostatistical Analysis*. 4th edition. Prentice Hall, Upper Saddle River, New Jersey.

Chapter 9
Craniofacial Strain Patterns During Premolar Loading: Implications for Human Evolution

David S. Strait, Barth W. Wright, Brian G. Richmond, Callum F. Ross, Paul C. Dechow, Mark A. Spencer, and Qian Wang

Contents

9.1 Introduction 173
    9.1.1 Hypotheses 174
9.2 Materials and Methods 175
    9.2.1 Solid Model Creation 175
    9.2.2 Mesh Creation 176
    9.2.3 Muscle Forces 177
    9.2.4 Constraints 178
    9.2.5 Elastic Properties 179
    9.2.6 Validity of the Model 180
    9.2.7 Modeling Experiments 181
    9.2.8 Evaluation of Experiments 181
9.3 Results 185
9.4 Discussion 190
    9.4.1 Applications of FEA to Mandibular Biomechanics 194
9.5 Conclusions 195
    References 195

9.1 Introduction

"Robust" australopiths exhibit enlarged cheek teeth and almost certainly possessed hypertrophied masticatory muscles, among other highly derived features of facial morphology (e.g., Robinson, 1954a, b; Tobias, 1967). These features have long been thought to be adaptations for feeding on resistant food items (e.g., Robinson 1954b, 1962, 1963, 1967; Jolly, 1970; Grine, 1981; Rak, 1983, 1985; Teaford and Ungar, 2000; Ungar, 2004; Scott et al., 2005). Among these features are some that have led researchers to suggest that premolar loading may have been an important component of feeding behavior in these species (Rak, 1983, 1985). Specifically, "robust" australopiths have expanded, "molarized" premolars exhibiting extra

---

D.S. Strait
Department of Anthropology, University at Albany, 1400 Washington Ave. Albany, NY, 12222
e-mail: dstrait@albany.edu

cusps (e.g., Robinson, 1956; Wood, 1991), and certain "robust" facial features (e.g., anterior pillars, an anteriorly placed zygomatic root) are thought to have played a role in resisting elevated premolar loads (Rak, 1983, 1985). This interpretation of anterior facial morphology assumes that the stresses and strains produced by premolar loading are sufficiently different from those produced by molar loading as to induce morphological adaptation in the craniofacial skeleton. In particular, premolar loading should induce elevated stresses in the anterior rostrum (Rak, 1983, 1985). The validity of this assumption has yet to be tested experimentally. However, the assumption can be tested using finite element analysis (FEA), an engineering method used to examine how objects of complex design respond to external loads (e.g., Huiskes and Chao, 1983; Cook et al., 1989; Richmond et al., 2005).

9.1.1 Hypotheses

Enlarged premolars are one of several derived australopithecine features that have long been thought to have dietary significance. For example, when comparing *Australopithecus africanus* to *Paranthropus robustus,* Robinson (1954b: 328) noted that "*Australopithecus,* with less disparity in size between anterior and posterior elements of the dentition, with appreciably larger canines and smaller premolars and molars than *Paranthropus,* probably had a more nearly omnivorous diet." Robinson (1954b: 328) continued by positing a biomechanical relationship between post-canine tooth size, mastication, and anterior facial morphology, "As Benninghoff ... and others have shown, the face skeleton is highly organized with regard to the forces of mastication. It is interesting to compare the skulls of *Australopithecus* and *Paranthropus* in light of this work. Both are stressed forms with the result that the nasal region has much the same shape and structure; the nasal region does not protrude; and the margin of the pyriform aperture is thick. The flattened shape ... results from this region being stressed. With relaxation of the forces affecting this region, as a result of reduced dental size in descendants, the nasal region would be more protuberant and without the strong buttressing on either side of the pyriform aperture."

The biomechanics in Robinson's (1954b) hypothesis are quite general and do not focus specifically on premolars, but Rak (1983, 1985) provides a much more detailed explanation for the relationship between premolar loading and facial morphology. Rak's (1983, 1985) hypothesis focuses on three variables, the location of the bite point, the location of the root of the zygoma, and the structural rigidity of the anterior rostrum. Rak (1983, 1985) notes that in *Praeanthropus afarensis* (more commonly referred to as *Australopithecus afarensis,* but see Strait et al., 1997; Strait and Grine, 2004), the premolars are not molarized, and thus the bite point was presumably habitually positioned on the molars. The molars in this species are found directly underneath a posteriorly positioned zygomatic root, which would have acted to resist superiorly directed bite forces. However, if the bite point was instead positioned on the premolars, and if the premolars were anterior to the root, then the anterior rostrum would experience elevated stresses, due, primarily, to sagittal bending, shear, axial compression, or some combination thereof. Rak (1983, 1985) proposed that these stresses induced two evolutionary changes in the australopithecine face. First, anterior pillars evolved to act as struts in the anterior rostrum to reduce the stress. Second, the zygomatic root migrated anteriorly in order to add support and, presumably, to minimize the moment of the bite force relative to the zygoma. Anterior pillars were needed so long as the premolars remained anterior to the root (as in *A. africanus* and *P. robustus*), but were unnecessary once the zygoma completed its anterior migration and is found directly above the premolars (as in *Paranthropus boisei*). The key prediction of this hypothesis is that loading on bite points anterior to the zygomatic root induces appreciably higher strains in the anterior rostrum.

An alternative hypothesis states that craniofacial strain patterns induced by molar and premolar loading do not differ appreciably. Rather, derived australopith features might be the adaptations for withstanding elevated molar loads associated with eating resistant foods, they may be adaptations for consuming large volumes of food using bites that incorporate the entire post-canine tooth row (Walker, 1981), or they may have evolved for reasons unrelated to feeding (e.g., developmental constraints; McCollum, 1999). There are at least a few reasons to suspect that the loading regimes produced by molar and premolar chews are similar. First, in most catarrhine primates, the positions of the premolars and first molar may differ by as little as a few millimeters to a couple of centimeters, a distance that may be smaller than the food item being ingested. Thus, it is conceivable that routine "molar" mastication may in fact be associated with some routine premolar loads. Moreover, although dorso-ventral bending moments in the anterior rostrum may be greater for premolar as opposed to molar loads, the bite forces produced at the premolars are generally expected to be less than those at the molars, given an equivalent muscle force (Du Brul, 1977; Smith, 1978; Greaves, 1978; Spencer, 1995, 1998). Thus, these two factors may have the effect of roughly canceling each other out.

This study uses a finite element model of a macaque to assess how craniofacial strain varies by bite point location along the premolars and molars, and tests the hypothesis that the stresses and strains in the face induced by premolar loading differ from those induced by molar loading.

9.2 Materials and Methods

9.2.1 Solid Model Creation

The finite element model was based on the skull of a male, wild-shot *Macaca fascicularis* (Specimen # 114505, National Museum of Natural History). This species was chosen because it has been the subject of prior studies of masticatory electromyography (EMG) and bone strain that provide data for creating and validating FE models (e.g., Hylander, 1979, 1984; Hylander et al., 1991; Borrazzo et al., 1994; Hylander and Johnson, 1997), and because it is appropriate for evaluating Rak's (1983, 1985) hypothesis. Rak suggests that strains in the anterior rostrum will be elevated when premolar bite points are employed by a species that has a posteriorly positioned zygoma. *M. fascicularis* exhibits this configuration, and although macaques are typically more prognathic than early hominids, some hominids (e.g., *Pr. afarensis* and some specimens of *A. africanus*) have faces that project fairly strongly.

Sixty-one 2 mm thick CT scans were obtained from the specimen. Scans were digitized using commercially available architecture and design software (Solid-Works), and a virtual solid model was created. The model was then divided into 53 parts, each of which could be assigned its own set of elastic properties, and then reassembled (Fig. 9.1A). A validation study has shown that modeling elastic properties using this level of precision produces results that closely match in vivo strain results (Strait et al., 2005). Included among the 53 parts are 12 parts representing regions of trabecular bone (Fig. 9.1B).

9.2.2 Mesh Creation

During mesh creation, a complex object is modeled as a virtual mesh of many small, simple elements. These elements generally take the form of bricks or tetrahedra (Richmond et al., 2005). The elements are linked at their corner points, called nodes, and as the nodes are displaced, strain is generated. The finite element model (FEM) of the macaque skull was constructed using ALGOR FEMPRO software. The model consisted of 311,057 polyhedral elements containing between four and eight corner nodes each, as well as mid-side nodes (Fig. 9.2). When constructing the FEM, the skull was aligned such that the occlusal plane was horizontal (i.e., in the X–Z plane).

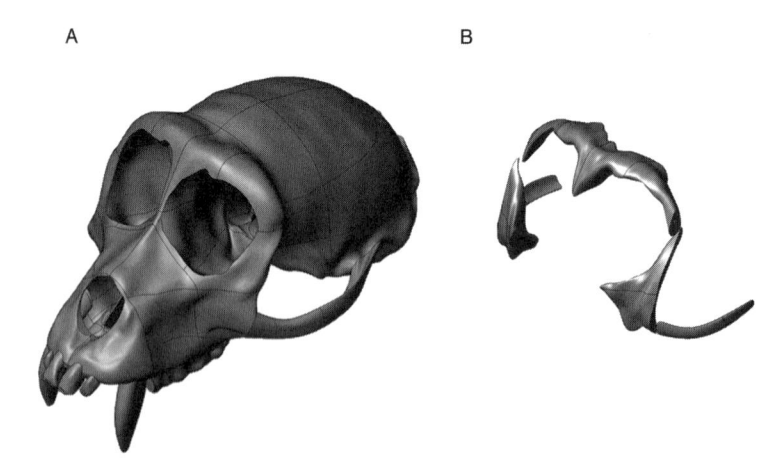

Fig. 9.1 Solid model of *M. fascicularis* skull. (**A**) Parts of skull representing cortical bone. Lines represent boundaries between parts of the model assigned different elastic properties. (**B**) Parts of skull representing trabecular bone in the supraorbital torus, postorbital bar, zygomatic body, and zygomatic arch

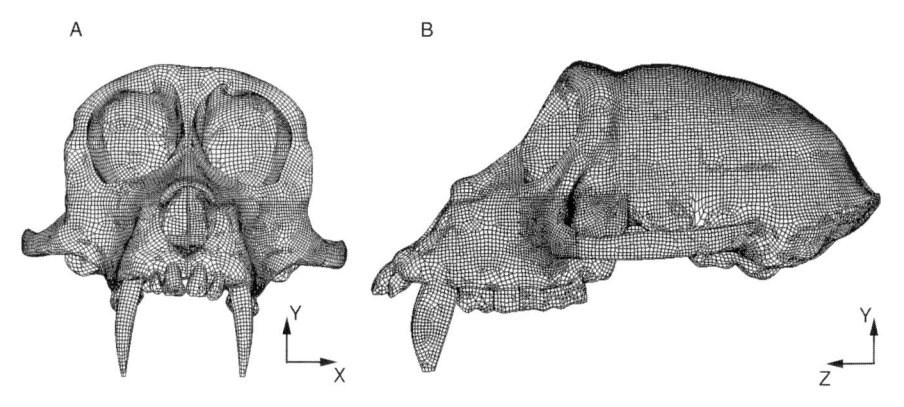

Fig. 9.2 Finite element mesh consisting of 311,057 polyhedral elements in (**A**) frontal and (**B**) lateral view. Arrows indicate directions of x, y, and z axes

9.2.3 Muscle Forces

Eight muscle forces were applied to the mesh, representing the right and left anterior temporalis, superficial masseter, deep masseter, and medial pterygoid. These muscles are principally responsible for jaw elevation during mastication. Muscle force magnitudes and orientations are summarized in Table 9.1. Force orientation was estimated by measuring the relative positions of muscle origins and insertions, and by examining muscle maps based on dissections (Antón, 1993). Muscle force magnitude was estimated by combining data on muscle activity and physiological cross sectional area. Within vertebrates, myofibrillar cross-sectional area is the closest correlate of force-generating capacity (Murphy, 1998). Area and force are related such that approximately 300 kN are produced for every square meter of striated muscle (Murphy, 1998). Area data were obtained from Antón (1993). However, Antón (1993) did not collect data for the anterior temporalis in *M. fascicularis*. Thus, the data of Antón (1993) for *Macaca mulatta* were used instead for that muscle. *M. mulatta* is larger than *M. fascicularis*, and as a result the muscle force

Table 9.1 Muscle forces applied to finite element model

Muscle	Magnitude in Newtons	Orientation vector (x, y, z)[1]
Working-side superficial masseter	70.627	(−0.2, −1, −0.2)
Balancing-side superficial masseter	34.682	(0.2, −1, −0.2)
Working-side deep masseter	22.591	(−0.6, −1, 0)
Balancing-side deep masseter	8.214	(0.6, −1, 0)
Working-side medial pterygoid	34.794	(0.75, −1, 0)
Balancing-side medial pterygoid	6.904	(−0.75, −1, 0)
Working-side anterior temporalis	36.592	(0.1, −1, −0.1)
Balancing-side anterior temporalis	15.147	(−0.1, −1, −0.1)

[1] X-direction is positive to the model's left (working) side. Y-direction is positive superiorly. Z-direction is positive anteriorly.

magnitudes employed in the model somewhat overestimate the forces actually generated by *M. fascicularis*. However, our previous work (Ross et al., 2005) suggests that a slight overestimate in anterior temporalis force is unlikely to have a substantial impact on strain patterns.

Cross-sectional area measurements do not by themselves provide reliable estimates of muscle force, because at any given moment different muscles may have very different levels of activity. The relative force magnitudes exerted by each muscle at or near centric occlusion were calculated by assuming that force production is proportional to the magnitude of muscle activity as measured by the root mean square (rms) of electromyography (EMG) data collected during chewing experiments (Hylander and Johnson, 1989). The highest standardized rms EMG activity recorded from each electrode during an experiment is assigned a value equal to 100% of the cross-sectional area; when the muscle is acting at less than peak activity (e.g., at 50% of peak), then force is proportional to a corresponding percentage of cross-sectional area. EMG data gathered simultaneously from all eight muscles enable relative force magnitudes to be generated (Ross, 2001; Ross et al., 2003). The EMG data used here are taken from a single power stroke that was representative of other power strokes recorded during the same in vivo chewing experiment (Ross et al., 2003).

In FEA, loads are translated into strains instantaneously. When investigating chewing, a logical instant to model is the moment at which bite force is maximized. Bone strain magnitudes recorded from the lateral aspect of the mandibular corpus below M_{1-2} in macaques are highly correlated with the magnitude and timing of bite force during isometric biting on a force transducer ipsilateral to a strain gage (Hylander, 1986), so the timing of peak bite force was estimated using the timing of peak strain in the mandibular corpus. Root mean square EMG activity in the masseter (Hylander and Johnson, 1989) and temporalis muscles precedes the force generated by those muscles by approximately 20 msec. Using the muscle data at a 20 msec latency performed better than the muscle data with no latency, in terms of producing strain results that more closely match experimental in vivo strain (Ross et al., 2005). Consequently, the muscle forces entered into the FEA were calculated from rms EMG activity 20 msec prior to the instant of peak corpus strain. EMG and strain data from the corpus and elsewhere were recorded during the same set of experiments. In summary, muscle force magnitude is calculated as: F = (cross-sectional area) × (300 kN/m^2) × (% of peak activity 20 msec prior to peak corpus strain).

9.2.4 Constraints

Three sets of constraints were applied to the model. Nodes at the right and left articular eminences and at a bite point were fixed in place. The position of the bite point varied in each of the analyses (see below). When muscle forces are applied to a model with these constraints, the model is pulled inferiorly onto the fixed points. Reaction forces are generated at each location, simulating the contact between the mandibular condyles and the articular eminences, and between the teeth and a food item. Obviously, in life, the masticatory muscles act principally to move the mandible rather than the cranium. However, in a free-body diagram, the masticatory muscle forces act to pull the skull down onto a bite point, producing strains in the face equivalent to those produced by pulling a mandible up onto a resistant food item, and then having the item contact a bite point on the upper tooth row. In either case, bite force is a reaction force at the bite point.

9.2.5 Elastic Properties

Elastic properties refer to the force–displacement relations of the substance being modeled. These relations are summarized by several variables, including the elastic modulus, the shear modulus, and Poisson's ratio. The elastic modulus (E) is defined as stress/strain measured in simple extension or compression. It, therefore, numerically describes the stiffness of a material. For example, rubber will strain (deform) far more than steel under a given amount of stress and it has a correspondingly lower E. The shear modulus (G) is analogous to the elastic modulus in that it describes the stiffness of a material under shear. Poisson's ratio (v) is the lateral strain divided by axial strain, thus representing how much the sides of a material will contract as it is tensed (or, conversely, how the material will expand as it is compressed). E, G, and v are expressed along axes (or within planes defined by axes), and those axes have orientations that can be considered variables as well.

Bone presents a formidable modeling challenge for a number of reasons (see review in Currey, 2002). First, the elastic properties of bone vary in different regions across the skull (Peterson and Dechow, 2003; Wang and Dechow, 2006). For example, bone in the postorbital bar in macaques is 51% stiffer than the bone in the adjacent supraorbital torus (as reflected by the elastic modulus in the axis of maximum stiffness; Wang and Dechow, 2006). Moreover, bone is anisotropic, meaning that its elastic properties are not the same in all directions. Specifically, many regions of craniofacial bone are approximately orthotropic, meaning that bone exhibits three orthogonal material axes, each of which has its own set of properties. A further complication is that the orientation of the material axes may vary according to the shape of the bone. In most regions of cortical bone that have been investigated, including the facial skeleton, two of the three material axes are approximately parallel to the bone's surface, while the third axis is normal to the surface. Thus, if the surface of the bone is curved (as are many surfaces in the face), then the orientations of the material axes may vary with the curvature.

Each region in the face corresponding to cortical bone was assigned its own set of orthotropic elastic properties (Wang and Dechow, 2006) (Table 9.2). Cortical regions in the neuro- and basicranium were assigned isotropic elastic properties based on an average of values obtained from all parts of the skull ($E = 17.3$ GPa, $G = 5.5$ GPa, $v = 0.28$; Wang and Dechow, 2006). Trabecular bone in the supraorbital torus, postorbital bar, zygomatic body, and zygomatic arch was also modeled isotropically ($E = 0.64$ GPa, $G = 0.13$ GPa, $v = 0.28$;

Table 9.2 Elastic properties employed in finite element analysis[1]

Region	$E_1{}^{2,3}$	$E_2{}^{2,3}$	$E_3{}^{2,3}$	$G_{12}{}^2$	$G_{13}{}^2$	$G_{23}{}^2$	v_{12}	v_{13}	v_{23}
Premaxilla	10.0	13.9	18.5	4.4	5.2	7.3	0.29	0.18	0.15
P³-M¹ alveolus	9.9	12.1	16.7	4.3	5.8	7.4	0.33	0.24	0.17
M²-M³ alveolus	12.6	15.4	20.6	4.9	6.4	7.9	0.35	0.24	0.22
Anterior palate	7.5	8.8	15.3	2.6	2.8	3.6	0.41	0.36	0.26
Posterior palate	6.4	7.5	18.8	2.2	2.5	3.3	0.48	0.26	0.23
Dorsal rostrum	12.2	14.0	19.9	5.0	6.9	8.9	0.32	0.21	0.14
Lateral rostrum	11.5	14.4	18.1	4.7	5.3	7.3	0.37	0.24	0.15
Root of zygoma	8.9	10.9	17.9	3.7	5.3	8.6	0.53	0.30	0.18
Anterior zygomatic arch	8.6	12.4	20.8	4.2	4.6	8.6	0.39	0.28	0.22
Posterior zygomatic arch	8.2	10.0	12.5	3.1	3.8	4.9	0.34	0.27	0.24
Medial orbital wall	7.1	11.5	14.6	3.6	4.2	9.0	0.46	0.40	0.23
Postorbital bar	11.3	13.1	19.8	4.4	6.4	8.0	0.44	0.22	0.15
Frontal torus	10.2	11.2	13.1	4.3	5.1	6.0	0.32	0.24	0.19
Glabella	9.2	9.7	14.4	3.3	4.8	5.1	0.46	0.14	0.21
Frontal squama	7.9	11.0	14.9	3.4	4.3	7.1	0.49	0.27	0.18

[1] Data from Wang and Dechow (2006).
[2] Values in gigpascals (GPa).
[3] By convention, axis 3 is the axis of maximum stiffness. Axis 2 is perpendicular to axis 3 within the plane of the bone's surface. Axis 1 is perpendicular to the bone's surface. For each region, the orientations of these axes are derived from Wang and Dechow (2006).

Ashman et al., 1989). Although this procedure is considerably more precise than that employed in most other finite element analyses of vertebrate crania, it is not without drawbacks. Principally, there may be unrealistic shifts in elastic properties at the boundaries between regions. Moreover, the material axes in each region receive only a single set of orientations. In regions with strongly curved surfaces, these orientations will not be accurate across an entire surface. Finally, the model does not incorporate information about the material properties of sutures. It is well known that patent sutures can affect strain patterns in the skull (Herring, 1972; Herring and Mucci, 1991; Herring et al., 1996, 2001; Rafferty and Herring, 1999; Herring and Teng, 2000; Shibazaki et al., 2007, Wang et al., 2008), and recent work demonstrates that craniofacial sutures in *Macaca* can remain at least partially unfused well into adulthood (Wang et al., 2006). It is clear, therefore, that the accuracy of finite element analysis would be improved by including sutures in the model. However, to do so would require direct information about suture elastic properties and how those change during ontogeny. Although we intend to collect these data (Wang, pers. comm.), they are not yet available in *Macaca*. Thus, for all of the reasons described above, it is critical to assess the validity of the model by comparing the results of FEA to those of experimental studies.

9.2.6 Validity of the Model

A prior study (Strait et al., 2005) has demonstrated that when the model employed here is loaded as described above and constrained at the LM¹, the resulting patterns of strain are very similar to those observed during in vivo chewing experiments. In most regions for which experimental data are available, maximum shear strains from FEA (Strait et al., 2005) are typically within or just beyond two standard deviations of mean experimental strains (Table 9.3), and the orientations of maximum principal strain in the FE model match the experimental results in most comparisons (Hylander et al., 1991; Hylander and Johnson, 1997; Ross et al., 2002). This suggests that the FE model deforms in a broadly realistic fashion, and can be used to make biologically meaningful interpretations of masticatory biomechanics.

9.2.7 Modeling Experiments

To examine the influence of variation in bite point on model deformation patterns, five modeling experiments were performed. In Experiment 1, the bite point was defined by constraining nodes on the surfaces of the LP^3 and LP^4 tooth crowns. In Experiment 2, the bite point was set at LM^1. In Experiment 3, the bite point was set at LM^2 and LM^3. In Experiment 4, the bite point was set at all of the left upper molars. Finally, in Experiment 5, the bite point was set at all of the left upper cheek teeth. Thus, Experiments 1–3 simulate bites on a relatively small food item at different locations along the tooth row. In contrast, Experiments 4 and 5 simulate bites on larger food items that either contact or do not contact the premolars, respectively.

Other than the location of the bite point, all variables and boundary conditions in each of the five experiments were equivalent. As a result, differences in strain patterns are exclusively a consequence of bite point position. Although this experimental design facilitates interpretation by altering and examining the influence of only one variable (bite point location), it is not entirely realistic. Biomechanical models predict and empirical observations during biting indicate (Greaves, 1978; Spencer, 1995, 1998) that muscle force magnitudes vary substantially during biting at different points along the tooth row. This may relate to the need to prevent distraction (inferior dislocation) of the working-side temporomandibular joint, or to differences in tooth root morphology among the teeth (Spencer, 2003). These variations in muscle forces were not incorporated into these finite element analyses, because the EMG data used to calculate muscle forces were collected during chewing experiments in which the subject used its post-canine teeth, but precise bite points are not known (Ross, 2001; Ross et al., 2003).

9.2.8 Evaluation of Experiments

Comparison among the results of the five experiments is complicated by the fact that statistical tests (such as analysis of variance) are not applicable in a straightforward manner. Statistical tests typically assess the probability that two or more groups of

Table 9.3 Validation of FE model

Region	FEA strains from Strait et al. (2005) Maximum shear strain[1]	Principal strain ratio (Max / Min)	Strains recorded from in vivo chewing experiments Mean maximum shear strain ± 2 standard deviations[1]	Mean principal strain ratio ± 2 standard deviations[2]	Experiment	Reference
Dorsal interorbital	139	4.4	169 ± 94	2.1 ± 0.4	5 A (W)	Hylander et al. (1991)
			266 ± 82	2.3 ± 0.2	5 A (B)	Hylander et al. (1991)
			185 ± 78	4.0 ± 0.4	6 (W)	Hylander et al. (1991)
			182 ± 68	4.0 ± 0.6	6 (B)	Hylander et al. (1991)
			139 ± 110	1.8 ± 0.2	2 A (W)	Hylander et al. (1991)
			129 ± 42	1.7 ± 0.2	2 A (B)	Hylander et al. (1991)
			86 ± 38	2.4 ± 0.6	2 B (W)	Hylander et al. (1991)
			117 ± 56	3.1 ± 0.6	5 B (W)	Hylander et al. (1991)
			240 ± 116	2.6 ± 0.4	5 C (W)	Hylander et al. (1991)
			200 ± 84	2.1 ± 1.0	5 C (B)	Hylander et al. (1991)
Working-side dorsal orbital	184	0.6	100 ± 62	0.5 ± 0.2	5 A	Hylander et al. (1991)
			85 ± 44	0.7 ± 0.2	6	Hylander et al. (1991)
Balancing-side dorsal orbital	159	1.3	147 ± 60	1.4 ± 0.2	5 A	Hylander et al. (1991)
			105 ± 29	1.4 ± 0.2	6	Hylander et al. (1991)
Working-side infraorbital	683	2.0	325 ± 174	1.4	2 C	Hylander et al. (1991)
			613 ± 256	1.1	5 C	Hylander et al. (1991)
			180 ± 128	0.1 ± 0.5[3]	7	Ross et al. (2002)

9 Craniofacial Strain Patterns During Premolar Loading

Table 9.3 (continued)

Region	FEA strains from Strait et al. (2005)		Strains recorded from in vivo chewing experiments			
	Maximum shear strain	Principal strain ratio (Max / Min)	Mean maximum shear strain ± 2 standard deviations	Mean principal strain ratio ± 2 standard deviations[1]	Experiment	Reference
Balancing-side infraorbital	269	2.3	199 ± 144	2.2	2 C	Hylander et al. (1991)
			295 ± 234	2.4	5 C	Hylander et al. (1991)
			192 ± 160		7	Ross et al. (2002)
Working-side mid-zygomatic	952	0.9	661 ± 414	1.0	2 A	Hylander et al. (1991)
			569 ± 244	0.9	2 B	Hylander et al. (1991)
			250 ± 104	0.7	5 B	Hylander et al. (1991)
			857 ± 360	0.6	2	Hylander and Johnson (1997)
			614 ± 274	0.7	5	Hylander and Johnson (1997)
			398 ± 204	0.7	7	Hylander and Johnson (1997)
			391 ± 72	0.7	9	Hylander and Johnson (1997)
Balancing-side mid-zygomatic	368	0.7	352 ± 238	0.9	2 A	Hylander et al. (1991)
			578 ± 212	0.5	2	Hylander and Johnson (1997)
			440 ± 254	0.6	5	Hylander and Johnson (1997)
			349 ± 262	0.6	7	Hylander and Johnson (1997)
			202 ± 168	0.6	9	Hylander and Johnson (1997)
Working-side postorbital bar	185	1.2	135 ± 113	1.0 ± 0.5	46	Ross et al. (unpubl.)
			194 ± 197		47	Ross et al. (unpubl.)
			142 ± 129		48	Ross et al. (unpubl.)

[1] Measured in microstrain.
[2] Standard deviations and, in some cases, means were not reported for all experiments.
[3] Data from this experiment were highly skewed, with most chews exhibiting high compression and low tension. However, the highest value observed in this experiment was 1.8.

randomly sampled, independent variates could have been drawn from a single statistical population. However, strain data derived from finite element analyses are not independent because the degree of deformation recorded at a given node is not independent of the degree of deformation at adjacent nodes. Moreover, the strain data at the given nodes are not randomly sampled. Rather, they are determined a priori by the variables (e.g., elastic properties, loads, constraints) incorporated into each analysis. Thus, the present study is not asking whether or not the model behaves differently during the five experiments, because clearly it must. Nor does the present study determine whether or not the magnitudes of the differences between experiments are sufficient to induce evolutionary adaptation in the craniofacial skeleton. This is not a question that can be answered simply because the precise relationship between strain and evolutionary adaptation is not fully understood. Although high strains can induce bone remodeling (Rubin and Lanyon, 1984; Burr et al., 1989; Cowin, 1993), low strains can do so as well under a repetitive loading regime (McLeod et al., 1998; Rubin et al., 2001). For example, the supraorbital torus experiences low strains during feeding (Picq and Hylander, 1989; Hylander et al., 1991), but our studies (e.g., Peterson, 2002) indicate that edentulous humans (who generate reduced masticatory loads) exhibit supraorbital thinning and a loss of bone mass. Thus, masticatory loads clearly influence supraorbital morphology to some degree, despite habitually low feeding strains in non-pathological subjects (i.e., individuals who have all of their teeth). Insofar as mastication is a repetitive loading regime, it is therefore not possible at present to identify a strain threshold below which differences between the FE models can be considered adaptively insignificant. Such a threshold might exist, but its value is not known. Regardless, the question of what causes bone remodeling in an individual is not equivalent to the question of what is responsible for evolutionary change.

The present study therefore assumes that even low-magnitude strain differences may be adaptively significant, but that higher-magnitude differences are likely to be more significant, where relative significance refers to the importance of the loading regime for determining evolution of the facial skeleton.

Differences between the results of the analyses were compared in three ways. First, a qualitative visual inspection of strain was performed. The shape of the FE model prior to loading was compared to that recorded after loading. Moreover, the magnitudes of maximum and minimum principal strains were mapped onto the model so as to identify the concentrations of strain. In addition, the percentage difference in maximum shear strain (a convenient summary of peak strain [Hylander et al., 1991; Hibbeler, 2000]) between models was calculated at 1,325 evenly spaced nodes on the surface of the model. These differences were summarized by reporting the proportion of the nodes whose shear strain values differed in a given pair of analyses by more than 10%, 20%, 30%, 40%, and 50%. Finally, strain values (maximum shear strain, the ratio of maximum to minimum principal strain, and the orientation of maximum principal strain) in a selected number of nodes in the anterior rostrum that correspond to observable strain concentrations were examined and compared.

9.3 Results

Results of the five experiments are summarized visually in Figs. 9.3 and 9.4. With respect to overall patterns of deformation, all experiments are similar in that the anterior rostrum exhibits some torsion around an antero-posterior axis, the ridge along the dorso-lateral aspect of the rostrum experiences mediolateral bending, the

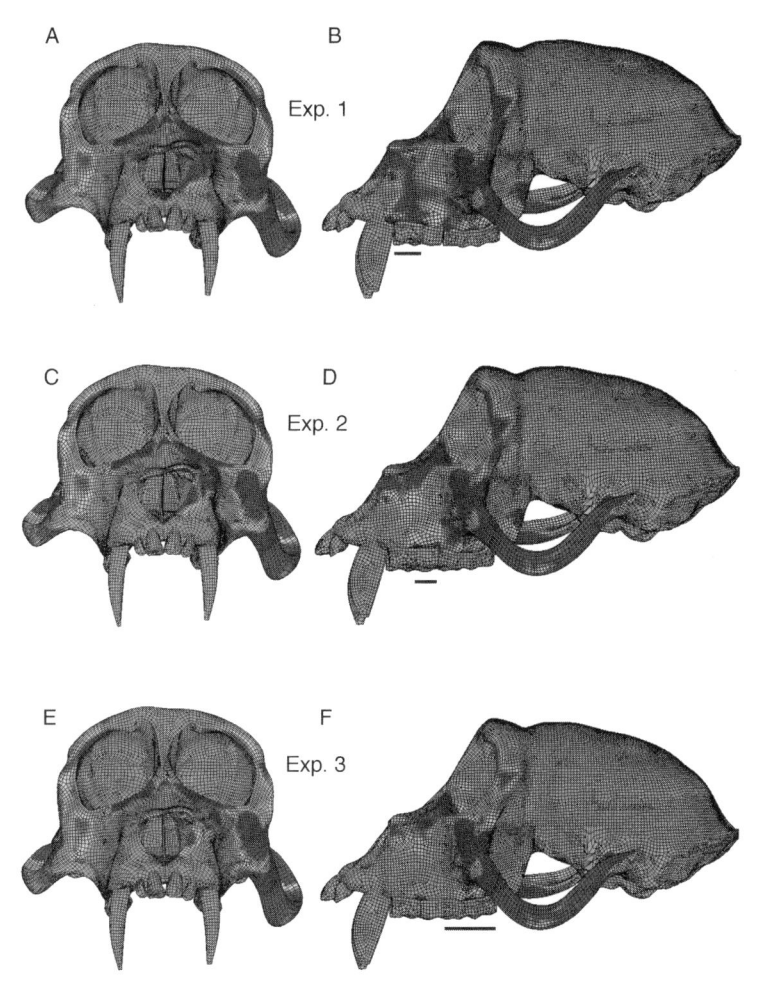

Fig. 9.3 Maximum principal strain in the finite element model as induced by a left-side chew. Please see color insert. Color mapping indicates the magnitude of strain. Horizontal bars indicate the size and location of the bite point. Gross deformations of the model are evident through comparisons with Fig. 9.2. (**A**, **B**) Experiment 1 (P^3–P^4 bite point) in frontal and lateral view. (**C**, **D**) Experiment 2 (M^1 bite point) in frontal and lateral view. (**E**, **F**) Experiment 3 (M^2–M^3 bite point) in frontal and lateral view. (**G**, **H**) Experiment 4 (M^1–M^3 bite point) in frontal and lateral view. (**I**, **J**) Experiment 4 (P^3–M^3 bite point) in frontal and lateral view (*See* Color Insert)

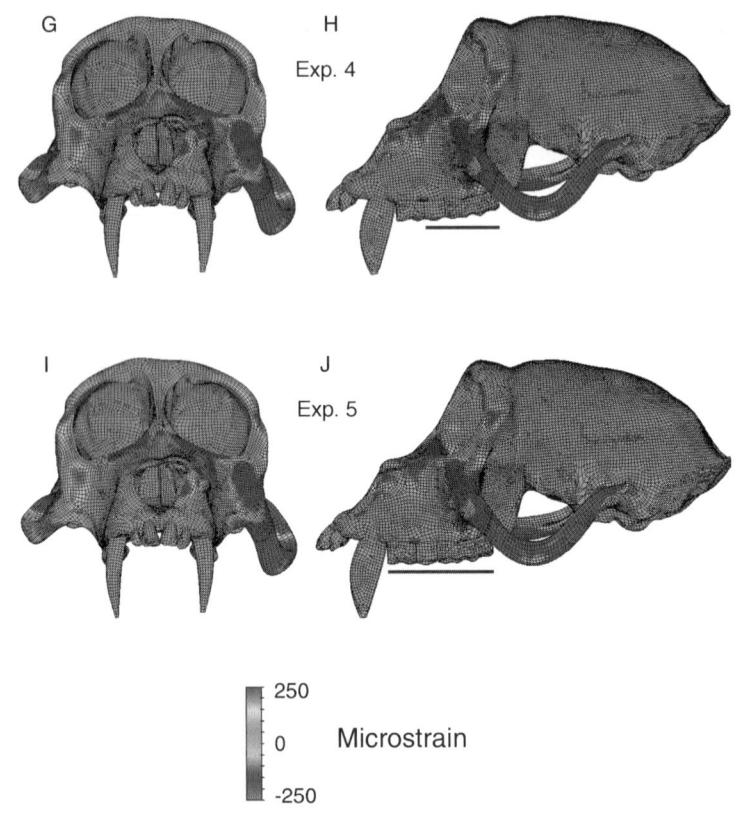

Fig. 9.3 (continued)

rostrum is bent or shears dorsally, the orbits (particularly on the working side) are compressed infero-superiorly, and the working-side zygomatic arch is displaced inferiorly to a greater extent than the balancing-side arch. However, torsion, mediolateral bending, and dorsal bending/shear of the anterior rostrum are most pronounced in Experiment 1, in which the bite point is located at the premolars. These deformations are moderately pronounced in Experiment 2, in which the bite point is set at M^1. Deformations are least pronounced in Experiments 3–5, in which the bite point is set at the distal-most two molars, all of the molars, and all of the molars and premolars, respectively.

Similar results are obtained when considering the distribution of strain concentrations (Figs. 9.3 and 9.4). The FE model exhibits strain concentrations in the circumorbital and zygomatic regions that are broadly comparable in all experiments, although close inspection reveals that these concentrations are more extensive in Experiments 1 and 2 (premolar and M^1 bites, respectively) than in the other experiments. However, notable differences between the experiments are observed in the antero-lateral and dorsal aspects of the rostrum, where high strains are observed in Experiment 1 (premolar bite point), moderate strains are seen in Experiment 2

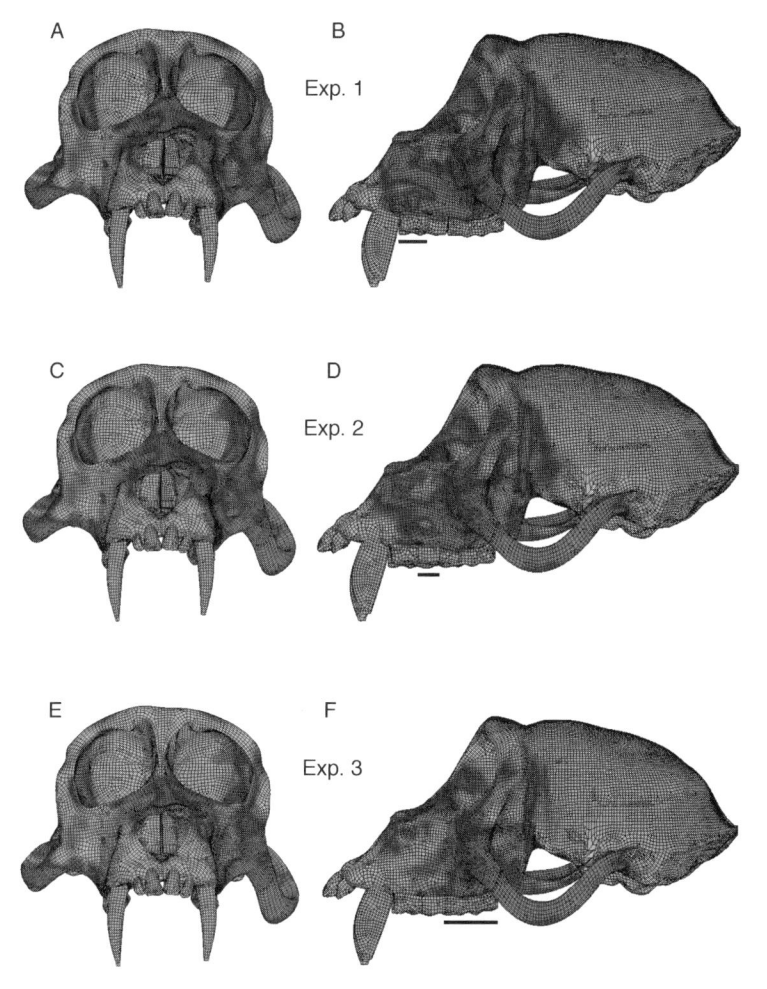

9 Craniofacial Strain Patterns During Premolar Loading

Fig. 9.4 Minimum principal strain in the finite element model as induced by a left-side chew. Please see color insert. Color mapping indicates the magnitude of strain. Horizontal bars indicate the size and location of the bite point. Gross deformations of the model are evident through comparisons with Fig. 9.2. (**A**, **B**) Experiment 1 (P^3–P^4 bite point) in frontal and lateral view. (**C**, **D**) Experiment 2 (M^1 bite point) in frontal and lateral view. (**E**, **F**) Experiment 3 (M^2–M^3 bite point) in frontal and lateral view. (**G**, **H**) Experiment 4 (M^1–M^3 bite point) in frontal and lateral view. (**I**, **J**) Experiment 4 (P^3–M^3 bite point) in frontal and lateral view (*See* Color Insert)

(M^1 bite point), and lower strains are recorded in Experiments 3–5 (bite points at M^2–M^3, all molars, all cheek teeth, respectively).

Quantitative results from 1,325 evenly spaced nodes on the surface of the face are consistent with the qualitative observations described above (Table 9.4). Several patterns are evident. First, shear strains are remarkably similar in Experiments 3–5, in which the bite point was set at M^2–M^3, all of the molars, and all of the cheek teeth, respectively. On average, maximum shear strains are only approximately

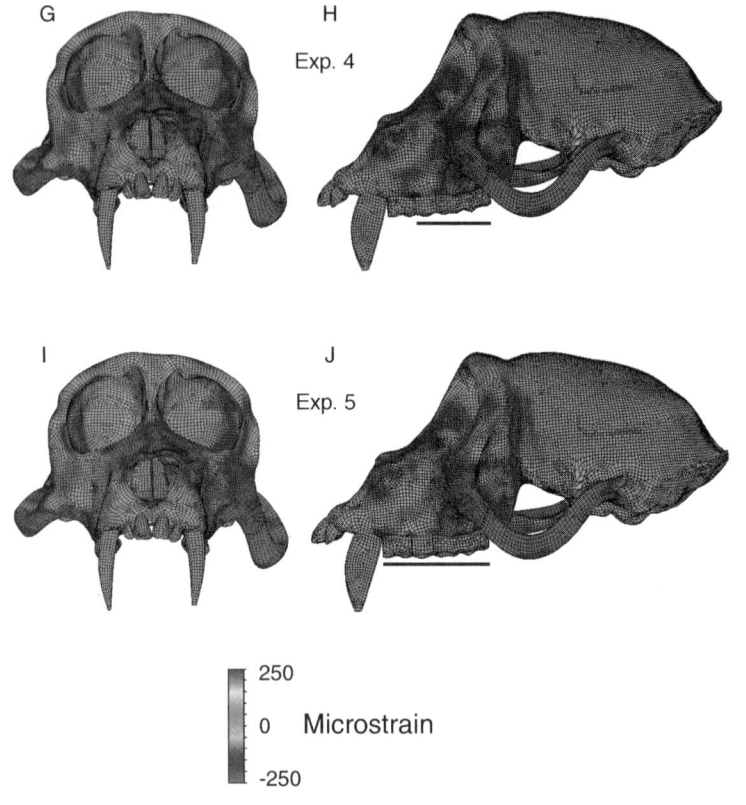

Fig. 9.4 (continued)

1–2% lower in Experiments 3 and 4 than in Experiment 5. When comparing strains node-by-node in each of these analyses, very few of these nodes (less than one-fifth) differ with respect to shear strain magnitude by more than 10%. A second notable result is that shear strains are markedly elevated in Experiment 1, in which the bite point was set at the premolars, relative to all other experiments. On average, shear strains in Experiment 1 exceed those in Experiments 3–5 by approximately 50%. In one-third of the selected nodes, strains in Experiment 1 exceed those in 3–5 by more than 50%. Strain magnitudes were more similar in Experiments 1 and 2, but on average, strains were nonetheless approximately 15% greater in Experiment 1. Experiment 2, therefore, exhibits an intermediate level of strain. Maximum shear strains in Experiment 2 are lower than in Experiment 1, but are, on average, approximately 25–30% greater than in Experiments 3–5. The actual magnitudes of maximum shear strain at the selected nodes fall reasonably close to the range of values observed in primate in vivo chewing experiments (Hylander et al., 1991; Hylander and Johnson, 1997; Ross, 2001; Ross et al., 2003). In all five experiments, approximately 97% of the selected nodes exhibit shear strain magnitudes of less than 1,000 microstrain.

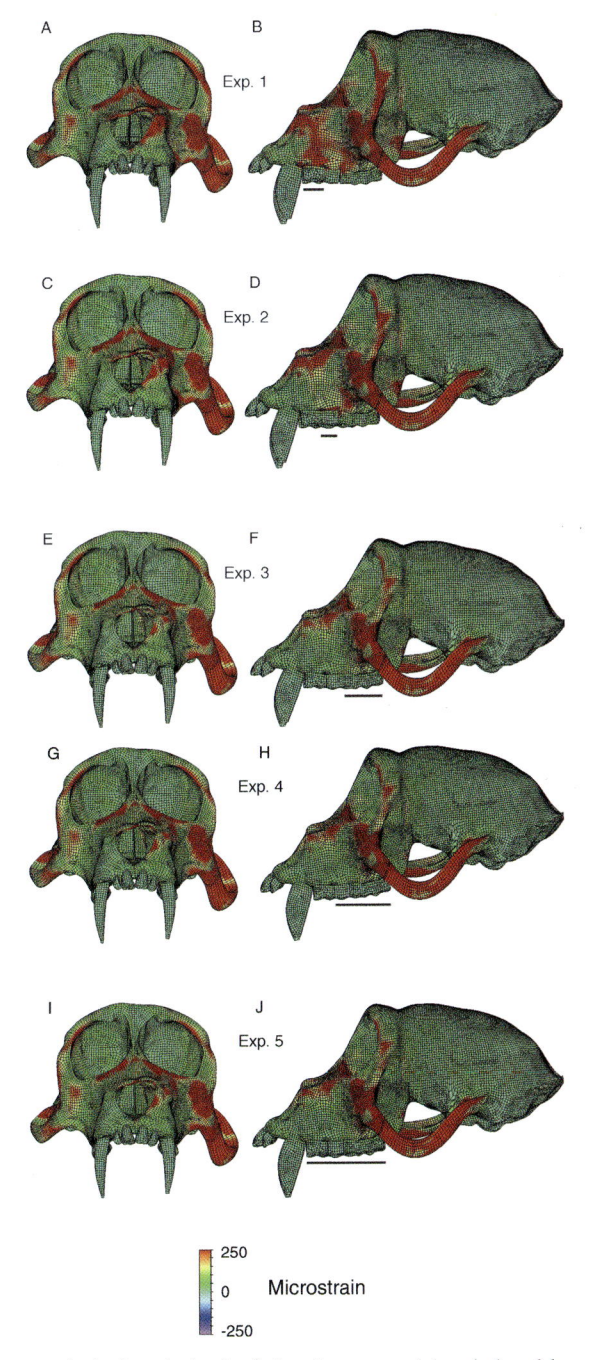

Fig. 9.3 Maximum principal strain in the finite element model as induced by a left-side chew. Color mapping indicates the magnitude of strain. Horizontal bars indicate the size and location of the bite point. Gross deformations of the model are evident through comparisons with Fig. 9.2. (**A, B**) Experiment 1 (P^3–P^4 bite point) in frontal and lateral view. (**C, D**) Experiment 2 (M^1 bite point) in frontal and lateral view. (**E, F**) Experiment 3 (M^2–M^3 bite point) in frontal and lateral view. (**G, H**) Experiment 4 (M^1–M^3 bite point) in frontal and lateral view. (**I, J**) Experiment 4 (P^3–M^3 bite point) in frontal and lateral view

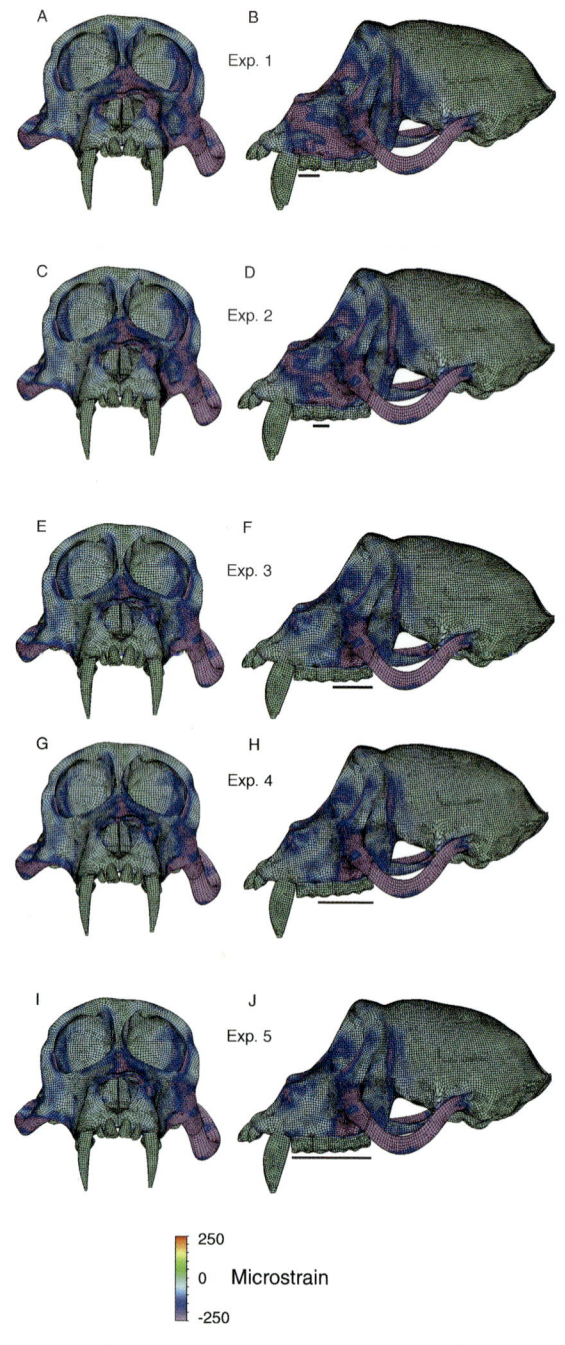

Fig. 9.4 Minimum principal strain in the finite element model as induced by a left-side chew. Color mapping indicates the magnitude of strain. Horizontal bars indicate the size and location of the bite point. Gross deformations of the model are evident through comparisons with Fig. 9.2. (**A, B**) Experiment 1 (P^3–P^4 bite point) in frontal and lateral view. (**C, D**) Experiment 2 (M^1 bite point) in frontal and lateral view. (**E, F**) Experiment 3 (M^2–M^3 bite point) in frontal and lateral view. (**G, H**) Experiment 4 (M^1–M^3 bite point) in frontal and lateral view. (**I, J**) Experiment 4 (P^3–M^3 bite point) in frontal and lateral view

9 Craniofacial Strain Patterns During Premolar Loading

Table 9.4 Comparisons of maximum shear strains between experiments[1]

Comparison	Mean % diff. (s.d.)	\multicolumn{6}{c}{Percentage of nodes differing in shear strain by more than:}
		10%	20%	30%	40%	50%	100%
Experiments 1 vs. 2 (P^{3-4} vs. M^1)	14.7 (36.0)	48.5	31.2	16.3	9.4	5.6	1.5
Experiments 1 vs. 3 (P^{3-4} vs. M^{2-3})	52.4 (99.9)	64.7	54.3	46.0	41.5	34.6	14.4
Experiments 1 vs. 4 (P^{3-4} vs. M^{1-3})	50.5 (106.6)	64.6	54.3	46.6	41.4	33.8	13.6
Experiments 1 vs. 5 (P^{3-4} vs. P^3-M^3)	48.5 (105.8)	63.7	53.0	44.6	38.0	30.3	10.2
Experiments 2 vs. 3 (M^1 vs. M^{2-3})	29.0 (58.7)	57.8	45.7	32.8	21.9	16.5	5.8
Experiments 2 vs. 4 (M^1 vs. M^{1-3})	27.5 (61.6)	56.7	44.8	28.8	19.7	14.4	4.6
Experiments 2 vs. 5 (M^1 vs. $P^3 - M^3$)	26.6 (69.5)	58.0	42.4	26.6	17.3	10.8	6.3
Experiments 3 vs. 4 (M^{2-3} vs. M^{1-3})	−1.2 (5.3)	5.9	1.6	0.5	0.2	0.1	0.0
Experiments 3 vs. 5 (M^{2-3} vs. P^3-M^3)	−2.4 (11.4)	17.6	6.9	3.7	1.7	0.9	0.0
Experiments 4 vs. 5 (M^{1-3} vs. $P^3 - M^3$)	−1.4 (9.4)	9.7	3.5	1.6	1.1	0.7	0.1

[1] In each experiment, strain is recorded at 1,325 surface nodes. In order to compare experiments, a percentage difference (% diff.) in strain is calculated for each node. For example, when comparing Experiments 1 and 2, % diff. = ($strain_{Analysis\,1}$ − $strain_{Analysis\,2}$) × $100/strain_{Analysis\,2}$. Positive values indicate that maximum shear strain is greater in Experiment 1, while negative values indicate that strain is greater in Experiment 2. Values greater than 10 or less than −10 indicate that the difference in strain at a given node is greater than 10%. This table indicates that in a comparison of Experiments 1 and 2, 48.5% of the selected nodes differ in shear strain by more than 10%, 31.2% of nodes differ by more than 20%, and so on.

Similar patterns of strain differences are observed in the anterior rostrum. Eleven representative nodes were selected, which sample strain concentrations in Experiment 1 (premolar bite point). These nodes are found along the nasal margin, the dorsal rostrum, the lateral rostrum above the premolars, and along a ridge of bone at the interface between the lateral and dorsal rostrum (Fig. 9.5). Strains at these nodes are compared in Table 9.5. As previously observed, Experiments 3–5 (bite points at M^2-M^3, all molars, all cheek teeth, respectively) exhibit very similar patterns of strain. Relative to Experiments 3–5, maximum shear strains in Experiment 1 are approximately 50–100% higher in the nasal margin and dorsal rostrum, 300–900% higher in the lateral rostrum, and 40–50% higher in the rostral ridge. Shear strain values in the anterior rostrum in Experiment 1 are broadly comparable to those observed in the zygomatic arch and infraorbital regions in in vivo chewing experiments (Table 9.3). Experiment 1 also differs from Experiments 3–5 with respect to principal strain ratio and strain orientation (particularly in the lateral rostrum and

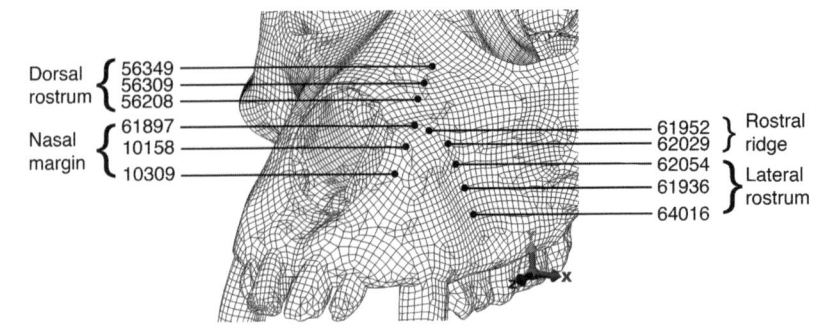

Fig. 9.5 Nodes in the anterior rostrum corresponding to high strain concentrations during premolar loading

rostral ridge), indicating that these experiments differ with respect to not only the magnitude but also the nature of the strains (Table 9.5).

9.4 Discussion

When the bite point was fixed at the premolars, strains were elevated in the model overall, and strain concentrations were observed that were not seen when the bite point was fixed at other locations. Thus, one can reject the hypothesis that craniofacial strain patterns induced by molar and premolar loading do not differ appreciably. One can further conclude that strains are elevated only when the premolars are loaded in isolation, rather than when they are loaded in association with the molars. This is perhaps unsurprising given that when many teeth are loaded simultaneously, occlusal loads are distributed across a large surface area, thereby minimizing occlusal pressure for any given set of muscle forces. However, it is notable that this effect is found not only in the alveolar region, close to where these pressure differences actually occur, but also in other more distant parts of the face. The observation that, compared with molar loading, premolar loading produces elevated strain magnitudes in many parts of the face is consistent with the notion that increased premolar loading might necessitate modifications to facial skeletal morphology. This provides indirect support for the hypothesis that the derived craniofacial features in "robust" australopiths are adaptations to withstand premolar loads (Rak, 1983, 1985). One feature that seems particularly likely to resist the sagittal bending/shear, mediolateral bending, and torsion imposed by premolar loading would be an anteriorly placed root of the zygomatic arch. Note, however, that the present study has not specifically tested the functional relationships of this and other features. Rather, this study has tested the premise of those functional hypotheses, namely, that premolar bites influence strain patterns in the facial skeleton so as to induce morphological adaptation. To fully establish that features such as anterior pillars and other traits are the adaptations to resist premolar mastication, one would

9 Craniofacial Strain Patterns During Premolar Loading

Table 9.5 Strains in the anterior rostrum

Type of strain	Region	Node	P^3-P^4 bite	M^1 bite	M^2-M^3 bite	M^1-M^3 bite	P^3-M^3 bite
Maximum shear strain (in microstrain):							
	Nasal margin	10309	527	404	244	257	277
	Nasal margin	10158	652	528	326	343	371
	Nasal margin	61897	504	424	269	280	296
	Dorsal rostrum	56208	636	501	360	370	390
	Dorsal rostrum	56309	529	421	299	307	322
	Dorsal rostrum	56349	489	417	326	331	341
	Lateral rostrum	64016	380	319	120	134	85
	Lateral rostrum	61936	568	307	85	81	67
	Lateral rostrum	62054	437	288	129	132	127
	Rostral ridge	62029	468	444	309	316	303
	Rostral ridge	61952	440	444	305	314	317
Principal strain ratio (max/min):							
	Nasal margin	10309	1.4	1.6	1.6	1.6	1.6
	Nasal margin	10158	0.9	1.0	1.1	1.0	1.0
	Nasal margin	61897	1.4	1.6	1.7	1.7	1.6
	Dorsal rostrum	56208	0.7	0.8	0.8	0.8	0.8
	Dorsal rostrum	56309	0.8	1.0	1.1	1.1	1.1
	Dorsal rostrum	56349	1.6	2.1	2.5	2.5	2.4
	Lateral rostrum	64016	2.7	0.6	0.4	0.3	0.4
	Lateral rostrum	61936	0.8	0.7	0.4	0.4	0.4
	Lateral rostrum	62054	1.5	1.5	1.3	1.4	1.5
	Rostral ridge	62029	1.3	1.4	1.6	1.6	1.7
	Rostral ridge	61952	2.2	2.6	2.7	2.7	2.6

Table 9.5 (continued)

Type of strain	Region	Node	P³-P⁴ bite	M¹ bite	M²-M³ bite	M¹-M³ bite	P³-M³ bite
Orientation of maximum principal strain (x, y, z):							
	Nasal margin	10309	(.75, −.44, −.54)	(.72, −.49, −.50)	(.72, −.49, −.49)	(.72, −.49, −.49)	(.72, −.49, −.49)
	Nasal margin	10158	(.77, −.34, −.48)	(.75, −.37, −.55)	(.75, −.39, −.54)	(.75, −.39, −.54)	(.76, −.38, −.53)
	Nasal margin	61897	(.46, −.04, −.89)	(.38, −.10, −.92)	(.35, −.12, −.93)	(.35, −.12, −.93)	(.37, −.10, −.92)
	Dorsal rostrum	56208	(.77, .26, −.58)	(.73, .29, −.61)	(.74, .28, −.61)	(.74, .28, −.61)	(.74, .29, −.61)
	Dorsal rostrum	56309	(.83, .18, −.52)	(.82, .22, −.52)	(.83, .22, −.51)	(.83, .22, −.51)	(.83, .29, −.51)
	Dorsal rostrum	56349	(.83, .45, −.33)	(.82, .50, −.27)	(.82, .52, −.23)	(.82, .52, −.23)	(.82, .52, −.23)
	Lateral rostrum	64016	(.16, .35, −.92)	(.34, .35, .87)	(.92, −.39, .02)	(.93, −.36, .07)	(.93, −.37, .07)
	Lateral rostrum	61936	(.15, .24, −.96)	(.17, .05, .98)	(.96, −.29, .02)	(.94, −.35, −.03)	(.93, −.36, −.06)
	Lateral rostrum	62054	(.37, .34, −.86)	(.62, −.07, .78)	(.67, .33, .66)	(.66, .28, .70)	(.61, .26, .75)
	Rostral ridge	62029	(.31, .32, −.90)	(.62, −.09, −.78)	(.96, −.11, −.25)	(.92, −.15, −.37)	(.78, −.18, −.60)
	Rostral ridge	61952	(.38, .04, −.93)	(.20, −.25, −.95)	(.15, −.32, −.93)	(.16, −.32, −.94)	(.17, −.29, −.94)

have to determine that those features reduce the strains induced by premolar bites. This could be done by building and loading FE models of fossil hominids, and by altering the geometry of the existing macaque model by adding to it features of "robust" australopiths. Thus, the possibility that some derived facial features in "robust" australopiths are the adaptations to resist premolar loads remains a viable, if not yet fully tested hypothesis.

Other results from this study have implications for human evolution. Strains are elevated in the anterior rostrum when the bite point is fixed at only the premolars, but strains are much lower when the premolars are loaded in conjunction with the molars. Thus, premolar loading is a reasonable explanation of "robust" craniofacial form only in situations in which the premolars are loaded in isolation or with high bite forces. This possibility is consistent with the notion that "robust" australopiths may have been feeding on small objects, which would only contact one or two teeth during each bite. Small, hard-object feeding has long been considered a potentially important adaptation of "robust" australopiths (e.g., Jolly, 1970; Grine, 1981), so these results are consistent with this hypothesis. However, the biomechanics of mastication do not implicate small-object feeding as the source of premolar loading. Biomechanical models (Du Brul, 1977; Smith, 1978; Greaves, 1978; Spencer, 1995, 1998) predict that bite forces at the premolars will be lower than or, at best, equal to those produced at the molars. Thus, given a small, resistant food object that can be easily positioned at any point along the tooth row, there is no obvious advantage to chewing that object on the premolars, especially in light of the fact that such chews elevate strains in the face. Thus, although small-object feeding might be a source of premolar loading, such an explanation does not make sense biomechanically.

An alternative source of premolar loading might be large, hard-object feeding. In this context, a large object is simply a food item that cannot be placed easily inside the oral cavity without first being processed by the teeth. Typically, anthropoids ingest large objects by processing them with their incisors, as when taking a bite out of a fruit (e.g., Hylander, 1975; Ungar, 1994). However, if that food item is so resistant that the incisors would not be able to withstand the forces needed to induce failure in the object, then the cheek teeth might be needed to crush the item (it is noteworthy that the incisors of "robust" australopiths are characterized by a low incidence of microwear features [Ungar and Grine, 1991]). The molars would provide the highest bite force, but they would not be available for use because the food item would be too large relative to the subject's gape to allow contact with the distal-most teeth. Because the premolars are positioned more mesially, near the orifice of the mouth, they could be used to crush the item. The crushing would either provide a bite-sized portion of the food item or crack open an outer casing (e.g., a shell) to facilitate extraction of the edible portions of the item. Lucas (2004) argues that cheek-tooth preparation of large, resistant objects would be most effective on food items that are not only resistant but also brittle (Lucas, 2004), or when used in combination with manual manipulation. *Cebus apella*, well known for its masticatory adaptations for resistant foods (Wright, 2005), has been observed to extract embedded ingestible tissues from hard brittle fruits, and insects from tough branches using its premolars and hands (Wright pers. obs.). Australopith premolars might be less useful in processing large food items that were tough (rather than brittle), because their occlusal morphology does not seem well designed for propagating cracks through tough foods (Teaford and Ungar, 2000). However, premolars might be useful in processing tough items if such items were gripped with the teeth and pulled or twisted with the hands, as is observed in some primates such as *C. apella*.

Living primates that feed on large, hard objects typically employ teeth other than the incisors during the initial stages of feeding. Pithecines and *Cebus apella* often use their canines for this purpose (Izawa and Mizuno, 1977; Struhsaker and Leland, 1977; van Roosmalen et al., 1988; Kinzey and Norconk, 1990), but premolars probably also play a role (Cole, 1992; Wright, 2005). Naturally, canines can be used to puncture an item, but canines are extremely reduced in "robust" australopiths, so puncturing was obviously not an important aspect of their food preparation. Rather, "robust" australopiths may have used their premolars to habitually crack open or crush large, resistant food items. This Large Object Feeding Hypothesis does not preclude the possibility that "robust" australopiths habitually ate small food items. Rather, it merely states that small food items do not provide the best explanation for the evolution of craniofacial features functionally related to premolar biting. It has also been hypothesized that "robust" australopiths used their expanded cheek teeth to process large quantities of food at one time (Walker, 1981). This, too, is a plausible interpretation of "robust" australopith feeding behavior, but likewise does not explain the evolution of features that resist premolar loading. "Large quantity" feeding presumably involves chewing with many teeth at one time, but those loading regimes were found here to induce minimal strains in the anterior rostrum. Recent studies (Wood and Strait, 2004; Sponheimer et al., 2005; Scott et al., 2005) have suggested that "robust" australopiths may have been ecological and dietary generalists. The Large Object Feeding Hypothesis is fully consistent with such dietary reconstructions, but notes that some of the items consumed in the diet may have specific morphological correlates. Indeed, insofar as such morphological features might have allowed "robust" australopiths to process food items that might have otherwise been inaccessible, adaptations for premolar loading may have played a key role in facilitating a generalized diet. Indeed, Wright (2005) has noted that similar features in *Cebus apella* (anteriorly placed, zygomatic, enlarged premolars) have the effect of expanding dietary breadth. Studies of premolar microwear in living primates might provide a comparative data set that would allow a means of testing the Large Object Feeding Hypothesis in australopiths.

9.4.1 Applications of FEA to Mandibular Biomechanics

Although this paper has focused on craniofacial morphology, FEA can be applied in a similar fashion to questions about the evolution of mandibular morphology in australopiths. With respect to mandibular biomechanics, Hylander has shown that the mandibular corpus and symphysis in "robust" australopiths are broad and deep, and he suggested that premolar loading may have accentuated twisting moments in the corpus, thereby contributing to the evolution of a transversely thick corpus in these species (Hylander, 1979, 1988). Other variables, such as bite force direction, relative muscle activity, and food material properties, may also have influenced mandibular morphology (Hylander, 1979, 1988). Although the results presented here do not speak directly to hypotheses regarding mandibular evolution, the finite element methods employed here should, in principle, be able to test aspects of these hypotheses.

9.5 Conclusions

Finite element analysis reveals that the location and size of the bite point can have a considerable influence on strain patterns in the primate face. Strains were highest when the bite point was restricted to the premolars. These results are consistent with hypotheses suggesting that some derived craniofacial features in "robust" australopiths may be the adaptations for resisting premolar loads. Although a diet of small, hard objects cannot be ruled out, an hypothesis that "robust" australopiths were using their premolars in the initial stages of ingesting large, resistant food items might better explain the evolution of these features.

Acknowledgments This contribution is dedicated to W. L. Hylander, whose seminal work on primate feeding biomechanics provides the framework underlying this and many other studies. We thank the editors for inviting us to participate in this volume, and we thank Fred Grine and Bernard Wood for useful discussions about australopith feeding adaptations. We also thank the editors and two anonymous reviewers for constructive comments that considerably improved this paper. Thanks also to the curatorial staff at the National Museum of Natural History. In vivo macaque data were collected using funding from an NSF Physical Anthropology grant to C. Ross and X. Chen (SBR 9706676). Finite element analyses were supported by an NSF Physical Anthropology grant to D. Strait, P. Dechow, B. Richmond, C. Ross, and M. Spencer (BCS 0240865) and NSF HOMINID grants (BCS 0725219, 0725183, 0725147, 0725141, 0725136, 0725126, 0725122, 0725078). Support was also provided by The Henry Luce Foundation, and the New York College of Osteopathic Medicine.

References

Antón, S. C. (1993). Internal masticatory muscle architecture in the Japanese macaque and its influence on bony morphology. *Am. J. Phys. Anthropol. Suppl.* 16:50.
Ashman, R. B., Rho, J. Y., and Turner, C. H. (1989). Anatomical variation of orthotropic elastic moduli of the proximal human tibia. *J. Biomech.* 22:895–900.
Borrazzo, E. C., Hylander, W. L., and Rubin, C. T. (1994). Validation of a finite element model of the functionally loaded zygomatic arch by in vivo strain gage data. Proc. Amer. Soc. Biomech. 18:51–52.
Burr, D. B., Schaffler, M. B., Yang, K. H., Lukoschek, M., Sivaneri, N., Blaha, J. D., and Radin, E. L. (1989). Skeletal changes in response to altered strain environments: is woven bone a response to elevated strain? *Bone*. 10:223–233.
Cole, T. M. (1992). Postnatal heterochrony of the masticatory apparatus in *Cebus apella* and *Cebus albifrons*. *J. Hum. Evol.* 23:253–282.
Cook, R. D., Malkus, D. S., and Plesha, M. E. (1989). *Concepts and Applications of Finite Element Analysis*. John Wiley & Sons, New York.
Cowin, S. C. (1993). Bone stress adaptation models. *J Biomech Eng*. 115:528–533.

Currey, J. D. (2002). *Bones: Structures and Mechanics.* Princeton University Press, Princeton.

Du Brul, E. L. (1977). Early hominid feeding mechanisms. *Am. J. Phys. Anthropol.* 47:305–320.

Greaves, W. S. (1978). The jaw lever system in ungulates: a new model. *J. Zool. (London)* 184:271–285.

Grine, F. E. (1981). Trophic differences between 'gracile' and 'robust' australopithecines: a scanning electron microscope analysis of occlusal events. *S. Afr. J. Sci.* 77:203–230.

Herring, S. W. (1972). Sutures–a tool in functional cranial analysis. *Acta Anat. (Basel)* 83:222–47.

Herring, S. W., and Mucci, R. J. (1991). *In vivo* strain in cranial sutures: the zygomatic arch. *J. Morphol.* 207:225–239.

Herring, S. W., and Teng, S. (2000). Strain in the braincase and its sutures during function. *Am. J. Phys. Anthropol.* 112:575–593.

Herring, S. W., Teng, S., Huang, X., Mucci, R. J., and Freeman J. (1996). Patterns of bone strain in the zygomatic arch. *Anat. Rec.* 246:446–57.

Herring, S. W., Rafferty, K. L., Liu, Z. J., and Marshall, C. D. (2001). Jaw muscles and the skull in mammals: the biomechanics of mastication. *Comp. Biochem. Physiol. A Mol. Integr. Physiol.* 131:207–19.

Hibbeler, R. C. (2000). *Mechanics of Materials.* Prentice Hall, Upper Saddle River, New Jersey.

Huiskes, R., and Chao, E. Y. S. (1983). A survey of finite element analysis in orthopedic biomechanics: the first decade. *J. Biomech.* 16:385–409.

Hylander, W. L. (1975). Incisor size and diet in anthropoids with special reference to Cercopithecidae. *Science* 189:1095–1098.

Hylander, W. L. (1979). The functional significance of primate mandibular form. *J. Morph.* 160:223–240.

Hylander, W. L. (1984). Stress and strain in the mandibular symphysis of primates: a test of competing hypotheses. *Am. J. Phys. Anthropol.* 64:1–46.

Hylander, W. L. (1986). *In vivo* bone strain as an indicator of masticatory bite force in *Macaca fascicularis. Archs. Oral Biol.* 31:149–157.

Hylander, W. L. (1988). Implications of *in vivo* experiments for interpreting the functional significance of "Robust" Australopithecine jaws. In: Grine, F.E. (Ed.) *Evolutionary History of the "Robust" Australopithecines.* New York, Aldine de Gruyter. pp. 55–83.

Hylander, W. L., and Johnson, K. R. (1989). The relationship between masseter force and masseter electromyogram during masticcation in the monkey *Macaca fascicularis. Archs. Oral Biol.* 34:713–722.

Hylander, W. L., and Johnson, K. R. (1997). *In vivo* bone strain patterns in the zygomatic arch of macaques and the significance of these patterns for functional interpretations of craniodental form. *Am. J. Phys. Anthropol.* 102:203–232.

Hylander, W. L., Picq, P. G., and Johnson, K. R. (1991). Masticatory-stress hypotheses and the supraorbital region of primates. *Am. J. Phys. Anthropol.* 86:1–36.

Izawa, K., and Mizuno, A. (1977). Palm-fruit cracking behavior of wild black-capped capuchin (*Cebus apella*). *Primates* 14:773–792.

Jolly, C. J. (1970). The seed eaters: a new model for hominid differentiation based on a baboon analogy. *Man* 5:5–26

Kinzey, W. G., and Norconk, M. A. (1990). Hardness as a basis of fruit choice in two sympatric primates. *Am. J. Phys. Anthropol.* 81:5–15.

Lucas, P. W. (2004). *Dental Functional Morphology.* Cambridge University Press, Cambridge.

McCollum, M. A. (1999). The robust australopithecine face: a morphogenetic perspective. *Science* 284:301—305.

McLeod, K. J., Rubin, C. T., Otter, M. W., and Qin, Y. X. (1998). Skeletal cell stresses and bone adaptation. *Am. J. Med. Sci.* 316:176–183.

Murphy, R. A. (1998). Skeleta muscle. In: Berne, R. M. and Levy, M. N. (eds.), *Physiology.* Mosby, St. Louis, pg. 294.

Peterson, J. (2002). Cortical material properties of the human dentate and edentulous cranial complex. Ph.D. thesis, Baylor College of Dentistry, Texas A & M University System Health Science Center, Dallas, TX.

Peterson, J., and Dechow, P. C. (2003). Material properties of the human cranial vault and zygoma. *Anat. Rec.* 274A:785–797.
Picq, P.G. and Hylander, W.L. (1989). Endo's stress analysis of the primate skull and the functional significance of the supraorbital region. *Am. J. Phys. Anthropol.* 79:393–398.
Rafferty, K.L. and Herring, S.W. (1999). Craniofacial sutures: morphology, growth, and *in vivo* masticatory strains. *J. Morph.* 242:167–179.
Rak, Y. (1983). *The Australopithecine Face.* Academic Press, New York.
Rak, Y. (1985). Australopithecine taxonomy and phylogeny in light of facial morphology. *Am J. Phys. Anthropol.* 66:281–287.
Richmond, B. G., Wright, B. W., Grosse, I., Dechow, P. C., Ross, C. F., Spencer, M. A., and Strait, D. S. (2005). Finite element analysis in functional morphology. *Anat. Rec. A* 283A:259–274.
Robinson, J. T. (1954a). The genera and species of the Australopithecinae. *Am. J. Phys. Anthropol.* 12:181–200.
Robinson, J. T. (1954b). Prehominid dentition and hominid evolution. *Evolution* 8:324–334.
Robinson, J. T. (1956). The dentition of the Australopithecinae. *Mem. Transvaal Mus.* 9:1–179.
Robinson, J. T. (1962). The origin and adaptive radiation of the australopithecines. In: Kurth, G. (Ed.) *Evolution und Hominisation.* Stuttgart, Fischer, pp. 120–140.
Robinson, J. T. (1963). Adaptive radiation in the Australopithecines and the origin of man. In Howell, F. C. and Boulière, F. (Eds.) *African Ecology and Human Evolution.* Cambridge, Cambridge University press, pp. 385–416.
Robinson, J. T. (1967). Variation and the taxonomy of the early hominids. In: Dobzhansky, T., Hecht, M. K., and Steere, W. C. *Evolutionary Biology.* New York, Meredith, pp. 69–100.
Ross, C. F. (2001). In vivo function of the craniofacial haft: The interorbital "pillar". *Am. J. Phys. Anthropol.* 116:108–139.
Ross, C. F., Strait, D. S., Richmond, B. G., and Spencer, M. A. (2002). *In vivo* bone strain and finite-element modeling of the anterior root of the zygoma in *Macaca* (abstract). *Am. J. Phys. Anthropol. Suppl.* 34:133.
Ross, C. F., Strait, D. S., Richmond, B. G., and Spencer, M. A. (2003). *In vivo* bone strain and finite-element modeling of the anterior root of the zygoma in *Macaca. Am. J. Phys. Anthropol. Suppl.* 34:133.
Ross, C. F., Patel, B. A., Slice, D. E., Strait, D. S., Dechow, P. C., Richmond, B. G., and Spencer, M. A. (2005). Modeling masticatory muscle force in finite element analysis: sensitivity analysis using principal coordinates analysis. *Anat. Rec. A* 283A:288–299.
Rubin, C. T., and Lanyon, L. E. (1984). Regulation of bone formationby applied dynamic loads. *J. Bone Joint Surg.* 66:397–402.
Rubin, C. T., Sommerfeldt, D. W., Judex, S., and Qin, Y. (2001). Inhibition of osteopenia by low magnitude, high-frequency mechanical stimuli. *Drug Discov. Today.* 6(16):848–858.
Scott, R. S., Ungar, P. S., Bergstrom, T. S., Brown, C. A., Grine, F. E., Teaford M. F., and Walker, A. (2005) Dental microwear texture analysis shows within-species diet variability in fossil hominins. *Nature* 436:693–695.
Shibazaki, R., Dechow, P. C., Maki, K. and Opperman, L. A. (2007). Biomechanical strain and morphological changes with age in rat calvarial bone and sutures. *Plastic Reconstr. Surg.* 119: 2167–2178.
Smith, R. J. (1978). Mandibular biomechanics and temporomandibular joint function in primates. *Am. J. Phys. Anthropol.* 49:341–350.
Spencer, M. A. (1995). Masticatory system configuration and diet in anthropoid primates. Ph.D. Dissertation, State University of New York at Stony Brook.
Spencer, M. A. (1998). Force production in the primate masticatory system: electromyographic tests of biomechanical hypotheses. *J. Hum. Evol.* 34:25–54.
Spencer, M. A. (2003). Tooth root form and function in platyrrhine seed-eaters. *Am. J. Phys. Anthropol.* 122:325–335.
Sponheimer, M., Lee-Thorp, J., de Ruiter, D., Codron D., Codron, J., Baugh, A. T., and Thackeray, F. (2005). Hominins, sedges and termites: new carbon isotope data from the Sterkfontein valley and Kruger National Park. *J. Hum. Evol.* 48:301–312.

Strait, D. S., and Grine, F. E. (2004). Inferring hominoid and early hominid phylogeny using craniodental data: the role of fossil taxa. *J. Hum. Evol.* 47:399–352.

Strait, D. S., Grine, F. E., and Moniz, M. A. (1997). A reappraisal of early hominid phylogeny. *J. Hum. Evol.* 32:17–82.

Strait, D. S., Wang, Q., Dechow, P. C., Ross, C. F., Richmond, B. G., Spencer, M. A., and Patel, B. A. (2005). Modeling elastic properties in finite-element analysis: how much precision is needed to produce an accurate model? *Anat. Rec. A* 283A:275–285.

Struhsaker, T. T., and Leland, L. (1977). Palm-nut smashing by *Cebus apella* in Colombia. *Biotropica* 9:124–126.

Teaford, M. F., and Ungar, P. S. (2000). Diet and the evolution of the earliest human ancestors. *Proc. Nat. Acad. Sci.* 97:13506–13511.

Tobias, P. V. (1967). *The cranium and maxillary dentition of* Australopithecus (Zinjanthropus) boisei. *Olduvai Gorge.* Volume 2. Cambridge University Press, Cambridge.

Ungar, P. S. (1994). Patterns of ingestive behavior and anterior tooth use differences in sympatric anthropoid primates. *Am. J. Phys. Anthropol.* 95:197–219.

Ungar, P. S. (2004). Dental topography and diets of *Australopithecus afarensis* and early *Homo. J. Hum. Evol.* 46:605–622.

Ungar, P. S., and Grine, F. (1991). Incisor size and wear in *Australopithecus africanus* and *Paranthropus robustus. J. Hum. Evol.* 20:313–340.

Van Roosmalen, M. G. M., Mittermeier, R. A. et al. (1988) Diet of the northern bearded saki (*Chiropotes satanas chiropotes*): a neotropical seed predator. *Am. J. Primatol.* 14:11–35.

Walker, A. C. (1981). Diet and teeth. Dietary hypotheses and human evolution. *Phil. Trans. Roy. Soc. (London)* 292:57–64.

Wang, Q., and Dechow, P. C. (2006). Elastic properties of external cortical bone in the craniofacial skeleton of the rhesus monkey. *Am. J. Phys. Anthropol.* 131:402–415.

Wang, Q., Dechow, P. C., Wright, B. W., Ross, C. F., Strait, D. S., Richmond, B. G., and Spencer, M. A. (in press, this volume). Surface strain on bone and sutures in a monkey facial skeleton: an *in vitro* approach and its relevance to Finite Element Analysis. In: Vinyard, C. J., Ravosa, M. J., and Wall, C. E. (Eds.), *Primate Craniofacial Function and Biology.* New York, Springer

Wang, Q., Strait, D. S. and Dechow, P. C. (2006). Fusion patterns of craniofacial sutures in rhesus monkey skulls of known age and sex from Cayo Santiago. *Am. J. Phys. Anthropol.* 131: 469–485.

Wood, B. (1991). *Koobi For a Research Project Volume 4 Hominid Cranial Remains.* Oxford, Clarendon Press.

Wood, B., and Strait, D. S. (2004) patterns of resource use in early *Homo* and *Paranthropus. J. Hum. Evol.* 46:119–162.

Wright, B. W. (2005). Craniodental biomechanics and dietary toughness in the genus. *Cebus. J. Hum. Evol.* 48:473–492.

Part IV
Jaw-Muscle Architecture

While the functional significance of jaw muscles for mastication is obvious, our knowledge of jaw-muscle architecture in primates lags behind other areas of craniofacial research. We presently have little insight into how jaw-muscle architecture relates to size, diet, and muscle activity patterns during mastication in primates. In this part, three chapters begin to address this shortcoming by examining the scaling patterns and functional correlates of jaw-muscle architecture in primates.

Anapol et al. examine the scaling patterns for reduced physiological cross-sectional area (RPCA) of the jaw-closing muscles across primates. They identify a general pattern of positive allometry for jaw-muscle RPCAs relative to body size. This pattern potentially fits the predictions from the geometric similarity or metabolic scaling models, but does not support the predictions of the fracture scaling hypothesis. Furthermore, Anapol et al. observe that catarrhines tend to have higher slopes for RPCA when compared to other primate groups. While the authors caution about the preliminary nature of these comparisons and voice important concerns related to interpreting interspecific data, they provide some interesting initial clues into how the force-producing abilities of the jaw muscles change with size in primates.

Perry and Wall also conduct a scaling analysis examining how jaw-muscle architectural parameters change with body size in prosimian primates. They find that muscle mass, physiological cross-sectional area (PCSA), and fiber lengths each scale close to isometry with PSCA, potentially showing slight positive allometry relative to body mass. When tested against the predictions of the fracture scaling model, the scaling pattern for prosimian PCSA does not fit the predicted negatively allometric slope. The discussion turns to the details of food fracture and how additional factors can influence the scaling of PSCA in prosimians.

In the final chapter of the part, Taylor and Vinyard compare jaw-muscle fiber architecture in tree-gouging versus non gouging callitrichids to assess the functional changes in jaw muscles related to the derived tree biting behavior of marmosets. Among callitrichids, marmosets feed differently from tamarins by biting trees to initiate exudate flow for later consumption. This gouging behavior appears to involve relatively large gapes, and marmoset jaw muscles exhibit relatively elongate fibers to facilitate this extreme jaw opening. As a structural consequence, marmosets have relatively reduced force-producing capabilities in their jaw muscles compared to tamarins. Collectively, these results demonstrate the potential for significant functional changes in jaw-muscle architecture related to novel feeding behaviors in primates.

Chapter 10
Scaling of Reduced Physiologic Cross-Sectional Area in Primate Muscles of Mastication

Fred Anapol, Nazima Shahnoor, and Callum F. Ross

Contents

10.1	Introduction	201
10.2	Materials and Methods	203
10.3	Results	204
10.4	Discussion	210
	10.4.1 Data Issues	210
	10.4.2 RPCA Scaling	210
	10.4.3 Geometric Similarity, Metabolic Scaling, or Fracture Scaling	212
	10.4.4 Intergroup Comparisons	214
	References	215

10.1 Introduction

In mammals, the functions of mastication (chewing) are to increase the surface area of food so as to increase the rate of digestion, to mechanically break down cell walls so that enzymes can digest intracellular contents, increasing the efficiency of digestion, and to break food into pieces that can be swallowed (Lucas, 2004). Food is broken down by the application of forces generated by the chewing muscles and transmitted by the bones of the skull to the teeth. The magnitude of the stress generated in the food (bite stress) is a function of a number of factors, including the size and arrangement (i.e., muscle and tooth lever arms) of the muscles and teeth, the location of the bite point, tooth size and shape, and the extent to which muscle force is recruited by the central nervous system (CNS). The relative importance of these various influences on bite stress in different primates is unknown. Primarily this is because, although the scaling of dental and skeletal dimensions is relatively well documented, the scaling of muscle physiological parameters and general principles of masticatory muscle force recruitment by the CNS are unknown.

F. Anapol
Department of Anthropology, University of Wisconsin-Milwaukee, P.O. Box 413, Sabin Hall, Milwaukee, WI 53211
e-mail: fred@uwm.edu

Here we present our data on the scaling of one reasonable estimator of muscle force potential: the physiologically related morphological variable, *reduced physiologic cross-sectional area* (RPCA). This variable combines muscle mass with the length and angle of pinnation of the constituent fibers to derive a representative estimate of the maximum force deliverable by a quantity of muscle. In this study we ask whether RPCA changes with body size according to the predictions of geometric similarity, metabolic demands, or fracture scaling.

Geometric similarity: If the primate feeding system scales with geometric similarity, then muscle RPCA will scale relative to body mass (M) to the power of 0.67; i.e., RPCA $\alpha M^{.67}$, lever arms will scale $\alpha M^{0.33}$, tooth area will scale $\alpha M^{0.67}$, and bite stress (force/area) will be independent of size (muscle force $\alpha M^{0.67}$/tooth area α $M^{0.67} = M^{0.0}$). Thus, if the feeding system scales with geometric similarity, relative muscle cross-sectional area and relative tooth area will increase at a lesser rate than increasing body mass, although bite stress will be invariant with respect to body mass.

Metabolic scaling: Resting metabolic rate scales $\alpha M^{0.75}$ (Kleiber, 1961; Peters, 1983; Calder, 1984; Schmidt-Nielsen, 1984; Savage et al., 2004; West and Brown, 2005), and it has been suggested that dental dimensions of mammals should scale with metabolic rate, i.e., $\alpha M^{0.75}$ in order for them to acquire the necessary amount of food per unit time (Pilbeam and Gould, 1974, 1975; Gould, 1975). This hypothesis assumes that the volume of food processed per chew increases isometrically with tooth surface area; i.e., $\alpha M^{0.75}$, if that is indeed how tooth surface area scales. However, subsequent work has revealed that geometric similarity and not positive allometry of dental dimensions is the rule for mammals (α $M^{0.67}$) (Kay, 1975; Fortelius, 1985; Lucas, 2004; Vinyard and Hanna, 2005), demanding an explanation for how larger animals acquire enough energy. Larger animals might spend relatively more time feeding every day, take in more energetically rich foods, process more food per chew (e.g., food volume processed might scale $\alpha M^{1.0}$), and/or extract energy more efficiently to meet their metabolic needs (Kay, 1985; Fortelius, 1985). One way to extract energy more quickly and efficiently would be to increase bite stress as body size increases, and this could be achieved by increasing muscle force faster than tooth area: e.g., RPCA α tooth area$^{>1.0}$, where tooth area $\alpha M^{>0.67}$.

Fracture scaling: Lucas (2004) has suggested an alternative model of scaling of the feeding system based on the proposition that larger particles of food fracture at smaller stresses than smaller particles. Consequently, if larger animals eat larger food particles than smaller animals, the amount of stress needed to fracture the larger particles will decrease. Specifically, Lucas predicts that bite forces, and therefore RPCA, need only increase α $M^{0.5}$ in geometrically similar animals, an exponent substantially less than the geometric scaling prediction of $M^{0.67}$. In essence, according to the fracture scaling, larger animals require proportionally smaller masticatory muscles to eat larger particles of the same kind of foods as their smaller counterparts.

In this paper, we present an analysis of scaling patterns for RPCA of chewing muscles in a wide variety of primates. The geometric similarity, metabolic and fracture scaling hypotheses are tested using both published body weights and a directly measured size surrogate, craniobasal length, as independent variables. We also evaluate the possibility that scaling patterns of RPCA differ between prosimians, platyrrhines, and catarrhines.

10.2 Materials and Methods

The sample consisted of 27 species of primates—prosimians, platyrrhines, cercopithecoid monkeys, and lesser apes. Three jaw elevator muscles—masseter, temporalis, and medial pterygoid—were examined. The muscle samples used in this study were dissected by CFR from preserved specimens in various museum and private collections between 1991 and 1993 and, although stored in 30% ethanol when received, were grossly titrated into buffered formalin (10%) at University of Wisconsin-Milwaukee (UWM) (see below). The skull, or craniobasal, length was measured on each specimen. Descriptions of the gross anatomy of the muscles in most of the species were published by Ross (1995), along with a preliminary account of the scaling of muscle mass.

The specimens were subsequently moved to UWM where RPCA was determined (Shahnoor, 2004). Muscle sampling and measurement protocols have been detailed previously in Anapol (1984), Anapol and Jungers (1986), Anapol and Barry (1996), Anapol and Gray (2003), and Anapol et al. (2004). Measurements of in situ sarcomere lengths, derived using histologic techniques, enabled restoration of in situ fiber lengths and pinnation angles to putative resting values (Anapol and Barry, 1996; Anapol and Gray, 2003; Anapol et al., 2004). These corrected variables were combined with muscle weights to calculate RPCA, using the formula of Schumacher (1961), after Weber (1851), and adjusted by Haxton (1944):

$$\text{RPCA (cm}^2) = [\text{mass (gm)} \times \cos\theta]/[l_\text{f} \text{ (cm)} \times \text{specific density}],$$

where l_f and θ represent resting fiber length and angle of pinnation of that fiber, respectively, and the specific density of muscle is $1.0564\,\text{gm/cm}^3$ (Murphy and Beardsley, 1974).

Following (natural) log transformation of the data, Pearson's product—moment correlation coefficients r, and least squares linear regression (LSR) equations (Sokal and Rohlf, 1981a) were performed for RPCA on both the craniobasal length (CBL) measured on the specimens from which the muscles were dissected (Ross, 1995) and the cube root of body weight (M) taken from the literature (Smith and Jungers, 1997). M and RPCA were converted to $M^{1/3}$ and RPCA$^{1/2}$, so that isometry values would equal 1.0. Reduced major axis (RMA) slopes were calculated by dividing each LSR slope by its corresponding correlation coefficient. Confidence intervals were computed for LSR slopes and intercepts by multiplying each standard error by the table (Students' T-test; Sokal and Rohlf, 1981b) value for 0.025 at the error degrees of freedom. The standard error of the RMA slope approximates that of the LSR slope (Sokal and Rohlf, 1981), enabling use of the same confidence interval. After testing for homogeneity of slopes, ANCOVAs were performed on RPCA$^{1/2}$s for each muscle separately and all muscles pooled with CBL or $M^{1/3}$ as the covariate. Groups were prosimians, platyrrhines, and catarrhines, or prosimians and anthropoids. Most results were computed using Statistical Analysis Sofware (SAS, SAS Institute, Cary, NC) on the University of Wisconsin-Milwaukee IBM mainframe UWM-3270 computer. ANCOVAs were performed using SPSS.

10.3 Results

The mean variables used in the regressions in this study are presented in Table 10.1. This includes published body weights (Smith and Jungers, 1997) and means of craniobasal lengths and RPCA of each muscle (masseter, temporalis, medial pterygoid) separated by species. In Table 10.2, Pearson's product-moment correlation coefficients (r), plus the probability that r is significantly different from 0, are shown for $M^{1/3} \times$ CBL, RPCA$^{1/2} \times M^{1/3}$, and RPCA$^{1/2} \times$ CBL. Published body weights are highly correlated with measured CBLs across all primates studied, and across prosimians and platyrrhines, but are not significantly correlated in catarrhines ($r = 0.63; P \leq 0.18$) (Table 10.2a). With the exception of catarrhines, correlation coefficients are similar for RPCA$^{1/2} \times$ either $M^{1/3}$ or CBL. In catarrhines, correlations between $M^{1/3}$ and RPCA$^{1/2}$ of muscles as a whole, masseter, and medial pterygoid are as strong or stronger than those between RPCA$^{1/2}$ and CBL (Table 10.2b and c), despite the fact that CBL was measured directly on the skulls upon which these muscles had been attached and M was taken from the literature. For RPCA$^{1/2} \times$ CBL, the probability that r is greater than zero (P $\neq 0.00$) exceeds 0.05 in catarrhines only for muscles as a whole, masseter, and medial pterygoid.

Table 10.3 contains regression statistics (isometry = 1.0) for CBL vs $M^{1/3}$, and for RPCA$^{1/2}$ vs $M^{1/3}$, and RPCA$^{1/2}$ vs CBL. Results are presented for the three muscles separately and for the RPCA pooled within a species. The data were analyzed across all primates and subdivided into prosimians, platyrrhines, and catarrhines. The confidence intervals are broad, such that, even when differences between groups are dramatic, their statistical significance cannot be proven. Nevertheless, discussion of some general patterns is possible and provides hypotheses for future testing.

When CBL is regressed on $M^{1/3}$ (Table 10.3a), both LSR and RMA slopes indicate negative allometry (< 1.0) for primates as a whole, prosimians, platyrrhines, and catarrhines. Thus, in this primate sample, skull length increases at a slower rate than body weight.

When RPCA is regressed on published body mass, regression slopes across primates as a whole suggest negative allometry for the RPCA of each muscle individually and for the RPCA of all muscles pooled (Fig. 10.1, Table 10.3b). In contrast, within prosimians, platyrrhines, and catarrhines, pooled muscle RPCAs—and in most cases individual muscles—scale with positive allometry. The difference between intra-group and all-primate results is due to transposition of intra-group lines, with prosimians having the higher and platyrrhines the lower intercept values (Fig. 10.1). Slopes for all primates pooled suggest positive allometry for all muscles combined and for individual muscles. In prosimians, slope estimates are

Table 10.1 Mean variables of species used in the regressions. Each genus comprises >1 species, except *Callithrix* and *Colobus* where two separate species are listed

Genera	TYPE	M (g)	N	CBL (mm)	MXSEC (cm^2)	TXSEC (cm^2)	MPXSEC (cm^2)
Eulemur	PRO	2215	3	91.77	1.36	2.09	0.64
Galago	PRO	62	4	35.98	0.17	0.2	0.09
Hapalemur	PRO	945	2	66.4	1.22	1.15	0.63
Otolemur	PRO	1150	3	81.93	3.53	5.2	1.09
Potto	PRO	833	3	62.57	1.3	1.01	0.57
Tarsier	PRO	123	2	38.15	0.33	0.35	0.23
Alouatta	PLA	5950	2	108.2	1.94	2.18	1.44
Aotus	PLA	775	4	63.00	0.66	0.63	0.23
Ateles	PLA	8775	4	122.08	2.56	3.01	1.63
Callicebus	PLA	988	4	62.45	0.26	0.4	0.14
Callimico	PLA	484	3	53.9	0.45	0.7	0.2
Callithrix argentata	PLA	345	4	45.45	0.19	0.27	0.09
Callithrix jacchus	PLA	372	5	43.7	0.27	0.39	0.1
Cebuella	PLA	116	3	34.13	0.24	0.31	0.11
Cebus	PLA	3085	4	87.93	1.67	2.83	0.64
Chiropotes	PLA	2740	5	83.66	2.00	4.21	1.45
Lagothrix	PLA	7150	4	107.95	1.47	1.9	0.43
Leontopithecus	PLA	609	2	57.8	0.73	1.17	0.28
Pithecia	PLA	1760	2	77.65	1.33	1.63	0.84
Saguinus	PLA	411	3	47.7	0.31	0.25	0.25
Saimiri	PLA	721	5	62.6	0.11	0.24	0.09
Colobus badius	CAT	8285	4	103.78	4.36	5.18	1.74
Colobus polykomos	CAT	9100	3	113.93	3.43	2.82	1.53
Hylobates	CAT	5795	1	118.4	1.91	2.95	1.03
Miopithecus	CAT	1250	4	71.33	0.16	0.31	0.17
Presbytis	CAT	17000	3	89.57	2.35	2.12	1.16
Symphalangus	CAT	11300	1	128.5	1.29	3.52	1.1

Abbreviations= PRO, *prosimians;* PLA, *platyrrhines,* CAT, *catarrhines*; M, mass; CBL, craniobasal length; MXSEC, reduced physiologic cross-sectional area of masseter; TXSEC, reduced physiologic cross-sectional area of temporalis; MPXSEC, reduced physiologic cross-sectional area of medial pterygoid.

positively allometric for all muscles combined, masseter, and temporalis, while negatively allometric for the medial pterygoid. In platyrrhines, slopes for all muscles pooled, masseter, and temporalis are close to the values for prosimians, but that of medial pterygoid is much greater than in prosimians. In catarrhines, slopes of muscles pooled and taken individually are strongly positive with masseter exhibiting a steeper slope than temporalis, which in turn scales with a steeper slope than the medial pterygoid.

When RPCA$^{1/2}$ is regressed on CBL (Table 10.3c), all slopes are higher than when RPCA$^{1/2}$ is regressed on $M^{1/3}$ and all are positively allometric (Fig. 10.2, Table 10.2b). In prosimians, platyrrhines, and catarrhines, the absolute values of the slopes differ, but the slope values for masseter, temporalis, and medial pterygoid are more or less the same for both LS and RMA regressions, regardless of which independent variable is chosen. Slope estimates relative to both CBL and M are higher across catarrhines than for both prosimians and platyrrhines.

Table 10.2 Pearson's product–moment correlation coefficients between (**a**) body weight$^{1/3}$ (M) × craniobasal length (CBL), (**b**) RPCA$^{1/2}$ × body weight$^{1/3}$, and (**c**) RPCA$^{1/2}$ × craniobasal length

a.

	Primates	PRO	PLA	CAT
$M \times$ CBL	0.96	0.98	0.99	0.63
($Pr = 0.0$)	(0.00)	(0.00)	(0.00)	(0.18)

b.

	Mm.	Masseter	Temporalis	Med Pterygoid
Primates	0.81	0.77	0.81	0.82
($Pr = 0.0$)	(0.00)	(0.00)	(0.00)	(0.00)
PRO	0.91	0.91	0.89	0.93
($Pr = 0.0$)	(0.01)	(0.01)	(0.02)	(0.01)
PLA	0.85	0.83	0.83	0.83
($Pr = 0.0$)	(0.00)	(0.00)	(0.00)	(0.00)
CAT	0.86	0.84	0.82	0.89
($Pr = 0.0$)	(0.03)	(0.04)	(0.04)	(0.02)

c.

	Mm.	Masseter	Temporalis	Med Pterygoid
Primates	0.84	0.79	0.85	0.84
($Pr = 0.0$)	(0.00)	(0.00)	(0.00)	(0.00)
PRO	0.93	0.91	0.93	0.91
($Pr = 0.0$)	(0.01)	(0.01)	(0.01)	(0.01)
PLA	0.84	0.83	0.83	0.83
($Pr = 0.0$)	(0.00)	(0.00)	(0.00)	(0.00)
CAT	0.77	0.66	0.84	0.76
($Pr = 0.0$)	(0.07)	(0.15)	(0.04)	(0.08)

For abbreviations see Table 10.1.

The rank order of slope values among taxa is the same, regardless of the independent variable used: for muscles pooled, masseter, and temporalis: platyrrhines ≤ prosimians < catarrhines; while for medial pterygoid: prosimians < platyrrhines < catarrhines. However, when RPCA pooled across all muscles is regressed against M or CBL, tests of homogeneity of slopes reveal the slopes are not significantly different, allowing ANCOVA to test for differences in elevation of the regression lines. ANCOVA reveals that there is an effect of taxon membership on the mean RPCA when controlling for body mass, with prosimians having larger pooled RPCA than platyrrhines or catarrhines, but these effects are not significant for RPCA vs CBL. When platyrrhines and catarrhines are pooled together as anthropoids, homogeneity of slopes is confirmed for RPCA vs body mass, and ANCOVA again confirms that prosimians have larger pooled RPCA than anthropoids when controlling for body mass.

ANCOVAs were also performed on the data from individual muscles to ask whether prosimians have larger RPCAs than anthropoids when CBL and M are controlled. These analyses revealed that prosimians do not have larger RPCAs than anthropoids at similar CBLs, but that prosimians do have larger RPCAs for temporalis, masseter, and medial pterygoid than the anthropoids with similar M.

Table 10.3 Regression statistics for linear regressions of (a) craniobasal length on body weight$^{1/3}$, (b) RPCA$^{1/2}$ on body weight$^{1/3}$, and (c) RPCA$^{1/2}$ on craniobasal length

a.

	Primates				PRO				PLA				CAT			
	Slope		Intercept	r^2	Slope		Intercept	r^2	Slope		Intercept	r^2	Slope		Intercept	r^2
	LSR	RMA	LSR		LSR	RMA	LSR		LSR	RMA	LSR		LSR	RMA	LSR	
Estimate	0.75	0.78	2.45	0.92	0.81	0.83	2.40	0.96	0.87	0.87	2.15	0.99	0.45	0.71	3.31	0.40
Std. Err.	(0.04)		(0.11)		(0.08)		(0.17)		(0.03)		(0.07)		(0.28)		(0.82)	
95% CI	Slope ± 0.08		±0.23		Slope ± 0.22				Slope± 0.06				Slope ± 0.78			

b.

	All muscles				Masseter				Temporalis				Med Pterygoid			
	Slope		Intercept	r^2	Slope		Intercept	r^2	Slope		Intercept	r^2	Slope		Intercept	r^2
	LSR	RMA	LSR		LSR	RMA	LSR		LSR	RMA	LSR		LSR	RMA	LSR	
Primates	0.86	1.06	−1.64	0.66	0.84	1.08	−2.14	0.60	0.87	1.07	−2.06	0.66	0.85	1.04	−2.48	0.67
(std. err.)	(0.12)		(0.30)		(0.14)		(0.34)		(0.13)		(0.31)		(0.12)		(0.29)	
CI	Slope ± 0.25				Slope ± 0.29				Slope ± 0.27				Slope ± 0.25			
PRO	1.06	1.16	−1.78	0.83	1.06	1.17	−2.27	0.82	1.12	1.26	−2.33	0.79	0.89	0.96	−2.29	0.86
(std. err.)	(0.24)		(0.52)		(0.25)		(0.52)		(0.29)		(0.61)		(0.18)		(0.38)	
CI	Slope ± 0.67				Slope ± 0.69				Slope ± 0.81				Slope ± 0.50			
PLA	1.00	1.18	−2.05	0.72	0.99	1.19	−2.56	0.69	1.00	1.20	−2.41	0.70	1.03	1.24	−2.99	0.69
(std. err.)	(0.17)		(0.42)		(0.18)		(0.44)		(0.18)		(0.43)		(0.19)		(0.45)	
CI	Slope ± 0.37				Slope ± 0.39				Slope ±0.39				Slope ± 0.41			
CAT	1.44	1.67	−3.46	0.74	1.65	1.97	−4.65	0.70	1.36	1.65	−3.61	0.68	1.25	1.41	−3.71	0.79
(std. err.)	(0.43)		(1.27)		(0.54)		(1.59)		(0.46)		(1.37)		(0.33)		(0.96)	
CI	Slope ± 1.19				Slope ± 1.50				Slope ± 1.28				Slope ± 0.92			

Table 10.3 (continued)

c.

	All muscles			Masseter			Temporalis			Med Pterygoid		
	Slope	Intercept	r^2	Slope	Intercept	r^2	Slope	Intercept	r^2	Slope	Intercept	r^2
	LSR RMA	LSR		LSR RMA	LSR		LSR RMA	LSR		LSR RMA		
Primates	1.13 1.35	−4.39	0.70	1.10 1.40	−4.78	0.62	1.16 1.37	−4.91	0.72	1.11 1.33	−5.16	0.70
(std. err.)	(0.15)	(0.63)		(0.17)	(0.73)		(0.15)	(0.62)		(0.14)	(0.62)	
CI	Slope ± 0.31			Slope ± 0.35			Slope ± 0.31			Slope ± 0.29		
PRO	1.31 1.41	−4.92	0.86	1.29 1.42	−5.34	0.83	1.42 1.52	−5.79	0.87	1.05 1.16	−4.73	0.82
(std. err.)	(0.26)	(1.08)		(0.29)	(1.18)		(0.28)	(1.13)		(0.25)	(1.01)	
CI	Slope ± 0.72			Slope ± 0.81			Slope ± 0.78			Slope ± 0.69		
PLA	1.14 1.35	−4.46	0.71	1.12 1.36	−4.94	0.68	1.13 1.36	−4.82	0.69	1.18 1.41	−5.51	0.70
(std. err.)	(0.20)	(0.84)		(0.21)	(0.89)		(0.21)	(0.88)		(0.22)	(0.91)	
CI	Slope ± 0.43			Slope ± 0.45			Slope ± 0.45			Slope ± 0.48		
CAT	1.82 2.35	−7.64	0.60	1.83 2.76	−8.25	0.44	1.93 2.29	−8.57	0.71	1.50 1.97	−6.99	0.58
(std. err.)	(0.74)	(3.44)		(1.03)	(4.79)		(0.63)	(2.90)		(0.64)	(2.95)	
CI	Slope ± 2.05			Slope ± 2.86			Slope ± 1.75			Slope ± 1.78		

Abbreviations: LSR, least squares; RMA, reduced major axis; std. err., standard error; 95% CI, confidence interval. For additional abbreviations see Table 10.1.

10 Scaling of Reduced Physiologic Cross-Sectional Area

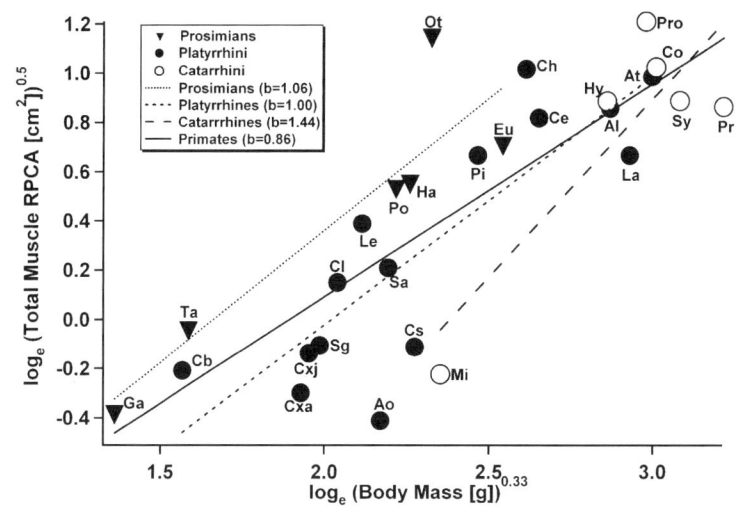

Fig. 10.1 Bivariate plot of $\log_e$ of the square root of the sum of all muscle's RPCA cm^2 against $\log_e$ cube root of body mass (g). Least squares regression lines for prosimians, platyrrhines, catarrhines, and all primates are shown. The slope values are given in the figure legend. ANCOVA reveals an effect of taxon group on relative RPCA: prosimians have larger RPCAs than the anthropoids at equivalent body masses

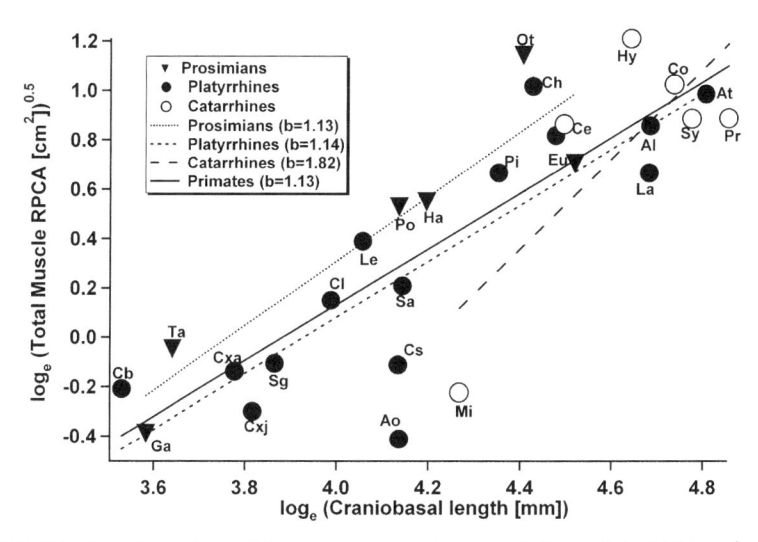

Fig. 10.2 Bivariate plot of $\log_e$ of the square root of the sum of all muscle's RPCA cm^2 against $\log_e$ of craniobasal length (CBL). Least squares regression lines for prosimians, platyrrhines, catarrhines, and all primates are shown. The slope values are given in the figure legend. ANCOVA reveals no effect of taxon group on relative RPCA: prosimians do not have larger RPCAs than the anthropoids at equivalent CBLs

10.4 Discussion

10.4.1 Data Issues

Certain limitations of the data set presented here must be acknowledged. Sample size and taxonomic sampling in studies of muscle anatomy are rarely as satisfactory as one would like. The current array is no exception, being skewed strongly toward platyrrhine species, and having low sample sizes within species. Cercopithecine primates in particular are poorly represented, and catarrhines are only represented by a cluster of similarly sized large animals and one small species, *Miopithecus talapoin* (Fig. 10.1). In addition, the specimens from which these muscles were collected come from diverse sources with diverse histories, including length of time since death of the animal. Moreover, some of the specimens lived in captivity and some lived in the wild. Whether these factors affect the results presented here cannot be determined at present. Finally, reduced physiologic cross-sectional area (RPCA) is a composite variable, derived from several measurements extracted from muscle tissue. Although great care is taken to standardize control measurement protocols, each of these raw variables—sarcomere length, muscle weight, fasciculus length, and angle of pinnation—is subject to measurement error. These limitations should be addressed in future work.

10.4.2 RPCA Scaling

RPCA is a physiologically relevant morphological variable in that it provides an estimate of maximum muscle force for a whole muscle, or a specific portion from in situ measurements (Gans and Bock, 1965; Gans, 1982). To estimate RPCA, muscle weight is converted to a volume by dividing it by the specific gravity of muscle. The cross-sectional area of the muscle volume is then estimated by dividing the muscle volume by the length of muscle fibers or fasciculi. Multiplying this value by cosine of the angle of pinnation isolates the proportion of the cross-sectional area contributing to whole muscle shortening (an angular adjustment that is not required for parallel-fibered muscles.)

In this study, in situ measurements of fasciculus (rather than fiber) length were used as the functional contractile length by which muscle weight was divided (Alexander, 1974). We argue that this is appropriate because, as other authors have demonstrated, what is ordinarily considered to be a muscle fiber extending between attachment sites (either tendon or bone) is often a series of shorter fibers with overlapping ends (Huxley, 1957; Fawcett, 1986; Gaunt and Gans, 1992). If the muscle is chemically treated so as to free-up individual cells from their fasciculi, the released individual fibers would often be shorter than the inter-attachment length, artificially inflating the calculation of cross-sectional area by reducing l_f in the equation. Moreover, chemical digestion of the muscle precludes measurement of the angle of pinnation. (We note that if, in an attempt to preserve pinnation angles, the muscle was sectioned before maceration, fibers would likely be severed, reducing their lengths and also inflating the calculated RPCA.) Thus, by using fasciculus rather than fiber length, not only more accurate estimates of cross-sectional area are produced but also pinnation angles can be measured, allowing the calculation of a more physiologically relevant RPCA.

In general, our results suggest positive allometry of RPCA within primate subgroups: i.e., prosimians, platyrrhines, and catarrhines. When all LSR slopes are examined, only the slope of masseter $RPCA^{1/2}$ against $M^{1/3}$ in platyrrhines (slope = 0.99) and the slope of medial pterygoid $RPCA^{1/2}$ against $M^{1/3}$ in prosimians (slope = 0.89) are less than 1.0. Among RMA slopes, only the slope of medial pterygoid $RPCA^{1/2}$ against $M^{1/3}$ in prosimians (slope = 0.96) is less than 1.0. All regression slopes of $RPCA^{1/2}$ against $CBL^{1/3}$ are greater than 1.0.

The regression equations for catarrhines suggest stronger positive allometry than the other groups, but examination of Figs. 10.1 and 10.2 suggest that these equations should be viewed with skepticism. The RPCAs for the smallest catarrhine for which we have data, *Miopithecus talapoin*, fall well below the regression lines for the platyrrhines and prosimians, yet the larger catarrhines fall in amongst the larger platyrrhines in our sample. Although we cannot say for sure that the *Miopithecus* data are invalid for some reason, we suspect that they are. Because *Miopithecus* is so much smaller than the rest of the catarrhine sample, its low RPCA is driving the catarrhine regression values. We retained *Miopithecus* in the calculation of the primate regression equations because its body mass and CBL are close to the mean values of the sample, so its seemingly anomalous RPCAs have little effect on the slopes of these regression lines.

Different scaling relationships were documented depending on whether we examined scaling of the RPCA relative to body mass data from the literature (M) or skull length measured from the specimens (CBL). Regressions of RPCA on CBL were associated with steeper, more positive slopes than the regressions on body mass because CBL scales with negative allometry against body weight across primates as a whole, as well as within prosimians, platyrrhines, and catarrhines. The finding that skull length is increasing at a slower rate than body weight across groups and for primates as a whole is no surprise, since brain and related braincase size scale negatively allometric to body size (Martin, 1990). The vast departure in slope values by catarrhines may reflect different scaling coefficients at larger body sizes (catarrhines have, on average, larger body sizes than the other groups). Alternately, the slope value calculated for the catarrhines studied here may not be representative of the true scaling relationships within the group. Not only are CBL and M less tightly correlated in catarrhines ($r = 0.63$) than in prosimians ($r = 0.98$) and platyrrhines ($r = 0.99$), but catarrhines show higher correlation coefficients between RPCA and body mass than between RPCA and CBL, the reverse of the situation in prosimians and platyrrhines, and the confidence intervals of the slope values for all scaling analyses within catarrhines are much larger than those for the other groups.

The allometry of skull length against body mass begs the question: which independent variable should RPCA be regressed against to test the hypotheses presented in the Introduction? If body mass is seen as a logical choice for evaluating the metabolic scaling hypothesis, then the observation that pooled RPCA$^{1/2}$ scales at slopes higher than 1.00 within prosimians, platyrrhines, and catarrhines (Table 10.3b, Fig. 10.1) suggests that increasing relative bite force might indeed be one way that primates compensate for the fact that tooth area scales more slowly relative to body mass than the metabolic rate. However, RPCA is converted to bite stress via the masticatory system, which lies within the head, suggesting that scaling of RPCA against CBL might more accurately reflect the effect of RPCA scaling on bite stress. The observation that RPCA scales with strong positive allometry relative to CBL in all primate groups (Table 10.3c, Fig. 10.2) confirms the suspicion that positive allometry of RPCA relative to body size and size of the feeding system might be a strategy whereby primates increase food energy intake rate.

Whether the scaling relationships of the chewing muscles are computed relative to CBL or to M has little effect on the interpretation of the inter-specific scaling results: the relative values among taxonomic groups essentially are the same. The one exception would be in the scaling relationships of masseter between prosimians and platyrrhines. Relative to $M^{1/3}$, masseter RPCA$^{1/2}$ scales with almost identical slopes (1.17 and 1.19) in the two clades; using CBL, platyrrhines have the lowest allometric slope of the three taxa, although slope ranges for prosimians and platyrrhines are slightly overlapping.

10.4.3 Geometric Similarity, Metabolic Scaling, or Fracture Scaling

Geometric similarity. The data currently available, e.g., that presented in Table 10.4, do not definitively exclude the possibility that jaw adductor RPCA scales with geometric similarity relative to body mass. Confidence intervals are very broad, so that whether LSR or RMA is used, slopes of 1.00 lie within the 95% CI. However, we believe that an argument for positive allometry of RPCA relative to both body mass and skull length is more likely to be correct in the long term. RMA slopes are most

Table 10.4 Review of previous published scaling studies relevant to the data presented in this paper

Variable	Taxon	Expectation of isometry in listed papers	Published slopes of variable vs. body mass	Published slopes converted to isometric slope value = 1.0	Reference
Chewing Frequency	Mammals	No expectation	−0.13 (RMA −0.13)	No expectation	Druzinsky, 1993
Jaw Length	Mammals	0.33	0.31	0.93	Druzinsky, 1993
Muscle PCA	Mammals	0.67	0.50 Theoretical	0.75	Lucas, 2005
Jaw-Muscle PCA	Humans	0.67	0.78	1.17	Weijs and Hillen, 1985
Total jaw adductor RPCA	Prosimians	0.67	0.77 (RMA 0.83)	1.16 (RMA 1.25)	Perry and Wall, 2005
Metabolic Rate	Mammals	1.00	0.75 Empirical	0.75	Kleiber, 1961
Total Jaw Adductors	Rodents	0.67	0.78 (RMA 0.79)	1.17 (RMA 1.19)	Druzinsky, 1993

appropriate because the body masses upon which these analyses are based derive from the literature, and not from the specimens from which muscles were taken, introducing an error in the body mass estimates of unknown relative magnitude. If RMA is accepted to be the most appropriate regression model, then the only muscle that exhibits slope values less than one for RPCA$^{1/2}$ vs $M^{1/3}$ is the medial pterygoid of prosimians (slope = 0.96). However, even if LSR is regarded as the more appropriate regression model, only the masseter RPCA$^{1/2}$ of platyrrhines, and the regressions calculated across all primates, scales against $M^{1/3}$ with a slope < 1.0. Positive allometry is also consistent with other studies of physiologic cross-sectional area, reduced (Perry and Wall, 2005) or otherwise (Weijs and Hillen, 1985), which indicates positive allometry of muscle cross-sectional area to body mass. Thus, we argue that positive allometry of jaw-muscle RPCA is a reasonable hypothesis at present.

Metabolic scaling. Positive allometry of jaw-muscle RPCA is consistent with the metabolic scaling models summarized above, in which tooth area $\alpha M^{0.67}$, metabolic rate $\alpha M^{0.75}$, and positive allometry of RPCA compensates for this discrepancy by increasing bite stress with body size, allowing more food to be processed per chew. If positive allometry of RPCA did indeed compensate for decreases in tooth surface area relative to metabolic rate, then RPCA would scale $\alpha M^{1.126}$ (= $\alpha\ M^{0.75/0.67}$), a figure close to the slope values found among prosimians and platyrrhines in this study, and for prosimians by Perry and Wall (2005). However, despite the similarities of these slope values, we hesitate to claim that this provides strong support for the idea that positive allometry of RPCA relative to body mass compensates for decreasing tooth surface area relative to metabolic rate. First, the confidence intervals of the slope values are very large, including a wide range of plausible slopes. Second, when scaling of RPCA relative to skull length is considered, the slope values are much higher than 1.12. The feeding system is largely housed in the head, suggesting that relative RPCA might be increasing faster than required by this metabolic model, generating even higher bite stresses than predicted. Two possible explanations for such strong positive allometry of RPCA (and estimated bite stress) are worth considering. First, it has been argued that larger primates eat "tougher" or "more obdurate" foods than smaller primates. Larger (>1000 g) primates obtain their protein from eating leaves (Kay and Simons, 1980), a behavior usually associated with increased molar occlusal surface area, e.g., as observed in *Alouatta* (Anapol and Lee, 1994). Since body size increases to length cubed, while surface area increases by length squared, the physiological cross-sectional area of masticatory muscle in animals on a comparatively tougher folivorous diet would have to increase faster than body size in order to provide enough chewing energy for the increasing diet of a larger animal. Second, size-related decreases in chewing speed also might be related to positive allometry of RPCA. Chewing cycle duration, the inverse of chewing frequency, scales positively with jaw length across mammals as a whole (Druzinsky, 1993) and within primates. Variation in chewing speed is influenced by many factors, including food type and, presumably, interspecific differences in fiber type composition of the masticatory muscles, but the primary determinant of chewing speed is body size: larger primates chew more slowly than smaller primates (Ross et al., 2008). Larger animals must therefore meet increasing demands for energy in the face of decreasing chewing speed. One way to compensate for slowing chew speed would be to increase bite stress and use it to process more food per chew.

Fracture scaling. Our data definitively reject the fracture scaling model of Lucas (2004). Lucas hypothesizes that the cross-sectional area of the chewing muscles scales relative to $M^{0.50}$; i.e., with negative allometry. Using $RPCA^{1/2}$ and $M^{1/3}$, fracture scaling predicts a slope of $M^{0.75}$. In contrast, our data suggest that RPCA increases more rapidly relative to body mass than predicted by fracture scaling, with RMA slopes ranging from 0.96 to 1.97, and LSR slopes ranging from 0.85 to 1.65, depending on the muscle and taxon. The 95% confidence intervals of the catarrhine and prosimian slopes include 0.75 (all muscles combined or separate), as do those of the LSR and platyrrhine and primate LSR equations. However, we have already given our reasons for being suspicious of the catarrhine slopes and for believing that the RMA slope calculations are more appropriate than LSR. Moreover, Perry and Wall (2005) also report higher slopes for the scaling of jaw adductor muscle mass relative to body mass than predicted for fracture scaling. We therefore think it unlikely that fracture scaling explains the scaling primate jaw-muscle RPCAs. There are various assumptions underlying the fracture scaling model, any of which might be undermining the applicability of the model: that food toughness remains constant at different body sizes, that foods are homogeneous, and that foods have linear stress–strain curves up to yielding. Violation of any of these assumptions could result in a deviation from the predictions of fracture scaling.

10.4.4 Intergroup Comparisons

Taken at face value, our regression results suggest clade-level differences in RPCA scaling, with catarrhine slopes being steeper than platyrrhine and prosimian slopes, and prosimian slopes being steeper than those of platyrrhines. However, as discussed above, we are hesitant to attach biological significance to the steep catarrhine slopes at present; and, moreover, tests for homogeneity of slopes do not reveal that these slopes are significantly different. The ANCOVA results do reveal that prosimians have larger RPCAs than anthropoid primates when controlling for body mass, but not when controlling for CBL. The reason for this is revealed by ANCOVA of CBL vs M, which reveals that prosimians have larger CBLs for their M than anthropoids. Prosimians do have larger masticatory muscle RPCAs relative to body mass than anthropoids, but this effect is not separable from the effect of general increases in head length in prosimians vs anthropoids.

Acknowledgments We heartily thank Chris Vinyard, Matt Ravosa, and Chris Wall for inviting us to contribute to this volume in honor of Bill Hylander—a great colleague, excellent scientist, and perpetually (sic; except, of course, in 2005) frustrated, yet faithful, fan of the Chicago White Sox. We also tremendously appreciate the generosity of Paula Jenkins and the British Museum of Natural History and of Bill Stanley and the Field Museum of Natural History, the Duke University

Primate Center, and the Neil Tappen Collection at the University of Wisconsin-Milwaukee. Many thanks are owed to Frank Stetzer for assistance with the statistical analysis.

References

Alexander, R. McN. (1974) The mechanics of jumping by a dog (*Canis familiaris*). *J. Zool. Lond.* 173:549–573.
Anapol, F. (1984) Morphological and functional diversity within the quadriceps femoris in *Lemur fulvus*: Architectural, histochemical, and electromyographic considerations. Ph. D. dissertation, State University of New York at Stony Brook.
Anapol, F., and Barry, K. (1996) Fiber architecture of the extensors of the hindlimb in semiterrestrial and arboreal guenons. *Amer. J. Phys. Anthropol.* 99(3):429–447.
Anapol, F., and Gray, J. P. (2003) Fiber architecture of the intrinsic muscles of the shoulder and arm in semiterrestrial and arboreal guenons. *Amer. J. Phys. Anthropol.* 122:51–65.
Anapol, F., and Jungers, W. L. (1986) Architectural and histochemical diversity within the quadriceps femoris of the brown lemur (*Lemur fulvus*). *Am. J. Phys Anthropol.* 69:355–375.
Anapol, F., and Lee, S. (1994) Morphological adaptation to diet in platyrrhine primates. *Am. J. Phys. Anthropol.* 94:239–261.
Anapol, F., Shahnoor, N., and Gray, J. P. (2004) Fiber architecture, muscle function, and behavior: Gluteal and hamstring muscles of semiterrestrial and arboreal guenons. In: Anapol, F., German, R.Z., and Jablonski, N. eds. *Shaping Primate Evolution* Cambridge: Cambridge University Press, Chapter 6.
Calder, W. A. (1984) *Size, Function, and Life History*. Dover, Mineola NY
Druzinsky, R. E. (1993) The time allometry of mammalian chwing movements: Chewing frequency scales with body mass in mammals. *J. theor. Biol.* 160:427–440.
Fawcett, D. W. (1986) *Bloom and Fawcett, A Textbook of Histology*, 11th ed. Philadelphia: Saunders.
Fortelius, M. (1985) Ungulate cheek teeth: Developmental, functional and evolutionary interratlations. *Acta Zool. Fenn.* 180:1–76.
Gans, C. (1982) Fiber architecture and muscle function. *Exerc. Spt. Sci. Rev.* 10:160–207.
Gans, C., and Bock W. F. (1965) The functional significance of muscle architecture—a theoretical analysis. *Ergeb. Anat. Entwicklungsgesch.* 38:115–142.
Gaunt, A. S., and Gans C. (1992) Serially arranged myofibers: An unappreciated variant in muscle architecture. *Experientia* 48:864–868.
Gould, S. J. (1975) On the scaling of tooth size in mammals. *Am. Zool.* 15:351–362.
Haxton, H. A. (1944) Absolute muscle force in the ankle flexors of man. *J. Physiol.* (London). 103:267–273.
Huxley, A. G. (1957) Muscle structure and theories of contraction. *Prog. Biophy. Biophy. Chem.* 4:255–312.
Kay, R. F. (1975). Allometry and early hominids. *Science* 189:61–63.
Kay, R. F. (1985). Dental evidence for the diet of *Australopithecus*. *Ann. Rev. Anthropol.* 14:315–341.
Kay, R. F., and Simons E. L. (1980) The ecology of Oligocene African Anthropoidea. *Int. J. Primatol.* 1(1):21–37.
Kleiber, M. (1961) *The Fire of Life*. New York, Wiley.
Lucas, P. W. (2004) *Dental Functional Morphology*. Cambridge University Press, Cambridge.
Martin, R. D. (1990) *Primate Origins and Evolution: A Phylogenetic Reconstruction*. Princeton University Press, Princeton.
Murphy, R. A., Beardsley, A. C. (1974) Mechanical properties of the cat soleus muscle in situ. *Am. J. Physiol.* 227:1008–1013.
Perry, J. M. G., and Wall, C. E. (2005) Scaling patterns of physiological cross-sectional area of the chewing muscles in prosimians. *Am. J. Phys. Anthropol.* Suppl. 40:170 (abst.).

Peters, R. H. (1983) *The Ecological Implications of Body Size?* Cambridge University Press, Cambridge.
Pilbeam, D., and Gould, S. J. (1974) Size and scaling in human evolution. *Science* 186:892–901.
Pilbeam, D., and Gould, S. J. (1975) Allometry and early hominids. *Science* 189:64.
Ross, C. F. (1995) Muscular and osseous anatomy of the primate anterior temporal fossa and the functions of the postorbital septum. *Am. J. Phys. Anthropol.* 98:275–306.
Ross, C. F., Reed, D. A., Washington, R. L., Eckhardt, A., Anapol, F., and Shahnoor, N. (2008) Scaling of chew cycle duration in Primates. *Am. J. Phys. Anthropol.* (in press).
Savage, V. M., Gillooly, J. F., Woodruff, W. H., West, G. B., Allen, A. P., Enquist, B. J., and Brown, J. H. (2004). The predominance of quarter-power scaling in biology. *Funct. Ecol.* 18:257–282.
Schmidt-Nielsen, K. (1984) *Scaling: Why is Animal Size So Important.* Cambridge University Press, Cambridge.
Schumacher, G H. (1961) *Funktionelle Morphologie der Kaumuskulatur.* Gustav Fischer, Jena. (Trans. By Z. Muhl).
Shahnoor, N. (2004) *Morphological Adaptations to Diet in Primate Masticatory Muscles.* Ph. D. dissertation, University of Wisconsin-Milwaukee.
Smith, R. J., and Jungers, W. L. (1997) Body mass in comparative primatology. *J. Hum. Evol.* 32:523–559.
Sokal, R. R., and Rohlf, F. J. (1981a) *Biometry, 2nd ed.* WH Freeman:San Francisco.
Sokal, R. R., and Rohlf, F. J. (1981b) Statistical Tables, 2nd ed. W. H. Freeman:San Francisco.
Vinyard, C. J., and Hanna, J. (2005). Molar scaling in strepsirrhine primates. *J Hum Evol* 49:241–269.
Weijs, W. A., and Hillen, B. (1985) Cross-sectional areas and estimated intrinsic strength of the human jaw muscles. *Acta Morphol. Neerl. Scand.* 23:267–274.
Weber EF (1851) Uber die Langenverhaltnisse der Fleischfasern der Muskeln im allgemeinen. Ber. K. sachs. Ges. Wiss. Nat. phys. K1:64–86.
West, G. B., and Brown, J. H. (2005). The origin of allometric scaling laws in biology from genomes to ecosystems: towards a quantitative unifying theory of biological structure and organization. *J. Exp. Biol.* 208:1575–1592.

Chapter 11
Scaling of the Chewing Muscles in Prosimians

Jonathan M.G. Perry and Christine E. Wall

Contents

11.1	Introduction	217
	11.1.1 Fiber Architecture	218
	11.1.2 Scaling Predictions	220
11.2	Materials and Methods	221
	11.2.1 The Chewing Muscles	222
	11.2.2 Muscle Mass	223
	11.2.3 Measuring Fiber Length	223
	11.2.4 Calculating PCSA	223
	11.2.5 Measuring Pinnation	224
	11.2.6 Statistical Analysis	225
11.3	Results	225
11.4	Discussion	232
	11.4.1 Cross-Sectional Area, PCSA, RPCSA, and Pinnation	233
	11.4.2 Fiber Length	235
	11.4.3 Muscle Mass	235
	11.4.4 Fracture Scaling	236
11.5	Conclusions	237
	References	238

11.1 Introduction

The jaw adductor muscles provide the input force for breaking down food. How much force a muscle produces is a function of many variables. Two of the most significant variables are the moment arm (or leverage) of a muscle and its cross-sectional area (Hylander, 1975; Weijs and Hillen, 1985). Muscle forces during mastication have been studied from a theoretical perspective by modeling the mandible as a third-class lever during a static bite (e.g., Hylander, 1975; Greaves, 1978; Smith, 1978; Spencer, 1999). Many studies have compared the leverages of the jaw

---

J.M.G. Perry
Department of Biological Anthropology and Anatomy, Box 3170, Duke University Medical Center, Durham, NC 27710
e-mail: jmp20@duke.edu

adductor muscles (e.g., Arendsen de Wolff-Exalto, 1951; Smith and Savage, 1959; Smith and Redford, 1990; Vizcaíno et al., 1998), and electromyographic (EMG) data provide a measure of the timing of activity and the relative force of the jaw adductors during mastication (e.g., Hylander and Johnson, 1985; 1994; Hylander et al., 2004; Wall et al., 2006).

Despite a growing data set on muscle leverages and on EMG activity, there are surprisingly few data on cross-sectional area and fiber architecture for the jaw adductor muscles of primates. Such data are critical for integrating theoretical, morphometric, and in vivo work to better understand the muscle forces produced during mastication. Mass has been reported for the chewing muscles of some nonhuman primates (Schumacher, 1961; Cachel, 1979; 1984; Antón, 1999; 2000; Taylor and Vinyard, 2004). Values for cross-sectional area and fiber length have been reported for three species of macaque (Antón, 1999; 2000) and in a comparison of two callitrichid species (Taylor and Vinyard, 2004). Shahnoor et al. (2005) and Anapol et al. (Chapter 9 in this volume) have reported data on the cross-sectional area for 27 primate species. Their data set includes a few prosimians, but prosimians are not the focus of their study.

Here we report the results of a study of the cross-sectional area, mass, fiber length, and pinnation of the jaw adductor muscles of prosimian primates. We measured these variables in 16 prosimian species that span a large range of body sizes and diets (Table 11.1). The goal of this study is to use these data to understand the scaling patterns of the jaw adductor muscles relative to body mass. This is a first step in comparing the force-producing abilities of the jaw adductors across prosimians. In the future, this data set will be expanded to include several other small- and medium-sized prosimians. The complete data set will focus on biomechanical scaling and will also be used to compare feeding adaptations in the jaw muscles and skulls of extant prosimians to Eocene adapids and subfossil lemurs.

As part of the present study, we are comparing the scaling patterns observed in this sample of prosimians to a set of scaling predictions (based on the fracture properties of food) for how the chewing muscles might scale to body mass (Lucas, 2004). The fracture scaling model of Lucas (2004) is an important first attempt to move beyond the assumptions of geometric similarity in predicting the scaling patterns of the chewing muscles relative to body mass (e.g., Cachel, 1979, 1984).

11.1.1 Fiber Architecture

The force a muscle can produce is proportional to its cross-sectional area. A general formula for cross-sectional area is as follows.

$$\text{Area} = \text{muscle mass}/(\text{average fiber length} \times \text{muscle density}) \quad (11.1)$$

Because much of the mass of a muscle is taken up by sarcomeres in series within a fiber, and because additional length in a muscle fiber does not provide additional force, muscle mass is an inaccurate estimator of muscle force. Even though muscle

Table 11.1 Specimens included in this study

Species	Specimen[a]	Side	Preservation	BM[b](g)	Diet[c]
Avahi laniger laniger	AMNH 170501	L		1085	mostly leaves
Cheirogaleus medius	DUPC 651m	R	70% ethanol	354	fruit and insects
Daubentonia madagascariensis	AMNH 185640	R		742	insect larvae, nuts
Galago senegalensis braccatus	ASU Ef	L	10% formalin	193	mostly insects
	ASU Jf	L			
Galagoides demidoff	NCZP 510f	R		60	mostly insects
	NCZP 507f	L			
Hapalemur griseus	DUPC 1353f	L	Fresh-frozen	940	shoots, stems, leaves
Lemur catta	BAA C2f	L		2207	fruit and leaves
Lepilemur mustelinus leucopus	AMNH 170790m	R	70% ethanol	760	mostly leaves
Mirza coquereli	DUPC 363f	R	Fresh-frozen	320	insects, insect secretions, fruit
Nycticebus coucang	BAA C6m	R		679	mostly insects
Otolemur crassicaudatus	ORPC 1f	L,R	10% formalin	1258	fruits, insects, gum
	ORPC 2f	L,R			
Otolemur garnettii	ASU 1m	L		834	fruits, insects
	Haines Am	L			
Perodicticus potto	AMNH 200640f	R	70% ethanol	1100	fruit and insects
Propithecus verreauxi coquereli	DUPC 6110f	L	Fresh-frozen	3700	leaves, stems, seeds
	DUPC 6560m	R		2780	
Tarsius syrichta	DUPC 89m	R	70% ethanol	140	insects and vertebrates
	AMNH 150143f	R		98	
Varecia variegata rubra	AMNH 201395f	R		3865	mostly fruit

[a] AMNH: American Museum of Natural History Mammalian Collection, ASU: Arizona State University, BAA: Biological Anthropology & Anatomy collection Duke University, DUPC: Duke University Primate Center, Haines: from the collection of Duane Haines, NCZP: North Carolina Zoological Park. All galago specimens courtesy of Dr. N.N. Cordell who dissected the animals. For galagos, the letters following the collection abbreviation are specimen identifiers. The use of 'f' and 'm' designates 'female' and 'male', respectively.

[b] Body Mass. Galago weights are species means for that sex (O.c., O.g., and G.s. from Bearder, 1987; G.d. from Charles-Dominique, 1972). Weights for *Lemur catta* and *Nycticebus coucang* are also species means for that sex (from Sussman, 1991 and Bearder, 1987 respectively). Weights for *Cheirogaleus medius*, *Propithecus verreauxi*, and *Tarsius syrichta* are last-known weights of the individual animals.

[c] Diets for prosimians are generalized and are taken from Nowak, 1999; Sussman, 1999; and Mittermeier et al., 2006.

mass is often the variable reported in publication on muscle anatomy (e.g., Turnbull, 1970; Cachel, 1984), muscle cross-sectional area is the anatomical variable of interest when one wants to estimate muscle forces (e.g., Antón, 1999). However, measuring cross-sectional area is complicated if, as is often the case, fiber architecture is complicated.

Pinnation increases a muscle's ability to generate force by increasing its cross-sectional area compared to that of a parallel-fibered muscle of the same shape and

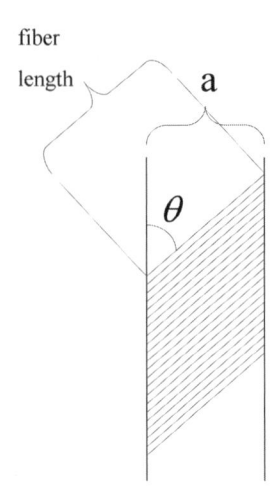

Fig. 11.1 Method of calculating pinnation. Thick vertical lines represent the planes of the surfaces of origin and insertion of the fibers. These surfaces are usually tendinous sheets and are anchored to bone at each end (therefore, the bony attachments of this muscle would be toward the top and bottom of this figure.). Diagonal lines delimit fibers. The angle of pinnation is represented by θ and is calculated as: Sin θ = a/fiber length. Figure modified from Anapol and Barry (1996). Distance "*a*" is the perpendicular distance between tendinous sheets of fiber attachment (mean of at least three measurements at different sites on the muscles). Fiber length was measured after chemical dissolution of connective tissue, and is the mean length of at least 30 measured fibers. We measured fibers from every major region of each muscle

volume (Gans, 1982; Gans and De Vree, 1987). A long, thin muscle in which each fiber originates on a bony surface and inserts on another distant bony surface must have long fibers if it is not pinnate (see Fig. 11.1). However, many muscles are bounded by tendon sheets anchored to (and running perpendicular to) the bony origin and insertion. Generally, in these muscles, many fibers insert on or originate from tendon and the muscle is considered pinnate. In this way, without increasing its volume, the muscle will have more fibers stacked in parallel, and thus it can generate more force.

11.1.2 Scaling Predictions

Previous research suggests that chewing muscle mass in prosimians scales isometrically with body mass (Cachel, 1979, 1984). However, Cachel's sample included only three prosimian species with one specimen each. She concluded that because anthropoids and prosimians together yield a pattern of isometry in chewing muscle mass, and because removal of the prosimians from the regression causes little change in the slope, prosimian chewing muscles scale in the same way as those of anthropoids (isometrically). More data on the prosimians are required to test this conclusion.

Recently, Lucas (2004) generated predictions about how the chewing muscles might scale to body mass. Lucas' "fracture scaling" is a set of predictions about the scaling patterns of the chewing muscles that move beyond assumptions of geometric similarity. Fracture scaling hinges on the fact that particles of food of different sizes fail at different stresses, even if they are homogeneous, equally tough, and deform in a linear-elastic fashion. "Failure" here refers to fragmentation of the particle as cracks propagate through it. Fragmentation is the phenomenon of interest because, in most cases, the purpose of mastication is to break food into smaller pieces.

Lucas compared the fracture properties of two food items of different sizes[1] in a three-point bending model. If the larger item is λ times larger than the smaller one for any linear dimension, it only takes $\lambda^{1.5}$ times as much force to fracture the larger item than the smaller item. Therefore, an $8\,cm^3$ ($=2^3$) cube of food requires only about 3 ($=2^{1.5}$) times as much force to be fractured as does a $1\,cm^3$ ($=1^3$) cube of food. If ingested food size is isometric to the size of the animal chewing the food, then this analysis has implications for primate scaling.

Geometric scaling predicts that the chewing muscles should be able to produce λ^2 more force in a larger animal (where the larger animal is λ times larger than the smaller one for any linear dimension). If ingested food size scales isometrically with body size, then a primate that is $8\,m^3$ in size needs to chew with only about 3 times as much force as a $1\,m^3$ primate [as opposed to 4 (2^2) times the amount of force – the value predicted by geometry].

Lucas used this relationship, along with the assumption of bite size isometry, to predict that chewing muscle force should scale to the 0.5 power of mass ($M^{0.5}$), and therefore chewing muscle cross-sectional area should scale to the 0.5 power of body mass. This slope is lower than the 0.67 slope predicted by geometric similarity.

From the prediction that fiber length is isometric ($M^{0.33}$) and that PCSA is negatively allometric ($M^{0.5}$), Lucas predicted slight negative allometry for muscle mass ($M^{0.83}$).

11.2 Materials and Methods

One of us (Perry) dissected the masticatory muscles of 16 prosimian species, represented by a total of 22 specimens (see Table 11.1). Some were preserved in 10% formalin, some in 70% ethanol, and others had been frozen since the time of death. Minimal fiber shrinkage is expected in preserved specimens as they were preserved whole with the muscles intact. All specimens had similar degrees of gape after death: most had a small part of the tongue interposed between the incisors (equivalent to 2–5° of gape at the central incisors).

---

[1] Here we are referring to the size of a food item as it is presented to the molars for mastication. This is probably roughly equivalent to the size of a food item as bitten off by the anterior dentition, but very different from the sizes of the food particles that result from the action of mastication.

The chewing muscles of the galago specimens had been examined and removed from the cadavers for previous work (Cordell, 1991). These animals had been preserved with mouth postures similar to those of the other prosimians in this sample. As these muscles were ex situ and muscle orientation could not be determined, it was impossible to accurately measure muscle thickness perpendicular to the line of action. Therefore, no pinnation correction was performed on them (see below).

11.2.1 The Chewing Muscles

Three groups of masticatory adductor muscles were examined: (1) the temporalis group, (2) the masseter group, and (3) the pterygoid group. The pterygoid group includes the medial and lateral pterygoid. The latter is primarily a jaw opener and was not included in this study.

Table 11.2 compares the nomenclature used by various authors to describe the mandibular adductors. The temporalis and masseter group each can be divided into

Table 11.2 Synonyms for masticatory muscle terminology used here

Muscle	Authors in agreement	Synonym
Superficial Masseter (SM)	De Gueldre and De Vree, 1988; Cordell, 1991	masseter pars superficialis (Minkoff, 1968; Schön, 1968), masseter superficialis & masseter intermedius (Gaspard et al., 1973b)
Deep Masseter (DM)	De Gueldre and De Vree, 1988; Cordell, 1991	masseter pars superficialis (Minkoff, 1968; Schön, 1968), superficial part of maxillo-mandibularis (Gaspard et al., 1973b,c)
Zygomatico-mandibularis (ZM)	Fiedler, 1953; De Gueldre and De Vree, 1988; Cordell, 1991	deepest portion of masseter (Starck, 1933), masseter pars profundus (Minkoff, 1968; Schön, 1968), masseter profundus (Gaspard et al., 1973b,c), deep masseter (Taylor and Vinyard, 2004)
Superficial Temporalis (ST)	Minkoff, 1968; De Gueldre and De Vree, 1988; Cordell, 1991	pars temporalis lamina superficialis & pars orbitalis (Gaspard et al., 1973c)
Deep Temporalis (DT)	Minkoff, 1968; Cordell, 1991	pars temporalis lamina profunda (Gaspard et al., 1973c), medial temporalis & deep temporalis (De Gueldre and De Vree, 1988)
Zygomatic Temporalis (ZT)	Fiedler, 1953; Cordell, 1991	zygomatico-mandibularis (Starck, 1933), zygomatico-mandibularis & deep part of maxillo-mandibularis (Gaspard et al., 1973c), suprazygomatic temporalis & posterior zygomaticomandibularis (De Gueldre and De Vree, 1988), zygomatico-mandibularis & maxillo-mandibularis (Ross, 1995)
Medial Pterygoid (MP)	as considered here, there is no disagreement about the medial pterygoid

individual layers of muscle fibers bounded wholly or partly by layers of connective tissue or bone. The temporalis group consists of the zygomatic temporalis, superficial temporalis, and deep temporalis. The masseter group consists of the superficial masseter, deep masseter, and zygomatico-mandibularis. The medial pterygoid muscle is very complex. Connective tissue sheets divide it into four major compartments of fibers (Gaspard et al. 1973a; pers. obs.). This muscle is impossible to divide into unipinnate layers and was treated differently (see below 'Correcting for Pinnation').

11.2.2 Muscle Mass

Following the removal of a muscle, it was patted gently with a paper towel to remove excess water and wet weight was taken using an electronic scale accurate to 0.001 g. Wet weights were preferred to dry weights for two reasons: first, they better reflect the hydrated condition of muscles in living animals; second, once a muscle is dried, it is impossible to chemically dissect its fibers for information on internal architecture and fiber dimensions (Antón, 1999).

11.2.3 Measuring Fiber Length

To determine the average fiber length for each muscle, we modified a protocol used by Rayne and Crawford (1972) for chemical dissolution of connective tissue. If much connective tissue remains, fibers are very likely to break during extraction and fiber lengths will be greatly underestimated.

To measure fiber length, each muscle was immersed in 10% sulfuric acid and cooked in an oven at 60°C. Cooking time varied between 1 and 6 h based on muscle mass and the amount of connective tissue present. Formalin-preserved, ethanol-preserved, and fresh-frozen muscle fibers were equally easy to extract and measure. However, formalin-preserved muscles took considerably more time to cook.

Muscles have to be monitored carefully as overcooking can cause the muscle to become very fragile and positional relationships within the muscle can be lost. We cooked the muscles in Petrie dishes and, to preserve positional relationships, we recorded the position and orientation of the muscle before placing it in the oven. Muscles were also checked frequently to make sure they were not overcooked.

Once enough connective tissue had been dissolved, the muscle was removed from its dish. Then it was blotted gently between water-soaked paper towels and it was examined under a dissecting microscope equipped with an ocular micrometer (accurate to 0.0825 mm at 25×). We measured 30–60 fibers from each muscle, ensuring that each region of the muscle was represented. Mean fiber length was recorded and used in subsequent equations.

11.2.4 Calculating PCSA

The cross-sectional area of each muscle was calculated from the following formula (Schumacher, 1961):

$$\text{PCSA (cm}^2\text{)} = \text{muscle mass (g)/(fiber length (cm)} \times \text{specific density)} \quad (11.2)$$

PCSA stands for physiological cross-sectional area. The specific density of muscle is taken to be 1.0564 g/cm^3 (Murphy and Beardsley, 1974).

For comparative purposes, the reduced physiological cross-sectional area (RPCSA)[2] of each muscle was calculated from the following formula (Anapol and Barry, 1996):

$$\text{RPCSA (cm}^2\text{)} = \text{muscle mass (g)} \times \cos\theta/(\text{fiber length (cm)} \times \text{specific density)} \quad (11.3)$$

The angle of pinnation θ is the angle between the fiber orientation and the line of action of the muscle.[3] RPCSA is used when one wants to measure a single component of the force that a muscle produces (Anapol and Barry, 1996). The effect of using RPCSA instead of PCSA is to remove the component of the muscle mass that produces a force vector that is normal to the line of action. Thus, RPCSA is always smaller than PCSA.

Anapol and Barry (1996) measured fiber length in limb muscles. The limbs in their specimens were in a variety of flexed and extended postures. To correct for inter-individual variability in the length of muscle fibers due to this variable preservation, they measured sarcomere lengths for their muscles and adjusted their average fiber length value based on a standard value for vertebrate sarcomere length (2.5 microns). Because all of our specimens were preserved in a similar mouth posture, we did not adjust the fiber length measurements to reflect a standard vertebrate sarcomere length.

11.2.5 Measuring Pinnation

The angle of pinnation θ was calculated using the following formula (Anapol and Barry, 1996, see also Fig. 11.1):

$$\sin\theta = a/\text{mean fiber length for that unit} \quad (11.4)$$

To calculate θ, it was necessary to obtain mean fiber length for each muscle. Fiber length can be measured directly in a cut section of muscle (Anapol and Barry, 1996; Taylor and Vinyard, 2004), but this can be problematic because it is difficult to know that one is cutting a muscle parallel to its fibers. We had already measured

---

[2] 'R' is for 'reduced'. This follows Anapol and Barry (1996) who use the abbreviation 'RPCA' for the same variable.

[3] Here, line of action refers to the direction of pull of a muscle. It can be defined in many different ways. By the method of Anapol and Barry (1996), the line of action is parallel to the tendon sheets of fiber attachment. This may be slightly different from a line drawn from the center of a muscle's bony origin to the center of its bony insertion.

fiber length in the chemically dissected muscles, so we calculated Equation (11.4) for each muscle using the mean of its measured fiber lengths.

The system of classification of the chewing muscles used here (and described above) is based on the presence of fascial planes. This facilitated the pinnation correction because "a" could be measured as the thickness of a muscle. The average thickness of each unit was measured in the coronal plane, perpendicular to the direction of pull of the muscle (bony origin to bony insertion). This thickness constitutes the perpendicular distance "a" between the surface of *fiber* origin and the surface of *fiber* insertion.

It was impossible to physically separate the thin and twisted sheets of the medial pterygoid muscle for measuring muscle weight and thickness without destroying the muscle. Instead, we measured total medial pterygoid weight and average fiber length for the entire muscle. We then used total medial pterygoid thickness divided by four as "a" in Equation (11.4) to get a value for pinnation. The four layers of medial pterygoid are roughly equal in thickness.

It was not possible to calculate RPCSA for the galagos in this sample. Therefore, the comparison of PCSA to RPCSA excludes the galagos.

11.2.6 Statistical Analysis

Species means for muscle mass, mean fiber length, PCSA, and body mass were derived for each muscle. For each individual prosimian, total PCSA is the sum of the PCSA values for all the muscles. Total adductor mass is the sum of the masses of all the muscles. Total mean fiber length is the grand mean of the mean fiber lengths for the muscles.

In most cases, the last known living weight was used as body mass. However, for the galagos, species mean weights were used because living weights were not known. For most of the American Museum of Natural History specimens, cadaver weights were used because living weights were not known. For the AMNH *Avahi* specimen, species mean weight was used because only the head was preserved.

Chewing muscle variables were regressed against body mass in log space using the reduced major axis (RMA) form of Model II regression. The (S)MATR (Version 1, Falster DS, Warton DI & Wright IJ http://www.bio.mq.edu.au/ecology/SMATR) software package was used to perform the regressions and generate confidence intervals. The F-test in (S)MATR was used to test the observed slopes against the slopes hypothesized by geometric scaling and by fracture scaling.

11.3 Results

Tables 11.3 through 11.6 provide the data for PCSA, RPCSA, muscle mass, and fiber length for the seven mandibular adductor muscles of the 16 prosimian species. Table 11.7 provides the pinnation values for all species as well as the ratio of PCSA to body mass and the ratio of RPCSA to body mass.

Table 11.3 Values of physiological cross-sectional area (PCSA) for sixteen species of prosimians (mm^2)

	Al[a]	Cm	Dm	Gd	Gs	Hg	Lc	Lm
SM[b]	52.02[c]	45.40	70.09	10.73	41.19	147.71	155.81	34.61
DM	77.09	20.05	70.09	2.28	7.99	41.55	49.66	64.75
ZM	49.94	21.25	69.44	2.35	10.74	49.34	35.32	23.39
Masseter Mean	179.05	86.70	209.63	15.36	59.91	238.61	240.79	122.75
ZT	29.10	7.67	80.08	2.50	7.13	44.42	67.38	22.60
ST	53.53	38.14	112.16	6.33	31.65	85.82	84.42	60.90
DT	48.38	37.13	127.68	15.25	58.60	107.41	183.32	50.25
Temporalis Mean	131.01	82.94	319.93	24.08	97.38	237.66	335.12	133.74
MP	90.75	32.63	65.76	6.92	23.12	163.61	118.45	54.70
Sum of Adductors	400.80	202.27	595.31	46.35	180.42	639.88	694.36	311.19

	Mc	Nc	Oc	Og	Pp	Pv	Ts	Vv
SM	46.92	92.89	222.86	209.32	81.50	127.72	45.20	97.61
DM	8.29	64.40	37.11	80.12	48.80	119.40	23.38	94.14
ZM	26.25	40.78	34.10	32.10	32.04	152.81	20.59	126.37
Masseter Mean	81.46	198.07	294.07	321.54	162.34	399.93	89.17	318.13
ZT	with ST[d]	29.73	46.31	26.43	48.64	100.60	9.20	77.13
ST	41.66	90.89	179.82	161.39	54.78	139.07	28.33	111.73
DT	41.62	154.35	244.94	283.73	96.93	211.03	42.05	244.39
Temporalis Mean	83.27	274.97	471.07	471.54	200.35	450.69	79.57	433.25
MP	22.35	88.45	103.74	95.40	70.89	260.07	53.70	117.44
Sum of Adductors	187.08	561.49	868.87	888.49	433.59	1110.69	222.44	868.81

[a] Species name abbreviations are as follows: Al (*Avahi laniger*), Cm (*Cheirogaleus medius*), Dm (*Daubentonia madagascariensis*), Gd (*Galagoides demidoff*), Gs (*Galago senegalensis*), Hg (*Hapalemur griseus*), Lc (*Lemur catta*), Lm (*Lepilemur mustelinus*), Mc (*Mirza coquereli*), Nc (*Nycticebus coucang*), Oc (*Otolemur crassicaudatus*), Og (*Otolemur garnettii*), Pp (*Perodicticus potto*), Pv (*Propithecus verreauxi*), Ts (*Tarsius syrichta*), and Vv (*Varecia variegata rubra*).
[b] Muscle name abbreviations are the same as in Table 11.2.
[c] These values are not corrected for pinnation.
[d] It was not possible to separate the ZT from the ST for *Mirza coquereli*. Therefore, these muscles were analyzed as a single unit for the measurement of PCSA and mass. However, it was possible to isolate fibers of ZT from the ST-ZT muscle mass once the muscle had been cooked. Therefore, an average fiber length value could be obtained for each of ZT and ST.

Tarsius syrichta, Daubentonia madagascariensis, and the galagos have high PCSA:body mass ratios (Table 11.7). *D. madagascariensis* specimen has very massive chewing muscles (~1% of body weight) (see Table 11.5 and Fig. 11.3). However, this specimen was a sub-adult. It had all of its adult dentition, but it weighed 742 g, less than half of the average weight of an adult aye-aye (~2.5 kg, Sterling et al., 1994). Any conclusions regarding aye-aye muscle scaling are tentative without a fully adult specimen.

Four folivores are included in this study (*Hapalemur griseus, Propithecus verreauxi, Avahi laniger,* and *Lepilemur mustelinus*). *H. griseus* has relatively large PCSA relative to body mass. *P. verreauxi, A. laniger,* and *L. mustelinus* have low

Table 11.4 Values of reduced physiological cross-sectional area (RPCSA) for twelve species of prosimians (mm^2)

	Al[a]	Cm	Dm	Hg	Lc	Lm
SM[b]	50.47[c]	42.61	69.86	134.09	145.67	33.76
DM	71.56	19.64	69.86	39.47	48.76	60.69
ZM	46.63	20.97	68.61	43.11	34.54	22.42
Masseter Mean	168.67	83.23	208.33	216.67	228.97	116.86
ZT	26.70	7.52	77.83	37.89	62.81	20.89
ST	52.92	37.65	111.52	81.83	83.52	60.40
DT	47.81	36.90	126.05	99.01	180.75	49.78
Temporalis Mean	127.42	82.07	315.41	218.73	327.08	131.06
MP	90.12	32.51	65.69	156.44	118.33	54.23
Sum of Adductors	386.21	197.81	589.43	591.84	674.38	302.15
	Mc	Nc	Pp	Pv	Ts	Vv
SM	45.18	89.70	80.28	125.03	40.07	97.08
DM	8.11	59.68	48.18	111.57	22.81	91.46
ZM	25.09	38.80	31.64	135.17	18.42	120.07
Masseter Mean	78.38	188.18	160.10	371.78	81.30	308.61
ZT	with ST	28.61	47.05	98.23	9.03	74.87
ST	41.12	86.53	54.22	136.30	27.72	110.75
DT	40.45	141.92	95.41	202.92	40.48	240.09
Temporalis Mean	81.56	257.06	196.68	437.46	77.23	425.72
MP	22.20	88.33	70.57	230.07	53.48	117.03
Sum of Adductors	182.14	533.57	427.35	1039.31	212.01	851.36

[a] Species name abbreviations are the same as in Table 11.3.
[b] Muscle name abbreviations are the same as in Table 11.2.
[c] These values are smaller than the PCSA values because they are corrected to reflect the fraction of muscle mass that contributes to force along the line of action of the muscle (parallel to the tendons of fiber attachment).

values of PCSA and RPCSA relative to body mass[4] (Table 11.7). These folivores have high angles of pinnation relative to other prosimians, so RPCSA is relatively much lower than PCSA (Table 11.7).

Relative to body mass, *Galagoides demidoff* has very long adductor fibers (Table 11.6 and Fig. 11.4). *T. syrichta, Galago senegalensis*, and *D. madagascariensis* also have long adductor fibers. Pinnation angles are relatively great in *T. syrichta*, but are relatively small in *D. madagascariensis*. *T. syrichta* has long adductor muscle fibers, but maintains large pinnation angles. *G. demidoff, G. senegalensis*, and *T. syrichta* are small-bodied insectivores.

On average, the deep temporalis, followed by the superficial masseter, generally has the greatest cross-sectional area (Tables 11.3 and 11.4). However, the medial pterygoid usually follows; it generally has a large cross-sectional area for its mass. On average, the deep temporalis is the largest part of the adductor muscle mass (Table 11.5). It is followed by the superficial masseter, then the superficial

---

[4] This is not with respect to a particular biological expectation, only relative to the other species in the sample.

Table 11.5 Muscle mass (g) for sixteen species of prosimians

	Al[a]	Cm	Dm	Gd	Gs	Hg	Lc	Lm
SM[b]	0.31[c]	0.20	0.75	0.05	0.25	1.83	1.61	0.20
DM	0.42	0.07	0.75	0.01	0.04	0.44	0.52	0.34
ZM	0.28	0.07	0.99	0.01	0.05	0.29	0.28	0.14
Masseter Mean	1.01	0.34	2.49	0.06	0.34	2.56	2.41	0.68
ZT	0.17	0.05	1.20	0.01	0.05	0.39	0.70	0.19
ST	0.42	0.22	1.65	0.03	0.22	0.87	1.01	0.58
DT	0.41	0.20	2.03	0.06	0.38	0.95	2.18	0.44
Temporalis Mean	1.00	0.47	4.88	0.09	0.64	2.21	3.89	1.21
MP	0.37	0.11	0.60	0.02	0.12	0.85	0.83	0.21
Sum of Adductors	2.38	0.92	7.97	0.18	1.09	5.62	7.13	2.10
	Mc	Nc	Oc	Og	Pp	Pv	Ts	Vv
SM	0.19	0.85	1.58	1.88	1.00	1.67	0.18	1.42
DM	0.03	0.43	0.25	0.71	0.26	1.11	0.09	0.72
ZM	0.08	0.28	0.22	0.12	0.32	1.05	0.07	1.12
Masseter Mean	0.30	1.56	2.04	2.71	1.58	3.83	0.35	3.26
ZT	with ST	0.30	0.39	0.30	0.46	0.89	0.07	1.04
ST	0.22	0.98	1.50	1.35	0.62	1.16	0.19	1.59
DT	0.14	1.59	1.79	1.11	1.25	1.98	0.23	3.62
Temporalis Mean	0.36	2.87	3.67	2.76	2.33	4.03	0.49	6.25
MP	0.09	0.44	0.58	0.52	0.44	1.51	0.19	1.10
Sum of Adductors	0.75	4.87	6.29	5.99	4.35	9.37	1.02	10.61

[a] Species name abbreviations are the same as in Table 11.3.
[b] Muscle name abbreviations are the same as in Table 11.2.
[c] Values are species means.

temporalis. The medial pterygoid has the shortest fibers of all the adductors, on average (Table 11.6). The three parts of the temporalis generally have the longest fibers.

A Mann-Whitney U-test (Zar, 1999) showed that RPCSA values are not significantly lower than PCSA values ($U = 87$, $p > 0.05$.). This result suggests that the fraction of the contractile mass that pulls normal to the line of action is not a statistically significant component of the total muscle cross-sectional area in this sample. However, the total cross-sectional areas for folivores drop considerably when they are corrected for pinnation using RPCSA.

Table 11.8 provides the results of the RMA regression analyses. The 95% confidence intervals for PCSA, muscle mass, and fiber length include isometry for all comparisons, except for the PCSA of the deep masseter and the zygomatic temporalis. The PCSA of each of these muscles is positively allometric relative to body mass and the F-test is significant (DM slope $= 0.953$, $p < 0.023$; ZT slope $= 0.917$, $p < 0.021$). The regression for zygomatico-mandibularis is almost significantly different from isometry and shows positive allometry (slope $= 0.849$, $p < 0.090$).

The observed slopes for most PCSA regressions are significantly higher than the slope of 0.5 predicted by fracture scaling (Lucas, 2004). The one exception is the regression for the superficial masseter (slope $= 0.692$, CI $= 0.479 - 1.001$).

11 Scaling of the Chewing Muscles in Prosimians

Table 11.6 Fiber length (cm) for sixteen species of prosimians

	Al[a]	Cm	Dm	Gd	Gs	Hg	Lc	Lm
SM[b]	0.56[c]	0.42	1.01	0.41	0.57	1.02	0.97	0.55
DM	0.52	0.33	1.01	0.33	0.49	0.86	1.00	0.50
ZM	0.53	0.31	1.35	0.36	0.42	0.49	0.75	0.57
Masseter Mean	0.54	0.35	1.13	0.37	0.49	0.79	0.91	0.54
ZT	0.55	0.62	1.42	0.42	0.62	0.72	0.98	0.86
ST	0.74	0.55	1.39	0.37	0.65	0.83	1.13	0.90
DT	0.80	0.51	1.51	0.36	0.61	0.72	1.13	0.83
Temporalis Mean	0.70	0.56	1.44	0.38	0.63	0.76	1.08	0.86
MP	0.39	0.32	0.86	0.30	0.48	0.43	0.66	0.36
Sum of Adductors	0.58	0.44	1.22	0.36	0.55	0.72	0.95	0.65
	Mc	Nc	Oc	Og	Pp	Pv	Ts	Vv
SM	0.38	0.87	0.67	0.85	1.16	1.24	0.38	1.38
DM	0.34	0.63	0.63	0.84	0.50	0.88	0.37	0.72
ZM	0.29	0.65	0.61	0.70	0.95	0.65	0.33	0.84
Masseter Mean	0.34	0.72	0.64	0.80	0.87	0.92	0.36	0.98
ZT	0.50	0.96	0.79	1.10	0.90	0.84	0.72	1.28
ST	0.50	1.02	0.79	0.79	1.07	0.79	0.62	1.35
DT	0.32	0.98	0.69	0.74	1.22	0.89	0.52	1.40
Temporalis Mean	0.44	0.98	0.76	0.88	1.06	0.84	0.62	1.34
MP	0.38	0.47	0.53	0.51	0.59	0.55	0.33	0.89
Sum of Adductors	0.39	0.80	0.67	0.79	0.91	0.83	0.47	1.12

[a] Species name abbreviations are the same as in Table 11.3.
[b] Muscle name abbreviations are the same as in Table 11.2.
[c] Values are species means.

Table 11.7 Pinnation values, PCSA values relative to body mass, and RPCSA values relative to body mass

Species	Mean θ[a]	Max θ	DM[b] θ	ZM θ	Sum PCSA/BM	Sum RPCSA/BM
A.l.[c]	14.93	23.44 (ZT)	21.82	20.98	0.343[d]	0.331
C.m.	10.41	20.16 (SM)	11.64	9.23	0.571	0.559
D.m.	7.09	13.61 (ZT)	4.67	8.88	0.694	0.689
G.s.	–	–	–	–	0.935	–
G.d.	–	–	–	–	0.778	–
H.g.	18.66	28.85 (ZT)	12.75	26.22	0.730	0.698
L.c.	12.23	21.22 (ZT)	10.93	12.08	0.315	0.306
L.m.	12.56	20.41 (DM)	20.41	16.58	0.380	0.370
M.c.	12.36	17.14 (ZM)	11.79	17.14	0.585	0.569
N.c.	16.41	23.15 (DT)	22.07	17.92	0.827	0.786
O.c.	–	–	–	–	0.691	–
O.g.	–	–	–	–	1.065	–
P.p.	9.53	14.69 (ZT)	9.13	9.13	0.350	0.346
P.v.	18.29	27.80 (ZM)	20.86	27.80	0.343	0.321
T.s.	15.77	27.56 (SM)	12.70	26.55	1.869	1.782
V.v.	10.70	18.17 (ZM)	13.72	18.17	0.205	0.201

[a] The symbol for the angle of pinnation is θ. Each mean pinnation value is the mean of the pinnation values for the seven adductors for that species (species means).
[b] Muscle name abbreviations are the same as in Table 11.2.
[c] Species name abbreviations are the same as in Table 11.3.
[d] Each PCSA and RPCSA value is the sum of the PCSA or RPCSA values for the seven adductors for that species (species means). PCSA, RPCSA, and body mass values are raw values (not logged).

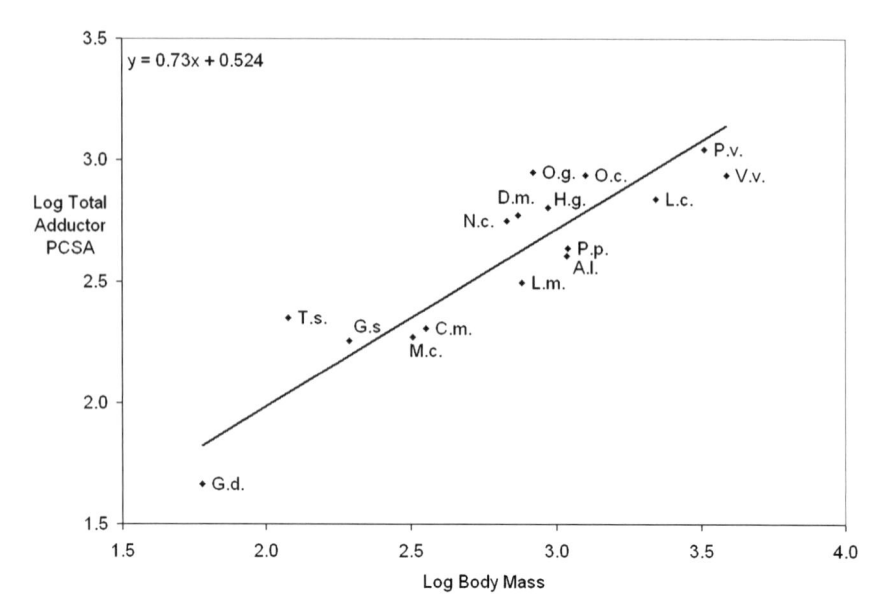

Fig. 11.2 Log of the sum of the PCSA values for all seven adductor muscles plotted against the log of body mass. Values plotted are species means. The line is the reduced major axis fit. Species name abbreviations are the same as in Table 11.3

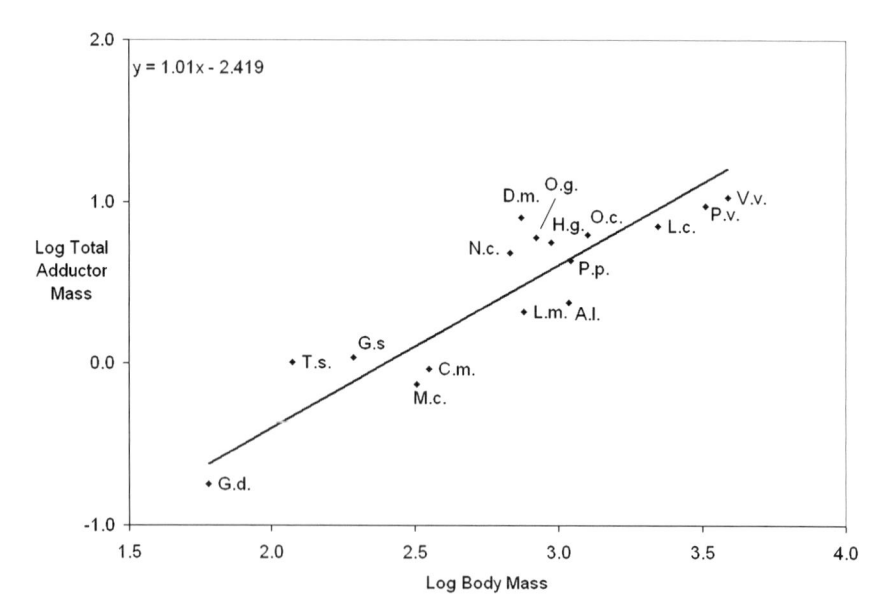

Fig. 11.3 Log of the sum of the muscle mass values for all seven adductor muscles plotted against the log of body mass. Values plotted are species means. The line is the reduced major axis fit. Species name abbreviations are the same as in Table 11.3

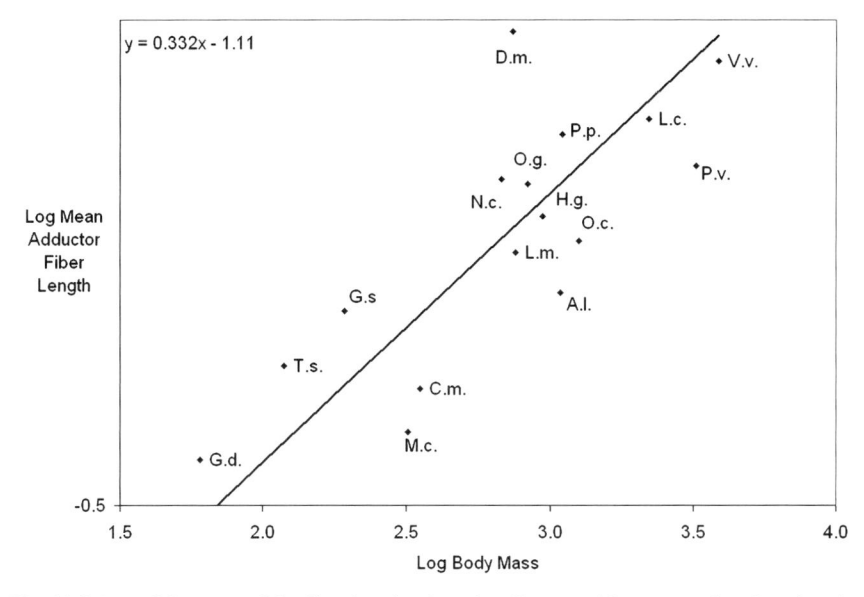

11 Scaling of the Chewing Muscles in Prosimians   231

Fig. 11.4 Log of the mean of the fiber length values for all seven adductor muscles plotted against the log of body mass. Values plotted are species means. The line is the reduced major axis fit. Species name abbreviations are the same as in Table 11.3

Table 11.8 Results of reduced major axis (RMA) regressions of muscle dimensions on body mass

Muscle	Slope	R^2	y-intercept	Lower 95% CL	Upper 95% CL	p-value for isometry H_o slope[a]	p-value for fracture scaling H_o slope[b]
Log (10) PCSA vs. Log (10) Body Mass							
DM[c]	0.953	0.72	−1.152	0.707	1.285	0.023*	0.0002**
DT	0.758	0.657	−0.191	0.545	1.053	0.444	0.016*
MP	0.786	0.746	−0.419	0.591	1.045	0.256	0.004**
SM	0.692	0.567	−0.101	0.479	1.001	0.855	0.081
ST	0.737	0.7	−0.292	0.541	1.004	0.524	0.017*
ZM	0.849	0.76	−0.9	0.643	1.12	0.090	0.001**
ZT	0.917	0.788	−1.156	0.706	1.19	0.021*	0.0001**
SUM	0.73	0.777	0.524	0.559	0.954	0.506	0.008**
Log (10) Muscle Mass vs. Log (10) Body Mass							
DM	1.22	0.755	−4.128	0.92	1.62	0.146	0.009**
DT	1.04	0.753	−3.094	0.78	1.37	0.785	0.112
MP	0.974	0.817	−3.244	0.764	1.242	0.819	0.183
SM	1	0.702	−3.102	0.74	1.37	0.982	0.212
ST	0.989	0.748	−3.075	0.745	1.314	0.936	0.210
ZM	1.11	0.781	−3.881	0.85	1.45	0.416	0.033*
ZT	1.13	0.781	−3.845	0.87	1.48	0.333	0.024*
SUM	1.01	0.812	−2.419	0.79	1.3	0.901	0.102

Table 11.8 (continued)

Muscle	Slope	R^2	y-intercept	Lower 95% CL	Upper 95% CL	p-value for isometry H$_o$ slope[a]	p-value for fracture scaling H$_o$ slope[b]
Log (10) Fiber Length vs. Log (10) Body Mass							
DM	0.355	0.543	−1.242	0.243	0.519	0.688	0.688
DT	0.404	0.569	−1.264	0.279	0.583	0.267	0.267
MP	0.296	0.5	−1.158	0.199	0.439	0.572	0.572
SM	0.4	0.659	−1.281	0.288	0.556	0.233	0.233
ST	0.323	0.596	−1.013	0.226	0.462	0.903	0.903
ZM	0.386	0.445	−1.343	0.255	0.584	0.444	0.444
ZT	0.301	0.438	−0.957	0.199	0.458	0.658	0.658
MEAN	0.332	0.628	−1.11	0.235	0.467	0.975	0.975

[a] The slope of isometry is 0.67 for PCSA, 1 for muscle mass, and 0.33 for fiber length.
[b] The slope predicted by fracture scaling is 0.5 for PCSA, 0.83 for muscle mass, and 0.33 for fiber length (Lucas, 2004). An F-test was used to evaluate the significance of the difference between observed slopes and predicted slopes. The *p*-values reported are for the F-test (as performed in (S)MATR).
[c] Muscle name abbreviations are the same as in Table 11.2.
* Significant difference at the 5% level.
** Significant difference at the 1% level.

The observed slopes for muscle mass relative to body mass are not significantly different from isometry. The observed slopes for some muscle mass regressions (deep masseter, zygomatico-mandibular, and zygomatic temporalis) are significantly different from the slope of 0.83 predicted by fracture scaling. The observed slopes for fiber length are isometric.

Galagoides demidoff, the smallest prosimian in this sample, has a very low total adductor PCSA for its body mass (Fig. 11.2). This datum point has a considerable effect on the regression slope. If it is removed from the total adductor PCSA–body mass regression, the slope drops to 0.648 and is no longer significantly different from either isometry or the fracture scaling slope of 0.5. The addition of other small-bodied prosimians and more *G. demidoff* specimens will help to firm up the slope estimates. *G. demidoff* does have long chewing muscle fibers on average relative to body mass (see Table 11.6 and Fig. 11.4), and this has a necessary negative effect on PCSA for a given muscle mass.

11.4 Discussion

These preliminary results suggest that mandibular adductor PCSA scales with isometry or slight positive allometry relative to body mass in prosimians. Positive allometry of mandibular adductor cross-sectional area is supported by a recent study of a sample that includes both prosimians and anthropoids (Shahnoor et al., 2005; Anapol et al., Chapter 9 in this volume). Lucas (2004:142–144) predicted strong negative allometry of cross-sectional area relative to body mass; the predicted slope is 0.5. No study supports the predicted pattern of strong negative allometry (e.g., $M^{0.5}$) for chewing muscle cross-sectional area (Perry and Wall, 2005; Shahnoor et al., 2005; Anapol et al., Chapter 9 in this volume; this study).

The 95% confidence intervals for the slope of our regressions of PCSA against body mass are very large (~0.4 for the sum of all adductors, 0.45–0.6 for the individual muscle regressions). Furthermore, the r^2 value is below 0.8 in every case. This suggests that much of the variation in PCSA fails to be explained by variation in body mass.

We plan to examine the biomechanical scaling of cross-sectional area, mass, fiber length, and pinnation of the jaw adductor muscle. Important independent variables to evaluate include the bite force load arms during incisal and molar biting and the moment arms of the jaw muscles (Hylander, 1985).

This sample of prosimians is small and, for most species, only one individual was available. Nevertheless, all families of strepsirrhines were sampled. Five out of nine genera of lorisids (Nowak, 1999; Sussman, 1999) and nine out of fifteen genera of Malagasy lemurs (Mittermeier et al., 2006) were sampled. *Indri*, at the top of the body size range, is very rare, and the only cadaver available was that of a juvenile (studied, but not included here). We will add several more prosimian species to this sample, including many small-sized (e.g., *Microcebus murinus*) and medium-sized (e.g., *Eulemur coronatus*) species. Furthermore, we plan to add an adult aye-aye, another *G. demidoff*, as well as other specimens of *Lemur catta*, *P. verreauxi*, and *Hapalemur griseus*. We expect that the confidence intervals for the regressions will be reduced as a consequence. The gummivores *Elegantulus* and *Phaner*, as well as the greater bamboo lemur *Hapalemur simus*, would make useful additions to this sample. However, cadavers of these species are difficult to obtain.

11.4.1 Cross-Sectional Area, PCSA, RPCSA, and Pinnation

Dietary behavior likely has a strong influence on PCSA. For example, species that chew tough foods may have especially large mandibular adductor PCSA for their body size. Also, species that use large gape may have long chewing muscle fibers and, given isometry in muscle mass, less cross-sectional area relative to body size (cf. Taylor and Vinyard, 2004).

PCSA is low relative to body mass in *Avahi* and *Lepilemur*. It is isometric or nearly so relative to body mass in *Propithecus*. Another way to view this is that the folivores have relatively large body mass compared to their adductor PCSA. Perhaps, this is in part because folivorous prosimians have large gastrointestinal systems that contribute considerable mass to their total body mass (Chivers and Hladik, 1980). Pinnation angles are relatively high in all of these folivores. These data together suggest that folivorous prosimians generate relatively less muscle force for their body mass compared to non-folivorous prosimians of the same size and that a large proportion of this force is normal to the line of action. The moment that the jaw adductors are capable of generating is a function of leverage and force.

The short faces and mandibles of *Propithecus*, *Avahi*, and – to a slightly lesser degree – *Lepilemur* and *Hapalemur*, suggest that muscle leverage is high relative to body mass. Estimation of moments generated by the jaw adductors may reveal that greater leverage in the masticatory system of folivorous prosimians (relative to non-folivores) compensates for the small cross-sectional areas of their chewing muscles. Another possibility is that the maximum force required by the mandibular adductors of folivores is smaller than the force required of these muscles in prosimians that exploit other kinds of food. Folivores must chew their food many times, but perhaps each chew demands little muscle force.

The adductor muscles of *Hapalemur griseus* have large PCSA values relative to body mass. This species feeds on bamboo shoots, stems, leaves, fruits, flowers, and fungi in the wild (Overdorff et al., 1997). *Hapalemur* is often considered a folivore for the purposes of studies of functional morphology (e.g., Kay, 1975). However, *Hapalemur* clearly has a different pattern of PCSA compared to *Propithecus*, *Avahi*, and *Lepilemur*.

Slopes were especially high for the PCSA regressions for the deep masseter, zygomatico-mandibularis, and zygomatic temporalis. This suggests that these muscles have especially large cross-sectional areas in the larger prosimians in this sample. If muscle leverage and activity pattern are equal across body sizes in prosimians, these muscles are capable of contributing a larger proportion of the total adductor force in larger prosimians.

The RPCSA of Anapol and Barry (1996) is a necessary modification of PCSA when one is interested in a single component of muscle force (parallel to the line of action), because it reduces muscle mass by the fraction of muscle that generates force *perpendicular* to the line of action. However, incorporating cross-sectional area into models of chewing muscle mechanics is complicated by several factors. First, the chewing muscles have different lines of action. For example, even if all of the fibers of the medial pterygoid were arranged in parallel to its line of action, and the same were true of the deep masseter, these two muscles would still be pulling against one another in the mediolateral plane. Second, many feeding behaviors presumably include mediolateral and anteroposterior muscle forces (e.g., the power stroke of mastication), in addition to vertical force. Pinnate fibers in these muscles might be oblique to the line of action for some masticatory actions, whereas for other actions the same fibers might be parallel to the line of action. Moreover, pinnation angle changes if fibers swivel during concentric or eccentric contractions. Third, chewing muscle force vectors affect food in the occlusal plane. This is a complication because, for example, a vertically oriented muscle vector has a different effect at a vertically oriented occlusal plane than at a horizontally oriented occlusal plane.

Gans (1982) suggested that a muscle layer (or array) should be considered pinnate only if its fibers swivel as they contract. It is unknown if this occurs during jaw adductor contraction in prosimians, though it is likely to occur when fibers originate or insert on connective tissue sheets. Most of the fibers of the medial pterygoid are attached to at least one connective tissue sheet. This is also true of the superficial and deep temporalis, and the superficial and deep masseter. Fewer than half of the fibers in the zygomatico-mandibularis are attached to connective tissue (most attachment is to bone). No fibers of the zygomatic temporalis are attached to a connective tissue sheet, though many converge on a strap-like tendon of insertion.

We are currently working on a modification of the Anapol and Barry (1996) method of correcting for pinnation. This modification will consider different hypothetical lines of action for different masticatory actions. For example, it is possible to correct for pinnation, assuming a vertical line of action for each chewing muscle. For a vertical line of action, pinnation would be measured in the same way as in Equation (11.4), except distance "*a*" would be measured in the horizontal (i.e., perpendicular to vertical) plane for each muscle, rather than as the distance perpendicular to the tendons of fiber attachment. In choosing the lines of action of interest, reference will be made to the direction of the power stroke of mastication and to the orientation of the occlusal plane.

11.4.2 Fiber Length

Fiber length scales isometrically with body mass in this sample. In this manner, our data conform to the predictions of both geometric and fracture scaling. Deviations from this pattern for individual species may reflect different gape requirements (c.f., Taylor and Vinyard, 2004).

The slopes of regressions of fiber length against body mass have wide confidence intervals and the r^2 values are very low. As with PCSA, much of the variation in fiber length fails to be explained by body mass variation. However, our data do not support the claim that variation in fiber length has *no* relationship to variation in body mass (Antón, 1999; 2000), as all regressions are significant.

It is not surprising that body mass variation explains only some of the variation in fiber length. Fiber length should bear some relation to the range of motion experienced by the lower jaw (Herring and Herring, 1974; Herring et al., 1979). This range of motion is expected to differ based on differences in several factors including jaw and face length, the structural properties of food, the manner of food acquisition, the degree and nature of food processing, and grooming behavior.

It is interesting that the prosimians in this sample, which have the relatively longest fibers, are small-bodied insectivores. *Tarsius* is capable of a large gape (60–70° for *Tarsius bancanus*) and routinely consumes very large animal prey (Jablonski and Crompton, 1994). Immature aye-ayes use gapes of approximately 40° to gnaw holes in wood for extracting insect larvae (Krakauer, 2005).

11.4.3 Muscle Mass

The mass of the mandibular adductors in this sample of prosimians scales isometrically to body mass. Cachel (1979, 1984) found this to be true of primates generally. R-squared values for regressions of muscle mass against body mass range from 0.7 to 0.81. Body mass variation does not explain all of the variation in jaw adductor muscle mass. The 95% confidence intervals for the slope are large for all muscle mass regressions. For all muscles, the confidence interval includes the slope of isometry and, for most, it also includes the slope of 0.83 predicted by fracture scaling. As with PCSA, the muscles that have the highest slopes for regressions of muscle mass against body mass are the deep masseter, zygomatico-mandibularis, and zygomatic temporalis.

Adductor muscle mass is especially great relative to body size for *Daubentonia*, *Tarsius*, and for the larger lorisids. Great adductor mass in these species is likely a result of selection for adductors that produce high forces but that can withstand high degrees of excursion. It is uncertain whether or not adult *Daubentonia* fit this pattern.

11.4.4 Fracture Scaling

Most of the results for scaling of fiber length and muscle mass support the predictions of fracture scaling (Lucas, 2004). However, the predictions regarding cross-sectional area are not supported by the data.

One explanation for the failure of the PCSA data to support the predictions of fracture scaling is that PCSA is an estimator of muscle force (input force), whereas fracture scaling predicts occlusal force (output force), the force needed to fracture foods. The output force differs from the input force because it is modulated by several factors such as muscle activity, muscle leverage, and occlusal morphology. Data on cross-sectional areas of chewing muscles cannot rule out the possibility that *output* muscle force scales with negative allometry relative to body mass.

Another explanation for the lack of correspondence between the observed PCSA scaling patterns and those predicted by fracture scaling is that one or more of the assumptions of fracture scaling is not met by this sample. Lucas (2004) specifically acknowledged that several of these assumptions, such as foods being homogeneous, linearly elastic, and of uniform toughness, are violated in most comparative samples of real primates. Some of these assumptions may be violated in a systematic fashion with respect to body size. For example, larger primates may *generally* feed on tougher foods (e.g., Kay, 1975). It may be difficult to assemble a sample of mammals that vary in body size, but that do not vary with respect to the material properties of the foods they eat.

The fracture scaling model assumes that foods are broken down in three-point bending. This model may be appropriate for many foods, such as nuts and seeds. However, leaves must be cut into many small pieces to be digested efficiently (e.g., Kay and Sheine, 1979). The nutritional contents of some foods, such as juicy fruits and larval insects, may be extracted primarily by tearing or puncturing an outer membrane. Alternately, this may be accomplished by compression/tension. If a different model of food breakdown were used in the fracture scaling equations, different predictions about chewing muscle scaling would result.

The predictions of fracture scaling require that chewing rate scale to the one-quarter power of body mass. However, there is a great deal of variation in the scaling exponents reported for chewing rate (Fortelius, 1985; Druzinsky, 1993). Moreover, chewing rate changes as a bolus of food is reduced (Hiiemae and Kay, 1973), and likely differs with the material properties of food.

Finally, fracture scaling assumes that the size of ingested items scales isometrically with body size. Currently, there are very few data available to evaluate this assumption. While ingested food size may scale isometrically with body mass for some foods, it is likely that there may be little relationship across folivorous primates between body size and the size of ingested leaf materials.

11.5 Conclusions

Based on this sample, the mandibular adductors of prosimians more or less fit the predictions of geometric scaling relative to body mass. Adductor PCSA scales isometrically or with slight positive allometry to body mass. Muscle mass and fiber length scale isometrically to body mass. For most individual adductors and for the sum of all adductors, the PCSA scaling pattern is significantly different from the negative allometry predicted by fracture scaling. The same is true of jaw adductor mass for the deep masseter, the zygomatico-mandibularis, and the zygomatic temporalis. Slopes for jaw adductor mass scaling are not significantly different from the prediction of fracture scaling for any of the other muscles.

Much of the variation in PCSA in prosimians is likely due to differences in food properties and food processing behavior. In this sample, general diet-related patterns of PCSA, muscle mass, and fiber length relative to body mass are not clear. However, the small-bodied folivores, *A. laniger* and *L. mustelinus*, have small PCSA values relative to body mass, whereas *H. griseus* has a high value of PCSA.

The prediction of fracture scaling regarding PCSA was not borne out in this sample. However, we cannot rule out the possibility that activity pattern and muscle leverage modulate input muscle force such that the output occlusal force *does* scale in the manner predicted by Lucas (2004). Regardless, the fracture scaling model has unique merit because its predictions are based on food properties.

As Lucas (2004) pointed out, food properties are likely to affect selection in the masticatory system. Because primate diets are complex, it may be fruitful to modify the fracture scaling model to reflect the complexity of primate chewing rhythms and the many different ways in which primate foods fail during mastication.

Scaling of PCSA to mechanical variables (e.g., load arm length) may yield further insights into adaptation in the jaw adductor muscles (Hylander, 1985). Data on additional prosimian species as well as on the scaling of ingested food size will improve the quality of this ongoing study.

Acknowledgments We thank the Duke University Primate Center (especially S. Combes and J. Ives), the Department of Biological Anthropology & Anatomy (Duke University), the American Museum of Natural History (especially E. Westwig), and Dr. Nancy Cordell for the use of prosimian cadaver material. We thank Drs. R.F. Kay and W.L. Hylander for the use of laboratory space and materials. We thank N.N. Cordell, R.C. Fox, A. Hartstone-Rose, W.L. Hylander, K.R. Johnson, R.F. Kay, A.B. Taylor, C.J. Vinyard, and A.C. Zumwalt for their helpful advice. We thank NSF, NSERC, Duke University, University of Alberta, and Sigma Xi for funding.

References

Anapol, F. and Barry, K. (1996). Fiber architecture of the extensors of the hindlimb in semiterrestrial and arboreal guenons. *Am. J. Phys. Anthropol.* 99:429–447.

Antón, S. C. (1999). Macaque masseter muscle: internal architecture, fiber length, and cross-sectional area. *Int. J. Primatol.* 20:441–462.

Antón, S. C. (2000). Macaque pterygoid muscle: internal architecture, fiber length, and cross-sectional area. *Int. J. Primatol.* 21:131–156.

Arendsen de Wolff-Exalto, E. (1951). On differences in the lower jaw of animalivorous and herbivorous mammals I and II. *Proc. K. Ned. Akad. Wet. C* 54:237–246, 405–410.

Bearder, S. K. (1987). Lorises, bushbabies, and tarsiers: diverse societies in solitary foragers. In: Smuts, B. B., Cheney, D. L., Seyfarth, R. M., Wrangham, R. W. and Struhsaker, T. T. (eds.), *Primate Societies*, The University of Chicago Press, Chicago, pp. 11–24.

Cachel, S. M. (1979). A functional analysis of the primate masticatory system and the origin of the anthropoid post-orbital septum. *Am. J. Phys. Anthropol.* 50:1–18.

Cachel, S. M. (1984). Growth and allometry in primate masticatory muscles. *Arch.Oral Biol.* 29:287–293.

Charles-Dominique, P. (1972). Écologie et vie sociale de *Galago demidovii* (Fisher 1808; Prosimii). *Z. Tierpsychol., Beiheft* 9:7–41.

Chivers, D. J. and Hladik, C. M. (1980). Morphology of the gastrointestinal tract in primates: comparisons with other mammals in relation to diet. *J. Morphol.* 166:337–386.

Cordell, N. N. (1991). *Craniofacial Anatomy and Dietary Specialization in the Galagidae*. Ph.D. Dissertation, University of Washington, Seattle.

De Gueldre, G. and De Vree, F. (1988). Quantitative electromyography of the masticatory muscles of *Pteropus giganteus* (Megachiroptera). *J. Morphol.* 196:73–106.

Druzinsky, R. E. (1993). The time allometry of mammalian chewing movements: chewing frequency scales with body mass in mammals. *J. Theor. Biol.* 160:427–440.

Falster, D. S., Warton, D. I., and Wright I. J., (S)MATR: standardised major axis tests and routines, Version 1.0, 2003, Sydney (August 15, 2006); http://www.bio.mq.edu.au/ecology/SMATR

Fortelius, M. (1985). Ungulate cheek teeth: developmental, functional and evolutionary interrelationships. *Acta Zool. Fennica* 180:1–76.

Gans, C. (1982). Fiber architecture and muscle function. *Exerc. Sport Sci. Rev.* 10:160–207.

Gans, C. and De Vree, F. (1987). Functional bases of fiber length and angulation in muscle. *J. Morphol.* 192:63–85.

Gaspard, M., Laison, F. and Mailland, M. (1973a). Organisation architecturale et texture des muscles ptérygoïdiens chez les primates supérieurs. *J. Biol. Buccale* 1:215–233.

Gaspard, M., Laison, F. and Mailland, M. (1973b). Organisation architecturale et texture du muscle masseter chez les primates et l'homme. *J. Biol. Buccale* 1:7–20.

Gaspard, M., Laison, F. and Mailland, M. (1973c). Organisation architecturale du muscle temporal et des faisceaux de transition du complexe temporo-massétérin chez les primates et l'homme. *J. Biol. Buccale* 1:171–196.

Greaves, W. S. (1978). The jaw lever system in ungulates: a new model. *J. Zool. (Lond.)* 184:271–285.

Herring, S. W., Grimm, A. F., and Grimm, B. R. (1979). Functional heterogeneity in a multipinnate muscle. *Am. J. Anat.* 154:563–576.

Herring, S. W. and Herring, S. E. (1974). The superficial masseter and gape in mammals. *Am. Nat.* 108:561–576.
Hiiemae, K. M. and Kay, R. F. (1973). Evolutionary trends in the dynamics of primate mastication. In: Zingeser, M. R. (ed.), *Symposia of the Fourth International Congress of Primatology, Volume 3: Craniofacial Biology of Primates.* Karger, Basel, pp. 28–64.
Hylander, W. L. (1975). The human mandible: lever or link? *Am. J. Phys. Anthropol.* 43:227–242.
Hylander, W. L. (1985). Mandibular function and biomechanical stress and scaling. *Am. Zool.* 25:315–330.
Hylander, W. L. and Johnson, K. R. (1985). Temporalis and masseter muscle function during incision in macaques and humans. *Int. J. Primatol.* 6:289–322.
Hylander, W. L. and Johnson, K. R. (1994). Jaw muscle function and wishboning of the mandible during mastication in macaques and baboons. *Am. J. Phys. Anthropol.* 94:523–547.
Hylander, W. L., Vinyard, C. J., Ravosa, M. J., Ross, C. F., Wall, C. E., and Johnson, K. R. (2004). Jaw adductor force and symphyseal fusion. In: Anapol, F., German, R. Z. and Jablonski, N. G. (eds.), *Shaping Primate Evolution*. Cambridge University Press, Cambridge, pp. 229–257.
Jablonski, N. G. and Crompton, R. H. (1994). Feeding behavior, mastication, and tooth wear in the Western Tarsier (*Tarsius bancanus*). *Int. J. Primatol.* 15:29–59.
Kay, R. F. (1975). The functional adaptations of primate molar teeth. *Am. J. Phys. Anthropol.* 43:195–216.
Kay, R. F. and Sheine, W. S. (1979). On the relationship between chitin particle size and digestibility in the primate *Galago senegalensis*. *Am. J. Phys. Anthropol.* 50:301–308.
Krakauer, E. B. (2005). *Development of Aye-aye (*Daubentonia madagascariensis*) Foraging Skills: Independent Exploration and Social Learning*. Ph.D. Dissertation, Duke University, Durham.
Lucas, P. W. (2004). *Dental Functional Morphology: How Teeth Work*. Cambridge University Press, Cambridge.
Minkoff, E. C. (1968). *A Study of the Head, Neck, and Forelimb Muscles in the Genus* Galago*(Primates, Lorisidae)*. Ph.D. Dissertation, Harvard University, Cambridge.
Mittermeier, R. A., Konstant, W. R., Hawkins, F., Louis, E. E., Langrand, O., Ratsimbazafy, J., Rasoloarison, R., Ganzhorn, J. U., Rajaobelina, S., Tattersall, I. and Meyers, D. M. (2006). *Lemurs of Madagascar*, 2nd ed. Conservation International, Washington, D.C.
Murphy, R. A. and Beardsley, A. C. (1974). Mechanical properties of the cat soleus muscle *in situ*. *Am. J. Physiol.* 227:1008–1013.
Nowak, R. M. (1999). *Walker's Primates of the World*. The Johns Hopkins University Press, Baltimore.
Overdorff, D. J., Strait, S. G., and Telo, A. (1997). Seasonal variation in activity and diet in a small-bodied folivorous primate, *Hapalemur griseus*, in southeastern Madagascar. *Am. J. Primatol.* 43:211–223.
Perry, J. M. G. and Wall, C. E. (2005). A study of the scaling patterns of physiological cross-sectional area of the chewing muscles in prosimians. *Am. J. Phys. Anthropol.* 40S:165.
Rayne, J. and Crawford, G. N. C. (1972). The relationship between fibre length, muscle excursion and jaw movement in the rat. *Arch. Oral Biol.* 17:859–872.
Ross, C. F. (1995). Muscular and osseous anatomy of the primate anterior temporal fossa and the functions of the postorbital septum. *Am. J. Phys. Anthropol.* 98:275–306.
Schön, M. A. (1968). The muscular system of the Red Howling Monkey. *U. S. Nat. Mus. Bull.* 273:1–185.
Schumacher, G. H. (1961). *Funktionelle Morphologie der Kaumuskulatur*. Fisher, Jena.
Shahnoor, N., Anapol, F., Ross, C. (2005). Scaling of reduced physiological cross-sectional area in primate masticatory muscles. *Am. J. Phys. Anthropol.* 40S:188.
Smith, J. M. and Savage, R. J. G. (1959). The mechanics of mammalian jaws. *Sch. Sci. Rev.* 141:289–301.
Smith, K. K. and Redford, K. H. (1990). The anatomy and function of the feeding apparatus in two armadillos (Dasypoda): anatomy is not destiny. *J. Zool.* (London) 222:27–47.
Smith, R. J. (1978). Mandibular biomechanics and temporomandibular joint function in primates. *Am. J. Phys. Anthropol.* 49:341–350.

Spencer, M. A. (1999). Constraints on masticatory system evolution in anthropoid primates. *Am. J. Phys. Anthropol.* 108:483–506.

Starck, D. (1933). Die Kaumuskulatur der Platyrrhinen. *Gegenbaurs Morphologisches Jahrbuch* 72:212–285.

Sterling, E. J., Dierenfeld, E. S., Ashbourne, C. J. and Feistner, A. T. C. (1994). Dietary intake, food composition and nutrient intake in wild and captive populations of *Daubentonia madagascariensis*. *Folia Primatol.* 62:115–124.

Sussman, R. W. (1991). Demography and social organization of free-ranging *Lemur catta* in the Beza Mahafaly Reserve, Madagascar. *Am. J. Phys. Anthropol.* 84:43–58.

Sussman, R. W. (1999). *Primate Ecology and Social Structure: Volume 1: Lorises, Lemurs and Tarsiers*. Pearson Custom Publishing, Needham Heights, MA.

Taylor, A. B. and Vinyard, C. J. (2004). Comparative analysis of masseter fiber architecture in tree-gouging (*Callithrix jacchus*) and nongouging (*Saguinus oedipus*) callitrichids. *J. Morphol.* 261:276–285.

Turnbull, W. D. (1970). Mammalian masticatory apparatus. *Fieldiana: Geol.* 18:147–356.

Vizcaíno, S. F., De Iuliis, G. and Bargo, M. S. (1998). Skull shape, masticatory apparatus, and diet of *Vassallia* and *Holmesina* (Mammalia: Xenarthra: Pampatheriidae): when anatomy constrains destiny. *J. Mammal. Evol.* 5:291–322.

Wall, C. E., Vinyard, C. J., Johnson, K. R., Williams, S. H., and Hylander, W. L. (2006). Phase II jaw movements and masseter muscle activity during chewing in *Papio anubis*. *Am. J. Phys. Anthropol.* 129:215–224.

Weijs, W. A. and Hillen, B. (1985). Physiological cross-section of the human jaw muscles. *Acta Anat.* 121:31–35.

Zar. J. H. (1999). *Biostatistical Analysis*, 4th ed. Pearson Education (Singapore) Pte. Ltd., Delhi, India.

Chapter 12
The Relationship Between Jaw-Muscle Architecture and Feeding Behavior in Primates: Tree-Gouging and Nongouging Gummivorous Callitrichids as a Natural Experiment

Andrea B. Taylor and Christopher J. Vinyard

Contents

12.1	Introduction	241
	12.1.1 The Importance of Fiber Architecture for Understanding Muscle Function	243
	12.1.2 Feeding Behavior in Tree-Gouging and Nongouging Callitrichids: A Natural Model of Behavioral Divergence	244
	12.1.3 Are Tree Gougers Maximizing Jaw-Muscle Force or Muscle Stretch?	245
12.2	Materials and Methods	247
	12.2.1 Samples	247
	12.2.2 Data Collection	248
	12.2.3 Data Analysis	250
12.3	Results	251
12.4	Discussion	254
	12.4.1 The Functional Significance of Jaw-Muscle Fiber Architecture in Tree-Gouging Marmosets	254
	12.4.2 Functional Partitioning of the Masseter and Temporalis Muscles	256
	12.4.3 Integrating Fiber Architecture with Bony Morphology in Studies of Feeding Behavior	258
	References	259

12.1 Introduction

Feeding behavior and diet are among the most important factors underlying variation in primate craniofacial morphology. Thus, primate biologists interested in craniofacial evolution have directed a great deal of effort toward understanding the function and evolution of the primate masticatory apparatus. Teeth, because they are

---

A.B. Taylor
Doctor of Physical Therapy Division, Department of Community and Family Medicine, Duke University School of Medicine, and Department of Evolutionary Anthropology, Duke University, Box 104002, Durham, NC 27708
e-mail: andrea.taylor@duke.edu

both in direct contact with foods and comprise a significant portion of the primate fossil record, have been well studied from functional perspectives (e.g., Mills, 1963; Kay, 1975, 1978; Hershkovitz, 1977; Szalay and Delson, 1979; Gingerich et al., 1982; Lucas, 1982, 2004; Oxnard, 1987; Plavcan, 1993; Kay and Williams, 1994; Ungar, 1998; Jernvall and Jung, 2000; Teaford et al., 2000; McCollum and Sharpe, 2001; Swindler, 2002). We therefore have a basic understanding of how primate tooth shapes relate to the ingestion and mechanical break down of foods. *In vivo* investigations, many of which were carried out by Hylander and colleagues (e.g., Hylander, 1979a,b, 1984, 1985; Hylander and Johnson, 1994, 1997; Hylander et al., 1987, 1992, 1998, 2000, 2005; Ross and Hylander, 2000; Vinyard et al., 2005; Wall et al., 2006), have significantly advanced our knowledge of how primates recruit their jaw muscles, and how these muscle activity patterns create internal loads in the masticatory apparatus during mastication and incision. Likewise, numerous comparative analyses have demonstrated associations between jaw form and diet within and among primate species (e.g., Bouvier, 1986; Cole, 1992; Daegling, 1992; Antón, 1996; Taylor, 2002, 2005, 2006a,b; Vinyard et al., 2003). Functional and adaptive hypotheses must draw on knowledge gained through these experimental investigations, biomechanical modeling, and comparative approaches if we are going to better understand the evolution of the masticatory apparatus in primates.

Recognition that jaw muscles are recruited in specific patterns, and knowledge of how these activity patterns generate the forces and movements necessary to facilitate specific feeding behaviors, is vital to understanding morphological adaptations of the primate masticatory apparatus. Even though it is well understood that jaw muscles are responsible for generating jaw forces and movements, we know surprisingly little about the architecture of these muscles in primates. This is because studies seeking to link primate feeding behavior and morphology have largely overlooked how jaw-muscle fiber architecture affects masticatory apparatus performance. Muscle fiber architecture plays a crucial role in modulating movements and forces during feeding. Yet despite this role of the jaw muscles in feeding performance (e.g., Taylor and Vinyard, 2004; Taylor et al., in press), studies have focused almost exclusively on bony form, without integrating information from jaw-muscle architecture. We suggest there are important insights to be gained from examining fiber architecture, which can be applied to comparative studies of masticatory function and adaptation in both living and fossil primates.

We begin this chapter with a brief description of muscle architecture and its importance for understanding muscle function. Subsequently, we illustrate the importance of integrating jaw-muscle fiber architecture and skull morphology in a study of feeding behavior using gummivorous callitrichids. Specifically, we compare fiber architecture of the jaw-closing muscles in tree-gouging common (*Callithrix jacchus*) and pygmy (*Cebuella pygmaea*) marmosets to that of nongouging cotton-top tamarins (*Saguinus oedipus*). Callitrichids provide a natural model for addressing morphological divergence as a function of feeding behavior both because these feeding behaviors are well documented from both field and laboratory research and because these taxa are closely related at the subfamily level (Callitrichinae:Primates), thereby limiting the influence of phylogenetic history. We demonstrate that tree-gouging marmosets exhibit architectural features of the masseter and temporalis muscles that can be functionally linked to tree-gouging based on the comparison to nongouging callitrichids. Thus, knowledge of jaw-muscle fiber architecture broadens and refines our understanding of the functional and/or adaptive correlates of this specialized feeding behavior. We end this chapter by exploring what this case study suggests for future research integrating jaw-muscle architecture with the existing data on primate masticatory apparatus form and function.

12.1.1 *The Importance of Fiber Architecture for Understanding Muscle Function*

At both macroscopic and microscopic levels, skeletal muscle is a highly organized tissue. Skeletal muscle is arranged hierarchically from whole muscle, to muscle fascicles, fibers, and myofibrils, down to the functional unit of contraction, the sarcomere. The sarcomere is comprised of smaller myofilaments, the thin actin and thicker myosin proteins that overlap each other and form cross-bridges, thereby providing the contractile properties of a muscle fiber (Lieber, 2002). The length–tension relationship describes the amount of isometric tension that a muscle can generate at varying amounts of overlap between the actin and the myosin filaments (Gordon et al., 1966). Maximum tetanic tension occurs at the plateau region of the length–tension curve, when the number of cross-bridges formed between the actin and the myosin filaments is maximized (Gordon et al., 1966).

Myofibrils are arranged in parallel, influencing the physiological cross-sectional area, while sarcomeres are arranged longitudinally in series (i.e., end-to-end), thereby modulating the distance through which a muscle fiber can move and, by extension, its contraction velocity. All other factors being equal, the thicker the muscle fiber, the greater its force-producing capability, while the longer the fiber, the greater the distance and speed through which a fiber can shorten (or lengthen). Thus, both fiber diameter and length influence how a muscle fiber functions.

If all fibers were arranged parallel to the force-generating axis of the muscle, then estimating the force-producing capability of a muscle would be a simple task of measuring a muscle's volume. This is because the maximum potential force a muscle can generate in a given direction is equal to the sum of the contractile forces produced by all of the individual fibers oriented in that direction. In reality, however, the directional arrangement of individual muscle fibers varies within a muscle. This variation can be extrapolated to differences in orientation both across species for a given muscle and across muscles within a given species.

Some muscles comprise parallel or nearly parallel fibers, such as the vastus medialis muscle in humans and guenons (Anapol and Barry, 1996; Lieber, 2002). Most muscle fibers, however, are angled relative to the force-generating axis of the muscle. Unipinnate muscles comprise fibers oriented at a single angle relative to the axis of force generation, while multipinnate muscles comprise fibers oriented at several angles. Pinnation angle for these muscles generally ranges from 0° to 30°; the force a muscle can generate drops precipitously beyond 60°; Gans, 1982. In rodents and other mammals, the jaw-closing muscles, and the masseter in particular, tend to be among the most complex and highly pinnate-fibered muscles (Herring, 1992; van Eijden et al., 1997). For example, angle of pinnation in pig masseter ranges from 0 to 25° (Herring et al., 1979). We note, however, that even in so-called parallel-fibered muscles, fibers rarely extend the entire length of the muscle (Lieber, 2002).

Because the directional arrangement of muscle fibers varies, muscle weight or volume is insufficient to inform us of a muscle's capacity to produce force or to move through a given range. We need additional information on fiber length (l_f), fiber pinnation angle, and physiological cross-sectional area (PCSA). Because the total shortening distance of a muscle fiber is equal to the sum of the shortening distances of the individual sarcomeres (Lieber, 2002), muscle fiber length (l_f) is proportional to a muscle's maximum potential range of motion (excursion) and contraction velocity. Physiological cross-sectional area is proportional to the maximum force that a muscle can generate, and is computed as:

$$\text{muscle mass (gm)} \times \cos\theta / l_f(\text{cm}) \times 1.0564\,\text{gm/cm}^3,$$

(Gans and Bock, 1965; Powell et al., 1984.) Here, θ represents the angle of pinnation, or the orientation of the fiber relative to the force-generating axis of the muscle. It has been both theoretically argued (e.g., Gans and Bock, 1965; Gans and de Vree, 1987) and empirically demonstrated for jaw muscles (e.g., Taylor and Vinyard, 2004; Perry and Wall, 2008; Shahnoor et al., 2008) that there is a trade-off between PCSA and fiber length. Thus, a muscle can be best suited to either force production or excursion, but not both.

Finally, muscle fibers will be shortened when a muscle is contracting concentrically. This shortening will increase angles of pinnation relative to their resting position. Stretching a muscle will produce the opposite effect. This means that the posture in which a muscle is measured will affect fiber length and, by extension, all variables involving fiber length. Ideally, fiber length and angle of pinnation should be measured with a muscle at its resting length (Lieber, 2002). In practice, however, this is not always possible. Comparative analyses of fiber architecture should minimally measure muscles from different individuals in similar functional positions if not at resting length (e.g., incisor tip-to-tip occlusion for the jaw muscles) (Taylor and Vinyard, 2004).

12.1.2 Feeding Behavior in Tree-Gouging and Nongouging Callitrichids: A Natural Model of Behavioral Divergence

Callitrichid monkeys have been described as representing an adaptive radiation of gum feeders (Sussman and Kinzey, 1984). Gums provide an important food source and some callitrichids, such as pygmy marmosets (*Cebuella pygmaea*), can spend up to 70% of their feeding time consuming tree exudates during certain times of the year (Garber, 1992 and references therein). While all callitrichids feed on gums, they vary in terms of how they gain access to these gums. Some callitrichids, such as the cotton-top tamarin (*Saguinus oedipus*), are opportunistic gum feeders, capitalizing on tree exudates that have been exuded through damage by insects or other means. By contrast, common (*Callithrix jacchus*) and pygmy (*C. pygmaea*) marmosets actively elicit exudate flow by mechanically injuring trees with their anterior dentition. We define this type of biting behavior as tree gouging. During tree gouging, marmosets anchor their upper jaw in the tree substrate while using their lower jaw to scrape holes in the tree bark to stimulate exudate flow (Coimbra-Filho and Mittermeier, 1977). They later return to eat these exudates.

This divergence in feeding behavior makes tree-gouging and nongouging callitrichids an exceptional natural model for evaluating the relationship between muscle fiber architecture, muscle function, and feeding behavior. This is because gouging marmosets differ from nongouging tamarins based on the presence or absence of this gouging behavior, rather than forming part of a continuum of feeding behaviors. Furthermore, apart from this distinction in gum-feeding behavior, tree-gouging marmosets (*C. jacchus* and *C. pygmaea*) and nongouging tamarins (*S. oedipus*) are closely related genera that have similar diets consisting of insects, fruits, and gums. Thus, gouging is likely to be a feeding behavior that uniquely differentiates marmosets from tamarins.

12.1.3 Are Tree Gougers Maximizing Jaw-Muscle Force or Muscle Stretch?

Numerous investigators (e.g., Szalay and Seligsohn, 1977; Szalay and Delson, 1979; Dumont, 1997; Spencer, 1999) have hypothesized that tree gouging requires the generation of large jaw forces. To the contrary, preliminary *in vivo* work (Vinyard et al., 2001, in press; Mork et al., 2004) suggests that common marmosets generate jaw forces during simulated tree gouging that are significantly less than their maximum biting ability. The *in vivo* work also reveals that maximum jaw gapes during gouging are significantly larger than those during chewing. Importantly, both laboratory and field studies indicate that common marmosets use jaw gapes during gouging which approach their maximum structural capacity for jaw opening (Vinyard et al., 2001, 2004; in press).[1] Increased gapes have been hypothesized to increase mandibular excursion and/or facilitate optimal incisor alignment for penetrating the tree substrate during a gouge (Vinyard et al., 2003). Collectively, these studies strongly suggest that the ability to generate wide jaw gapes is important for tree-gouging primates, and that any musculoskeletal features facilitating the production of wide gapes would be advantageous.

Comparative studies have shown that tree-gouging primates exhibit morphological features of the bony masticatory apparatus that distinguish them from closely related nongouging taxa. For example, the tree-gouging *Callithrix jacchus*

---

[1] Maximum jaw-opening ability was measured on anesthetized animals by passively opening the jaws to their widest gape.

and *Cebuella pygmaea* have incisiform canines and long, chisel-like incisors with very thick labial and thin lingual enamel compared to nongouging callitrichids (Coimbra-Filho and Mittermeier, 1977; Rosenberger, 1978). This morphology has been referred to as the "short-tusked" condition (e.g., Hill, 1957; Napier and Napier, 1967; Coimbra-Filho and Mittermeier, 1977), and has been linked to the use of the canines and incisors as a functional unit for gouging trees.

Tree-gouging primates, including common marmosets, are characterized by skull shapes that are theoretically advantageous for achieving wide jaw gapes when compared to closely related, nongouging taxa. These features include relatively low condyles positioned closer to the height of the tooth row, anteroposteriorly elongated condyles (correlated with greater condylar curvature) and temporal articular surfaces, and higher superficial masseter origin–insertion ratios (Vinyard et al., 2003). Masseter muscle position, as reflected by a masseter origin–insertion ratio that deviates from 1.0 and a low condylar height, has been geometrically linked to the capacity to generate large gapes via reduction in masseter stretching (Herring and Herring, 1974). Carnivores, for example, which require wide jaw gapes for feeding, display, and/or fighting, show high superficial masseter origin–insertion ratios and relatively low condyles compared to other mammals where gape requirements are less important (Herring and Herring, 1974). A relatively low condyle, as exhibited by *C. jacchus*, is one way of reducing masseter stretch for a given gape. Individuals with lower condyles have an increased included angle from the masseter origin – condyle – masseter insertion (all other things being equal) (Fig. 12.1). An origin–insertion ratio that deviates from 1.0, and/or an increase in the aforementioned included angle, is theoretically predicted to reduce superficial masseter stretching and hence potentially increase maximum gape. The available evidence from the bony parts of the masticatory apparatus, therefore, strongly suggests that tree-gouging marmosets should be optimizing muscle excursion over muscle force production.

Here we present an analysis of the superficial masseter and anterior temporalis muscles in tree-gouging and nongouging callitrichids. Based on a geometric model, stretching of the masseter and temporalis muscles is thought to potentially limit maximum jaw opening in most mammals (Herring and Herring, 1974). Empirical evidence from pig masseter demonstrates that fiber elongation during gape, which may be as great as 50% of muscle resting length, is greatest at the anterior margin of the masseter relative to other regions along the muscle (Herring et al., 1979). Moreover, the longest masseter fibers are observed to lie in the anterior portion of the muscle in pigs (Herring et al., 1979), as well as the marmosets and tamarins examined here (Taylor and Vinyard, 2004).[2]

---

[2] Fiber lengths change depending on the position of the jaws at the time of fixation. We observed longer fibers in the posterior region of the masseter muscle in common marmosets and cotton-top tamarins when the jaws were fixed with the incisors in tip-to-tip occlusion (Taylor and Vinyard, 2004). However, when adjusted for muscle resting length, the anterior fibers were observed to be the longest (unpublished data).

Fig. 12.1 Schematic representation of the upper and lower jaws in lateral view depicting (**a**) the distance from the condyle to both the masseter origin and insertion (*solid lines*). Idealized length of the most anterior masseter fibers is represented as the dark dotted line, while gape is shown between the upper and the lower incisors. Positioning the condyle closer to the tooth row (**b**) compared to a condyle positioned farther from the toothrow (**c**) results in a larger gape for the same amount of stretch in the anterior masseter fibers. (Note that the length of the dark dotted line is similar in (**b**) and (**c**).) Thus, with the condyle positioned closer to the tooth row, a wider gape can be achieved with less stretch of the masseter for a given degree of angular rotation of the mandible

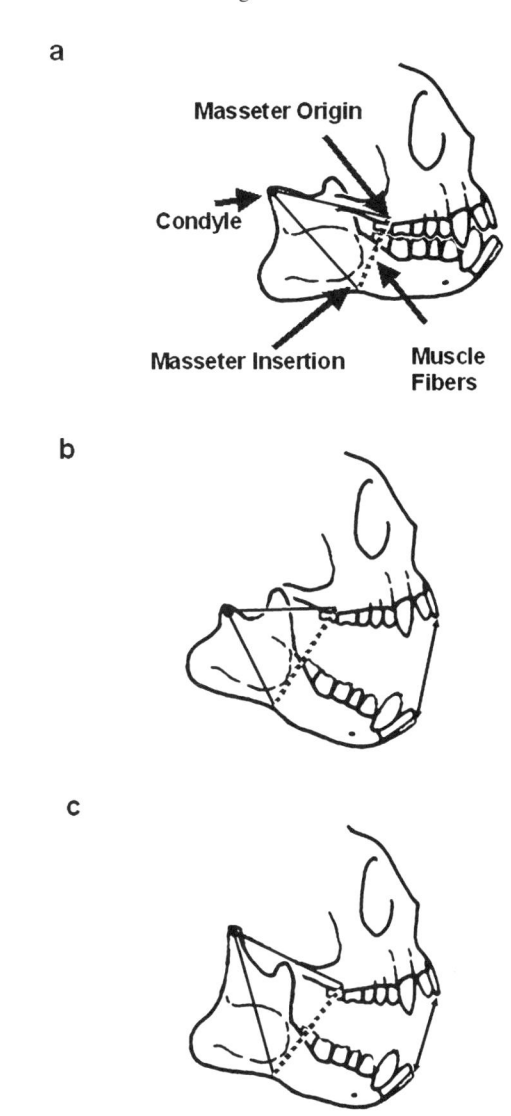

12.2 Materials and Methods

12.2.1 Samples

We analyzed the superficial masseter and anterior temporalis muscles of 15 *Callithrix jacchus*, 5 *Cebuella pygmaea*, and 9 *Saguinus oedipus*. We determined adult status based on the patterns of tooth eruption and occlusion, as well as husbandry records documenting age in years. Cadavers were provided courtesy of the Wisconsin National Primate Research Center, New England Primate Research Center, and Southwest National Primate Research Center.

12.2.2 Data Collection

We removed the superficial fat and fascia overlying the jaw muscles and photographed the specimens in lateral view. We measured masseter and temporalis muscle lengths from intact muscles with calipers to the nearest 0.01 mm. We then dissected the masseter and temporalis muscles free from their bony attachments, trimmed them of excess tendon and fascia, blotted them dry, and weighed them to the nearest 0.0001 g. We measured jaw length (to the nearest 0.01 mm) from the posterior edge of the condyle to prosthion on each specimen.

We measured fasciculus length and angle of pinnation for both muscles, following Taylor and Vinyard (2004). Briefly, we bisected the masseter into a superior

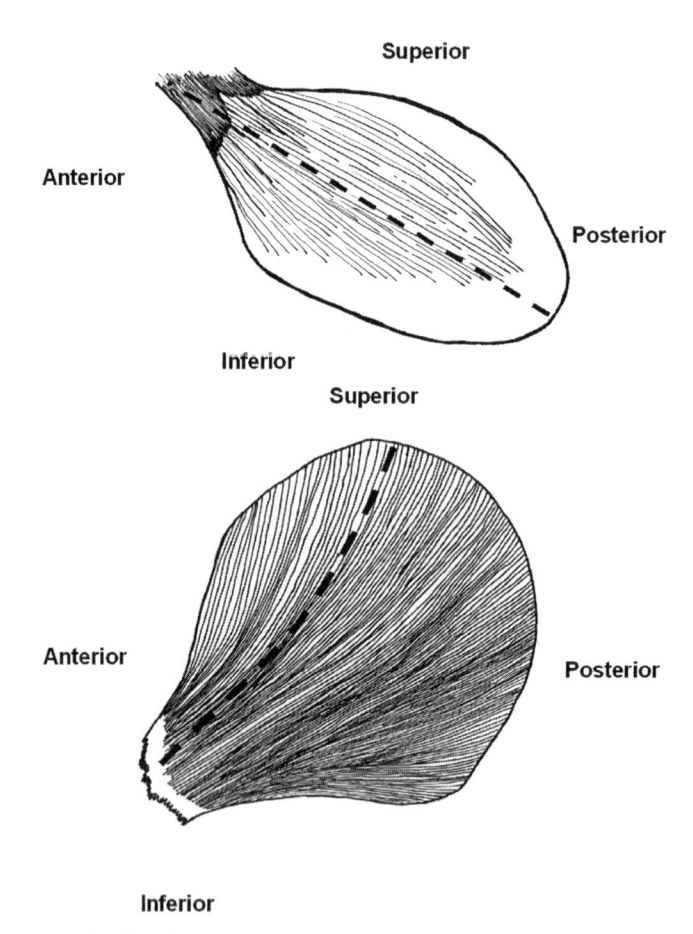

Fig. 12.2 Schematic of a left masseter muscle (**a**) and temporalis muscle (**b**) of *Callithrix jacchus*. The masseter muscle was bisected from superficial to deep, as depicted by the dotted line (**a**). The inferior portion was used for data analysis. The temporalis muscle was bisected into anterior and posterior portions, as illustrated by the dotted line, and the anterior portion used for data analysis (**b**) (Figure 12.2**a** reprinted with permission from John Wiley & Sons, Inc.)

and inferior portion along its length, roughly approximating the muscle's line of action (Fig. 12.2a). We analyzed the inferior portion in order to avoid the deep masseter, which has a different fiber orientation and muscle action than the superficial masseter. We bisected the temporalis muscle into anterior and posterior portions and analyzed the anterior portion (Fig. 12.2b). We oriented each segment so as to view the fibers in cross section, pinned the segment to a styrofoam block, and then visualized the proximal and distal attachments to tendon of individual fasciculi under a magnification light. Because of the exceptionally small size of the *Cebuella* muscles, we employed a pair of Zeiss (×2.3) binocular glasses in addition to the magnification light during data collection.

Depending on the position of the jaws at the time of fixation, muscle fibers may be either elongated or contracted relative to their resting lengths. To control for the effects of position at fixation on whole muscle and fiber lengths across individuals (and by extension, all variables involved), we only included specimens whose jaws were fixed in a standardized jaw posture. Following Taylor and Vinyard (2004), the incisors were in tip-to-tip occlusion in all specimens.

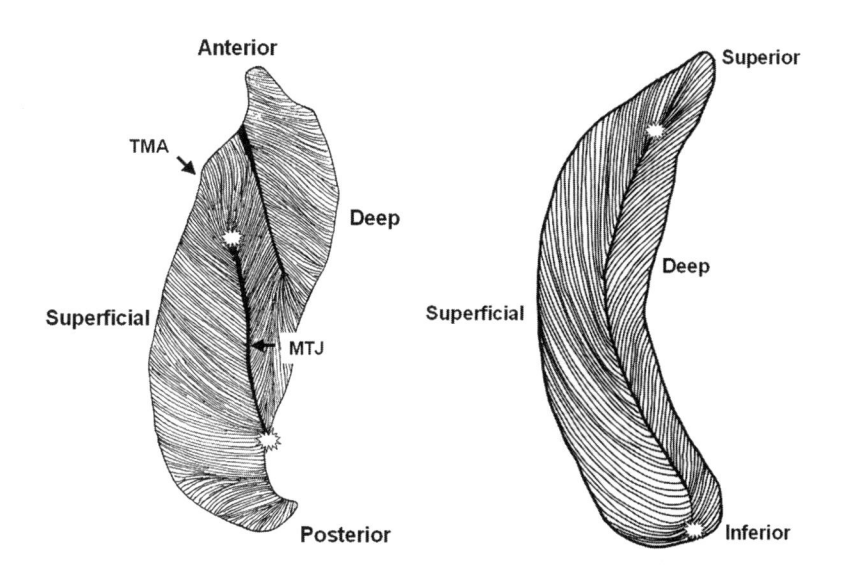

Fig. 12.3 Superior view of the internal architecture of marmoset masseter (**a**) and temporalis (**b**). The masseter is multipinnate in *Callithrix jacchus*, *Cebuella pygmaea*, and *Saguinus oedipus*. The fibers of the superficial portion of the masseter are bipinnate, and fasciculus measurements were taken from the myotendinous junction (MTJ) to the tendon of muscle attachment along the region of the zygomatic arch (TMA). The stars at the anterior and posterior ends of the MTJ mark the sampling sites for anterior and posterior fibers, respectively. The temporalis is bipinnate in all taxa. Fasciculus measurements were taken from the MTJ to the tendon of muscle attachment along the temporal bone. The stars at the superior and inferior ends of the MTJ mark the sampling sites for superior and inferior fibers, respectively (Figure 12.3a reprinted with permission from John Wiley & Sons, Inc.)

We selected anterior and posterior sampling sites for measurements along the length of the superficial masseter, whereas we chose proximal and distal sampling sites for the anterior temporalis muscle (Fig. 12.3). At each sampling site, we measured a maximum of six adjacent fasciculi. For each fasciculus, we measured the following: (1) fasciculus length, between the proximal and the distal myotendinous junctions (l_f); (2) the perpendicular distance from the tendon of insertion to the proximal attachment of the fasciculus (a); (3) the length of tendon from the proximal bony attachment to the proximal myotendinous junction (t_p); and (4) the length of tendon from the distal bony attachment to the distal myotendinous junction (t_d) (Table 12.1 and Fig. 12.4). We calculated the angle of pinnation (θ) as the arcsin of a/l_f (Anapol and Barry, 1996) (Table 12.1; Fig. 12.4).

12.2.3 Data Analysis

We used one-tailed Mann-Whitney U-tests to address the hypothesis that tree-gouging *Callithrix jacchus* and *Cebuella pygmaea* exhibit architectural properties of the masseter and temporalis muscles that are functionally linked to the production

Table 12.1 Muscle fiber architecture variables and predicted differences between tree-gouging marmosets (*Callithrix jacchus* and *Cebuella pygmaea*) and a nongouging tamarin (*Saguinus oedipus*)

Variable	Definition	Prediction
Fiber length for the superficial masseter and anterior temporalis muscles (l_f)	Calculated as the average fiber length of the six anterior and six posterior fasciculi for the superficial masseter, and the six proximal and six distal fasciculi for the anterior temporalis muscles.	l_f should be relatively greater in *C. jacchus* and *C. pygmaea* compared to *S. oedipus*.
Pinnation angle (θ) = arcsin a/l_f	Angle of fibers relative to the axis of force generation.	θ should be greater in *S. oedipus* compared to *C. jacchus* and *C. pygmaea*.
Maximum excursion of the anterior and distal tendons of attachment (h) for the masseter and temporalis muscles, respectively.	$l_f (\cos \theta - \sqrt{\cos \theta^2 + n^2 - 1})$, where θ represents the pinnation angle, and n is the coefficient of contraction: (fiber length after contraction/resting fiber length) = 0.767 (Anapol and Gray (2003) after Gans and Bock (1965) and Muhl (1982)).	h should be greater in *C. jacchus* and *C. pygmaea* compared to *S. oedipus*.
Physiological cross-sectional area (PCSA)	(muscle mass (g) × cos θ)/(l_f (cm) × 1.0564 g/cm^3), where θ represents the pinnation angle, and 1.0564 the specific density of muscle (Mendez and Keys, 1960; Powell et al., 1984).	PCSA should be relatively greater in *S. oedipus* compared to *C. jacchus* and *C. pygmaea*.
$[l_t/(l_f + l_t)]$	Ratio of total tendon length (l_t) to fasciculus + tendon length ($l_f + l_t$) (Anapol and Barry, 1996).	$[l_t/(l_f + l_t)]$ should be greater in *S. oedipus* compared to *C. jacchus* and *C. pygmaea*.

12 The Relationship Between Jaw-Muscle Architecture and Feeding Behavior 251

Fig. 12.4 Schematic of a bipinnate muscle depicting the measurements taken in this study (from Taylor and Vinyard, 2004). These measurements include: (1) fasciculus length (l_f); (2) the perpendicular distance from the proximal myotendinous junction to the tendon of anterior or distal muscle attachment (**a**); (3) length of tendon attaching to the most proximal end of the fasciculus (t_p); (4) length of tendon attaching to the anterior or distal-most end of the fasciculus (t_d); (5) angle of pinnation (θ)

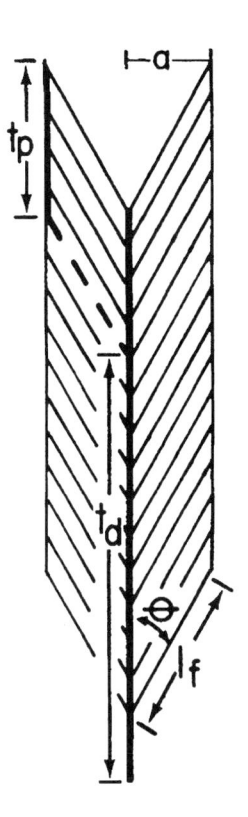

of wide jaw gapes (Table 12.1). We evaluated differences in masseter and temporalis fiber lengths and associated architectural variables by holding other functional aspects of jaw gape constant. Specifically, we created shape ratios by dividing fiber length and the square root of PCSA against jaw length. The maximum excursion of the anterior masseter and distal temporalis tendons of attachment (h) were standardized relative to whole muscle length (Taylor and Vinyard, 2004).

12.3 Results

Tree-gouging common (*C. jacchus*) and pygmy (*C. pygmaea*) marmosets exhibit architectural features of the masseter and temporalis muscles that are functionally linked to facilitating muscle stretch and muscle excursion when compared to cotton-top tamarins (*S. oedipus*) (Table 12.2). For example, *C. jacchus* has both absolutely (Fig. 12.5) and relatively (Fig. 12.6) longer fibers for both muscles compared to *S. oedipus*. While *C. pygmaea* has relatively longer masseter fibers than *S. oedipus*, the difference in temporalis fiber length between these two taxa only approaches significance following Bonferroni adjustment (Fig. 12.6). Both marmosets have significantly greater maximum excursions of the anterior and distal tendon attachments of the masseter and temporalis muscles, respectively, compared to cotton-top

Table 12.2 Comparison of masseter and temporalis fiber architecture between tree-gouging marmosets (*Callithrix jacchus* and *Cebuella pygmaea*) and a non-gouging tamarin (*Saguinus oedipus*)

	C. jacchus vs. *S. oedipus*	Direction of difference	*C. pygmaea* vs. *S. oedipus*	Direction of difference
Masseter muscle				
Fiber length (mm)	**0.0035**	Cj > So	**0.0200**	So > Cp
Fiber length (mm)/jaw length (mm)	**0.0005**	Cj > So	**0.0285**	Cp > So
h/masseter muscle length (mm)	**0.0015**	Cj > So	**0.0040**	Cp > So
PCSA (cm^2)/jaw length (mm)	**0.0005**	So > Cj	**0.0015**	So > Cp
Total tendon length per muscle fasciculus	**0.0055**	So > Cj	**0.0040**	So > Cp
Pinnation angle	0.3500	So = Cj	**0.0015**	So > Cp
Temporalis muscle				
Fiber length (mm)	**0.0000**	Cj > So	0.1430	So = Cp
Fiber length (mm)/jaw length (mm)	**0.0000**	Cj > So	0.0480	Cp > So
h/temporalis muscle length (mm)	**0.0000**	Cj > So	**0.0045**	Cp > So
PCSA (cm^2)/jaw length (mm)	**0.0000**	So > Cj	**0.0015**	So > Cp
Total tendon length per muscle fasciculus	**0.0015**	So > Cj	**0.0100**	So > Cp
Pinnation angle	0.2105	So = Cj	**0.0140**	So > Cp

All boldface p-values significant based on one-tailed Mann-Whitney U-tests and significant at $p < 0.05$.

tamarins (*h*/muscle length) (Table 12.2). By contrast, cotton-top tamarins exhibit significantly greater PCSAs and higher proportions of muscle tendon to fiber (Figs. 12.7 and 12.8). Pinnation angles are significantly smaller in *C. pygmaea* compared to *S. oedipus*, but *C. jacchus* and *S. oedipus* do not differ in pinnation angles (Table 12.2).

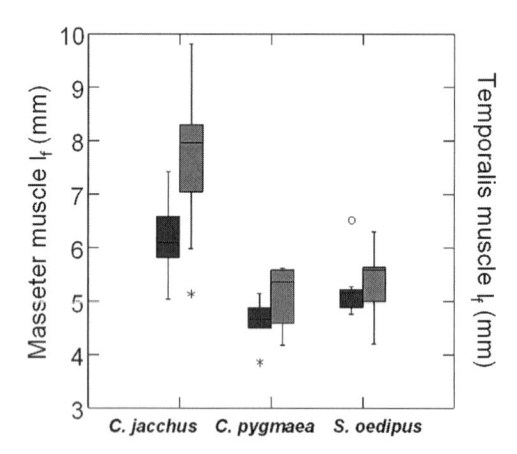

Fig. 12.5 Box plots comparing masseter and temporalis muscle fiber lengths (l_f). Masseter muscle fiber length depicted on the left (dark hatched box), temporalis fiber length on the right (light hatched box). *Callithrix jacchus* has significantly longer masseter and temporalis fibers. *Cebuella pygmaea* has fiber lengths that approach those of *Saguinus oedipus*, at roughly 25% of *S. oedipus*' body weight. Sample sizes for the masseter and temporalis muscles include *C. jacchus* (12, 15), *C. pygmaea* (5, 5) and *S. oedipus* (8, 9), respectively

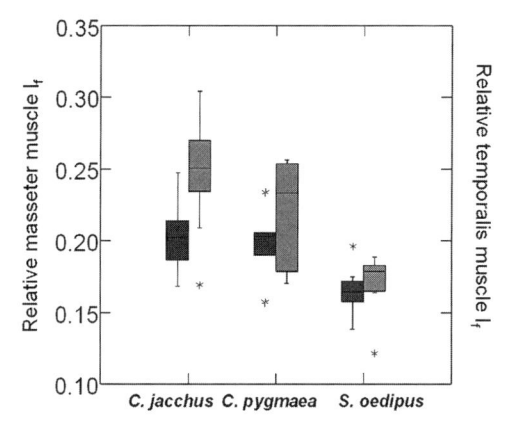

12 The Relationship Between Jaw-Muscle Architecture and Feeding Behavior 253

Fig. 12.6 Box plots comparing relative masseter and temporalis muscle fiber lengths (l_f/jaw length). Masseter muscle fiber length depicted on the left (dark hatched box), temporalis fiber length on the right (light hatched box). *Callithrix jacchus* and *Cebuella pygmaea* have relatively longer masseter and temporalis muscles compared to *Saguinus oedipus*, though the difference in relative temporalis fiber length only approaches significance between *C. pygmaea* and *S. oedipus* (p = 0.048) after Bonferroni adjustment. These data indicate that tree-gouging marmosets have jaw-closing muscles that facilitate increased muscle stretch during jaw opening, and thus the production of relatively wide jaw gapes. Sample sizes as in Fig. 12.2

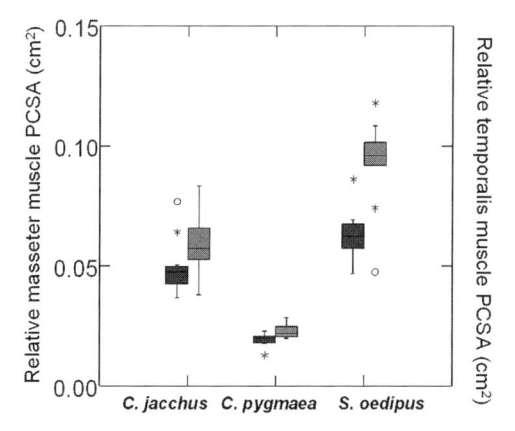

Fig. 12.7 Box plots comparing relative masseter and temporalis PCSAs (PCSA/jaw length). Masseter muscle PCSA depicted on the left (dark hatched box), temporalis PCSA on the right (light hatched box). *Saguinus oedipus* has significantly greater relative PCSAs for the masseter and temporalis muscles compared to *Callithrix jacchus* and *Cebuella pygmaea*. These data suggest that *S. oedipus* has the potential to generate relatively greater muscle and bite force compared to tree-gouging marmosets. Sample sizes as in Fig. 12.2

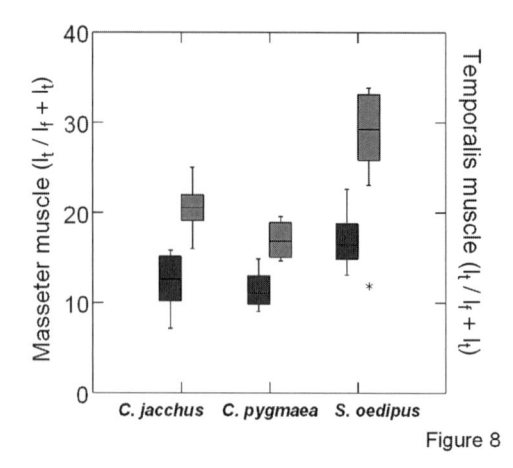

Fig. 12.8 Box plots comparing the ratio of total tendon length to fasciculus + tendon length $[l_t/(l_f + l_t)]$. Masseter muscle depicted on the left (dark hatched box), temporalis on the right (light hatched box). *Saguinus oedipus* has a significantly greater ratio of tendon to fiber for both the masseter and the temporalis muscles compared to *Callithrix jacchus* and *Cebuella pygmaea*. Lower ratios suggest the potential for greater neural control over jaw movement and jaw posture in marmosets. Sample sizes as in Fig. 12.2

12.4 Discussion

12.4.1 The Functional Significance of Jaw-Muscle Fiber Architecture in Tree-Gouging Marmosets

Results presented here demonstrate that tree-gouging marmosets have masseter and temporalis muscles that are well suited to facilitate muscle stretch. These features include relatively longer fibers and relatively greater maximum excursion capabilities. Elsewhere (e.g., Taylor and Vinyard, 2004; unpublished data), we have shown that common marmosets exhibit relatively longer fibers in both the anterior and the posterior portions of the masseter, and have higher ratios of muscle mass to effective maximal tetanic tension (M/P_0) for both the masseter and the temporalis. A higher ratio of M/P_0 (where the specific tension of muscle has been empirically determined to be 2.3 kg/cm^3; Spector et al., 1980; Sacks and Roy, 1982; Powell et al., 1984), reflects the dedication of a muscle to excursion and contraction velocity over its capacity to generate force (Sacks and Roy, 1982; Wottiez et al., 1986; Weijs et al., 1987; Anapol and Barry, 1996).

Marmosets also have lower ratios of tendon length to muscle fiber + tendon length $[l_t/(l_f + l_t)]$ for both the masseter and the temporalis muscles (Table 12.1; Fig. 12.8). This ratio contrasts the energy cost required for generating tension (Anapol and Gray, 2003) versus the relative amount of neural control in the muscle–tendon unit. Lower $l_t/(l_f + l_t)$ ratios indicate less noncontractile tendon relative to contractile fiber, suggesting higher energy cost because more of the muscle force is actively generated through hydrolysis of adenosine triphosphate (ATP) rather than

12 The Relationship Between Jaw-Muscle Architecture and Feeding Behavior 255

by passive tension stored in the tendon (Lieber, 2002). However, because muscle fibers have increased neural control relative to tendon (Biewener and Roberts, 2000; Anapol and Gray, 2003), a greater proportion of contractile tissue indicates increased neural control of the masseter and temporalis muscles throughout their ranges of motion. We hypothesize that tree gouging is a highly modulated behavior that would benefit from greater, more deliberate neural control of muscle stretch as a means of minimizing the risk of injury to the masticatory apparatus.

For marmosets, the greater capacity for muscle excursion comes at the expense of a decrease in relative force production. Previous investigators have theoretically argued (Gans and Bock, 1965; Gans and de Vree, 1987) and empirically demonstrated (Anapol and Barry, 1996; Anapol and Gray, 2003; Taylor and Vinyard, 2004) that it is difficult for an individual muscle to simultaneously maximize excursion and force production. Our results demonstrate that there is an architectural trade-off in callitrichid jaw muscles between longer fibers, which facilitate muscle excursion, and shorter, sometimes more pinnate fibers, which improve force production

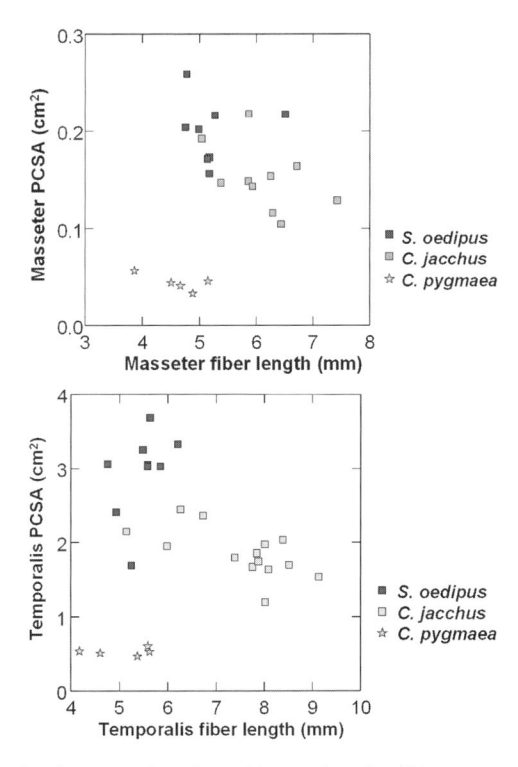

Fig. 12.9 Bivariate plot demonstrating the architectural trade-off between muscle force (PCSA) and muscle excursion (fiber length) for masseter (**a**) and temporalis (**b**) muscles. In general, marmoset masseter and temporalis, with their relatively low PCSAs and long fibers, are suited more for muscle excursion (i.e., production of wide gapes), while in *Saguinus oedipus* the larger PCSAs and shorter fibers make these muscles better suited for generating larger muscle forces with smaller excursions

(Fig. 12.9). Thus, *Saguinus oedipus*, with relatively higher PCSAs for the masseter and temporalis muscles, has the capacity to generate relatively greater maximal muscle force compared to marmosets. Furthermore, the higher ratios of tendon to fiber indicate that force production in *S. oedipus* is energetically more efficient compared to force production in marmosets. That said, we have no dietary or performance data to indicate that *S. oedipus* generates relatively greater maximum muscle or bite forces compared to marmosets. In lieu of such data, we conservatively interpret our finding of relatively greater PCSAs in *S. oedipus* as reflecting this structural trade-off between muscle excursion (fiber length) and muscle force (PCSA).

The combination of altered jaw-muscle fiber architecture shown here and muscle orientation (e.g., Vinyard et al., 2003) provide compelling evidence that tree-gouging common and pygmy marmosets are maximizing muscle stretch abilities. During the generation of wide jaw gapes, which involves active lengthening of the jaw-closing muscles, it is likely that marmosets are stretching their jaw-closing muscle well beyond their resting length. Bite forces in primates diminish as jaw-muscle fibers are stretched beyond their resting length (Dechow and Carlson, 1982, 1986, 1990), and decreases in bite force as gape increases are exacerbated by relatively inefficient jaw leverage for producing bite force at the anterior teeth. Alterations in both the bony and the soft tissue structures of the masticatory apparatus strongly suggest that tree-gouging marmosets are under pressure to reduce the amount of muscle stretch at a given gape in order to generate adequate bite forces at wide jaw gapes.

Collectively, relatively longer fibers of the masseter and temporalis muscles correspond with field and laboratory studies demonstrating that common marmosets gouge trees with relatively wide jaw gapes. The relatively lower muscle PCSAs support previous morphological and laboratory studies, which found no evidence that tree gouging requires generating relatively large bite forces (Vinyard et al., 2001, 2003, 2004, n.d.). The presence of relatively longer fibers, their capacity to enhance muscle stretch during jaw opening, and the performance data indicating that tree gouging involves the generation of wide jaw gapes, all suggest that longer masseter and temporalis fibers function to facilitate the production of wide jaw gapes during tree gouging in their natural environment (sensu Bock and von Wahlert, 1965).

12.4.2 Functional Partitioning of the Masseter and Temporalis Muscles

Most mammals have a basic masticatory muscle plan that includes four paired, recognizably distinct muscles: the masseter, temporalis, medial and lateral pterygoid muscles (e.g., Turnbull, 1970). Within this common organizational structure, the complexity of muscle architecture, the position of the muscles on the bony masticatory apparatus, and the proportional contribution to total muscle mass vary widely across taxa. Nevertheless, the primitive pattern for mammals, and one to which all three of these callitrichids conform, is characterized by a temporalis muscle that is relatively larger than the masseter and pterygoid muscles (Turnbull, 1970). Thus, while the force-generating capacity (PCSA) of the masseter and temporalis muscles are relatively lower in marmosets compared to tamarins, all three species have larger temporalis muscle weights (Eng et al., 2005) and maximum force-generating capacities relative to their masseters (Fig. 12.7).

If we consider the temporalis and masseter as part of a larger functional group involved in feeding, we can ask whether there are differences in the relative contributions of these two muscles to the overall force- and excursion-producing capabilities of the jaw-muscle group. If the PCSA of each muscle is expressed as a percentage of the combined PCSA for both muscles, it is apparent that there is less intermuscular diversity between the masseter and the temporalis muscles in marmosets compared to tamarins. That is, the force-generating capacities (PCSAs) of the two jaw-closing muscles are more similar within the marmosets than cotton-top tamarins (Fig. 12.10). Thus, while all taxa exhibit significant intermuscular diversity ($p < 0.05$), the disparity in % PCSA for the masseter and temporalis muscles is most pronounced in *S. oedipus*, suggesting a greater capacity for functional partitioning of these jaw-closing muscles in nongouging tamarins. We speculate that such partitioning may be more constrained in marmosets owing to their unique gape requirements for tree gouging. These observations on intersegmental differences in fiber architecture should be regarded as preliminary however, both because the masseter and temporalis muscles represent only two of the four muscles of mastication and because the masticatory muscles are architecturally complex and contain regionally differentiated, task-specific portions (e.g., Herring et al., 1979; Weijs and Dantuma, 1981; van Eijden et al., 1993; van Eijden and Turkawski, 2001), not all of which have been evaluated.

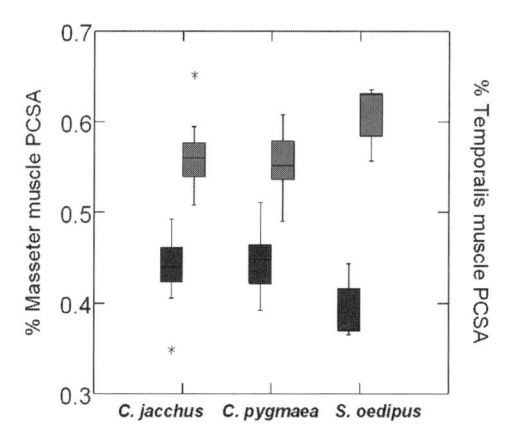

Fig. 12.10 The PCSA of the masseter depicted on the left (dark-hatched box) and temporalis on the right (light-hatched box), expressed as a percentage of the combined PCSA for both muscles. The disparity between the masseter and the temporalis muscles is greatest in *Saguinus oedipus*, suggesting a greater division of labor between these two jaw-closing muscles in terms of their dedication to generating force

12.4.3 Integrating Fiber Architecture with Bony Morphology in Studies of Feeding Behavior

Previous craniometric studies of the external morphology and internal geometry of tree-gouging marmosets suggest a masticatory apparatus that is well suited to facilitating the production of wide jaw gapes, but not necessarily the generation or dissipation of large jaw forces (Vinyard et al., 2003; Vinyard and Ryan, 2006). Based on the performance data collected during tree gouging in both laboratory and field settings, we argue that differences in skull form between tree-gouging and nongouging callitrichids are functionally linked to their divergent feeding behaviors. However, skull form alone is generally insufficient for determining skull function (Daegling, 1993; Daegling and Hylander, 2000; Vinyard and Ryan, 2006). Moreover, it is often impractical to obtain performance data on living primates in laboratory or natural environments, and impossible to obtain such data on extinct taxa.

We have shown a correspondence between variation in jaw-muscle fiber architecture and skull form in tree-gouging and nongouging callitrichids. Compared to nongouging tamarins, tree-gouging marmosets have relatively longer masseter and temporalis fibers that facilitate the production of wide jaw gapes. Common marmosets also have relatively longer mandibles, anteroposterior elongated glenoids and condyles, and lower condyles relative to the height of the mandibular toothrow, all of which assist in opening the jaw widely. Conversely, tree-gouging marmosets have relatively smaller PCSAs compared to tamarins, indicating less maximal force production capability. Likewise, common marmosets have no skull morphologies that can be functionally linked to generating or dissipating large jaw forces. To our knowledge, this is the first such study to demonstrate a relationship between jaw-muscle fiber architecture, skull form, and feeding behavior in a primate. Evaluation of fiber architecture, along with other anatomical and physiological parameters of muscle, provide an important means of corroborating functional hypotheses linking feeding behavior and diet to skull form.

Apart from providing data on muscle structure that is essential for building more robust interpretations of function and performance, an architectural analysis of muscle groups yields important information on how muscles may be functionally partitioned in various behavioral repertoires, information that cannot be readily gleaned from skeletal morphology (e.g., Anapol and Jungers, 1986, 1987; Anapol and Barry, 1996; Anapol and Gray, 2003). Knowledge of muscle structure aids our interpretation of jaw-muscle recruitment patterns, while muscle recruitment patterns inform our understanding of muscle fiber architecture and physiology. In this way, fiber architecture and jaw-muscle electromyography (EMG) are mutually informative. To date, we do not know whether jaw-muscle EMG and fiber architecture are correlated across primates. An improved understanding of this relationship will help us determine the extent to which muscle form reflects its function during various feeding and biting behaviors.

Acknowledgments Andrea B. Taylor gratefully acknowledges Matthew J. Ravosa, Christopher J. Vinyard, and Christine E. Wall for the invitation to contribute to this volume. ABT and CJV thank

Elizabeth Curran (NEPRC), Amanda Trainor (WPRC), Suzette Tardiff (SFBR), and Donna Layne (SFBR) for their help in providing the muscle tissue. This research was supported by NSF (BCS – 0412153) and (BCS-0094666).

References

Anapol, F.C., Jungers, W.L. (1986). Architectural and histochemical diversity within the quadriceps femoris of the brown lemur (*Lemur fulvus*). *Am. J. Phys. Anthropol*. 69:355–375.

Anapol, F.C., Jungers, W.L. (1987). Telemetered electromyography of the fast and slow extensors of the leg of the brown lemur (*Lemur fulvus*). *J. Exp. Biol*. 130:341–358.

Anapol, F., Barry, K. (1996). Fiber architecture of the extensors of the hindlimb in semiterrestrial and arboreal guenons. *Am. J. Phys. Anthropol*. 99:429–447.

Anapol, F., Gray, J.P. (2003). Fiber architecture of the intrinsic muscles of the shoulder and arm in semiterrestrial and arboreal guenons. *Am. J. Phys. Anthropol*. 122:51–65.

Antón, S.C. (1996). Cranial adaptation to a high attrition diet in Japanese macaques. *Int. J. Primatol*. 17:401–427.

Biewener, A.A., Roberts, T.J. (2000). Muscle and tendon contributions to force, work, and elastic energy savings: a comparative perspective. *Exerc. Sport. Sci. Rev*. 28:99–107.

Bock, W.J., von Wahlert, G. (1965). Adaptation and the form-function complex. *Evolution* 19: 269–299.

Bouvier, M. (1986). A biomechanical analysis of mandibular scaling in Old World monkeys. *Am. J. Phys. Anthropol*. 69:473–482.

Coimbra-Filho, A.F., Mittermeier, R.A. (1977). Tree-gouging, exudate-eating and the "short-tusked" condition in *Callithrix* and *Cebuella*, in: Kleiman, D.G. (ed.), The Biology and Conservation of the Callitrichidae, Smithsonian Institution Press, Washington DC, pp. 105–115.

Cole, T.M. III. (1992). Postnatal heterochrony of the masticatory apparatus in *Cebus apella* and *Cebus albifrons*. *J. Hum. Evol*. 23:253–282.

Daegling, D.J. (1992). Mandibular morphology and diet in the genus *Cebus*. *Int. J. Primatol*. 13:545–570.

Daegling, D.J. (1993). The relationship of *in vivo* bone strain to mandibular corpus morphology in *Macaca fascicularis*. *J. Hum. Evol*. 25:247–269.

Daegling, D.J., Hylander, W.L. (2000). Experimental observation, theoretical models, and biomechanical inference in the study of mandibular form. *Am. J. Phys. Anthropol*. 112:541–551.

Dechow, P.C., Carlson, D.S. (1982). Bite force and gape in rhesus macaques. *Am. J. Phys. Anthropol*. 57:179.

Dechow, P.C., Carlson, D.S. (1986). Occlusal force after mandibular advancement in adult rhesus monkeys. *J. Oral. Maxillofac. Surg*. 44:887–893.

Dechow, P.C., Carlson, D.S. (1990). Occlusal force and craniofacial biomechanics during growth in rhesus monkeys. *Am. J. Phys. Anthropol*. 83:219–237.

Dumont, E.R. (1997). Cranial shape in fruit, nectar, and exudate feeders: Implications for interpreting the fossil record. *Am. J. Phys. Anthropol*. 102:187–202.

Eng, C.M., Vinyard, C.J., Anapol, F., Taylor, A.B. (2005). Stretching the limits: Jaw-muscle fiber architecture in tree-gouging and nongouging callitrichid monkeys. *Am. J. Phys. Anthropol*. Supplement 40:103.

Gans, C. (1982). Fiber architecture and muscle function. *Exerc. Sport Sci. Rev*. 10:160–207.

Gans, C., Bock, W.J. (1965). The functional significance of muscle architecture: A theoretical analysis. *Ergeb. Anat. Entwicklungsgesch* 38:115–142.

Gans, C., de Vree, F. (1987). Functional bases of fiber length and angulation in muscle. *J. Morphol*. 192:63–85.

Garber, P.A. (1992). Vertical clinging, small body size, and the evolution of feeding adaptations in the Callitrichinae. *Am. J. Phys. Anthropol*. 88:469–482.

Gingerich, P.D., Smith, B.H., Rosenberg, K. (1982). Allometric scaling in the dentition of primates and prediction of body weight from tooth size in fossils. *Am. J. Phys. Anthropol.* 58:81–100.

Gordon, A.M., Huxley, A.F., Julian, F.J. (1966). The variation in isometric tension with sarcomere length in vertebrate muscle fibres. *J. Physiol. Lond.* 184:170–192.

Herring, S.W. (1992). Muscles of mastication: Architecture and functional organization, in: Davidovitch, Z. (ed.), *The Biological Mechanisms of Tooth Movement and Craniofacial Adaptation*, Ohio State University College of Dentistry, Ohio, pp. 541–548.

Herring, S.W., Herring, S.E. (1974). The superficial masseter and gape in mammals. *Am. Nat.* 108:561–576.

Herring, S.W., Grimm, A.F., Grimm, B.R. (1979). Functional heterogeneity in a multipinnate muscle. *Am. J. Anat.* 154:563–576.

Hershkovitz, P. (1977). *Living New World Monkeys (Platyrrhini)*. University of Chicago Press, Chicago.

Hill, W.C.O. (1957). *Primates: Comparative Anatomy and Taxonomy*. Edinburgh University Press, Edinburgh, and Wiley-Interscience, New York.

Hylander, W.L. (1979a). The functional significance of primate mandibular form. *J. Morphol.* 160:223–240.

Hylander, W.L. (1979b). Mandibular function in *Galago crassicaudatus* and *Macaca fascicularis*: An *in vivo* approach to stress analysis in the mandible. *J. Morphol.* 159:253–296.

Hylander, W.L. (1984). Stress and strain in the mandibular symphysis of primates: A test of competing hypotheses. *Am. J. Phys. Anthropol.* 61:1–46.

Hylander, W.L. (1985). Mandibular function and biomechanical stress and scaling. *Am. Zool.* 25:315–330.

Hylander, W.L., Johnson, K.R. (1994). Jaw muscle function and wishboning of the mandible during mastication in macaques and baboons. *Am. J. Phys. Anthropol.* 94:523–547.

Hylander, W.L., Johnson, K.R. (1997). In vivo bone strain patterns in the zygomatic arch of macaques and the significance of these patterns for functional patterns of craniofacial form. *Am. J. Phys. Anthropol.* 102:203–232.

Hylander, W.L., Johnson, K.R., Crompton, A.W. (1987). Loading patterns and jaw movements during mastication in *Macaca fascicularis*: A bone-strain, electromyographic, and cineradiographic analysis. *Am. J. Phys. Anthropol.* 72:287–314.

Hylander, W.L., Johnson, K.R., Crompton, A.W. (1992). Muscle force recruitment and biomechanical modeling: An analysis of masseter muscle function during mastication in *Macaca fascicularis*. *Am. J. Phys. Anthropol.* 88:365–387.

Hylander, W.L., Ravosa, M.J., Ross, C.F., Johnson, K.R. (1998). Mandibular corpus strain in primates: Further evidence for a functional link between symphyseal fusion and jaw-adductor muscle force. *Am. J. Phys. Anthropol.* 107:257–271.

Hylander, W.L., Ravosa, M.J., Ross, C.F., Wall, C.E., Johnson, K.R. (2000). Symphyseal fusion and jaw-adductor muscle force: An EMG study. *Am. J. Phys. Anthropol.* 112:469–492.

Hylander, W.L., Wall, C.E., Vinyard, C.J., Ross, C., Ravosa, M.J., Williams, S.H., Johnson, K. (2005). Temporalis function in anthropoids and strepsirrhines: An EMG study. *Am. J. Phys. Anthropol.* 128:35–56.

Jernvall, J., Jung, H.S. (2000). Genotype, phenotype, and developmental biology of molar tooth characters. *Yrbk. Phys. Anthropol.* 43:171–190.

Kay, R.F. (1975). The functional significance of primate molar teeth. *Am. J. Phys. Anthropol.* 43:195–215.

Kay, R.F. (1978). Molar structure and diet in extant Cercopithecidae, in: Butler, P.M., Joysey, K. (eds.), *Development, Function and Evolution of Teeth*, Academic Press, London, pp. 309–339.

Kay, R.F., Williams, B.A. (1994). Dental evidence for anthropoid origins, in: Fleagle, J.G., Kay, R.F. (eds.), *Anthropoid Origins*, Plenum Press, New York, pp. 361–445.

Lieber, R.L. (2002). *Skeletal Muscle Structure, Function, and Plasticity*, Second Edition. Lippincott, Williams and Wilkins, Baltimore, MD.

Lucas, P.W. (1982). Basic principles of tooth design, in: Kurten, B. (ed.), *Teeth: Form, Function and Evolution*, Columbia University Press, New York, pp. 154–162.

Lucas, P.W. (2004). *Dental Functional Morphology: How Teeth Work*. Cambridge University Press, Cambridge.

McCollum, M., Sharpe, P.T. (2001). Developmental genetics and early hominid craniodental evolution. *BioEssays* 23:481–493.

Mendez, J., Keys, A. (1960). Density and composition of mammalian muscle. *Metabolism* 9: 184–188.

Mills, J.R.E. (1963). Occlusion and malocclusion of the teeth of primates, in: Brothwell, D.R. (ed.), *Dental Anthropology*, Pergamon Press, New York, pp. 29–52.

Mork, A.L., Wall, C.E., Williams, S.H., Garner, B.A., Johnson, K.R., Schmitt, D., Hylander, W.L., Vinyard, C.J. (2004). The biomechanics of tree gouging in common marmosets (*Callithrix jacchus*). *Am. J. Phys. Anthropol*. Supplement 40:158.

Napier, J.R., Napier, P.H. (1967). *A Handbook of Living Primates*. Academic Press, New York.

Oxnard, C. (1987). *Fossils, Teeth and Sex*. Hong Kong University Press, Hong Kong.

Plavcan, J.M. (1993). Canine size and shape in male anthropoid primates. *Am. J. Phys. Anthropol*. 92:201–216.

Powell, P.L., Roy, R.R., Kanim, P., Bellow, M.A., Edgerton, V.R. (1984). Predictability of skeletal muscle tension from architectural determinations in guinea pig hindlimbs. *J. Appl. Physiol.: Respirat. Environ. Exerc. Physiol* 57:1715–1721.

Rosenberger, A.L. (1978). Loss of incisor enamel in marmosets. *J. Mamm*. 59:207–208.

Ross, C., Hylander, W.L. (2000). Electromyography of the anterior temporalis and masseter muscles of owl monkeys (*Aotus trivirgatus*) and the function of the postorbital septum. *Am. J. Phys. Anthropol*. 112:455–468.

Sacks, R.D., Roy, R.R. (1982). Architecture of the hind limb muscles of cats: Functional significance. *J. Morphol*. 173:185–195.

Spector, S.A., Gardiner, P.F., Zernicke, R.F., Roy, R.R., Edgerton, V.R. (1980). Muscle architecture and the force velocity characteristics of the cat soleus and medial gastrocnemius: Implications for motor control. *J. Neurophysiol*. 44:951–960.

Spencer, M.A. (1999). Constraints on masticatory system evolution in anthropoid primates. *Am. J. Phys. Anthropol*. 108:483–506.

Sussman, R.W., Kinzey, W.G. (1984). The ecological role of the Callitrichidae: A review. *Am. J. Phys. Anthropol*. 64:419–449.

Swindler. D.R. (2002). *Primate Dentition: An Introduction to the Teeth of Non-human Primates*. Cambridge University Press, Cambridge

Szalay, F.S., Delson, E. (1979). *Evolutionary History of the Primates*. Academic Press, New York.

Szalay, F.S., Seligsohn, D. (1977). Why did the strepsirhine tooth comb evolve? *Folia Primatol*. 27:75–82.

Taylor, A.B. (2002). Masticatory form and function in the African apes. *Am. J. Phys. Anthropol*. 117:133–156.

Taylor, A.B. (2005). A comparative analysis of temporomandibular joint morphology in the African apes. *J. Hum. Evol*. 48:555–574.

Taylor, A.B. (2006a). Diet and mandibular morphology in the African apes. *Int. J. Primatol*. 27:181–201.

Taylor, A.B. (2006b). Feeding behavior, diet, and the functional consequences of jaw form in orangutans, with implications for the evolution of *Pongo*. *J. Hum. Evol*. 50:377–393.

Taylor, A.B., Vinyard, C.J. (2004). Comparative analysis of masseter fiber architecture in tree-gouging (*Callithrix jacchus*) and nongouging (*Saguinus oedipus*) callitrichids. *J. Morphol*. 261:276–285.

Taylor, A.B., Eng, C.M., Anapol, F.C., Vinyard, C.J. (n.d.). The functional significance of jaw-muscle fiber architecture in tree-gouging marmosets, in: Davis, L.C., Ford, S.M., Porter, L.M. (eds.), *The Smallest Anthropoids: The Marmoset/Callimico Radiation*, Springer-Verlag, in press.

Teaford, M.F., Smith, M.M., Ferguson, M.W.J. (2000). *Development, Function and Evolution of Teeth*. Cambridge University Press, Cambridge.

Turnbull, W.D. (1970) Mammalian masticatory apparatus. *Fieldiana Geol*. 18:1–356.

Ungar, P. (1998). Dental allometry, morphology, and wear as evidence for diet in fossil primates. *Evol. Anthropol.* 6:205–217.

van Eijden, T.M.G.J., Turkawski, S.J. (2001). Morphology and physiology of masticatory muscle motor units. *Crit. Rev. Oral. Biol. Med.* 12:76–91.

van Eijden, T.M.G.J., Blanksma, N.G., Brugman, P. (1993). Amplitude and timing of EMG activity in the human masseter muscle during selected motor tasks. *J. Dent. Res.* 72:599–606.

van Eijden, T.M.G.J., Korfage, J.A.M., Brugman, P. (1997). Architecture of the human jaw-closing and jaw-opening muscles. *Anat. Rec.* 248:464–474.

Vinyard, C.J., Ryan, T. (2006). Cross-sectional bone distribution in the mandibles of gouging and non-gouging platyrrhines. *Int. J. Primatol.* 27:1461–1490.

Vinyard, C.J., Wall, C.E., Williams, S.H., Hylander, W.L. (2003). A comparative functional analysis of the skull morphology of tree gouging primates. *Am. J. Phys. Anthropol.* 120:153–170.

Vinyard, C.J., Wall, C.E., Williams, S.H., Schmitt, D., Hylander, W.L. (2001). A preliminary report on the jaw mechanics during tree gouging in common marmosets (*Callithrix jacchus*), in: Brooks, A. (ed.), *Dental Morphology 2001: Proceedings of the 12th International Symposium on Dental Morphology*, Sheffield Academic Press Ltd., Sheffield, UK, pp. 283–297.

Vinyard, C.J., Wall, C.E., Williams, S.H., Mork, A.L., Garner, B.A., de Oliveira Melo, L.C., Valença-Montenegro, M.M., Valle, Y.B.M., Monteiro da Cruz, M.A.O., Lucas, P.W., Schmitt, D., Taylor, A.B., Hylander, W.L. The evolutionary morphology of tree gouging in marmosets, in: Davis, L.C., Ford, S.M., Porter, L.M. (eds.), The Smallest Anthropoids: The Marmoset/*Callimico* Radiation, Springer, New York, in press.

Vinyard, C.J., Williams, S.H., Wall, C.E., Johnson, K.R., Hylander, W.L. (2005). Jaw-muscle electromyography during chewing in Belanger's treeshrews (*Tupaia belangeri*). *Am. J. Phys. Anthropol.* 127:26–45.

Wall, C.E., Vinyard C.J., Johnson, K.R., Williams, S.H., Hylander, W.L. (2006). Phase II jaw movements and masseter muscle activity during chewing in *Papio anubis*. *Am. J. Phys. Anthropol.* 129:215–224.

Weijs, W.A., Dantuma, R. (1981). Functional anatomy of the masticatory apparatus in the rabbit (*Oryctolagus cuniculus* L.). *Neth. J. Zool.* 31:99–147.

Weijs, W.A., Brugman, P., Klok, E.M. (1987). The growth of the skull and jaw muscles and its functional consequences in the New Zealand rabbit (*Oryctolagus cuniculus*). *J. Morphol.* 194:143–161.

Wottiez, R.D., Heerkens, Y.F., Huijing, P.A., Rijnsburger, W.H., Rozendal, R.H. (1986). Functional morphology of the m. gastrocnemius medialis of the rat during growth. *J. Morphol.* 187:247–258.

Part V
Bone and Dental Morphology

Analyses of skull and tooth morphology account for the bulk of craniofacial research in biological anthropology. While Hylander has made important contributions to our understanding of dental and mandibular form, he has primarily advanced others' efforts through their applying his in vivo data as empirical support for metric approximations and in thoughtful evaluation of the mechanical relevance of mandibular measurements. In this final part, we have several contributions that span the range of what is current in morphological analyses of skull form. These chapters range from microstructural analyses of bone properties to studies of bone plasticity, interspecific morphological analyses of bone and dental function to several studies examining skull form and function in fossil primates.

In Chapter 13, Dechow et al. explore the elastic properties of mandibular and femoral cortical bone in order to relate observed patterns of elastic properties to the three-dimensional orientation of osteons. By applying confocal microscopy, ultrasonic techniques, and microCT, they provide an initial test of the hypothesis that bone elastic properties are primarily a function of osteon orientation in bone. Femoral bone showed a stronger relationship between osteonal orientation and elastic properties than mandibular bone, although some association between these parameters is present in the jaw.

In Chapter 14, Ravosa et al. document changes in cartilage and bone microstructure in the symphyses and temporomandibular joints (TMJ) of rabbits fed mechanically challenging diets and mice with genetically altered jaw muscles. Rabbits with resistant diets and mice with genetically enlarged jaw muscles both tended to exhibit larger jaw dimensions and increased biomineralization in their symphyses and TMJs. Histological analyses diverge in that rabbits tended to show evidence of degradative changes in soft tissue structures, while knockout mice exhibited increased resistance abilities in their soft tissue structures. Because the rabbits experienced the experimental dietary treatment for a relatively long period, these results suggest that the time course of loading history is important to studying both the plasticity of masticatory joints in mammals and jaw joints as integrated tissue systems.

Hogue conducts a comparative morphometric analysis of mandibular corpus form among marsupials to assess whether morphologies that provide increased resistance to bending and twisting correlate with dietary variation in this clade. While a significant amount of variation exists, some correlation is observed between dietary categories and corpus morphology. This result suggests a wide-ranging association between diet and jaw form in mammals.

In Chapter 16, Vinyard notes that one of Hylander's main contributions to morphometric studies of jaw function and scaling was his advocacy of using independent variables with explicit mechanical relevance to mastication. This chapter builds on Hylander's contribution by exploring how size adjustment methods and metric variables affect the functional interpretations of morphometric comparisons. In particular, the recent tendency to accept a geometric mean of all characters as a size variable is critiqued from a functional perspective.

In Chapter 17, Yamashita relates dental morphology in ring-tailed lemurs and sifakas to the mechanical properties of kily eaten by both species. Ring-tailed lemurs exhibit relatively long crests that correlate with the relatively tougher parts of the kily plant they consume. Based on this correspondence, Yamashita argues that food properties are driving changes in ring-tailed lemur occlusal morphology. Alternatively, the hypothesis is put forth that sifaka dental form is constraining their behavioral choice of foods. In a predominately laboratory-based volume, this chapter reminds us of the importance of integrating naturalistic observations of animal behaviors with morphological analyses of jaw and tooth forms for understanding primate adaptations.

Simons describes the orbital convergence and frontation of Fayum anthropoids. In comparison to other primates, he notes that the Fayum anthropoids demonstrate a gradual acquisition of modern degrees of convergence and frontation during primate evolution. He discusses these results in the context of current hypotheses explaining primate and anthropoid origins. Simons concludes by proposing a new "synergy theory," where he integrates ideas from historic and current hypotheses of primate and anthropoid origins into a composite model.

Rak and Hylander consider the functional implications of the tall mandibular ramus in robust australopithecine, suggesting that this morphology is part of a suite of features demonstrating a unique grinding pattern during mastication. Specifically, they propose that the tall ramus and relatively short jaw facilitate an increase in anterior movements at the end of the power stroke. When combined with increased mediolateral movements, a "rotatory" type of motion is hypothesized for the robust australopithecine chewing cycle. This interpretation is supported by the relatively flat postcanine teeth in robust australopithecines and may help explain the relatively robust corpora in this fossil taxa.

In the final chapter, Antón examines the morphological evidence for a foraging shift between *Homo habilis* (*sensu lato*) and *H. erectus* (*s.l.*). Given prior suggestions that *H. erectus* acquired a higher-quality, softer diet, it is predicted that these fossils will exhibit relatively smaller teeth and jaws. To more effectively compare fragmentary fossils, Antón develops a chimera approach that combines fossil cranial and postcranial remains for comparison. The general result of these comparisons focuses on the substantial overlap in jaw and dental form between these two groups of hominids. Based on this overlap, she suggests that any foraging shift in the *Homo* lineage likely occurred prior to *H. habilis*.

Chapter 13
Relationship Between Three-Dimensional Microstructure and Elastic Properties of Cortical Bone in the Human Mandible and Femur

Paul C. Dechow, Dong Hwa Chung, and Mitra Bolouri

Contents

13.1	Introduction	265
13.2	Materials and Methods	267
	13.2.1 Determination of the Orientation of the Axis of Greatest Stiffness in the Cortical Plane	269
	13.2.2 Studies of Cortical Structure Using the Confocal Microscope	270
	13.2.3 Studies of Cortical Structure Using Micro Computed Tomography	271
13.3	Results	272
	13.3.1 Determination of the Orientation of the Axis of Greatest Stiffness in the Cortical Plane	272
	13.3.2 Studies of Cortical Structure Using the Confocal Microscope	272
	13.3.3 Studies of Cortical Structure Using Micro Computed Tomography	278
13.4	Discussion	281
	13.4.1 Orientation of Elastic Properties in Cortical Bone Tissue	281
	13.4.2 Orientation of Cortical Bone Tissue Structure	283
	13.4.3 Osteonal Morphology in Cortical Bone	285
	13.4.4 Elastic Anisotropy, Bone Tissue Structure, and Adaptation	285
	References	288

13.1 Introduction

Observations that bone has grain, or differences in mechanical properties by direction, predate our ability to measure these characteristics. Studies of patterns of trabecular orientation in the femoral head were seminal in forming early hypotheses

---

P.C. Dechow
Department of Biomedical Sciences, Baylor College of Dentistry, Texas A&M Health Science Center, 3302 Gaston Avenue, Dallas, TX 75254 USA, 214-370-7229
e-mail: pdechow@bcd.tamhsc.edu

about structure/function relationships in the skeleton (Koch, 1917), and remain of great relevance today. Modern techniques have allowed some studies on craniofacial tissues including investigations into the trabecular structure of the mandibular condyle in pigs (Teng and Herring, 1995, 1996), and in humans (Giesen and van Eijden, 2000). Advances in stereology and, more recently, microCT imaging have provided tools for documenting orientation and structure in trabecular bone.

Material orientation in cortical bone has been more difficult to assess systematically. Gebhardt in 1906 proposed that the three-dimensional functional behavior of mature cortical bone is primarily dependent on the three-dimensional organization of osteons. However, attempts to determine cortical material organization relied on techniques such as split-line patterns on the periosteal surface (for example, Dempster, 1967; or Tappen, 1970), which were influenced by osteon orientation among other factors (Buckland-Wright, 1977). The split-line patterns were thought to reflect directions in which the microstructure of cortical bone was oriented to resist stress. Subsequent research has shown this hypothesis to be incorrect (Buckland-Wright, 1977;Bouvier and Hylander, 1981).

Although there is a large amount of literature on the material properties of cortical bone (for summaries, see Evans, 1973; Currey, 1984; Cowin, 1989; Martin et al., 1998; Guo, 2001), information on material properties in three dimensions is limited. Most studies use mechanical methods to measure the material properties unidirectionally, primarily in the axial direction, or presumed direction of greatest stiffness or strength. Acoustic microscopes and the development of nanoindentation techniques have allowed some characterization of bone material properties at a microstructural and even subosteonal level (Guo, 2001) in multiple directions. However, ultrasonic techniques (Lees, 1982; Katz and Yoon, 1984; Ashman, 1989; Dechow et al., 1993; Kohles et al., 1997; Schwartz-Dabney and Dechow, 2003) using 2–20 MHz transducers remain the best method for the characterization of the anisotropy of apparent elastic moduli of cortical bone at an intermediate or tissue level.

Much controversy exists about which structural features are most important for producing anisotropy in the mechanical characteristics of bone. This problem stems from deficiencies in our knowledge of (1) the basic structure of bone at the molecular, ultrastructural, cellular, and microstructural (tissue) levels and (2) associated developmental processes. At the microstructural level, this problem is important because of the necessity for understanding the link between the bone structure and the mechanical behavior of the skeleton. Yet few tools have been developed for (1) quantifying three-dimensional aspects of bone microstructure and (2) modeling the mechanical impact of variations in this structure. To understand bone anisotropy, questions must be addressed with consideration of the hierarchical structure of bone. At the tissue level, cortical microstructural features that determine anisotropy may include both patterns of (1) porosity and (2) the intrinsic anisotropy of the bone matrix itself.

The link between skeletal function, cortical bone structure, and remodeling must take two factors into consideration. The first is the material properties of bone at a tissue or microstructural level, as these determine bone deformation in response to load. The second is the tissue-level structure of cortical bone, as this structure is essential for understanding how these patterns reflect chronic loading patterns and how osteocytes arranged in this structure or bone matrix are able to transduce mechanical signals into cellular and tissue response.

Previous investigations suggest that bulk elastic properties of remodeled femoral diaphyseal cortical bone are transversely isotropic with the principal material axis (axis of greatest stiffness) aligned with the anatomical long axis of the bone and the predominant direction of the osteons (Yoon and Katz, 1976; Katz et al., 1984; Lipson and Katz, 1984; Katz and Meunier, 1987). While these conclusions seem reasonable, it is of interest to note that previous investigations have not actually attempted to quantify orientation of the elastic orthotropic axes for femoral cortical bone, although much data are now available for the human mandible (Schwartz-Dabney and Dechow, 2003). Likewise, no work has attempted to quantify the average osteon orientation in cortical bone from any region. Existing references have provided qualitative descriptions of Haversian canal orientations in a few postcranial bones (Cooper et al., 2003; Stout et al., 1999).

The goal of this chapter is to begin an exploration of the relationship between tissue-level structure in cortical bone and the three-dimensional elastic properties of the tissue. To do this, ultrasonic techniques were used to determine the orientation of the principle orthotropic axes in three samples of cortical bone. Techniques of confocal microscopy and microcomputerized tomography (microCT) were used to examine osteonal structure, as reflected by Haversian canal structure, in these same bone samples. The hypothesis is a modification of that of Gebhardt (1906), specifically that the three-dimensional elastic structure of mature cortical bone is primarily dependent on the three-dimensional organization of osteons.

13.2 Materials and Methods

Two disc-shaped specimens (10-mm diameter) were collected from the corpora of a human mandible (Fig. 13.1) on the buccal surface inferior to the first molar. The mandible was from a 62-year-old unembalmed frozen female cadaver. An additional specimen was taken from the flat medial surface of the midshaft of a femur from a 71-year-old embalmed male cadaver. Although embalming has been shown to have a small effect on mechanical properties, such as yield stress, there is even less effect on elastic modulus, especially when measured ultrasonically (Guo, 2001; Currey et al., 1994). In any case, the goal of our study is to compare the orientation of the presumed orthotropic elastic axes to overall osteon orientation in the specimens. While embalming in the femur might have minor effects on elastic properties, it is reasonable to assume that these are unlikely to affect patterns of anisotropy or material orientation (Fig. 13.2) in the bone specimens, although data are not available on this point in the literature.

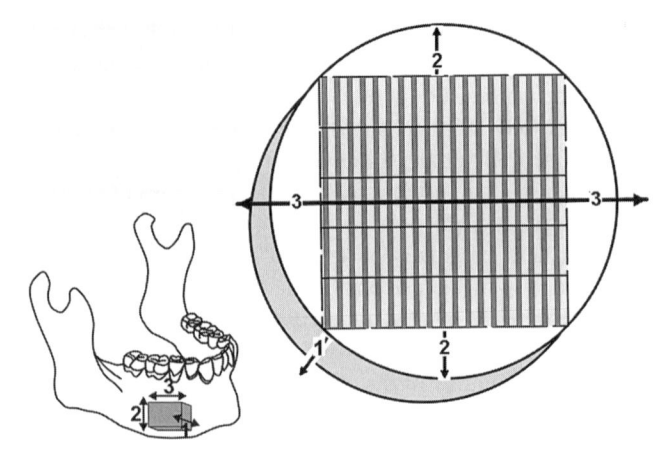

Fig. 13.1 Region of bone sampling for human mandibular bone specimens. The approximate location on the mandibles from which the bone cylinders were removed is illustrated by the box. Following measurement of longitudinal ultrasonic velocities through the diameter and around the perimeter of each cylinder, the axes of minimum and maximum velocity in the cortical plane (which correspond to axes of minimum and maximum stiffness) were used to orient the specimens for reshaping as a block. Both illustrations show the orientation of the axes of the specimens. The "3" direction is that of greatest stiffness in the cortical plane and was approximately parallel to the lower border of the mandible. The "2" direction is that of least stiffness in the cortical plane, and is perpendicular to the "3" direction in cortical plane. The "1" direction is perpendicular to the cortical plane and is in the direction of cortical thickness. The squared structure in the cylinder illustrates the orientation of the serial sections of bone cut with the vibratome. The horizontal lines dividing each vertical white or gray bar in the square into five regions illustrates the relative position of the regions imaged and reconstructed with the confocal microscope. In all, 19 sections (vertical white and gray bars) were cut and five regions or volumes were imaged and reconstructed in each section

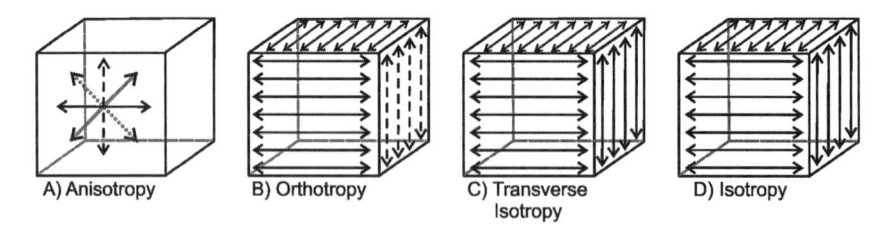

Fig. 13.2 Graphical explanation of material anisotropy. (**A**) In anisotropy, elastic constants differ in all directions with no planes of symmetry. (**B**) In orthotropy, elastic constants vary between three perpendicular planes of symmetry. (**C**) In transverse isotropy, elastic constants are similar in two perpendicular planes, but differ from the third plane. (**D**) In isotropy, elastic properties are the same in all directions

13.2.1 Determination of the Orientation of the Axis of Greatest Stiffness in the Cortical Plane

Longitudinal ultrasonic velocities were measured around the perimeter of each specimen. Prior to harvesting, cortical bone sites were referenced with a graphite line for orientation. The lower border of the mandible and the long axis of the femoral diaphysis were used as reference axes.

Cancellous bone on the inner surface of the specimens was removed with grinding wheels on a water-cooled Tormek grinder or by gentle hand sanding. Bone was cooled continuously with a water drip during preparation. Bone cylinders were harvested from the cortex using a rotary tool (Dremel 732) and 10.0 mm inner-diameter trephine burrs (Ace Dental Implant System). Specimens were stored in a solution of 95% ethanol and isotonic saline in equal proportions. This media has been shown to maintain the elastic properties of cortical bone over time (Ashman et al., 1984; Zioupos and Currey, 1998).

Longitudinal ultrasonic waves were generated by Panametrics transducers (V312-N-SU) resonating at 10 MHz. The transducers were powered with a pulse generator (Hewlett Packard Model 214A). Pulse delays induced by passage of ultrasonic waves through the bone were read on a digitizing oscilloscope (Tektronix TDS420). Velocities were calculated by dividing pulse delays by the specimen width.

First, the bone specimen was attached to the flat top of a dowel (3 mm in diameter) with cyanoacrylate. This dowel was attached to a 4" Rotary Table (P/N 3700, Sherline Products, Inc.), capable of accurate rotations to a tenth of a degree. We then measured the velocities of longitudinal ultrasonic waves passed through the cortical bone for each specimen at 1° angular intervals from 0° to 180°.

A plot of ultrasonic velocities by orientation in an orthotropic material yields a curve resembling a sinusoid if two of the three material axes are in the plane of measurement. The direction corresponding to the apogee is that of greatest stiffness in that plane, and the direction corresponding to the nadir should always be at 90° to the apogee and is the direction of least stiffness. The third material axis would be orthogonal to the first two and thus normal or radial to the plane of the cortical plate (Schwartz-Dabney and Dechow, 2002; Schwartz-Dabney and Dechow, 2003; Ashman et al., 1984; Yoon and Katz, 1976). Tests of ultrasonic velocities through test bone specimens with and without the attachment of the dowel did not show a difference in velocities.

Following collection of the ultrasonic velocities, the data set was fitted to the sinusoidal model using the sinfit function (a least squares fit) in Mathcad Professional (Mathsoft Engineering and Education, Cambridge MA) to generate parameter coefficients for each sine curve. The coefficient of interest here is the angle of the principal axis from the reference line. This describes the direction of the axis of greatest stiffness.

Additional details of the methods for measuring ultrasonic velocities for cortical bone samples have been described previously (Schwartz-Dabney and Dechow, 2003).

13.2.2 Studies of Cortical Structure Using the Confocal Microscope

The three-dimensional microstructure of a cortical bone specimen from the mandible was examined through the use of a confocal microscope (Leica SP2). Attempts to visualize internal structure of undecalcified cortical specimens were unsuccessful because the material was too dense to allow adequate transmission of light, although surface structures can be visualized well without staining.

Before decalcification, a Sherline miniature lathe was used to machine the 10-mm diameter bone cylinder into a squared specimen which was 7 mm × 7 mm × cortical thickness. The square was oriented such that the faces were perpendicular to the material axes, as determined ultrasonically.

After decalcification with EDTA, the unembedded specimen was attached to a block with cyanoacrylate and sectioned at a thickness of 200 μm[1] with a vibrotome (Series 1000 Sectioning System, Technical Products International, Inc., St. Louis, MO). There was little distortion if the specimens remained wet. The sections were immediately attached to a slide with glycerol, placed under a coverslip, and imaged on the confocal microscope. Decalcification did not appear to remove the autofluoresence inherent in the bone tissue, and thus the details were sufficient to reconstruct Haversian canal orientations, and to see some details of osteonal structure.

A series of images at increasing depths using a 10× objective allowed imaging of a volume of 1500 × 1500 × 150 μm^2 (volume of 1024 × 1024 × 102 voxels)[2]. These volumes were then reconstructed with a voxel size of 1.47 μm^3, visualized, and measured with Analyze software (AnalyzeDirect, Lenexa, KS). These reconstructions allowed measurement of the orientation of all Haversian canals over the depth of the reconstruction.

The Analyze software package was used to study the splitting patterns, length, and orientations of all Haversian canals within each volume. The canal structures were confirmed as Haversian canals by the visualization of a matrix resembling osteonal structure around each canal. An attempt was made to follow osteons

---

[1] After trials on a number of test bone specimens, in which section thicknesses ranged from 50 μm to 400 μm, a 200 μm thickness for the ultramicrotomed section was found to be optimal for producing a maximum depth for three-dimensional reconstruction. Thicker sections blocked too much light, while thinner specimens restricted the depth of the reconstruction. The vibrotome caused some rippling on the surface of the specimen, which showed up as parallel lines on images taken near the top of the specimen. If the microscope was used to image below the surface, these artifacts were not visible.

[2] A single 1500 × 1500 × 150 μm confocal volume was about 0.28% of the volume of the 7 mm × 7 mm × 2.5 mm cortical bone samples. Each section cut from the sample contained five such adjacent volumes. In total, 19 200 μm thick sections were cut from the bone sample, resulting in 95 reconstructed confocal volumes. These 19 sections and 95 volumes together sampled 35% of the volume of the bone sample, although only 26% could be visualized. There was 9% wastage due to problems associated with reconstructing near the surface of the confocal volumes.

between each of the 19 serial sections. This was successful in many cases, but could not be relied on consistently to reconstruct the entire osteonal tree because of wastage between adjacent sections.

Variation in canal orientation was quantified by determining the angle of each canal within each volume. All angles for all volumes were then combined for study. Thus, if a canal continued on in more than one volume, its angles of orientation would be included as separate observations in the total summary of angles, and by this means weighting for long osteons over the whole specimen was included.

13.2.3 Studies of Cortical Structure Using Micro Computed Tomography

The three-dimensional microstructure of one cortical bone specimen from the mandible and one from the femur were examined through the use of a micro computed tomography (μCT) scanner (μCT 40, ScanCo) at 9 μm nominal isotropic resolution with a 2048 × 2048 pixels image matrix. A miniature lathe (Sherline) was used to machine the 10-mm diameter bone cylinders into rectangular specimens, which were approximately 7 mm × 4 mm × (cortical thickness). The rectangle was oriented such that the 7 mm long axis was in the direction of greatest material stiffness, as determined ultrasonically.

The μCT scanner allowed imaging of the canal structure throughout each cortical bone specimen. Volumetric data from the scanner was imported into Analyze software. Within the software, the specimens were oriented, trimmed, and segmented. Thresholds for segmentation were the same for each specimen and were determined visually so that the maximum amount of canal structure could be seen with a minimum of noise. The software was then used to measure the length and orientation of all canals within each specimen. No attempt was made to measure the diameter of the canals because, to do this, more attention and additional testing would be needed to determine repeatable threshold levels.

Orientations of canals were quantified per unit length of canal. All canals were first divided up manually into approximately straight canal segments. The length and orientation of each of these canal segments were measured within the Analyze environment. For example, for the canal illustrated in Fig. 13.3, the angle and length of the canal relative to the orthogonal axes were determined for the projection of the canal in each of the three planes 12, 13, and 23. Within each plane, the angle was recorded and the length of the canal within the planar projection was used for weighting. In this way, a histogram of the relative orientation of the canals was constructed within each plane. Likewise, canal segments were treated as vectors and trigonometric functions were used to determine the proportion of the vector along each axis (see Fig. 13.3 and lengths of lines 1,2, and 3). These values were then summed for all canal segments to create a comparison of canal orientation along the material axes of the specimens.

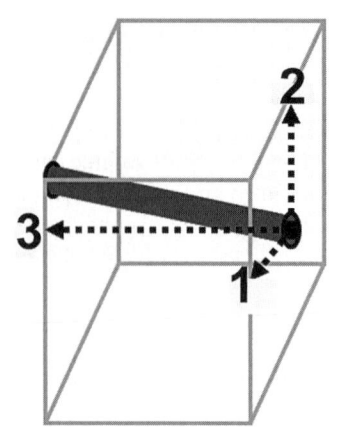

Fig. 13.3 Hypothetical Haversian canal segment within a box oriented to the orthogonal material axes of the cortical bone specimen. The canal segment was treated as a vector and trigonometric functions were used to determine the proportion of the vector along each axis (length of lines 1, 2, and 3). These values were then summed for all canal segments to create a comparison of canal orientation along the material axes of the specimens

13.3 Results

13.3.1 Determination of the Orientation of the Axis of Greatest Stiffness in the Cortical Plane

Plots of ultrasonic velocity by angle in the cortical plane of mandibular and femoral bone specimens reveal a sinusoidal pattern (Fig. 13.4). This pattern is expected in a material that can be approximated as orthotropic.

The primary difference between the femoral and the mandibular specimens in Fig. 13.4 is in the deviation from the idealized curve in the mandibular specimen. At the minimum and especially at the maximum velocities, the curves are flatter and slightly broader than those found in the femoral specimen, indicating a difference in internal microstructure.

13.3.2 Studies of Cortical Structure Using the Confocal Microscope

Confocal microscopy of decalcified mandibular cortical bone thick sections allows unique patterns of visualization of cortical bone microstructure (Figs. 13.5, 13.6). Two hundred and seventy nine osteonal segments were examined in the 95 volumes. These segments varied in length between 0.2 mm (found in two serial sections at most) and 3.8 mm (found in all 19 serial sections) (Fig. 13.6). Actual length could have been longer, because only length in the 3 direction was considered here. Because there were gaps of about 50 μm between serial sections or volumes, it could not be determined if all osteonal canals were connected. However, it is likely that

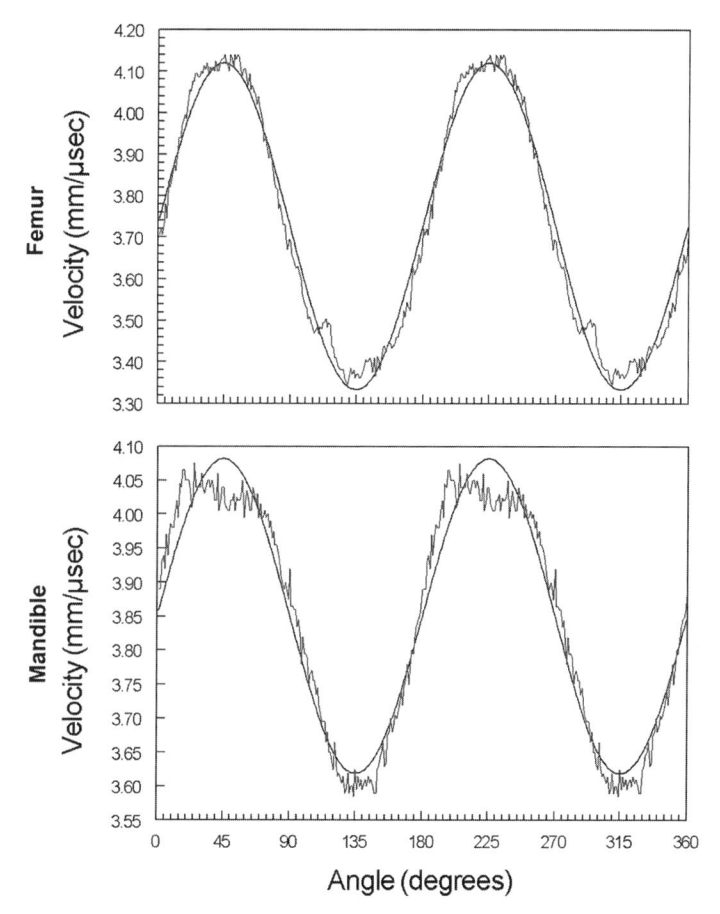

Fig. 13.4 Ultrasonic longitudinal velocities (10 MHz) versus wave direction in the femoral and mandibular specimens used in the μCT portion of the study. Readings were taken at intervals of 1° around the perimeter of the cortical cylinders so that the wave passed through the cortical plane. Peak velocity is the direction of maximum stiffness, and is registered at 45° and 225° for both graphs so that the shapes can be more easily compared. This direction is approximately parallel to the long axis of the femur and the lower border of the mandible respectively. Note the differences in shape in the peaks and the valleys between the femoral and the mandibular specimens

most were, even though these connections were not visualized, because they make up the structure of the vascular bed in the cortical bone tissue. Lamina of cortical bone, or osteonal bone matrix, could be seen around all canal structures, regardless of the orientation of the canal. Likewise, canals without surrounding osteonal matrix were not seen.

Of the 279 osteonal segments, 64 split. These 64 were the longer of the reconstructed canals. The amount of splitting varied from a single splitting in 26 of the 64 to 22 splittings in one osteonal structure. In all, 227 splittings were seen. Of these 207 were bifurcations, 16 were trifurcations and 4 were quadfircations. Thirty one

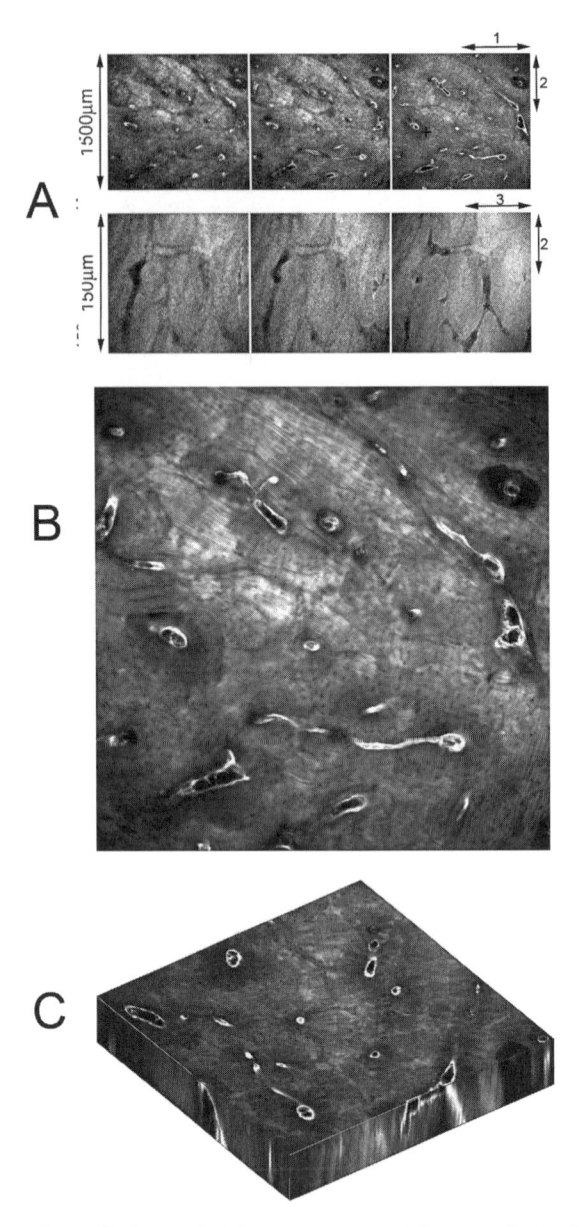

Fig. 13.5 Images of mandibular cortical bone reconstructed from confocal microscope scans. **A**. Reconstructed sections of a single mandibular cortical volume (1 of 95 volumes studied) to show the quality of confocal images taken from a decalcified unstained cortical bone specimen. The top three images are oriented parallel to the 1-2 plane and perpendicular to the 3 (longitudinal) axis. Each section is 1500 ×1500 μm, and the three serial sections are imaged at 15 μm intervals. From this angle, most osteons are seen in cross-section. Close examination of the images reveals changes in canal structure between images. Bone matrix around the canals can also be visualized and varies from faint with indistinct borders (location of cement lines) to very distinct (for example, see the dark osteon in the upper right portion of each photograph or on the left marked with a "+").

osteonal segments rejoined after previous bifurcation. Osteonal matrix surrounding the canals also surrounded the regions of splitting. Daughter branches were most likely to be angled along the long axis of the osteon rather than perpendicular to it. Thus, classic Volksman's canals, which are described in histology textbooks as being perpendicular to longitudinal Haversian canals, were not evident.

Longer osteonal segments were more likely to have multiple splittings than shorter osteonal segments. The correlation coefficient ($r = 0.689$) of the number of nodes with the longitudinal length of the osteonal segment was significant ($P < 0.05$) but moderate, because little splitting was seen in a few of the longer osteonal segments (Fig. 13.7). A diagram of the pattern of splitting in one complex osteonal segment (Fig. 13.8) shows four bifurcations, three trifurcations, and one rejoining.

Rosette histograms of the orientation of all osteonal segments in the three planes of each of the 95 reconstructed confocal volumes all showed a similar pattern. This pattern is illustrated cumulatively for all volumes (Fig. 13.9). The pattern showed that there is a difference between the projection of osteonal orientation in the 2-3 plane compared to the 1-3 plane. This difference can be also be seen in the 1-2 plane projection because more osteonal segments in that plane are oriented along the 2 axis than along the 1 axis. This pattern indicates that osteons vary more in orientation relative to the 3 axis by being oriented up or down (along the 2 axis), rather than in and out through the cortical thickness of the specimen (along the 1 axis).

The mean direction of osteon orientation in the 1-3 plane was 88.8°(Circular SD $= 20.6°$), which was not significantly different from 90°, or the orientation of maximum stiffness. The mean direction in the 2-3 plane was 95.1° (circular SD $= 27.0°$), which was significantly different from 90°. This small difference indicated that, on average, the osteons in this plane were oriented 5° above the orientation of maximum stiffness relative to an anterior to posterior vector. Most interesting in this comparison is the larger standard deviation for the angles in the 2-3 plane

Fig. 13.5 (Continued) The bottom three images are also serial sections at 15° µm intervals, but are smaller in size at 150 ×150 µm. These images are parallel to the 2-3 or cortical plane and perpendicular to the direction of cortical thickness. From this orientation, canals appear to be primarily split longitudinally, and faint shadows of osteonal bone matrix can be visualized around the canals. Osteonal splitting can be clearly seen ("*" on lower right figure. **B**. Collapsed confocal image. The region shown is the same as that in the top three images in part A, except here all 100 images of varying depths of the volume are collapsed (added and averaged) to create a single image. This is equivalent to examining an image of a 150 µm thick section, although details from all levels are overlain equally well. In an optical image, there would be a bias toward parts of the volume closer to the lens. In this image, it is much easier to visualize details than in the single sections in **A**. Bone matrix in the osteons is much more distinct. Laminae and osteocyte lacunae can be seen. **C**. Reconstructed volume of mandibular cortical bone. Here canal and osteon structures are also visible and can be seen from three orientations simultaneously. Within Analyze software, this volume can be visually modified to examine and measure internal morphology. The figure gives an indication of how canals and osteons can be followed through the volume and into adjacent volumes.

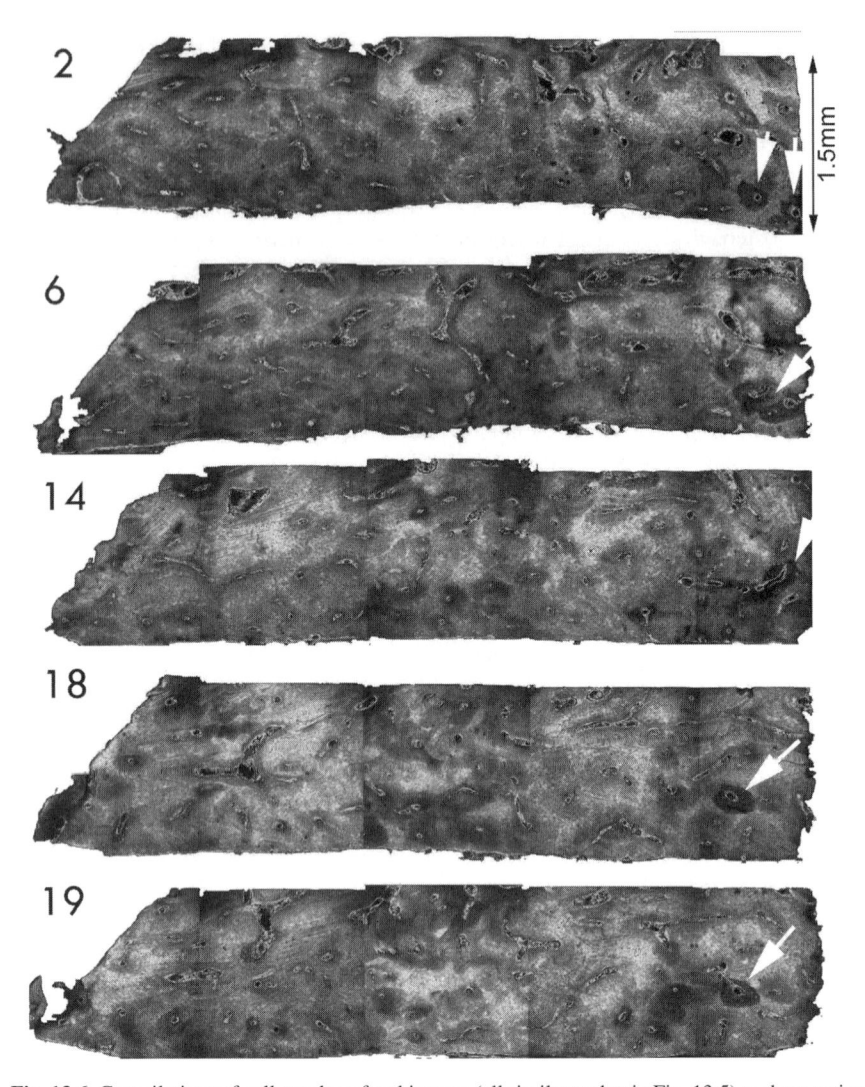

Fig. 13.6 Compilations of collapsed confocal images (all similar to that in Fig. 13.5) to show entire cross-sections of the bone specimen in the 2-3 plane or perpendicular to the 3 or longitudinal axis. Size of each section is about 7.5 × 1.5 mm. Details of canals and osteonal structure are seen in each section and can be followed to adjacent sections. Some osteons run through all 19 sections. For example, the dark osteon on the left (indicated by the white arrows) can be found in all sections. In sections 2 through 14, splitting of the osteon is seen

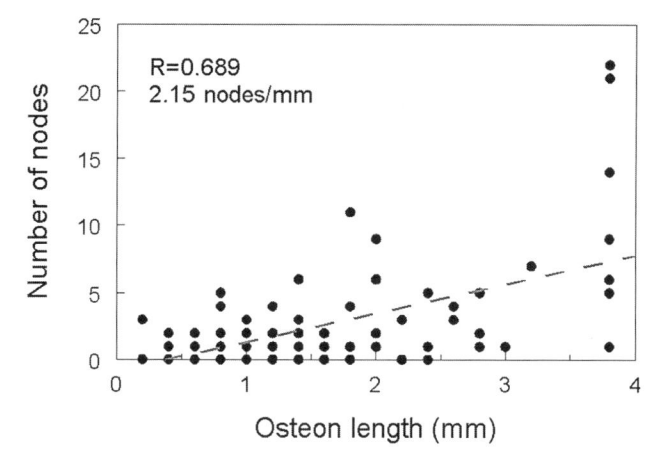

13 Relationship Between Bone 3D Microstructure and Elastic Properties

Fig. 13.7 The number of nodes correlates with the longitudinal length of the osteonal segments ($P < 0.05$). This correlation is moderate, and some osteons were found that had few branches despite their relatively long lengths

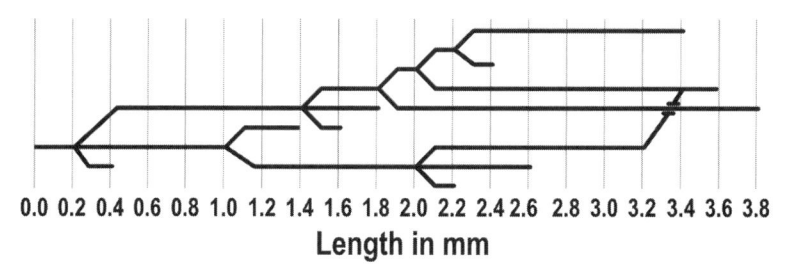

Fig. 13.8 Example of a splitting pattern in a complex osteonal segment. Note that this diagram does not give any details of tertiary structure. Splitting includes bifurcations and trifurcations as well as a rejoining. The Y-axis is not to scale

(Fig. 13.9). This difference is even more evident in other measures of circular statistics, such as concentration[3] (2.6 for 1-3 plane versus 1.7 for the 2-3 plane), and show a greater concentration of angles in the 1-3 plane.

---

[3] The concentration is a parameter specific to the von Mises distribution, and measures the departure of the distribution from a perfect circle (or a uniform distribution). It is related to the length of the mean vector. The value reported by Oriana Statistical Software is the maximum likelihood estimate of the population concentration, calculated using the formula in Fisher (1993, p. 88) and Mardia and Jupp (2000, pp. 85–6).

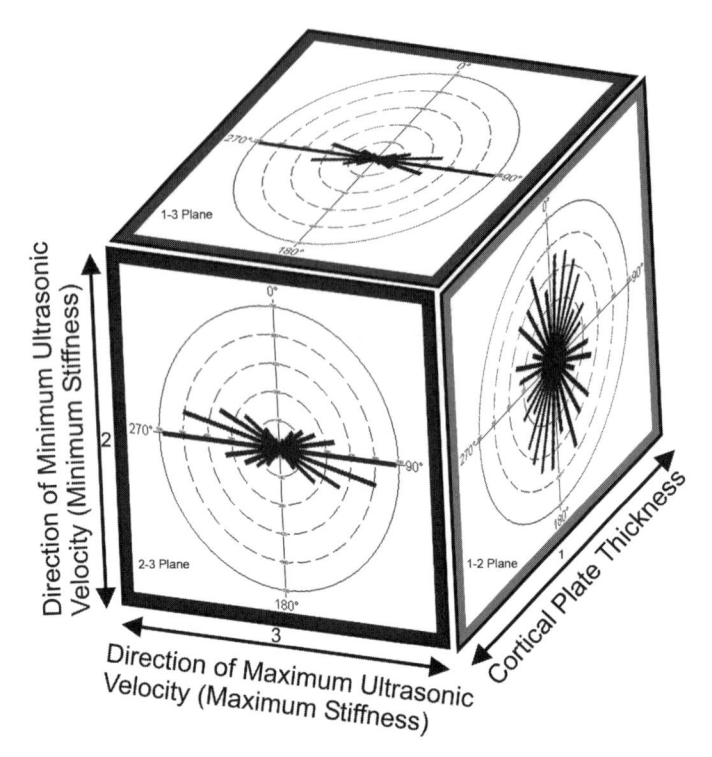

Fig. 13.9 Rosette histograms of angles of all osteonal segments in all 95 reconstructed confocal volumes. The scales for the three histograms are not equivalent and are shown here together in order to focus on the differences in the distributions. If osteonal structure is visualized in the 1-3 plane (histogram at the top of the box), most osteons will appear longitudinally oriented. In the 2-3 plane (histogram on the side of the box), most osteons also appear oriented along the long axis of the mandible, except there is significant variation seen; and many osteons appear oriented at an angle above or below the center line, indicating the direction of maximum stiffness in the cortical plane. When viewed in the 1-2 plane (histogram on the front of the box), osteons appear oriented in all directions, although there are more in the 2 or up-down direction

13.3.3 Studies of Cortical Structure Using Micro Computed Tomography

Reconstructions from μCT imaging showed a much denser canal network for the femoral specimen than for the mandibular specimen (Fig. 13.10). In the femoral specimen, the cumulative length of all canals was 153.0 mm, compared to 67.2 mm in the mandibular specimen. In per cubic millimeter of cortical bone, there were 12.4 mm of Haversian canal length for the femoral specimen versus 5.4 mm for the mandibular specimen.

In both μCT specimens, most osteons appear overall to be oriented in the direction of maximum stiffness, although much deviation is also apparent (Fig. 13.10). Unlike the numerous apparently independent osteonal fragments observed in the

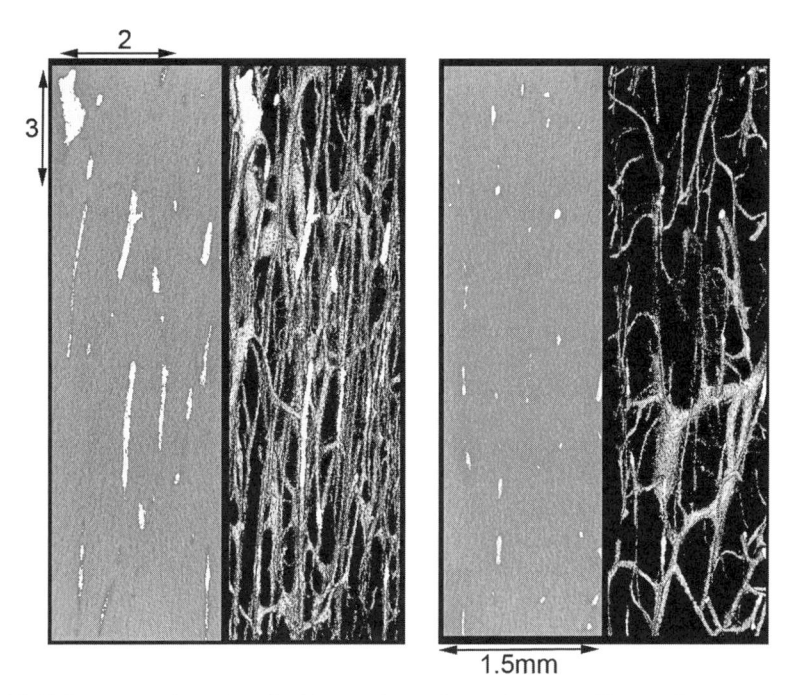

13 Relationship Between Bone 3D Microstructure and Elastic Properties

Fig. 13.10 Reconstructions of cortical bone volumes from μCT scans. The two figures on the left are from the femoral specimen and the two figures on the right are from the mandibular specimen. The left figure in each pair is a superficial surface image of the specimens, and the right figure shows the internal canal structures with the matrix removed. The up–down direction for each specimen corresponds with the orientation of maximum stiffness in the plane of the cortical plate (3 direction), and the right–left direction corresponds with the orientation of minimum stiffness in the plane of the cortical plate (2 direction). The size of the imaged regions is 5.5 mm (up–down) by 1.5 mm (*right–left*) by 1.5 mm (deep)

confocal microscope study, all canals examined in the femoral and mandibular μCT specimens were joined. This demonstrated that the networks of osteons in these specimens are each formed by a single vascular bed.

Statistical analysis of osteon orientations weighted by osteon length showed a similar pattern in the mandibular μCT specimen as that found in the confocal microscope mandibular specimen, (Fig. 13.11). In the mandibular μCT specimen, osteons were oriented on average along the axis of greatest stiffness in the 1-3 plane (mean = 94.1°, circular SD = 26.4°). In the 2-3 plane, the average orientation was 30° above an anterior–posterior vector aligned along the orientation of the axis of greatest stiffness (mean = 120.1°, circular SD = 45.9°). The much greater variance in the 2-3 plane compared to the 1-3 plane is evident in Fig. 13.11. This difference is also found in concentration (1-3 plane = 1.8; 2-3 plane = 0.6).

The pattern in the femoral μCT specimen was different. In both the 1-3 and the 2-3 planes, osteons were oriented in the direction of maximum stiffness with less variation in either plane than that found in the mandibular specimens (1-3 plane:

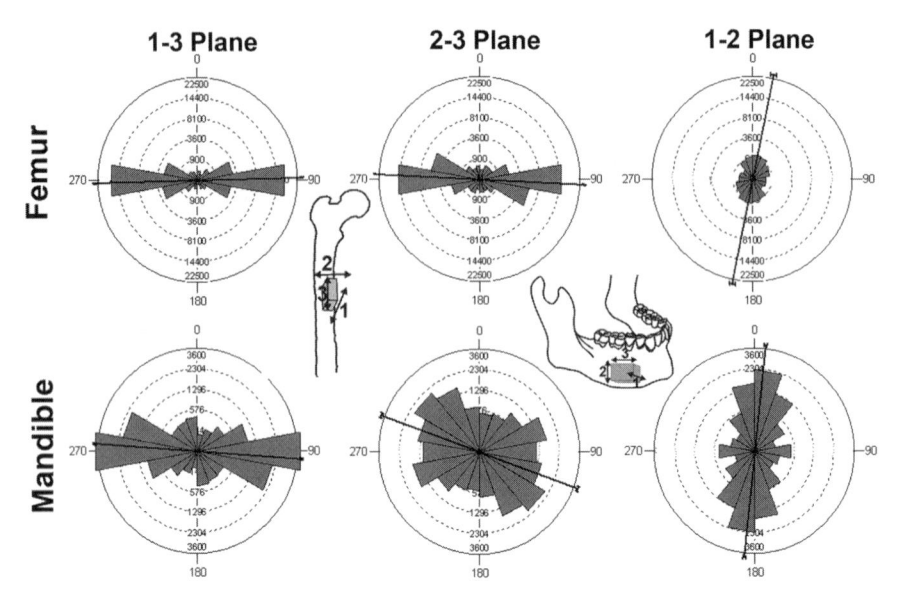

Fig. 13.11 Rosette histograms of the orientations of all osteonal segments per micron length relative to the plane of projection. Angles are weighted by the length of each osteonal segment, and the histograms for each projection within each bone specimen use the same scale. The scale is much larger for the femoral histograms because a much greater number of canals were found in a comparable volume of bone. In the femoral specimen, most osteons are oriented along the axis of greatest stiffness in the cortical plane (3 direction and longitudinal anatomical axis), although there is slightly more variation along the axis of minimum stiffness in the plane of the cortical plate (2 direction) than in the radial direction (1 direction). The mandibular specimen shows a similar pattern, except that the variation in the 1 and 2 directions are much larger relative to the 3 direction. This is especially so for the 2 direction, which corresponds with the tendency of many more osteons to be oriented at angles above or below the 3 axis

mean = 88.4°, circular SD = 13.3°; 2-3 plane: mean = 92.8°, circular SD = 18.4°). Note that there is still a difference in variance between the two planes, but this difference is relatively smaller than that found in the mandibular μCT specimen. The concentrations reflect the greater orientation of the femoral osteons (1-3 plane = 5.2; 2-3 plane = 3.0).

When the canal segments were treated as vectors and trigonometric functions were used to determine the proportion of the vectors along each axis (Fig. 13.3: length of lines 1, 2, and 3), the sum of these values for all canal segments showed a pronounced difference between the mandibular and the femoral μCT specimens. In the mandibular specimen, osteon orientation was 16% in the 1 direction (cortical thickness), 36% in the 2 direction (axis of minimum stiffness in the cortical plane), and 48% in the 3 direction (axis of maximum stiffness in the cortical plane). These values in the femoral specimen were respectively 11%, 16%, and 73%, indicating that osteons are more longitudinally oriented in the femoral specimen than in the mandibular specimen, and that there is a difference between the 1 and 2 axes in the mandibular specimen, which is much less apparent in the femoral specimen.

13.4 Discussion

The goal of this chapter is to explore the relationship between tissue-level structure and three-dimensional elastic properties in cortical bone tissue. The hypothesis is that the three-dimensional elastic structure of remodeled cortical bone at a tissue level of organization is primarily dependent on the three-dimensional organization of osteons (Hert et al., 1994; Skedros et al., 1994). Quantification of the three-dimensional structure of osteons has been an elusive goal due to the difficulty in visualizing internal structure in cortical bone. This study explores two novel technologies, confocal microscopy and μCT, to test how these imaging techniques can be used in quantifying the three-dimensional structure of cortical bone. A key component of this investigation is the use of ultrasound to determine the orientation of the material axes of the cortical bone material. This orientation can be compared to the mean orientation of osteons in a tissue specimen. Differences between regions in cortical bone elastic anisotropy can be compared to differences in the variation of osteonal orientation in these same regions.

13.4.1 Orientation of Elastic Properties in Cortical Bone Tissue

Tissue level elastic properties are necessary for both accurate mechanical models and to understand potential structural adaptations in the craniofacial skeleton (Hart et al., 1992; Korioth and Hannam, 1994a, b; Kabel et al., 1999; Dechow and Hylander, 2000; van Eijden, 2000; Vollmer et al., 2000). For example, local anisotropy and regional variations in skeletal elastic properties can have pronounced effects on the relationship between stress and strain patterns (Carter, 1978; Cowin and Hart, 1990; Cowin et al., 1991; Ricos et al., 1996, Dechow and Hylander, 2000).

Insights into the relationship between mandibular deformation, loading, and the structure of cortical bone can be gained by an examination of strain gage results in light of cortical elastic properties (Dechow and Hylander, 2000). Strain gage studies in the human mandible have demonstrated complex patterns of in vivo (Asundi and Kishen, 2000) and in vitro cortical bone strain (Andersen et al., 1991a, b; Throckmorton et al., 1992; Throckmorton and Dechow, 1994; Yamashita and Dechow, 2000). Studies using animal models suggest similar complexities (Bouvier and Hylander, 1996; Endo, 1973; Hylander, 1979a, b, 1984; Hylander et al., 1987; Marks et al., 1997; Teng and Herring, 1996), which result from variations in skeletal and muscular form and masticatory muscle contraction dynamics (Harper et al., 1997; Hylander and Johnson, 1994; Throckmorton et al., 1990; van Eijden et al., 1990). These variations may have differential effects on various structural and functional regions of the mandible (Dechow and Hylander, 2000; Schwartz-Dabney and Dechow, 2003).

This leads to the hypothesis that functional variations in some regions of the mandible are correlated with an increased variability of elastic properties. Confirmation of this may help elucidate a mechanistic explanation for why certain regions of the mandible have considerable variation in (1) the orientations of their axes of maximum stiffness and (2) the amount of anisotropy (Schwartz-Dabney and Dechow, 2003). Little is known about those variations in cortical microstructure, which might result in variations in three-dimensional elastic properties of cortical bone. This is especially important at the tissue level because it is the level most relevant to the studies of skeletal biomechanics. Knowledge of structural variation at this level may provide links to understanding functional adaptation within the levels of structural hierarchy of smaller dimension in cortical bone.

At the tissue or supraosteonal level, ultrasonic studies have shown that the elastic structure of mandibular cortical bone was unlike the presumed structure of femoral cortical bone. Mandibular cortical bone in the region of the corpus showed larger differences in ultrasonic velocities and calculated elastic moduli between each of its three orthotropic axes (Schwartz-Dabney and Dechow, 2003). This is a different pattern than has been suggested for the femur and the midshafts of other postcranial long bones.

In the midshafts of long bones, Sevostianov and Kachanov (2000) and Yeni et al. (2001 have shown that theoretical consideration of patterns of porosities (Haversian canals, Volkman's canals, canaliculi, and osteocyte lacunae) can explain much of the basic pattern of cortical bone anisotropy. These theoretical considerations suggest that cortical bone is dominated by longitudinally oriented osteons and can best be modeled as transversely isotropic with the plane of symmetry perpendicular to osteon orientation. Sevostianov and Kachanov (2000) note that studies by Katz et al. (1984) and Ashman et al. (1984) using ultrasonic methods, and Reilly and Burstein (1974) and Zioupos et al. (1995) using mechanical testing, have shown that the cortices of long bones are transversely isotropic as the measures of stiffness normal and tangential to the bone surface usually have values that vary by 10% or less. This is in contrast to elastic moduli in the longitudinal direction, which are 50% or more greater.

This picture differs from cortical anisotropy in the mandible. Our studies (Dechow et al., 1992, 1993; Dechow and Hylander, 2000, Schwartz-Dabney and Dechow, 2003) show orthotropy in the mandible, but not transverse isotropy. However, it is important to note that although the osteonal structure of femoral cortical bone strongly suggests that one axis of orthotropy parallels the long axis of the bone, this assumption has not been tested directly. In the mandible, studies of elastic property orientation in hundreds of cortical specimens (Schwartz-Dabney and Dechow, 2003) have shown variation both between and within regions in material orientation. The specimens in this study come from the lower border of the mandible in the molar region, where such variation is minimal. However, no previous studies have examined such variation in elastic orientation in the shaft of the femur. The specimen studied here showed that, as assumed in the literature, the femur is stiffest directly along its longitudinal axis. Studies on ten other femoral specimens in our laboratory (manuscript in preparation) have shown an identical pattern, and have indicated less variability in the orientation of cortical elastic structure than is found in the mandible.

13.4.2 Orientation of Cortical Bone Tissue Structure

Another problem with applying the theoretical considerations of Sevostianov and Kachanov (2000) is our lack of knowledge of the three-dimensional structure of bone at the tissue level. In particular, information on osteon orientation is both minimal and contradictory. Likewise, despite much research on the material properties of individual osteons, and investigations of variations of collagen structure in osteons, there is little information on the effects of three-dimensional osteon structure on tissue level material properties. Several studies point to contradictions in this area. Studies by Hert et al. (1994) and Petrtyl et al. (1996) review and support earlier work, suggesting that femoral osteons are oriented in opposite off-axis directions in the medial and lateral cortical walls. They suggest that these off-axis orientations, which vary $5°–15°$ from the long axis of the bone and are not found in atypical (unloaded?) femurs, are aligned with principal stresses resulting from the combination of the predominant compressive, bending, and torsional femoral loads. In contrast, Stout et al. (1999) used biomedical imaging technology (Analyze Software, Mayo Clinic) to create three-dimensional reconstructions of osteons from histological sections of dog femurs, previously described by Tappen (1977). They did not find a spiraling or even off-axis orientation of osteons, as reported in other investigations, but rather a predominately longitudinal organization, but with complex patterns of splitting. It is possible, although uninvestigated, that differences between these studies can be accounted for by several important factors, including probable functional differences between species, differences in visualization techniques, and the size or portion of the bones under study.

The results of this investigation clearly show that, in the femoral specimen, the average orientation of the canals in the cortical bone align very closely with the measured axis of maximum stiffness in the plane of the cortical plate, which in turn has the same orientation as the long axis of the bone. Likewise, variation in osteon orientation in the other two directions is relatively similar, as would be expected in a structure that can be approximated as transversely isotropic. The small difference between these two orientations involved a slightly greater variation in canal orientation in the 2-3 plane compared to the 1-3 plane, indicating more osteonal segments that diverged from the axis of maximum stiffness in the 2 direction than in the 1 direction. This greater variation suggests that, based on osteonal orientation alone, the bone should be slightly stiffer in the 2 direction than in the 1 direction (Ashman et al., 1984).

The difference in stiffness between the 1 and 2 directions in the femur (Ashman et al., 1984) is small compared to that found in the mandible (Schwartz-Dabney and Dechow, 2003). In both of our mandibular specimens, using different types of analysis in each, there was much greater variation in osteonal orientation in the 2-3 plane compared to the 1-3 plane. This difference was the greatest using the results of the mandibular µCT study. In the confocal microscopy study, the sampling was done by quantifying the osteonal segment angles in each of the 95 confocal volumes. Thus a longer osteonal segment in an individual volume that was not aligned in the direction of maximum stiffness was given the same weight as one aligned in the direction of maximum stiffness. Osteons in this later direction would by definition be shorter because of the orientation of the tissue sections. Thus, the measurement method in the confocal study might have created bias that underestimated the actual difference between the two planes in variability. In any case, the much greater variation in osteon segment orientation in the 2-3 plane compared to the 1-3 plane is what would be expected in a structure that is being modeled as orthotropic.

One problem here though is that the alignment between the axis of maximum stiffness and the mean orientation of the osteons does not correspond as well in the mandibular specimens as in the femoral specimen. In the femoral specimen, the difference is 1.6° in the 1-3 plane and 2.8° in the 2-3 plane, compared respectively to 1.2° and 4.9° in the confocal microscope mandibular specimen, and 4.1° and 30.1° in the µCT mandibular specimen. Even though the large deviance in the 2-3 plane of the µCT mandibular specimen is offset by the large variance in this measurement (Fig. 13.11), it is also clear that the majority of osteons are oriented in an anteroinferior to posterosuperior direction, while the orientation of greatest elastic stiffness is approximately parallel to the mandibular lower border.

It is also interesting in this regard that the ultrasonic curves (Fig. 13.4) for the mandibular specimen showed deviation from the idealized sine curve, especially by flattening at the apex. This pattern has not been found in most mandibular specimens in our laboratory (Schwartz-Dabney and Dechow, 2003), even though most show a pattern of orthotropy (rather than transverse isotropy). However, angular sampling in the 2003 study was insufficient to provide the more nuanced view of the shapes of these curves as described here. The greater density of osteons in the femoral specimen may be important, as the more prevalent non-osteonal interstitial bone may have more of an impact on the elastic structure in this mandibular specimen. Presumably, the interstitial bone is older than the osteonal bone and may reflect a different pattern of deposition and orientation than the osteonal matrix. This could result in increased variation in elastic structure in the mandibular cortical bone specimen.

If anisotropy in tissue elasticity in remodeled cortical bone is primarily dictated by osteonal structure, and osteonal structure has significant splitting, including off-axis, and horizontal components, as shown in this study and also in other investigations, such as those by Stout et al. (1999) in dog femur and in so-called drifting osteons in the midshafts of human metatarsals and baboon fibulas (Robling and Stout, 1999), then it is reasonable to question the limits of the assumption of orthotropy in mandibular cortical bone at a tissue level. For instance, some regions of cortical bone may not have a maximum axis of stiffness, but rather have microstructural components arrayed in a way so that there is equivalent stiffness over a range of angles. This may be the significance of the pattern of variation in the mandibular µCT specimen. But analyses of more specimens are required before any consistent pattern can be suggested. Findings from ultrasonic studies (Schwartz-Debney and Dechow, 2003, and unpublished data) suggest this may be true in some mandibular specimens and is also likely to vary regionally.

13.4.3 Osteonal Morphology in Cortical Bone

It is beyond the scope of this study to comment in detail on the implications for understanding of osteonal morphology. Yet the results do point out some important features. The results show that splitting patterns can be variable with some osteons splitting frequently and having complex networks, while other osteonal segments may travel for longer distances with no apparent splitting.

The results also call into question the significance of so-called Volksman's Canals, or channels that are usually perpendicular to the predominant long axis of osteons. The results show that splitting in osteonal trees occurs over a continuous range of angles from the axis of maximum stiffness to angles perpendicular to this axis. Study of matrix around all canals with confocal microscopy did not reveal channels at any angle that lacked a surrounding distinct matrix or osteonal structure.

Results from the μCT studies called into question the independence of the osteonal segments visualized using confocal microscopy. The canal structures in the μCT specimens were all continuous.

In total, this information suggests that it is important to view osteons in cortical bone as a structured vascular bed, rather than as a series of parallel longitudinal straws, or a series of osteons of variable complexity. The shape of this vascular bed may be affected by biomechanical forces, which likewise affect the three-dimensional aspects of material properties in cortical bone. In this sense, the internal walls of the canal system truly represent a third envelope for resorption, disposition, and potential response to mechanical changes in loading.

It is also logical to ask how the patterns of the vascular beds in cortical bone compare not only between regions but also with such beds in other tissues, which may or may not be subjected to significant loading. There is little data in the literature to even begin making such comparisons.

13.4.4 Elastic Anisotropy, Bone Tissue Structure, and Adaptation

Most studies of the mechanical properties of cortical bone have been concerned with structures at smaller levels of organization, with only a minor emphasis on anisotropy or three-dimensional structure. Variations within osteons and lamellar structures, both within and outside osteons, and a wide array of microarchitectural structures have been suggested to play an important role in mechanical properties (Martin and Burr, 1989; Martin et al., 1998), including the orientation of collagen fibers (Carando et al., 1991; Currey et al., 1994; Riggs et al., 1993) and mineral crystallites (Bacon and Goodship, 1991; Fratzl et al., 1992; Wenk and Heidelbach, 1999). Other investigators have discussed the importance of relative proportions of variable lamellae, containing distinct orientations of collagen fibers and crystallites (Ascenzi, 1988; Fratzl et al., 1993; Turner et al., 1995; Takano et al., 1996; Rinnerthaler et al., 1999). Other matrix factors include mineralization and density (Schaffler and Burr, 1988), as well as the degree of collagen cross-linking (Lees et al., 1990).

Currey and Zioupos (2001) pointed out that anisotropy in cortical bone is dependent on both porous structures and anisotropies within the bone matrix itself. Indeed, most studies that attempt to relate skeletal function with microstructural variation concern themselves with collagen matrix organization, patterns of bone mineralization, and their relationship. However, at a tissue level, which can be related to the mechanics of whole bones, the hierarchical structure of bone must be considered in the assessments of material properties.

At a tissue level, the organization of cortex matrix components, such as the osteons, surface lamellae, and interstitial bone, must be considered in tandem with porous structures. Using X-ray pole figure analysis, Sasaki et al. (1989, 1991) found that anisotropy can be explained by the axial distribution of bone mineral. Hasegawa et al. (1994) came to a similar conclusion by comparing ultrasonic measurements of elastic constants in demineralized bone and in bone with the organic phase removed. Demineralized bone showed no anisotropy; bone without organic phase showed anisotropies proportionate to fresh bone. However, it is unclear how microarchitectural features contribute to the anisotropy of remodeled cortical bone at a tissue or supraosteonal level. Exploration of how such factors play a role requires much more information about their variation in actual bone specimens and the ability to model them within the range of possibilities of osteonal structure.

Katz et al. (1984) demonstrated how remodeling of cortical bone can change elastic properties and their anisotropies. They argued that plexiform bone, constructed of parallel lamellar units, is symmetrical around three mutually perpendicular axes (orthotropy), while haversian bone in transversely symmetrical, that is, elastic constants are identical about two of three axes (transverse isotropy). Remodeling leads to increased symmetry and decreased stiffness. While this is important in the midshafts of long bones, which undergo large amounts of remodeling, its effect in craniofacial bone, with less remodeling, is unclear. Our results here suggest that the lesser amounts of remodeling can lead to differences between the axis of maximum stiffness in bulk tissue specimens and the mean direction of osteon orientation.

Katz et al. (2005) described anisotropy in the mandible as transversely isotropic. On the face of it, this appears to be a contradiction with the findings of Schwartz-Dabney and Dechow, 2003. But it is important to realize that the study of Katz and colleagues measured three-dimensional elastic properties at a smaller level of organization. The study followed longitudinal osteonal structures through the mandible using an acoustic microscope. At the level of a small number of parallel osteonal segments, a finding of transverse isotropy is likely. But in order to get a better idea of tissue organization that is relevant at a whole bone level, it is important to consider the supraosteonal tissue organization itself. When larger bulk specimens are measured ultrasonically, greater differences in ultrasonic velocities between the 1 and 2 directions are apparent, suggesting greater orthotropy. A close examination of the data of Katz and colleagues (2005) does show higher velocities circumferentially than axially, but these differences are not as great as those found with larger specimens.

These findings suggest, but do not directly show, that cortical bone can alter its three-dimensional elastic structure through remodeling. Some investigations have centered on the effects of altered bone strain on bone mass, porosity, mineralization, and structural integrity of the whole bone (for summary, see Martin et al., 1998). These studies show that, under certain conditions, cortical bone will model over time to increased loads by greater cortical thickness or increased mass, resulting in increased strength. However, interesting questions remain, such as whether bone can adapt by shifts in material properties or by changes in the direction in which it is most resistant to deformation, and to what degree such adaptation might occur.

Lanyon and Rubin (1985) have argued, based on the findings by Woo et al. (1981), that "internal bone remodeling is not an adaptive response to improve the material properties of bone tissue," although it may increase its fatigue life. However, bone modeling on periosteal and endosteal surfaces forms lamellar bone or woven bone (Ascenzi, 1988; Martin and Burr, 1989; Turner et al., 1992), which has variations in microstructural elements, such as collagen fiber or crystallite orientation, that correlate with material properties. Interstitially, drifting osteons or formation and splitting of normal (Type 1) osteons (Robling and Stout, 1999) may also influence bone mechanical properties.

Our work on the material properties in edentulous mandibles (Schwartz-Dabney and Dechow, 2002) suggests that functional changes in the mandible result in changes in material properties. It is reasonable to hypothesize that such changes result from alterations in the pattern of cortical modeling and remodeling. Differences between dentate and edentulous mandibles were site-specific and suggested both increases and decreases in stiffness, altered ratios of anisotropy, and, at a few sites, altered orientations of maximum stiffness, but no changes in apparent density.

Another test of experimental changes in three-dimensional material properties was that of Takano et al. (1999). They used a surgical model in greyhounds in which portions of the ulna were removed to alter loads on the radius. In controls, bone strains were shown to correlate with an indirect measure of collagen orientation (LSI), indicating differences in collagen orientation based on loading pattern. Radii from osteotomized animals showed increases in bone strain in concert with changes in anisotropy ratios in both demineralized and deproteinized tissues. While this experiment showed a change, information was not collected directly on three-dimensional changes in bone matrix structure.

We hypothesize that the microstructural features that account for the differences in three- dimensional elastic properties between the mandible and the mid-diaphysis of bones like the femur include not only the orientation, size (relative area), and density of osteons but also may include other structural features, such as the relative amounts of periosteal lamellar structures. Likewise, regional variations in these features are the likely cause of the heterogeneity in the material properties of the mandible. We suspect that similar heterogeneity would probably be found in the cortices of other bones, both cranial and postcranial, if sufficient studies were attempted. Current work in our laboratory has shown as great or greater variation in material properties within and among bones of the midface and cranium (Peterson and Dechow, 2002, 2003; Peterson et al., 2006).

Acknowledgments Much inspiration for the work contained in this chapter and others from my laboratory have sprung from the publications and work of Bill Hylander. Bill's approaches have pointed to the necessity of many new directions of investigation in evolutionary and functional morphology. Bone strain as an experimental technique is essential for making sense of primate mastication, but it also calls into focus the material characteristics of skeletal tissue, which in turn raises many issues about development, growth, and adaptation of hard tissues. Bill's work also raises the inherently important issues, often ignored, of the regional and evolutionary significance of variations in function, skeletal structure, mechanics, and adaptation at the tissue level in bone. I thank Bill for his insights and friendship over the years. I am proud to work in a field to which he has added so much.

Thanks to Dr. Qian Wang for reading and commenting on an earlier version of this manuscript. Thanks also to Drs. Mathew Ravosa, Chris Vinyard, and Chris Wall for their comments on this manuscript and for organizing this outstanding symposium and book dedicated to the growing field of research that stems from the seminal studies of Dr. William Hylander.

References

Andersen KL, Mortensen HT, Pedersen EH, Melsen B. (1991a) Determination of stress levels and profiles in the periodontal ligament by means of an improved three-dimensional finite element model for various types of orthodontic and natural force systems. J Biomed Eng 13:293–303.

Andersen KL, Pedersen EH, Melsen B. (1991b) Material parameters and stress profiles within the periodontal ligament. Am J Orthod Dentofacial Orthop 99:427–40.

Ascenzi A. (1988) The micromechanics versus the macromechanics of cortical bone–a comprehensive presentation. Journal of Biomechanical Engineering 110:357–63.

Ashman RB. (1989) Experimental techniques. In: Cowin SC, (editors), Bone Mechanics, Boca Raton, Florida: CRC Press, Inc. pp. 75–96.

Ashman RB, Cowin SC, Van Buskirk WC, et al. (1984) A continuous wave technique for the measurement of the elastic properties of cortical bone. J Biomech 17:349–61.

Asundi A, Kishen, A. (2000) A strain gauge and photoelastic analysis of in vivo strain and in vitro stress distribution in human dental supporting structures. Arch Oral Biol 45:543–50.

Bacon GE, Goodship AE. (1991) The orientation of the mineral crystals in the radius and tibia of the sheep, and its variation with age. J Anat 179:15–22.

Bouvier M, Hylander WL. (1981) The relationship between split-line orientation and in vivo bone strain in galago (*G.crassicaudatus*) and macaque (*Macaca mulatta* and *M. fascicularis*) mandibles. Am J Phys Anthropol 56:147–56.

Bouvier M, Hylander WL. (1996) The mechanical or metabolic function of secondary osteonal bone in the monkey *Macaca fascicularis*. Arch Oral Biol 41:941–50.

Buckland-Wright JC. (1977) The nature of split-line formation in bone. Proceedings of the Anatomical Society of Great Britain and Ireland.

Carando S, Portigliatti-Barbos M, Ascenzi A, et al. (1991) Macroscopic shape of, and lamellar distribution within, the upper limb shafts, allowing inferences about mechanical properties. Bone 12:265–9.

Carter DR. (1978) Anisotropic analysis of strain rosette information from cortical bone. J Biomech 11:199–202.

Cooper DMI, Turinsky AL, Sensen CW, Hallgrimsson B. (2003) Quantitative 3D analysis of the canal network in cortical bone by micro-computed tomography. Anat Rec 274B: 169–179.

Cowin SC. (1989) The mechanical properties of cortical bone tissue. In: Cowin SC, (editor), Bone Mechanics, CRC Press, Inc. Boca Raton, Florida, pp. 97–128.

Cowin SC, Hart RT. (1990) Errors in the orientation of the principal stress axes if bone tissue is modeled as isotropic. J Biomech 23:349–52.

Cowin SC, Sadegh AM, Luo GM. (1991) Correction formulae for the misalignment of axes in the measurement of the orthotropic elastic constants. J Biomech 24:637–41.

Currey JD. (1984) The mechanical adaptations of bones. Princeton University Press, Princeton, NJ.

Currey JD, Brear K, Zioupos P. (1994) Dependence of mechanical properties on fibre angle in narwhal tusk, a highly oriented biological composite. J Biomech 27:885.

Currey JD, Zioupos P. (2001) The effect of porous microstructure on the anisotropy of bone-like tissue: a counterexample. J Biomech 34:707–10.

Dechow PC, Nail GA, Schwartz-Dabney CL, Ashman RB. (1993) Elastic properties of human supraorbital and mandibular bone. Am J Phys Anthropol 90:291–306.

Dechow PC, Schwartz-Dabney CL, Ashman RB. (1992) Elastic properties of the human mandibular corpus. In: Carlson DS, Goldstein SA, (editors), Bone Biodynamics in Orthodontic and Orthopedic Treatment, Craniofacial Growth Series, Volume 27, Center for Human Growth and Development, The University of Michigan: Ann Arbor, Michigan, pp. 299–314.

Dechow PC, Hylander WL. (2000) Elastic properties and masticatory bone stress in the macaque mandible. Am J Phys Anthropol 112:553–74.

Dempster WT. (1967) Correlation of types of cortical grain structure with architectural features of the human skull. Amer J Anat 120:7–32.

Endo B. (1973) Stress analysis on the facial skeleton of gorilla by means of the wire strain gauge method. Primates 14:37–45.

Evans FG. (1973) Mechanical Properties of Bone. Charles C. Thomas, Springfield, IL.

Fisher NI. (1993) Statistical Analysis of Circular Data. Cambridge University Press, Cambridge. 277pp.

Fratzl P, Fratzl-Zelman N, Klaushofer K. (1993) Collagen packing and mineralization. An x-ray scattering investigation of turkey leg tendon. Biophys J 64:260–6.

Fratzl P, Groschner M, Vogl G, et al. (1992) Mineral crystals in calcified tissues: a comparative study by SAXS. J Bone Min Res 7:329–34.

Gebhardt W. (1906) Über funktionell wichtige Anordnungsweisen der feineren und gröberen Bauelemente des Wirbeltierknochens. II. Spezieller Teil.: Der Bau der Haversschen Lamellensysteme und seine funktionelle Bedeutung. Arch Entw Mech Org 20:187–322.

Giesen EB, van Eijden TM. (2000) The three-dimensional cancellous bone architecture of the human mandibular condyle. J Dent Res 79:957–63.

Guo E. (2001) Mechanical properties of cortical bone and cancellous bone tissue. In Cowin SC (editor) Bone Mechanics Handbook, Second Edition, CRC Press, Boca Raton, pages 10–1 to 10–23.

Harper RP, de Bruin H, Burcea I. (1997) Muscle activity during mandibular movements in normal and mandibular retrognathic subjects. J Oral Maxillofac Surg 55:225–33.

Hart RT, Hennebel VV, Thongpreda N, et al. (1992) Modeling the biomechanics of the mandible – a three- dimensional finite element study. J Biomechanics 25:261–86.

Hasegawa K, Turner CH, Burr DB. (1994) Contribution of collagen and mineral to the elastic anisotropy of bone. Calcif Tissue Int 55:381–6.

Hert J, Fiala P, Petrtyl M. (1994) Osteon orientation of the diaphysis of the long bones in man. Bone 15:269–77.

Hylander WL. (1979a) An experimental analysis of temporomandibular joint reaction force in Macaques. Am J Phys Anthropol 51:433–56.

Hylander WL. (1979b) Mandibular function in Galago crassicaudatus and Macaca fascicularis: an in vivo approach to stress analysis of the mandible. J Morphol 159:253–96.

Hylander WL. (1984) Stress and strain in the mandibular symphysis of primates: a test of competing hypotheses. Am J Phys Anthropol 64:1–46.

Hylander WL, Johnson KR. (1994) Jaw muscle function and wishboning of the mandible during mastication in macaques and baboons. Am J Phys Anthropol 94:523–47.

Hylander WL, Johnson KR, Crompton AW. (1987) Loading patterns and jaw movements during mastication in Macaca fascicularis: a bone-strain, electromyographic, and cineradiographic analysis. Am J Phys Anthropol 72:287–314.

Kabel J, van Rietbergen B, Odgaard A, Huiskes R. (1999) Constitutive relationships of fabric, density, and elastic properties in cancellous bone architecture. Bone 25:481–6.

Katz JL, Meunier A. (1987) The elastic anisotropy of bone. J Biomech 20:1063–70.

Katz JL, Yoon HS. (1984) The structure and anisotropic mechanical properties of bone. IEEE Trans Biomed Eng 31:878–84.
Katz JL, Kinney JH, Spencer P, Wang Y, Fricke B, Walker MP, Friis EA. (2005) Elastic anisotropy of bone and dentitional tissues. J Mater Sci Mater Med. 16:803–6.
Katz JL, Yoon HS, Lipson S, Maharidge R, Meunier A, Christel P. (1984) The effects of remodeling on the elastic properties of bone. Calcif Tissue Int 36:Suppl 1:S31–6.
Koch JC. (1917) The laws of bone architecture. Am J Anat 21:177–297.
Kohles SS, Bowers JR, Vailas AC, Vanderby R Jr. (1997) Ultrasonic wave velocity measurement in small polymeric and cortical bone specimens. J Biomech Eng 119:232–6.
Korioth TWP, Hannam AG. (1994a) Deformation of the human mandible during simulated tooth clenching. J Dent Res 73:56–66.
Korioth TWP, Hannam AG. (1994b) Mandibular forces during simulated tooth clenching. J Orofac Pain 8:178–89.
Lanyon LE, Rubin CT. (1985) Functional adaptation in skeletal structures. In: Hildebrand M, Bramble DM, Liem KF, Wake DB, (editors), Functional Vertebrate Morphology, The Belknap Press of Harvard University Press: London, England, pp. 1–25.
Lees S. (1982) Ultrasonic measurements of deer antler, bovine tibia and tympanic bulla. J Biomech 15:867–74.
Lees S, Eyre DR, Barnard SM. (1990) BAPN dose dependence of mature crosslinking in bone matrix collagen of rabbit compact bone: corresponding variation of sonic velocity and equatorial diffraction spacing. Connect Tissue Res 24:95–105.
Lipson SF, Katz JL. (1984) The relationship between elastic properties and microstructure of bovine cortical bone. J Biomech 17:231–40.
Mardia KV, Jupp PE. (2000) Statistics of Directional Data. 2nd Edition. John Wiley & Sons, Chicester. 429pp.
Marks L, Teng S, Artun J, Herring S. (1997) Reaction strains on the condylar neck during mastication and maximum muscle stimulation in different condylar positions: an experimental study in the miniature pig. J Dent Res 76:1412–20.
Martin RB, Burr DB. (1989) Structure, function, and adaptation of compact bone. Raven Press, New York.
Martin RB, Burr DB, Sharkey NA. (1998) Skeletal Tissue Mechanics. Springer, New York.
Peterson J, Dechow PC. (2002) Material properties of the inner and outer cortical tables of the human parietal bone. Anat Rec 268:7–15.
Peterson J, Dechow PC. (2003) Material properties of the cranial vault and zygoma. Anat Rec 274A:785–797.
Peterson J, Wang Q, Dechow PC. (2006) Material properties of the dentate maxilla. Anat Rec, 288A:962–972.
Petrtyl M, Hert J, Fiala P. (1996) Spatial organization of the haversian bone in man. J Biomech 29:161–9.
Reilly DT, Burstein AH. (1974) The mechanical properties of cortical bone. J Bone Joint Surg (AM) 56-A:1001–22.
Ricos V, Pedersen DR, Brown TD, Ashman RB, Rubin CT, Brand RA. (1996) Effects of anisotropy and material axis registration on computed stress and strain distributions in the turkey ulna. J Biomech 29:261–7.
Riggs CM, Vaughan LC, Evans GP, et al. (1993) Mechanical implications of collagen fibre orientation in cortical bone of the equine radius. Anat Embryol 187:239–48.
Rinnerthaler S, Roschger P, Jakob HF, Nader A, Klaushofer K, Fratzl P. (1999) Scanning small angle X-ray scattering analysis of human bone sections. Calcif Tissue Int 64:422–9.
Robling AG, Stout SD. (1999) Morphology of the drifting osteon. Cells Tiss Org 164:192–204.
Sevostianov I, Kachanov M. (2000) Impact of the porous microstructure on the overall elastic properties of the osteonal cortical bone. J Biomech 33:881–8.
Sasaki N, Ikawa T, Fukuda A. (1991) Orientation of mineral in bovine bone and the anisotropic mechanical properties of plexiform bone. J Biomech 24:57–62.

Sasaki N, Matsushima N, Ikawa T, et al. (1989) Orientation of bone mineral and its role in the anisotropic mechanical properties of bone- transverse anisotropy. J Biomech 22:157–64.

Schaffler MB, Burr DB. (1988) Stiffness of compact bone: effects of porosity and density. J Biomech 21:13–6.

Schwartz-Dabney CL, Dechow PC. (2002) Edentulation alters material properties of mandibular cortical bone. J Dent Res 81:613–617.

Schwartz-Dabney CL, Dechow PC. (2003) Variations in cortical material properties throughout the human dentate mandible. Am J Phys Anthropol 120:252–77.

Skedros JG, Mason MW, Bloebaum RD. (1994) Differences in osteonal micromorphology between tensile and compressive cortices of a bending skeletal system: indications of potential strain-specific differences in bone microstructure. Anat Rec 239:405–13.

Stout SD, Brunsden BS, Hildebolt CF, Commean PK, Smith KE, Tappen NC. (1999) Computer-assisted 3D reconstruction of serial sections of cortical bone to determine the 3D structure of osteons. Calcif Tissue Int 65:280–4.

Takano Y, Turner CH, Burr DB. (1996) Mineral anisotropy in mineralized tissues is similar among species and mineral growth occurs independently of collagen orientation in rats: results from acoustic velocity measurements. J Bone Miner Res 11:1292–301.

Takano Y, Turner CH, Owan I, Martin RB, Lau ST, Forwood MR, Burr DB. (1999) Elastic anisotropy and collagen orientation of osteonal bone are dependent on the mechanical strain distribution. J Orthop Res 17:59–66.

Tappen NC. (1970) Main patterns and individual differences in baboon skull split-lines and theories of causes of split-line orientation in bone. Am J Phys Anthropol 33:61–72.

Tappen NC. (1977) Three-dimensional studies on resorption spaces and developing osteons. Am J Anat. 149:301–17.

Teng S, Herring SW (1995) A stereological study of trabecular architecture in the mandibular condyle of the pig. Arch Oral Biol 40:299–310.

Teng S, Herring SW. (1996) Anatomic and directional variation in the mechanical properties of the mandibular condyle in pigs. J Dent Res 75:1842–50.

Throckmorton GS, Dechow PC. (1994) In vitro strain measurements in the condylar process of the human mandible. Archs Oral Biol 39:853–67.

Throckmorton GS, Ellis E, III, Winkler AJ, Dechow PC. (1992) Bone strain following application of a rigid bone plate: an in-vitro study in human mandibles. J Oral Maxillofac Surg 50:1066–73.

Throckmorton GS, Groshan GJ, Boyd SB. (1990) Muscle activity patterns and control of temporo-mandibular joint loads. J Prosthet Dent 63:685–95.

Turner CH, Chandran A, Pidaparti RM. (1995) The anisotropy of osteonal bone and its ultrastructural implications. Bone 17:85–9.

Turner CH, Woltman TA, Belongia DA. (1992) Structural changes in rat bone subjected to long-term, in vivo mechanical loading. Bone 13:417–22.

van Eijden TM. (2000) Biomechanics of the mandible. Crit Rev Oral Biol Med 11:123–36.

van Eijden TM, Brugman P, Weijs WA, Oosting J. (1990) Coactivation of jaw muscles: recruitment order and level as a function of bite force direction and magnitude. J Biomech 23:475–85.

Vollmer D, Meyer U, Joos U, Vegh A, Piffko J. (2000) Experimental and finite element study of a human mandible. J Craniomaxillofac Surg 28:91–6.

Wenk HR, Heidelbach F. (1999) Crystal alignment of carbonated apatite in bone and calcified tendon: results from quantitative texture analysis. Bone 24:361–9.

Woo SL-Y, Kuel SC, Amiel DG, Hayes WC, White FC, Akeson WH. (1981) The effect of prolonged physical training on the properties of lone bone: a study of Wolff's law. J Bone Joint Surg (AM) 63-A:780–7.

Yamashita J, Dechow PC. (2000) Strain patterns of the human mandible during artificial loading. J Dent Res 79, Special Issue: Abstract 2833.

Yeni YN, Vashishth D, Fyhrie DP. (2001) Estimation of bone matrix apparent stiffness variation caused by osteocyte lacunar size and density. J Biomech Eng 123:10–7.

Yoon HS, Katz JL (1976) Ultrasonic wave propagation in human cortical bone-I. Theoretical considerations for hexagonal symmetry. J Biomech 9:407–12.

Zioupos P, Currey D. 1998. Changes in the stiffness, strength, and toughness of human cortical bone with age. Bone 22:57–66.

Zioupos P, Currey JD, Mirza MS, Barton DC. (1995) Experimentally determined microcracking around a circular hole in a flat plate of bone: comparison with predicted stresses. Philos Trans R Soc Lond B Biol Sci 347:383–96.

Chapter 14
Adaptive Plasticity in the Mammalian Masticatory Complex: You Are What, and How, You Eat

Matthew J. Ravosa, Elisabeth K. Lopez, Rachel A. Menegaz, Stuart R. Stock, M. Sharon Stack, and Mark W. Hamrick

Contents

14.1	Introduction and Background	293
	14.1.1 Prior Studies of Masticatory Loading, Adaptive Plasticity, and Disease Progression	294
14.2	Materials and Methods	297
	14.2.1 Experimental Models	297
	14.2.2 Morphometry of Masticatory Elements	299
	14.2.3 MicroCT Analysis of Skeletal Biomineralization	299
	14.2.4 Histology and Immunohistochemistry of Cartilage Composition	300
	14.2.5 Statistical Analysis and Predictions	302
14.3	Results	302
	14.3.1 Dietary Manipulation and Joint Loading in Rabbits	302
	14.3.2 Myostatin Deficiency and Joint Loading in Mice	309
	14.3.3 Preliminary Analysis of Joint Aging in Rabbits	313
14.4	Discussion and Conclusions	316
	References	321

14.1 Introduction and Background

Of late, adaptive plasticity has attracted considerable attention in myriad fields of biology (Gotthard & Nylin, 1995; Agrawal, 2001; Holden & Vogel, 2002; West-Eberhard, 2003). Adaptive plasticity refers to the ability of an organism to respond during the course of its ontogeny to an altered environmental condition(s) (Gotthard & Nylin, 1995). It is intimately related to the concept of functional

---

M.J. Ravosa
Department of Pathology and Anatomical Sciences, University of Missouri School of Medicine, Medical Sciences Building, One Hospital Drive DC055.07 Columbia, Missouri 65212
e-mail: ravosam@missouri.edu

adaptation, which typically refers to the dynamic coordinated series of cellular, tissue, and biochemical processes of modeling and remodeling, which occur to maintain a sufficient safety factor of a given element or system to routine stresses (Lanyon & Rubin, 1985; Biewener, 1993; Bouvier & Hylander, 1996a, b; Vinyard & Ravosa, 1998; Hamrick, 1999; Ravosa et al., 2000). In the case of cortical bone, the link between functional adaptation and altered loading patterns is reasonably well documented for vertebrate limb elements and the mammalian mandibular corpus (Bouvier & Hylander, 1981; Lanyon & Rubin, 1985; Biewener et al., 1986; Biewener & Bertram, 1993). Studies regarding the ontogeny of locomotor performance have been equally fundamental for identifying behavioral, anatomical, and physiological adaptations and constraints specific to particular ages (Carrier, 1996). A common goal of these and other investigations is to analyze, under naturalistic conditions, the range of behaviors an organism employs with a given morphology, as well as the role of adaptive plasticity in fine-tuning the fit between form and behavior during an organism's lifespan (Grant & Grant, 1989; Losos, 1990). In doing so, such analyses of the biological role of a feature or complex directly address one or more facets of the important inter-relationships among behavior, morphology, performance, fitness and evolution, information critical for understanding ontogenetic and interspecific variation in character–state transformations (Bock & von Walhert, 1965; Losos, 1990; Wainwright and Riley, 1994; Lauder, 1995).

For those interested in the evolution of cranial variation, an understanding of the short- and long-term effects of altered masticatory stress is critical for interpreting the behavioral and/or ecological correlates of fossil form, functional adaptation in routinely loaded systems/elements, as well as the onset and progression of temporomandibular joint (TMJ) disease and dysfunction. Given the evolutionary and clinical implications of research on adaptive plasticity, it is not surprising our knowledge of this subject as regards the mammalian masticatory complex has been greatly influenced by the work of Bill Hylander and his early students.

One purpose of the following chapter is to provide a brief overview of our knowledge of adaptive plasticity in the skull vis-à-vis the masticatory apparatus. Acknowledging that such studies are largely limited to mandibular corpus cortical bone or TMJ cartilage, the second goal is to offer new experimental data from diet-modified rabbits and myostatin-deficient mice on the nature of the plasticity response of hard *and* soft tissues in two cranial joints: the mandibular symphysis and TMJ. In identifying dynamic determinants of symphyseal and TMJ growth, form and function in two model organisms, this integrative research develops a hierarchical framework for comparative ontogenetic analyses of the important relationships among adaptive plasticity, mechanobiology, and performance in the mammalian skull and masticatory system.

14.1.1 Prior Studies of Masticatory Loading, Adaptive Plasticity, and Disease Progression

In vivo and comparative analyses indicate that the postnatal development of masticatory elements and tissues is influenced by variation in jaw-loading patterns, with weaning being a particularly important life-history stage (Ravosa & Hylander, 1994; Ravosa, 1996, 1999; Ravosa & Hogue, 2004). Once weaned, mammals ingest "adult" foods (Watts, 1985; Grieser, 1992; Tarnaud, 2004) and develop "adult" jaw-adductor activity patterns (Herring & Wineski, 1986; Weijs et al., 1987; Herring et al., 1991; Iinuma et al., 1991; Langenbach et al., 1991, 1992, 2001; Westneat & Hall, 1992; Huang et al., 1994), with corresponding soft-tissue and bony responses to "adult" jaw-loading regimes (Ravosa, 1991a, 1992, 1996, 1998, 1999, 2007; Cole, 1992; Biknevicius & Leigh, 1997; Vinyard & Ravosa, 1998; Taylor et al., 2006). Most evidence regarding dynamic plasticity in the mammalian masticatory complex is derived from the mandibular corpus and TMJ cartilage of alert organisms subjected to variation in jaw-loading patterns via the postweaning manipulation of dietary properties (Bouvier & Hylander, 1981, 1982, 1984, 1996a, b). This methodology has proven very beneficial because masticatory stresses due to jaw-adductor, bite and reaction forces are elevated during the processing of relatively tough and/or resistant items (Herring & Scapino, 1973; Luschei & Goodwin, 1974; Thexton et al., 1980; Weijs et al., 1987, 1989; Gans et al., 1990; Hylander et al., 1992, 1998, 2000, 2005), and mammals with such diets typically exhibit relatively larger corpora, symphyses, and TMJs (Freeman, 1979, 1981, 1988; Hylander, 1979b; Bouvier, 1986; Daegling, 1989, 1992; Ravosa, 1991a, b, 2000; Biknevicius & Ruff, 1992; Ravosa & Hylander, 1994; Spencer, 1995; Biknevicius & Van Valkenburgh, 1996; Hogue, 2004; Ravosa & Hogue, 2004). Such a naturalistic approach ensures that potential tissue, cellular, and biochemical responses do not result from aberrant behaviors and/or surgical artifacts, thus facilitating the identification of a range of physiological responses (i.e., norms of reaction) of bony and connective tissues to altered masticatory loads.

Early research by Bouvier and Hylander observed that growing monkeys raised on an over-use diet of "hard" or resistant items exhibit greater cortical bone remodeling as well as greater mandibular depth and cortical bone thickness (Bouvier & Hylander, 1981). Compared to the TMJ of "under-use" or "soft-diet" macaques, "over-use" or "hard-diet" macaques of the same age also develop a higher density of connective tissue and subchondral bone as well as thicker condylar articular cartilage (Bouvier & Hylander, 1982). Similar postnatal patterns characterize condylar/craniofacial proportions and articular cartilage thickness in rats fed differing diets, and TMJ articular disc thickness in over-use rabbits (Beecher & Corruccini, 1981; Bouvier & Hylander, 1984; Kiliardis et al., 1985; Bouvier, 1987, 1988; Bouvier & Zimny, 1987; Block et al., 1988).

More recent work provides considerable support for the hypothesis that cartilage of the mandibular condyle and TMJ articular disc is affected by local biomechanical effects. As chondrocytes are exquisitely sensitive to 3-D microenvironment and exhibit changes in differentiation status in response to environmental cues (Lemare et al., 1988; Goldring, 2004a, b), expression of cartilage extracellular elements likely evince regional variation, reflecting differential loading patterns in distinct joint regions (Bayliss et al., 1983; Nakano & Scott, 1989; Mow et al., 1990; Hamrick, 1999; Tanaka et al., 2000). Altering TMJ force application by varying masticatory loading regime, tooth extraction, unilateral bite raise, or corticotomy has been shown to result in gene expression changes and elevated glycosaminoglycan

(GAG) content in condylar cartilage (Copray et al., 1985; Carvalho et al., 1995; Holmvall et al., 1995; Pirttiniemi et al., 1996; Mao et al., 1998; Agarwal et al., 2001; Huang et al., 2002, 2003). Increased alkaline phosphatase activity associated with biomineralization of condylar cartilage and changes in osteoclastic and osteoblastic activity also have been noted (Bouvier, 1988; Kim et al., 2003). Changes in expression of type I and type II collagen vary in response to joint loads, further supporting the hypothesis that mechanotransduction may signal changes in gene expression, which alter tissue composition and function as a response to induced degeneration of the cartilage matrix (Mizoguchi et al., 1996; Pirttiniemi et al., 1996; Grodzinsky et al., 2000; Honda et al., 2000; Lee et al., 2000; Huang et al., 2003; Wong & Carter, 2003). In this regard, it is interesting that collagen- and proteoglycan-degrading proteinases have been reported in TMJ tissues and synovial fluids (Kiyoshima et al., 1993, 1994; Marchetti et al., 1999; Puzas et al., 2001; Srinivas et al., 2001).

As reviewed above, bony *and* cartilaginous structures of the TMJ respond developmentally to dynamic changes in masticatory loads with altered proliferation and changes in gene expression (Kim et al., 2003). Growth responses of the mandibular condyle following alteration of local biomechanical conditions (both increased and decreased loads) can lead to hyperplastic or hypoplastic changes in TMJ cartilage and bone (Bouvier & Hylander, 1984; Nicholson et al., 2006). Based largely on short experimental periods in growing mammals (<2 months), these studies support the hypothesis that altered, excessive, and/or repetitive forces induce secondary osteonal remodeling of mandibular cortical bone and chondroblastic activity of articular cartilage, a suite of physiological responses or "functional adaptations" that maintain a sufficient safety factor for the tissues of a cranial element or joint complex to routine masticatory loads (Lanyon & Rubin, 1985; Biewener, 1993; Bouvier & Hylander, 1996a, b; Vinyard & Ravosa, 1998; Ravosa et al., 2000). This work also suggests a minimum loading level and frequency are required for the growth and maintenance of normal adult skull form and function (Beecher, 1983; Bouvier & Hylander, 1984). Interestingly, the magnitude of such responses appears to be age-dependent and may be underlain by genetic and epigenetic factors that vary systemically and interspecifically (Bouvier, 1988; Bouvier & Hylander, 1996a, b; this study, below). Despite that an understanding of the performance and integrity of the mandibular symphysis and TMJ hinges on the ability of individual tissues of composite structures to adapt to applied stresses, no comprehensive comparative data exist regarding the dynamic cascade of anatomical, biochemical, and biomechanical responses of the bone, cartilage, and ligaments of cranial joints vis-à-vis altered masticatory loads.

To address this problem, TMJ and symphyseal tissues were analyzed for changes in joint proportions; biomineralization of articular, subarticular and cortical bone, and cortical bone thickness via microcomputed tomography (microCT); and histology and immunohistochemistry of articular cartilage extracellular matrix (ECM) composition. Rabbits and mice subjected to elevated masticatory loads are predicted to develop: relatively larger symphyses, corpora, condyles and jaw-adductor muscles; greater symphyseal cortical bone thickness; elevated bone-density levels along the symphysis and condyle; and increased collagen and proteoglycan expression of symphyseal fibrocartilage (FC) pad and TMJ articular cartilage. Such evidence on anatomical, structural, and biochemical patterns of variation also are used to address several outstanding issues regarding mammalian masticatory function: the lack of data on adaptive plasticity for cranial arthroses (symphysis), joints with highly disparate functional and structural constraints compared with synovial joints (TMJs) and syndesmoses (sutures); the correlational nature of the in vivo and anatomical support for models of symphyseal fusion, and a related claim that symphyseal strength is not a determinant of fusion; and the preponderance of experimental information on symphyseal fusion from only one mammal order (Primates). In presenting these preliminary analyses, it is argued that a vital component of current and future work on adaptive plasticity in the skull, and especially joints, should employ a multifaceted characterization of a given functional network, one that incorporates data on myriad tissues so as to evaluate the role of altered load versus differential tissue response on functional adaptation of such composite structures. As tissue degradation is the failure of the adaptive process to adequately respond to altered and/or excessive loading conditions, this hierarchical perspective is important for understanding TMJ disease progression.

14.2 Materials and Methods

14.2.1 Experimental Models

Two animal models were employed to evaluate the plasticity of masticatory elements vis-à-vis altered loading levels. One sample consisted of New Zealand domestic white rabbits (*Oryctolagus cuniculus*) subjected to variation in dietary material properties. The second mammalian sample consisted of myostatin-deficient and control (i.e., wild type) domestic mice (*Mus musculus*). To control for variation in genetics and thus ensure the response to loading modification was established postnatally, only siblings were chosen. A third, small sample of young (6 months old) and mature (3 years old) adult rabbits fed the same diet was also examined.

Twenty mixed-sex rabbits were obtained as weanlings (4 weeks old) and housed in the AALAC-accredited NU Center for Comparative Medicine for 15 weeks until attaining subadult status at 19 weeks old (Sorensen et al., 1968; Yardin, 1974). In accordance with an ACUC-approved protocol, two diet cohorts of 10 rabbits each were established by MJR to induce postweaning variation in jaw-adductor forces and masticatory loads. Weaning was chosen as the starting point for dietary manipulation, because plasticity may decrease with age (Bouvier, 1988) and because we sought to minimize the confounding influence of postweaning diets other than those utilized herein. Weanlings were fed ad lib either a "soft" diet of ground pellets to model **under-use (U)** of the chewing complex or a "tough/hard" diet of Harlan TekLad rabbit pellets supplemented daily with two 2-cm hay blocks to model **over-use (O)**. Resistant food (pellets) and tougher foods with higher

Table 14.1 Dietary properties of rabbit experimental foods. Ground pellets require minimal oral preparation, which reduces the amount of cyclical loading during unilateral mastication. Hay necessitates greater forces to process, which increases peak loading magnitudes during biting and chewing

Food items	Young's modulus (E, MPa)	Toughness (R, Jm⁻2)	Hardness (H, MPa)
Pellets ($n = 10$)	29.2 (17.0–41.0)	–	11.8 (6.3–19.9)
Wet Hay ($n = 15$)	277.8 (124.9–451.0)	1759.2 (643.6–3251.9)	–
Dry Hay ($n = 15$)	3335.6 (1476.8–6711.4)	2759.8 (434.0–6625.5)	–

elastic moduli (hay) require absolutely larger jaw-adductor forces and increased transverse occlusal forces during biting and chewing, and both factors affect jaw-loading patterns and ultimately TMJ and symphyseal form. With a portable food tester (Darvell et al., 1996; Lucas et al., 2001), the material properties of pellets and hay were assessed (Table 14.1: averages, samples) and monitored to assure consistency (Wainwright et al., 1976; Vincent, 1992; Lucas, 1994; Currey, 2002). The elastic, or Young's, modulus (**E**) is the stress/strain ratio at small deformations and characterizes the stiffness or resistance to elastic deformation. Toughness (**R**) is an energetic property describing the work performed propagating a crack through the material. Hardness (**H**) is used to quantify indentation. While the properties of crushed pellets may differ little from intact pellets, the latter entail greater repetitive loading due to a longer processing time. Thus, the sequence from crushed pellets to whole pellets (only) to pellets with hay tracks items with longer preparation time and progressively greater elastic moduli, hardness, and toughness (well known to result in increasingly elevated masticatory stresses). As the between-cohort comparisons accentuate the influence of processing time (i.e., crushed pellets have similar properties to whole pellets), **U**-diet rabbits are posited to more closely resemble normal/non-pathological loading conditions. The inclusion of pellets in the diet of all postweaning rabbits ensured adequate nutrition for normal growth. Behavioral analyses and observations indicate **U**-diet rabbits did not show failure to thrive nor did they develop incisor malocclusions; 90% of the **U**-diet sample is within the skull-length range for 10 similar-aged **O**-diet rabbits; and no differences are indicated between the sexes.

A second sample consisted of genetically similar, domestic myostatin-deficient (*Mstn* −/−) and control (*Mstn* +/+) mice (*Mus musculus*). In accord with an ACUC-approved protocol, 23 male mice were bred by MWH and kept in the AALAC-accredited MCG Laboratory Animal Facility for 6 months until attaining adulthood. From a skeletal perspective, half-year old adult mice exhibit peak bone mass and mechanical properties (Ferguson et al., 2003). Versus 11 control (CD-1) mice characterizing a normal range of adult phenotypic variation, 12 myostatin-deficient mice bred on the CD-1 background were used to model masticatory over-loading (Nicholson et al., 2006). To control for the effects of dietary properties on masticatory plasticity, mice in the over-use *and* normal-use loading cohorts were fed Harlan TekLad rodent chow ad lib. Myostatin is a negative regulator of skeletal muscle growth, and knockout (KO) mice lacking myostatin develop approximately twice the skeletal muscle mass of wild-type mice at both 2 and 10 months of age (McPherron et al., 1997). Myostatin KO mice also develop masseter and temporalis muscles over 50% larger in mass due to larger muscle fiber cross sections and increased muscle cells (McPherron et al., 1997; Byron et al., 2004). This elevated muscle mass, and physiological cross section increases contractile forces, greater maximal bite forces (when jaw adductors are stimulated to tetanus) and jaw-muscle attachment size ver-sus similar-sized normal mice (Byron et al., 2004, 2006; Nicholson et al., 2006). Myostatin deficiency is dose-dependent, as mice heterozygous for the disrupted *Mstn* sequence exhibit muscle masses intermediate between those of +/+ wild-type mice and mice homozygous −/− for the myostatin mutation (McPherron & Lee, 2002). In situ hybridization data for mouse embryos indicate that myostatin is first expressed in the myotome compartment of somites, and myostatin transcripts still can be detected in adults (McPherron et al., 1997; Ji et al., 1998). Myostatin KO mice do not differ from normal mice (vs. body mass) in metabolic rate, food consumption, or body temperature (McPherron & Lee, 2002). In addition, myostatin-deficient mice develop greater mineral density than the normal mice in the hindlimb and spine (Hamrick, 2003; Hamrick et al., 2003). Elevated bone density in KO mice appears due solely to increased relative muscle forces as the myostatin receptor, the type IIB activin receptor, is not expressed at significant levels in skeletal tissues (Shuto et al., 1997). Thus, we posit that variation in masticatory form between mouse groups results from differences in jaw-muscle forces.

The third sample consisted of young adult ($n = 3$) and mature adult ($n = 3$) *O. cuniculus* raised for 6 months and 3 years, respectively, on the same diet of Harlan TekLad rabbit pellets. These adult specimens were obtained opportunistically and thus provide only a small sample with which to preliminarily investigate the influence of aging on soft and hard tissues of craniomandibular joints.

14.2.2 Morphometry of Masticatory Elements

After euthanasia, mouse and rabbit skulls were detached at the vertebral column and jaw-adductor muscles exposed and carefully dissected from their attachments. Left and right mandibles were detached from the skull and fixed in 10% buffered formalin. All specimens were weighed (to 0.01 g), with digital calipers (rabbits) or an optical scope (mice) used to obtain mandible length/breadth, symphysis length/width, corpus height/width, and condyle width/length (Nicholson et al., 2006; Ravosa et al., 2007a). Metric data were used to control for size-related variation in the skull and masticatory apparatus in comparisons of loading cohorts (Bouvier & Hylander, 1981, 1982, 1984). Symphyseal and TMJ samples then were used for microCT analyses of biomineralization and cortical bone thickness followed by histology and immunohistochemistry.

14.2.3 MicroCT Analysis of Skeletal Biomineralization

The influence of variation in routine joint loading on symphyseal and TMJ structure was evaluated via microCT (Wong et al., 1995; Nuzzo et al., 2002; Patel et al., 2003; Stock et al., 2003; Morenko et al., 2004; Nicholson et al., 2006;

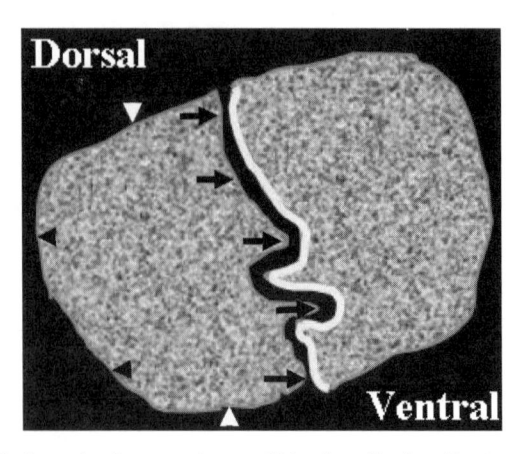

Fig. 14.1 MicroCT of symphysis proportions and biomineralization. Tracing of a coronal section of the middle joint site of a young adult rabbit. In each of 5 (rabbits) or 2 (mice) coronal sections, bone-density levels were evaluated with computer-assisted image analysis at 5 equidistant points along the articular surface (*arrows*) and 4 equidistant points along the lateral, superior, and inferior cortical bone surface (*arrowheads*). Also, joint height and width were quantified (not shown)

Ravosa et al., 2007a–c). MicroCT analyses were performed on fixed tissues to assess variation in the density or biomineralization of the joint articular surface, subarticular bone, and cortical bone. Northwestern's Scanco Medical MicroCT 40 was used to obtain samples in the coronal plane: five equidistant sites per rabbit symphysis (labial, anterior, middle, posterior, lingual – Fig. 14.1), two equidistant sites per mouse symphysis (anterior, posterior), and three equidistant sites per rabbit and mouse TMJ (anterior, middle, posterior – Fig. 14.2). At each joint site, 40 contiguous slices covering 0.31 mm were imaged, with one slice chosen to represent a given site. Values of linear attenuation (μ) were pooled for each specimen and used to characterize between-group variation in local tissue mineral density along the TMJ and symphysis (Nicholson et al., 2006; Ravosa et al., 2007a–c). For rabbit and mouse symphyseal sections, linear data on joint height and width as well as articular surface thickness and cortical bone thickness also were collected. In mouse TMJ sections, linear data on articular cartilage thickness also were obtained.

14.2.4 Histology and Immunohistochemistry of Cartilage Composition

Histological and immunohistochemical analysis of symphyseal and TMJ tissues followed standard methods (Scapino, 1981; Trevisan & Scapino, 1976a, b; Beecher, 1977, 1979; Hirschfeld et al., 1977; Bouvier & Hylander, 1982, 1984; Bouvier, 1987, 1988; Kiernan, 1999; Huang et al., 2002; Kim et al., 2003; Ravosa & Hogue, 2004). Joints were fixed in 10% neutral buffered formalin. Once analyzed via microCT, a specimen was decalcified using formic acid and sodium citrate. To verify the endpoint of decalcification, the oxalate test was performed. Subsequently,

Fig. 14.2 MicroCT of TMJ biomineralization. Tracing of the coronal section of the middle condylar site of a young adult rabbit. In each of 3 coronal sections, biomineralization levels were evaluated with computer-assisted image analysis at 5 equidistant points along the articular surface (*arrowheads*), 4 equidistant subchondral bone locations below the articular surface (*dots*), and 3 equidistant points along each side of the condylar neck below the articular surface (*arrows*)

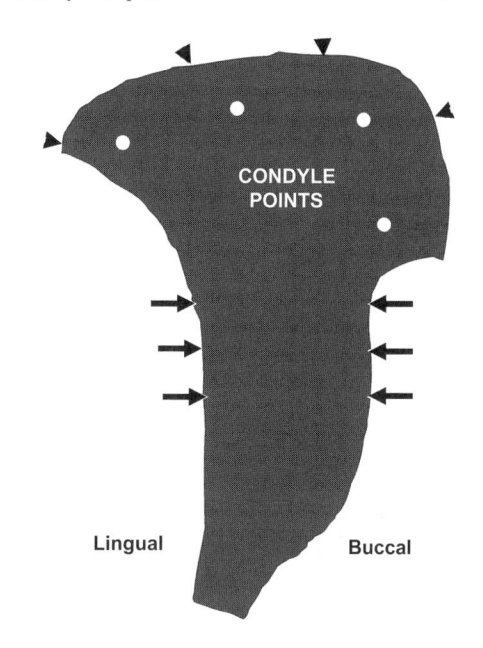

a joint was dehydrated in a graded concentration series of ethanol baths, washed in xylene, and then embedded in paraffin. Special care was exercised to maintain joint, and thus section, orientation parallel to the surface of the paraffin block. At five equidistant sites per symphysis (labial, anterior, middle, posterior, lingual) and at three equidistant sites per TMJ (anterior, middle, posterior), 4–6 μm sections were obtained with a Reichert-Jung autocut microtome in the coronal plane, i.e., orthogonal to craniofacial long axis. As sulfated GAGs are expressed in tissues regularly exposed to loads, and rat TMJ chondrocytes have been shown to increase GAG synthesis in response to mechanical force (Copray et al., 1985; Carvalho et al., 1995), the cationic dye safranin O was used to evaluate relative GAG content in TMJ articular/hyaline cartilage and symphysis fibrocartilage (Kiernan, 1999; Huang et al., 2002). Strong safranin O staining is indicative of chondroitin sulfate and keratan sulfate–containing proteoglycans, which in turn increases the viscoelastic ability of cartilage to resist compressive stresses. Type II collagen has a distinct fibrillar organization and associates strongly with proteoglycans and water, important for tissues subjected to compression, tension, and shear, such as the symphyseal FC pad and TMJ articular cartilage (Mizoguchi et al., 1996; Pirttiniemi et al., 1996; Benjamin & Ralphs, 1998; Tanaka et al., 2000). Thus, primary antibodies directed at variation in cartilage type II collagen were used to assess collagen and proteoglycan relative expression pattern (i.e., change in staining localization) as a function of masticatory loads (Type II Collagen Staining Kit – Chondrex Inc., WA). Tunel-staining was used to track variation in DNA fragmentation and chondrocyte apoptosis in response to joint loading (Apoptosis Detection Kit – Chemicon Inc., CA). Although not presented here, serial sections were H&E stained to distinguish the symphyseal FC pad and ligaments as well as the TMJ articular disc and articular cartilage layers.

Definitions of progressively deeper zones of TMJ articular cartilage are as follows: articular – filamentous network of elongate cells tangentially arranged and densely packed (high H_2O, low proteoglycan, collagen rich); proliferative – ovoid/round cells random in distribution (protein/proteoglycan production area); chondroblastic – large cell bundles arranged in columns (tidemark separates this from subjacent layer); hypertrophic chondrocyte/calcified – cells heavily encrusted in apatitic salts (Mankin et al., 1971; Newton & Nunamaker, 1985; Ostergaard et al., 1999).

14.2.5 Statistical Analysis and Predictions

The first step in the analysis of the linear data on symphyseal and TMJ proportions from morphometry and microCT was to adjust for variation in body/skull size between loading cohorts. This occurred by calculating the ratio of a given linear dimension, or cube root of a volumetric measure, versus jaw length (Bouvier & Hylander, 1981, 1982, 1984; Bouvier, 1986; Ravosa & Hylander, 1994; Ravosa & Hogue, 2004). To facilitate the comparison of specific masticatory parameters, to characterize the magnitude of difference between cohorts, and due to the more preliminary nature of the rabbit metric and microCT analyses, all between-group differences were investigated via non-parametric ANOVA (Mann-Whitney U-test, $p < 0.05$). Mouse morphometric analyses of size-adjusted masticatory proportions were examined via Mann-Whitney U-tests ($p < 0.05$). To provide an overall summary of between-group differences in bone-density levels between mouse loading cohorts, discriminant function analysis (DFA) was used (Nicholson et al., 2006; Ravosa et al., 2007b). DFA assessed if, based on a series of biomineralization parameters, a given joint was correctly identified as belonging to its cohort, thus offering a quantitative determination of morphological distinctness and adaptive plasticity among loading groups. As one can determine how individual variables load on a discriminant function, DFA facilitates identification of multivariate patterns of covariation among joint parameters (Nicholson et al., 2006). The underlying hypothesis is that dynamic alterations in jaw-loading regimes during mastication will translate into postnatal variation in gross proportions, soft-/hard-tissue anatomy, tissue properties, and biochemistry, as well as the resulting strength and integrity of the joint.

14.3 Results

14.3.1 Dietary Manipulation and Joint Loading in Rabbits

14.3.1.1 Morphometry

Rabbits were obtained as weanlings (4 weeks old) and for 15 weeks were fed either a diet of pulverized pellets to model under-use (**U**) of the masticatory system or a diet of intact pellets supplemented with hay blocks to model over-use (**O**).

Table 14.2 Comparison of size-adjusted means for load-resisting and force-generating elements between cohorts of 10 subjects each. As predicted, **O**-diet rabbits develop relatively larger masticatory proportions as well as thicker cortical bone along the symphysis articular and external surfaces (Mann-Whitney U-test, **=$p < 0.01$, *=$p < 0.05$)

Variable	O-Diet	U-Diet	% Difference
TMJ Condyle AP Length	0.178	0.150	18.7*
TMJ Condyle ML Width	0.074	0.069	7.3*
Corpus Height	0.242	0.237	2.1
Corpus Width	0.096	0.090	6.7*
Symphysis Length	0.396	0.351	12.8*
Symphysis Width	0.148	0.129	14.7*
Symphysis Articular Breadth	0.173	0.140	23.6*
Symphysis Superior Cortical Width	0.176	0.134	31.3*
Symphysis Lateral Cortical Width	0.089	0.058	53.4**
Masseter Muscle Mass	0.135	0.112	20.5*

Subsequently, the symphysis, TMJ and other masticatory elements were subjected to metric, microCT, histological, and immunohistochemical analyses of similar joint sections/sites (Ravosa et al., 2007a). ANOVAs indicate, as predicted, that size-adjusted measures of the symphysis, TMJ condyle, corpus, and jaw adductors are significantly larger in 10 **O**-diet versus 10 **U**-diet rabbits (Table 14.2).

14.3.1.2 MicroCT

MicroCT was used to visualize and quantify variation in biomineralization of the articular surface, subarticular bone, and cortical bone, along the symphysis and TMJ condylar head (Ravosa et al., 2007a). As predicted, significant variation develops in joint density and anatomy between **O**-diet and **U**-diet rabbits, with the former group exhibiting higher levels of biomineralization (Tables 14.3 & 14.4). ANOVAs also indicate the presence of significantly thicker cortical bone along the symphyseal outer and articular surfaces in **O**-diet rabbits (Fig. 14.3; Table 14.2). These findings

Table 14.3 Comparison of rabbit symphyseal mean biomineralization levels (μ) between cohorts of seven subjects each. As predicted, **O**-diet rabbits develop elevated bone-density levels at the symphyseal articular surface and external cortical bone (Mann-Whitney U-test, **= $p < 0.01$, *=$p < 0.05$)

Variable	O-Diet	U-Diet	% Difference
Symphysis Top	2.243	1.822	23.1**
Symphysis Upper	2.005	1.610	24.5**
Symphysis Middle	2.055	1.623	26.6**
Symphysis Lower	1.910	1.613	18.4**
Symphysis Bottom	1.904	1.673	13.8**
Inferior Corpus	2.163	1.762	22.8**
Inferior/Lateral Corpus	2.655	2.262	17.4**
Lateral Corpus	2.607	2.437	7.0*
Superior Corpus	2.611	2.281	14.5**

Table 14.4 Comparison of rabbit TMJ mean biomineralization levels (μ) between cohorts of eight subjects each. In addition to increases in external proportions, **O**-diet rabbits develop increased bone-density levels along the articular surface, subchondral bone, and cortical surface of the condylar neck (Mann-Whitney U-test, **= $p < 0.01$, *=$p < 0.05$)

Variable	O-Diet	U-Diet	% Difference
Outer 1	1.490	1.482	1.0
Outer 2	1.561	1.181	32.2**
Outer 3	1.485	1.314	13.0*
Outer 4	1.554	1.243	25.0*
Outer 5	1.618	1.425	13.3*
Inner 1	2.187	1.946	12.4*
Inner 2	2.180	1.787	22.0**
Inner 3	2.102	1.776	18.4**
Inner 4	2.111	1.953	8.1*
Neck 1	1.995	1.710	16.7*
Neck 2	2.009	1.746	15.1*
Neck 3	1.971	1.626	21.2**
Neck 4	2.051	1.807	13.5*
Neck 5	1.990	1.815	9.6*
Neck 6	2.008	1.648	21.8**

Fig. 14.3 Symphyseal cortical bone thickness. Coronal sections of "middle" joint sites from **U**-diet (**A**) and **O**-diet (**B**) subadult rabbits obtained via microCT. Comparisons of three size-adjusted measures of internal joint proportions indicate that, as predicted, **O**-diet rabbits develop significantly thicker cortical bone along the superior, lateral, and articular surfaces of the symphysis (*white lines with double arrowheads*) (see Table 14.2)

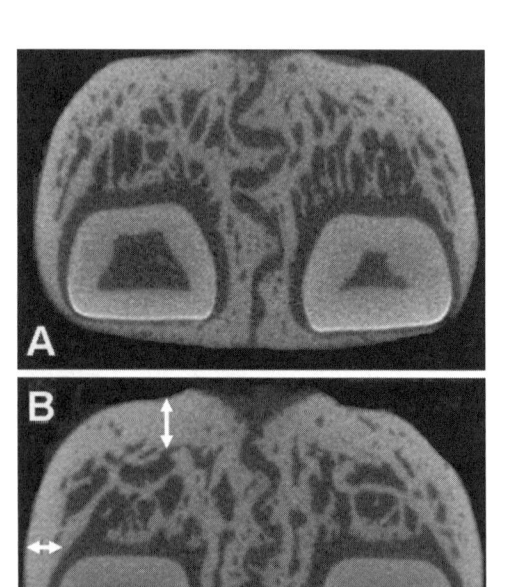

underscore a significant effect of dietary properties on adaptive plasticity in masticatory proportions, tissue structure, and bone mineral density levels.

14.3.1.3 Histology and Immunohistochemistry

Detached symphyses and TMJs were fixed, decalcified and sectioned, and then subjected to several histological and immunohistochemical staining protocols to identify variation in ECM components related to cartilage composition and properties (Figs. 14.4–14.9). Histological analyses of **U**-diet and **O**-diet subadults indicate more intense safranin O staining in the symphyseal FC pad (compare "A" vs. "B" in Fig. 14.4) and TMJ condylar articular cartilage (compare "A" vs. "B" in Fig. 14.5) of the **U**-diet rabbit. Lower proteoglycan content throughout the FC pad and in the condylar cartilage's lower two layers of **O**-diet rabbits mirrors findings for the articular surface of mammal limb elements, where age-related onset of osteoarthritis (OA) is linked to decreases in proteoglycan content (Mankin et al., 1971; Newton & Nunamaker, 1985; Haskin et al., 1995; Ostergaard et al., 1999). Due to the elevated viscoelasticity of proteoglycan-rich tissues in joints subjected to cumulatively low postnatal stresses (i.e., **U** diet), our analyses suggest that the articular cartilage and fibrocartilage of such organisms are able to resist greater compressive stresses than

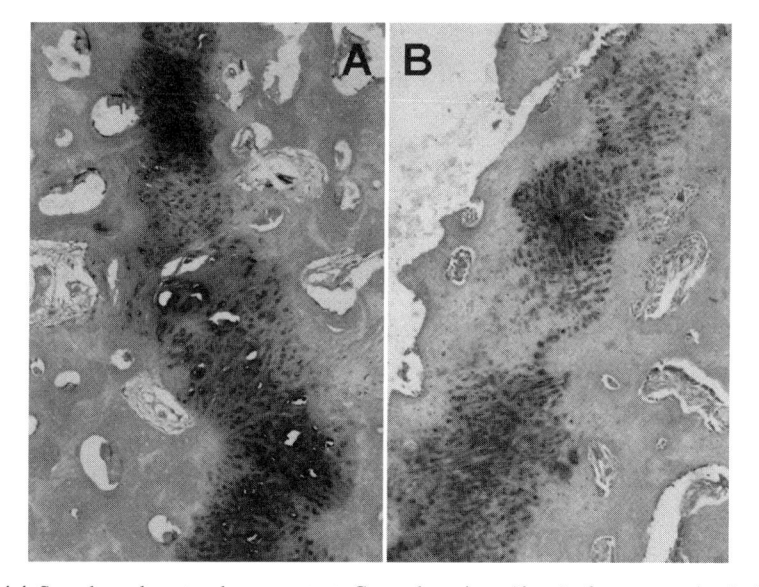

Fig. 14.4 Symphyseal proteoglycan content. Coronal sections (6 μm) of representative "middle" joint sites from **U**-diet (**A**) and **O**-diet (**B**) subadults stained with safranin O to identify GAG content in the FC pad. In the enlarged view of the ventral joint with the FC pad, darker staining in "A" vs. "B" indicates lower proteoglycan content and thus decreased FC pad viscoelasticity in **O**-diet rabbits. This suggests the FC pad of joints routinely subjected to greater stresses eventually exhibits ontogenetic reductions in the ability to counter compression between dentaries during mastication. In "B" note also the corresponding development of bony rugosities ("light grey" bone nearly completely traversing the "dark grey" FC pad)

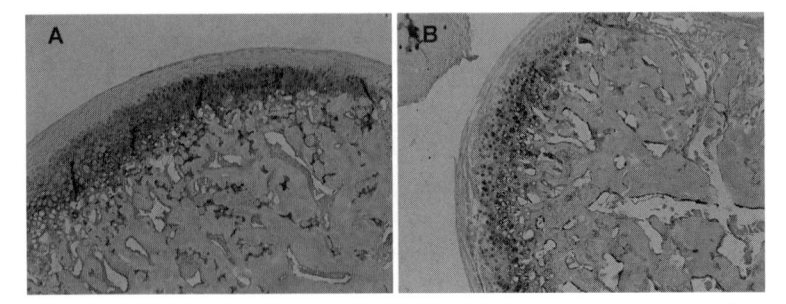

Fig. 14.5 TMJ proteoglycan content. Coronal sections (6 μm) of representative "middle" joint sites from U-diet (**A**) and O-diet (**B**) 4-month old subadults were stained with safranin O to identify GAG content in the articular cartilage. In views of the articular surface and underlying subchondral bone, more intense staining in "A" vs. "B" indicates lower proteoglycan content and thus diminished articular cartilage viscoelasticity in **O**-diet rabbits. This suggests that TMJs routinely subjected to higher loads develop postnatal decreases in the ability to resist compressive stresses during biting and chewing

that of repetitively over-loaded adult joints. As proteoglycan content is most pronounced in the two innermost layers of TMJ articular cartilage – chondroblastic and hypertrophic/ calcified chondrocyte – this suggests it is critical to account for regional variation in this and other ECM components in evaluating the biomechanical significance of variation in cartilage properties and proportions.

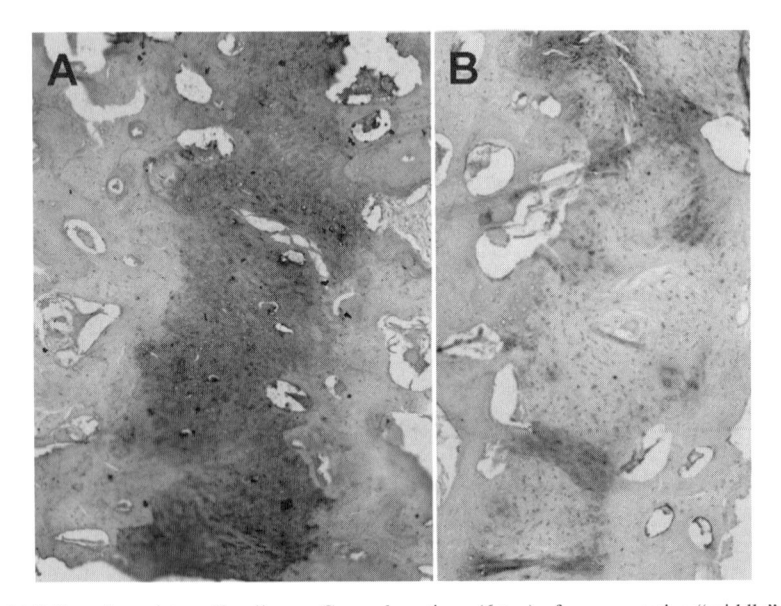

Fig. 14.6 Symphyseal type II collagen. Coronal sections (6 μm) of representative "middle" joint sites in 4-month old U-diet (**A**) and O-diet (**B**) subadults stained with a 1[ary] antibody directed against type II collagen. In the enlarged view of the ventral joint, darker staining in the FC pad of "A" vs. "B" demonstrates less type II collagen and thus lower viscoelasticity in **O**-diet rabbits. This suggests that the FC pad of joints regularly subjected to higher loading experiences postnatal decreases in the ability to resist compression during postcanine chewing and biting

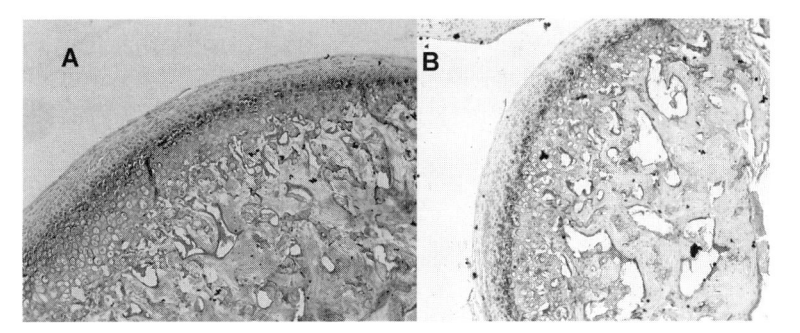

14 Adaptive Plasticity in the Mammalian Masticatory Complex

Fig. 14.7 TMJ type II collagen. Coronal 6 μm sections of representative "middle" sites in 4-month old **U**-diet (**A**) and **O**-diet (**B**) subadults were stained with a 1^{ary} antibody directed against type II collagen. More intense staining of TMJ articular cartilage in "A" vs. "B" indicates less type II collagen and thus diminished viscoelasticity in **O**-diet rabbits. This suggests that TMJ articular cartilage routinely subjected to increased stress experiences ontogenetic reductions in the ability to counter compressive loads during molar chewing and biting

Immunohistochemical data for **U**-diet versus **O**-diet subadult rabbits indicate a more widespread distribution of type II collagen in the symphyseal FC pad and TMJ condylar articular cartilage of **U**-diet rabbits (Figs. 14.6 & 14.7). Expression of type II collagen has been noted in the ECM of mature chondrocytes and inner cartilage

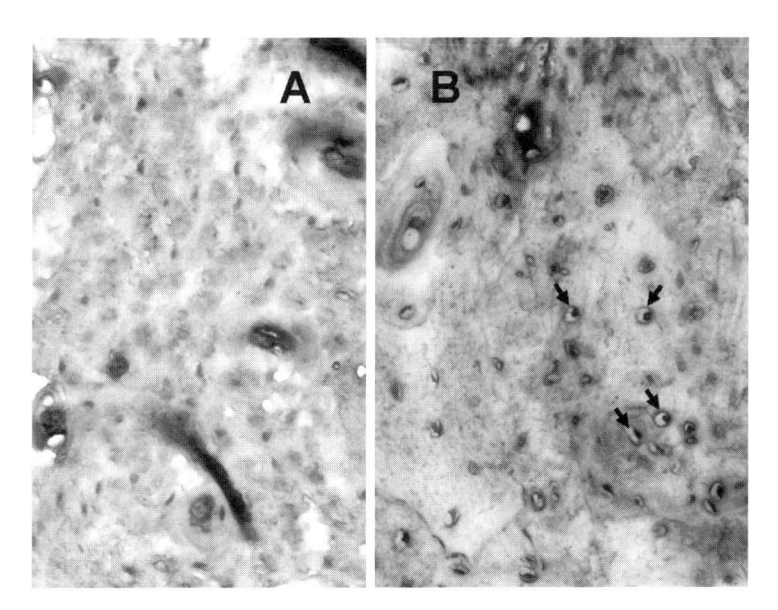

Fig. 14.8 Symphyseal apoptosis. Coronal sections (6 μm) of representative "middle" joint sites from **U**-diet (**A**) and **O**-diet (**B**) subadults tunel stained to identify fragmented DNA of apoptotic chondrocytes in the FC pad. In the enlarged view of the ventral joint with the FC pad, **O**-diet rabbits exhibit numerous apoptotic chondrocytes (*black arrows*). This suggests that the FC pad of joints routinely subjected to greater stresses eventually exhibits ontogenetic *increases* in cartilage cell degradation

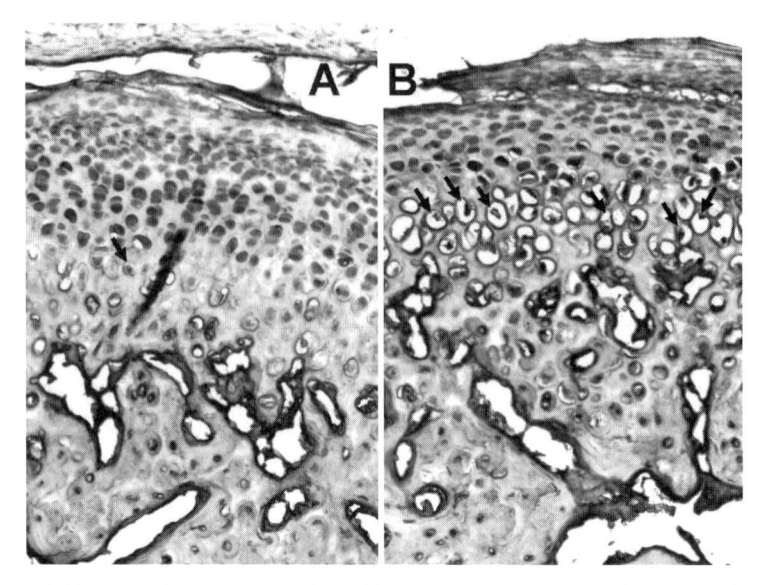

Fig. 14.9 TMJ apoptosis. Coronal sections (6 μm) of representative "middle" joint sites from U-diet (**A**) and O-diet (**B**) 4-month old rabbits were tunel stained to identify apoptosis in TMJ articular cartilage. In views of the articular surface and underlying subchondral bone, **O**-diet subadults exhibit elevated chondrocyte apoptosis versus **U**-diet rabbits (*black arrows*). This suggests that TMJs routinely subjected to higher loads develop postnatal increases in cartilage degeneration

layers, such as the TMJ's hypertrophic and chondroblastic zones. Type II collagen has a distinct fibrillar organization and associates more strongly with proteoglycans, and both ECM components are important in tissues subjected to compressive loads during biting and chewing. These comparisons suggest that, much as the case for the well-documented TMJ, symphyseal adaptive plasticity is characterized by similar of the patterns of postnatal variation in type II collagen *and* proteoglycan content (Figs. 14.3 & 14.4) (Ravosa et al., 2007a).

In the FC pad and TMJ articular cartilage, tunel staining indicates a significantly greater number of apoptotic chondrocytes in **O**-diet versus **U**-diet rabbits (Figs. 14.8 & 14.9). In fact, as symphyseal fibrocartilage is characterized by fewer chondrocytes than TMJ hyaline cartilage, it is exceedingly difficult to identify apoptotic cells in the **U**-diet symphysis. This pattern in both joints suggests that routine overloading induces accelerated cell death and increased cartilage degradation. In articular cartilage of the TMJ, **O**-diet rabbits appear to develop more hypertrophic chondrocytes (Fig. 14.9). In the growth plate of a joint, apoptosis is a normal terminal event for hypertrophic chondrocytes, and such cells express angiogenic factors initiating vascular invasion, erosion of the mineralized cartilage, and bone formation (Gerber et al., 1999). Thus, increased numbers of apoptotic, hypertrophic chondrocytes appear associated with advance of the subchondral mineralizing front. By selecting the same section/site samples for metric, microCT, histological, and immunohistochemical comparisons, this facilitated a characterization of the coordinated series of dynamic functional adaptations as well as the onset of degradative responses of cranial joint tissues vis-à-vis altered masticatory stresses.

14.3.2 Myostatin Deficiency and Joint Loading in Mice

14.3.2.1 Morphometry

After being raised from birth until 6 months old on a common diet of rodent chow, 23 mouse TMJs were subjected to morphometric, microCT, histological, and immunohistochemical analyses to evaluate the effects of joint over-loading on the response of TMJ hard and soft tissues (Nicholson et al., 2006). Metric analyses indicate, as predicted, that size-adjusted measures of the condyle, corpus, symphysis, and jaw adductors are mostly significantly larger in myostatin-deficient (n = 12) versus normal mice (n = 11) (Table 14.5).

14.3.2.2 MicroCT

MicroCT was used to visualize and quantify variation in biomineralization of the articular surface, subchondral bone, and cortical bone of the TMJ and symphysis (Nicholson et al., 2006; Ravosa et al., 2007b). Using linear attenuation coefficients for 15 condylar sites, DFA was performed for each of three slices (anterior, middle, posterior) to characterize overall patterns of variation in TMJ bone density between young adult KO and wild-type mice. As predicted, we found significant variation in joint density between KO and control mice, with the former group exhibiting greater levels of biomineralization (Table 14.6). Indeed, multivariate analyses of TMJ condylar values correctly identified 100% of the members of the wild-type cohort and 92% of the myostatin-deficient mice. The microCT analyses also highlight the degradation of TMJ subchondral bone among KO mice and corresponding

Table 14.5 Comparison of size-adjusted means of load-resisting and force-generating structures between loading cohorts. As predicted, myostatin-deficient mice typically develop relatively larger masticatory elements (Mann-Whitney U-test, *=$p < 0.05$)

Variable	Mstn −/−	Mstn +/+	% Difference
TMJ Condyle ML Width	0.067	0.061	9.8*
TMJ Condyle AP Length	0.134	0.136	−1.5
Corpus Height	0.301	0.292	3.1
Corpus Width	0.122	0.116	5.2*
Symphysis Length	0.339	0.330	2.8
Symphysis Width	0.239	0.232	3.0
Symphysis Articular Breadth	0.152	0.125	21.6*
Symphysis Sup. Cortical Width	0.160	0.140	14.3
Temporalis Insertion	0.324	0.296	9.5*
Masseter Insertion	0.327	0.290	12.8*
Masseter Muscle Mass	0.016	0.009	77.8*

Table 14.6 DFA comparison of mouse TMJ biomineralization levels (μ). In addition to increases in external proportions, myostatin-deficient mice develop elevated bone-density levels along the articular surface, subchondral region, and condylar neck. Accordingly, most subjects in each cohort are classified correctly

TMJ Slice	Anterior $Mstn+/+$	Anterior $Mstn-/-$	% Correct
Wild	10	0	100
KO	1	11	92
Total	11	11	95
TMJ Slice	Middle $Mstn+/+$	Middle $Mstn-/-$	% Correct
Wild	11	0	100
KO	0	12	100
Total	11	12	100
TMJ Slice	Posterior $Mstn+/+$	Posterior $Mstn-/-$	% Correct
Wild	10	0	100
KO	2	10	83
Total	12	10	91

increases in articular cartilage height (Fig. 14.10). These and other data above point to a significant effect of masticatory overloading on plasticity in TMJ proportions, tissue structure, and bone density. Data on TMJ external and internal structure in KO versus control mice respectively benefit our understanding of the effects of joint

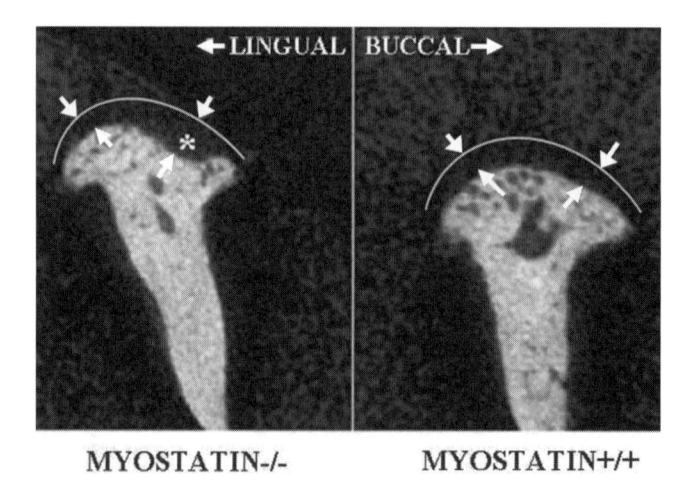

Fig. 14.10 TMJ anatomy and proportions. Myostatin-deficient mice develop adaptive changes in TMJ biomineralization, which are correlated with degradative changes in subchondral bone resorption along the junction between the articular cartilage and the subchondral bone ("*"). In turn, articular cartilage on the buccal side overlying the site of bony resorption is correspondingly thicker in 12 KO vs. 11 normal mice (mean of 0.20 vs. 0.14 mm: Mann-Whitney U-test, $p < 0.01$; comparisons of articular cartilage thickness at middle and lingual locations are not significantly different between loading cohorts). These findings underscore the need for characterizing adaptive and degradative patterns of change in craniomandibular joint bony and soft tissues. (Articular cartilage is identified between the arrows; the articular surface is denoted by the white line.)

14 Adaptive Plasticity in the Mammalian Masticatory Complex

Table 14.7 DFA of mouse symphysis bone-density levels that are similar to patterns for the TMJ

Symphysis Slice	Anterior Mstn+/+	Anterior Mstn−/−	% Correct	Posterior Mstn+/+	Posterior Mstn−/−	% Correct
Mstn +/+	9	1	90	10	0	100
Mstn −/−	1	8	89	0	10	100
Total	10	9	90	10	10	100

over-use versus normal use on TMJ disease progression. ANOVAs also indicate the presence of significantly thicker cortical bone along the symphysis articular surface in KO mice, whereas the thickness of the superior cortical bone only tends to be higher in this loading cohort (Table 14.5).

Discriminant function analyses highlight significant variation in biomineralization of the symphysis between the two loading groups (Table 14.7). Using linear attenuation coefficient values for 9 anterior and 12 posterior symphyseal locations, 95% of the myostatin-deficient mice (18 of 19) and 95% of the normal mice (19 of 20) were identified correctly, with no subject having a misidentified slice more than once (Ravosa et al., 2007b). Thus, similar to the TMJ findings, multivariate patterns of symphyseal biomineralization indicate that KO *and* normal mice cluster much more closely with members of their own loading group than with one another.

14.3.2.3 Histology and Immunohistochemistry

Detached TMJs were fixed, decalcified, embedded and sectioned, and then subjected to several histological and immunohistochemical staining protocols. Digital images of three condylar sites were sampled in the coronal plane (anterior, middle, posterior). Histological analyses of 6-month adult KO (*Mstn* −/−) and CD-1 control (*Mstn* +/+) mice indicate more intense safranin O staining in the condylar articular cartilage of the KO mice (compare "A" vs. "B" in Fig. 14.11). Higher proteoglycan content throughout the condylar cartilage of KO mice indicates the presence of adaptive changes in the ECM (unlike the case in the articular surface of older mammalian synovial joints, where the development of OA is linked to decreases in proteoglycan content). Due to the elevated viscoelasticity of proteoglycan-rich tissues in TMJs subjected to cumulatively higher postnatal stresses (i.e., KO mice), our analyses suggest that the articular cartilage of such organisms is able (at least during early adult stages) to resist greater compressive stresses than that of normally loaded adult joints. Note also that proteoglycan content is most pronounced in the two innermost layers of articular cartilage – chondroblastic and hypertrophic/calcified chondrocyte. Immunohistochemical data for KO versus control young adults indicate a more widespread distribution of type II collagen in condylar articular cartilage of KO mice (Fig. 14.12). Expression of collagen II has been noted in the ECM of mature chondrocytes and inner cartilage layers (hypertrophic and chondroblastic zones). As noted above, type II collagen has a distinct fibrillar organization and associates with proteoglycans. Both ECM components are vital for joint tissues subjected to compressive loads during biting and chewing.

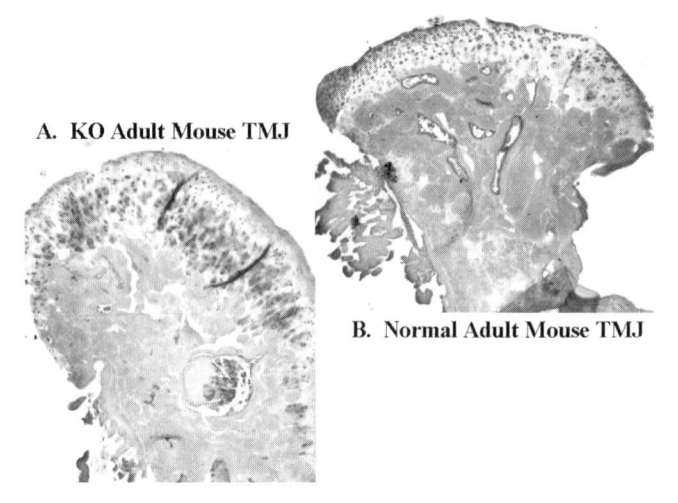

Fig. 14.11 TMJ proteoglycan content. Coronal sections (6 μm) of middle joint sites for KO (**A**) and control (**B**) adults were stained with safranin O to identify GAG content in articular cartilage. In views of the articular surface and subjacent subchondral bone, more intense staining in "A" vs. "B" indicates higher proteoglycan content and thus greater articular cartilage viscoelasticity in KO mice. This suggests that TMJs routinely subjected to greater stresses initially develop postnatal increases in the ability to counter compression during biting and chewing

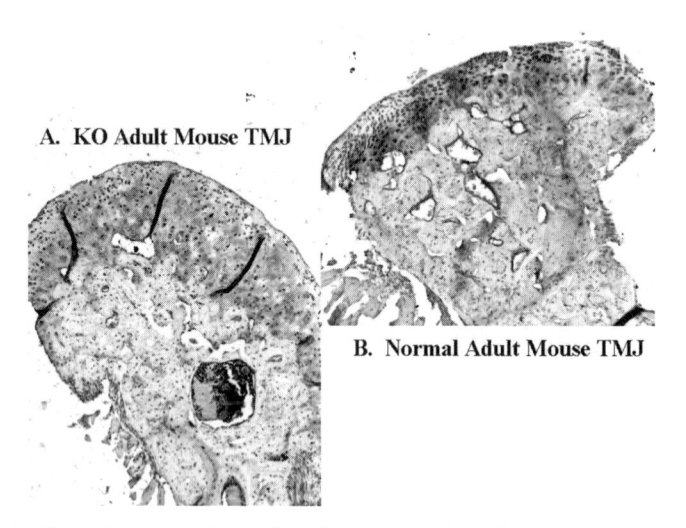

Fig. 14.12 TMJ type II collagen. Coronal sections (6 μm) of middle sites in 6-month old KO (**A**) and control (**B**) mice were stained with a 1[ary] antibody directed against type II collagen. More extensive staining of articular cartilage in "A" vs. "B" indicates more type II collagen and thus higher viscoelasticity in KO mice. This suggests that articular cartilage routinely subjected to higher masticatory loads initially develops postnatal increases in the ability to resist compressive stress during chewing and biting

Admittedly, variation in section location between KO and normal mice may play a role in our initial findings (i.e., the KO site is more posterior to the "standard" middle site of normal mice in Figs. 14.11 and 14.12). However, based on the fact that the posterior TMJ in mammals is thought to be loaded more in tension (Hylander, 1992), it is possible our initial findings may underemphasize differences between groups, as TMJ cartilage loaded in tension expresses higher type I collagen and lower type II collagen (Mizoguchi et al., 1996). Nonetheless, as noted in rabbits, the mouse comparisons underscore the important role of variation in ECM composition and cartilage proportions in determining joint biomechanics. In sampling roughly similar anatomical sites for a series of analyses between KO versus wild-type mice, we were able to further unravel the hierarchical suite of dynamic functional adaptations and degradative changes of TMJ soft and hard tissues in response to mechanical loads.

14.3.3 Preliminary Analysis of Joint Aging in Rabbits

14.3.3.1 MicroCT

MicroCT was used to evaluate variation in biomineralization of the articular surface, subarticular bone, and cortical bone, along the symphysis and TMJ condylar head in young adult (6 months old) and mature adult (3 years old) rabbits. (Quantitative data on age-related variation in external joint proportions were not obtained, although no differences were apparent.) Based on limited data, the symphysis of older rabbits generally exhibits higher mineral density than in younger adults (Table 14.8). In contrast, the TMJ of young adult rabbits is characterized by consistently *lower* levels of biomineralization than the mature adult rabbits (Table 14.9). These findings underscore a significant effect of aging on patterns of adaptive plasticity in bone-mineral density for different masticatory joints.

14.3.3.2 Histology and Immunohistochemistry

Histological analyses of younger and older adult rabbits indicate more intense safranin O staining in the symphyseal FC pad (compare "A" vs. "B" in Fig. 14.13)

Table 14.8 Comparison of rabbit symphyseal mean mineral density levels (μ) between cohorts of three subjects each. For the most part, older rabbits develop elevated bone-density levels at the symphysis articular surface and external cortical bone (Mann-Whitney U-test, $*=p < 0.05$)

Variable	Young adult	Mature adult	% Increase
Symphysis Top	1.936	1.990	2.8
Symphysis Upper	1.983	1.965	−1.0
Symphysis Middle	1.754	1.797	2.5
Symphysis Lower	1.824	2.065	13.2*
Symphysis Bottom	1.800	1.893	5.2
Inferior Corpus	1.937	2.152	11.1
Inferior/Lateral Corpus	2.450	2.209	−10.9
Lateral Corpus	2.549	2.708	6.2
Superior Corpus	2.823	2.813	≈ 0.0

Table 14.9 Comparison of rabbit TMJ mean biomineralization levels (μ) between cohorts of three subjects each. Opposite to findings for the symphysis of such older rabbits, the TMJ develops decreased bone-density levels along the articular surface, subchondral bone, and external cortical surface of the condylar neck (Mann-Whitney U-test, *=$p < 0.05$)

Variable	Young adult	Mature adult	% decrease
Outer 1	1.829	1.396	23.7*
Outer 2	1.475	1.355	8.1
Outer 3	1.842	1.282	30.4*
Outer 4	1.826	1.116	38.9*
Outer 5	1.855	1.400	24.5*
Inner 1	2.605	1.518	41.7*
Inner 2	2.815	1.673	40.6*
Inner 3	2.256	1.215	46.1*
Inner 4	2.203	1.227	44.3*

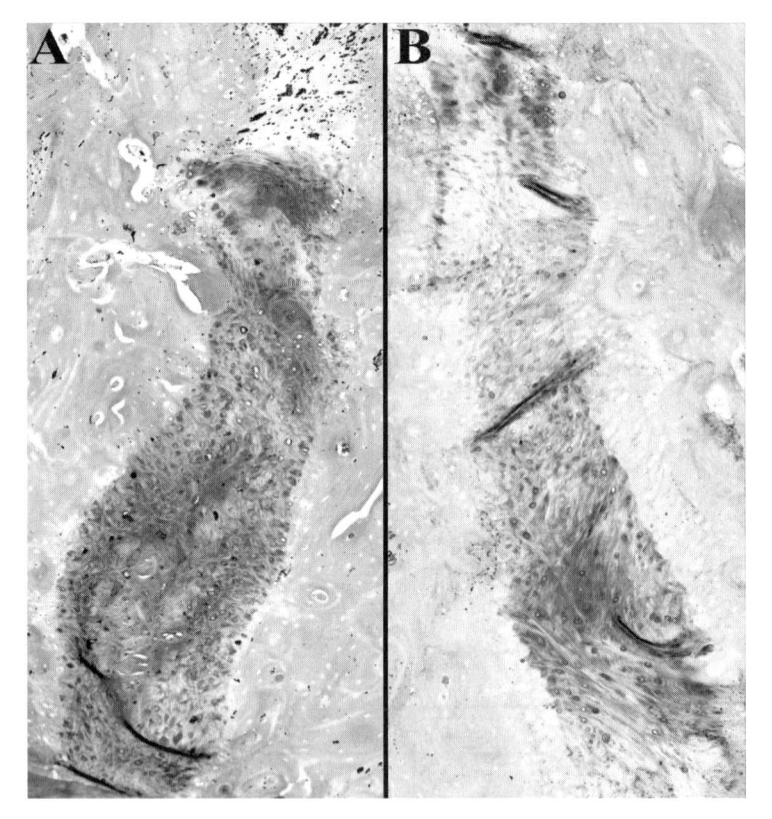

Fig. 14.13 Symphyseal proteoglycan content. Coronal sections (6 μm) of "middle" joint sites from young (**A**) and mature (**B**) adult rabbits stained with safranin O to identify GAG content in the FC pad. Darker staining in "A" vs. "B" indicates lower proteoglycan content and thus decreased FC pad viscoelasticity in older rabbits. This suggests the FC pad of older joints eventually exhibits ontogenetic reductions in the ability to counter compression between dentaries during mastication. In "B" note also the corresponding development of bony rugosities ("light grey" bone nearly traversing the "dark grey" FC pad)

14 Adaptive Plasticity in the Mammalian Masticatory Complex 315

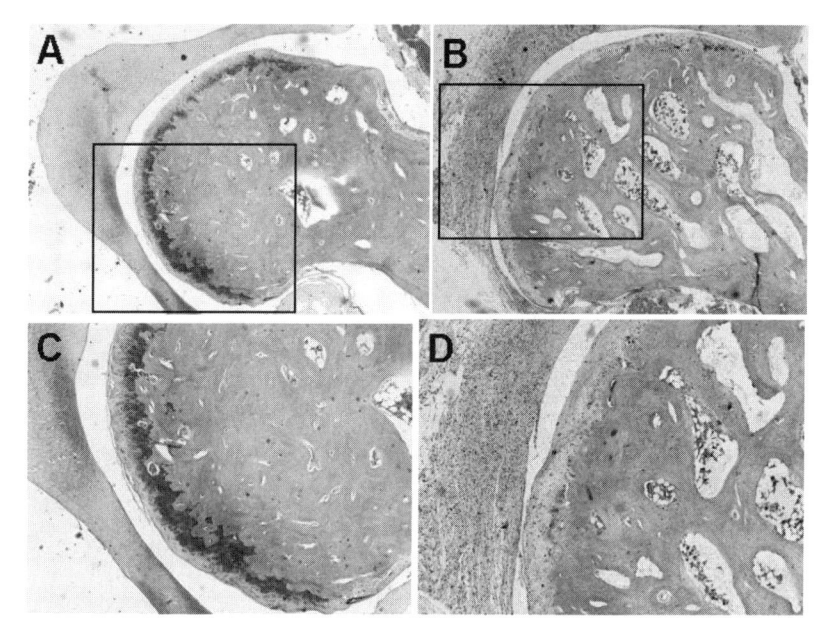

Fig. 14.14 TMJ proteoglycan content. Coronal sections (6 μm) of "middle" sites from young (**A**, **C**) and mature (**B**, **D**) adult rabbits were stained with safranin O to identify cartilage GAG content. In higher-power views of the upper panels, more intense staining of in "C" vs. "D" indicates higher proteoglycan content and thus diminished articular cartilage viscoelasticity in older rabbits. Therefore, older TMJs appear to evince decreases in the ability to counter compression during biting and chewing

and TMJ condylar articular cartilage (compare "A/C" vs. "B/D" in Fig. 14.14) of younger adults. Lower proteoglycan content throughout the FC pad and in the condylar cartilage's lower layers of older rabbits mirrors findings for cranial cartilage of diet-manipulated rabbits (as well as the limbs of mammals with joint disease – above). Due to the elevated viscoelasticity of proteoglycan-rich tissues in younger adult joints, our analyses suggest that the articular cartilage and fibrocartilage of such organisms is able to resist greater compressive stresses during biting and chewing than that of older joints. Much as noted previously, since proteoglycan content is most marked in the innermost layers of TMJ articular cartilage, this suggests it is important to account for regional variation in ECM composition in assessing functional and ontogenetic variation in cartilage proportions.

Immunohistochemical data for younger versus older adult rabbits indicate a more widespread distribution of type II collagen in the symphyseal FC pad and TMJ condylar articular cartilage of young adults (Figs. 14.15 & 14.16). These comparisons suggest that TMJ and symphysis aging are characterized by similar of patterns of variation in type II collagen *and* proteoglycan content, and both ECM components are important in tissues subjected to compressive loads during biting and chewing.

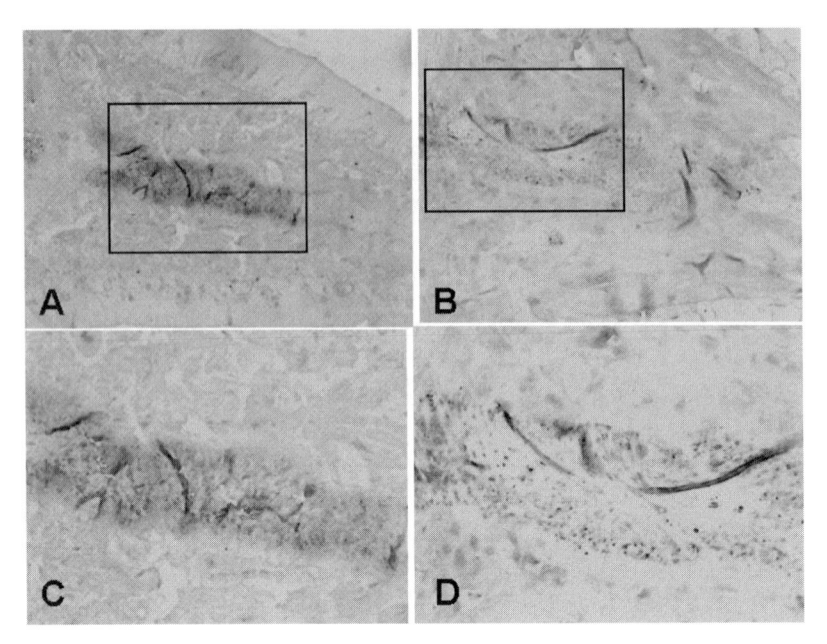

Fig. 14.15 Symphyseal type II collagen. Coronal sections (6 μm) of the "middle" joint site in young (**A**, **C**) and mature (**B**, **D**) adult rabbits stained with a 1^{ary} antibody directed against type II collagen (sections rotated 90 degrees in each panel). In the enlarged views of the ventral joint, darker staining in the FC pad of "C" vs "D" demonstrates less type II collagen and thus lower viscoelasticity in older rabbits. This suggests that the FC pad of older joints experiences postnatal decreases in the ability to resist compressive stresses during postcanine chewing and biting

Tunel staining indicates a greater number of apoptotic chondrocytes in the FC pad and TMJ articular cartilage of older adult rabbits (Figs. 14.17 & 14.18). This pattern suggests that aging in craniomandibular joints is characterized by elevated cell death and increased cartilage degradation. In TMJ articular cartilage, older rabbits also develop more hypertrophic chondrocytes (Fig. 14.18). As noted above, increased numbers of apoptotic, hypertrophic chondrocytes appear associated with advance of the subchondral mineralizing front. By selecting the same sections and sites for microCT, histological, and immunohistochemical analyses, this facilitated a preliminary characterization of the age-related responses of cranial joint tissues vis-à-vis altered masticatory stresses.

14.4 Discussion and Conclusions

The mammalian mandibular symphysis and TMJ are highly specialized joints capable of both rotational and translational movements, and thus encounter multidirectional compressive, shear, and tensile forces during biting and chewing (Rigler & Mlinsek, 1968; Beecher, 1977, 1979; Hylander, 1979a–c, 1992; Scapino, 1981; Ravosa & Hogue, 2004). In addition to cortical and trabecular bone, TMJs and

14 Adaptive Plasticity in the Mammalian Masticatory Complex

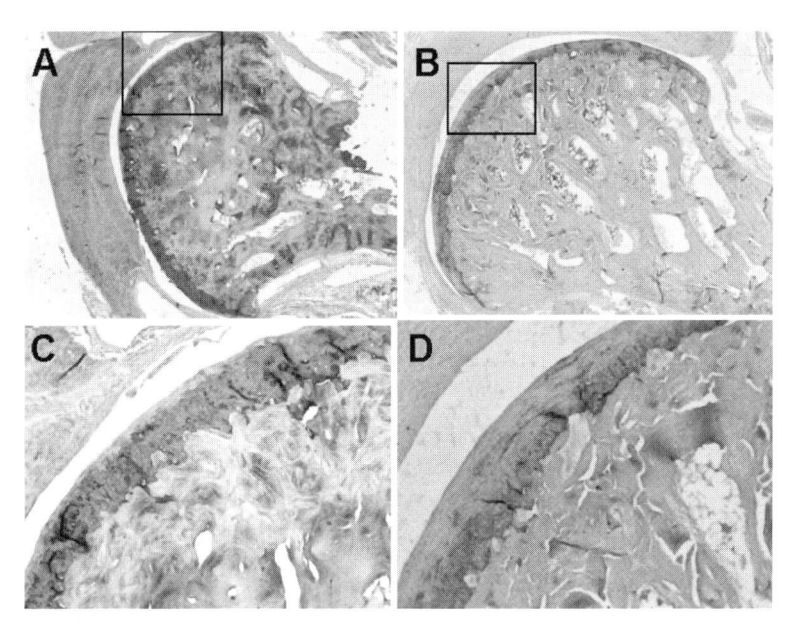

Fig. 14.16 TMJ type II collagen. Coronal sections (6 μm) of "middle" sites in young (**A, C**) and mature (**B, D**) adult rabbits were stained with a 1^{ary} antibody directed against type II collagen. In higher-power views of the upper panels, more intense staining of TMJ articular cartilage in "C" vs. "D" indicates less type II collagen and thus diminished cartilage viscoelasticity in older rabbits. This suggests that older TMJ cartilage experiences ontogenetic reductions in the ability to counter compressive loads during molar chewing and biting

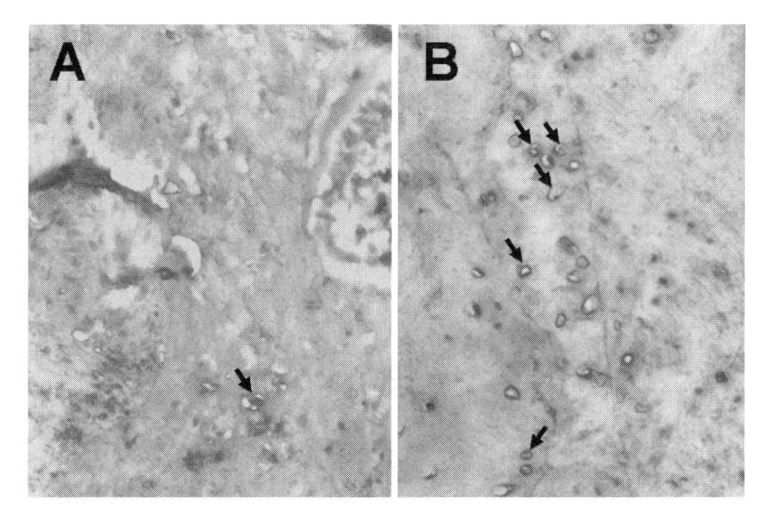

Fig. 14.17 Symphyseal apoptosis. Coronal sections (6 μm) of "middle" joint sites from young (**A**) and mature (**B**) adult rabbits tunel stained to identify fragmented DNA of apoptotic chondrocytes in the FC pad. In the enlarged view of the ventral joint with the FC pad, older rabbits exhibit numerous apoptotic chondrocytes (*black arrows*). This suggests that the FC pad of older joints develop postnatal increases in cartilage degradation

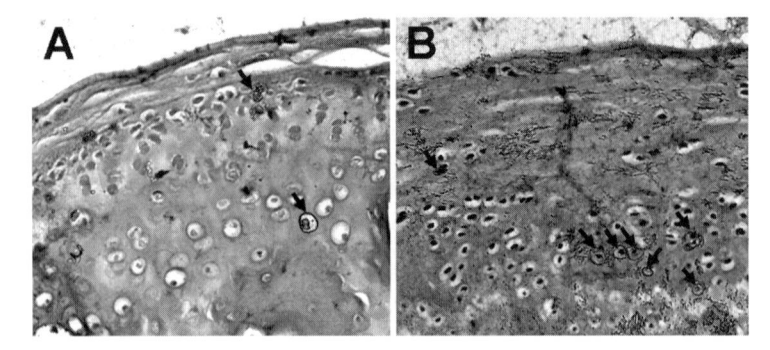

Fig. 14.18 TMJ apoptosis. Coronal sections (6 μm) of "middle" joint sites in young (**A**) and mature (**B**) adult rabbits were tunel stained to identify cellular apoptosis in articular cartilage. In views of the articular surface and underlying subchondral bone, older rabbits exhibit elevated chondrocyte apoptosis, and thus greater cartilage degradation vs. young adults (*black arrows*)

symphyses comprised cartilage, ligaments, and dense fibrous tissue containing collagens and proteoglycans (Figs. 14.3–14.18). As the symphyseal FC pad and TMJ articular cartilage are anchored into subarticular bone, their stress distributions are constrained respectively by movements between dentaries (symphysis) or between the mandible and temporal bone (TMJ). Employing animal models from clades for which in vivo data on feeding behavior are already available (Hiiemäe & Ardran, 1968; Weijs, 1975; Weijs & Dantuma, 1975; Weijs & de Jongh, 1977; Weijs et al., 1987, 1989; Langenbach et al., 1991, 1992, 2001; Kobayashi et al., 2002), we performed a series of integrative experiments to probe the longer-term dynamic hierarchical relationships among mechanical loading, adaptive plasticity, norms of reaction, and performance in mammalian masticatory joint systems.

Analyses of rabbits represent the first case where plasticity is assessed at two different joints in the same model organism. In subadults of this species, symphyses and TMJs of over-loaded joints develop larger joint proportions and higher bone-density levels coupled with lower proteoglycan content, lower type II collagen, and greater chondrocyte apoptosis. Thus, although symphyseal fibrocartilage is more acellular, it exhibits responses similar to that for hyaline cartilage of the TMJ articular surface. However, while the gross anatomical and bone biomineralization data are as predicted, findings for the ECM composition of joint cartilage seemingly contradict shorter-term experimental studies cited above. In light of this prior work, it is reasonable to interpret rabbit cartilage patterns as the result of degradative changes due to long-term joint over-loading. Thus, we do not and cannot refute the fact that cartilage exhibits a compensatory adaptive response to joint over-loading (e.g., see below for mice). Rather, the duration of dietary manipulation in our study typically exceeded that of earlier investigations, and it is well known that cartilage exhibits accelerated degradation in response to elevated and/or repetitive loading (Guerne et al., 1994, 1995; Bae et al., 1998). Such changes in cartilage composition reflect the early onset and progression of degenerative diseases like OA that compromise the structural integrity of a joint (Mankin et al., 1971; Newton & Nunamaker, 1985; Haskin et al., 1995; Kamelchuk & Major, 1995; Ishibashi et al., 1996; Ostergaard et al., 1999; Fujimura et al., 2006). This interpretation is consistent with the age-related patterns of change in rabbit TMJ connective tissues (Figs. 14.14, 14.16 & 14.18). In fact, it is likely that a component of the *adaptive* changes in rabbit TMJ proportions and biomineralization represents a compensatory mechanism to cartilage degradation that maintains the overall functional integrity of such composite joint systems. If age-related decreases in rabbit TMJ biomineralization are borne out by larger samples, then adaptive changes in TMJ skeletal tissues likely characterize only early ontogenetic stages, with later adult stages dominated by degradative patterns of change (Table 14.9).

In the case of the rabbit symphysis, the development of bony rugosities, larger joint surfaces due to thicker cortical bone and greater bone density, all represent adaptive responses to joint over-loading (Tables 14.2 & 14.3; Figs. 14.3, 14.4 & 14.6). However, repetitive joint overloading results in the FC pad eventually becoming *less* viscoelastic, which diminishes its ability to resist compressive stresses. As joint ossification clearly does not compromise symphyseal function as it would with the TMJ, the disparate long-term responses of symphyseal soft versus hard tissues may explain a common (but poorly understood) mammalian trend whereby older adults develop increased fusion (cf., Beecher, 1977, 1979, 1983; Scapino, 1981; Ravosa & Hylander, 1994; Ravosa, 1996, 1999; Hogue & Ravosa, 2001). Thus, age-related changes in fusion, especially among older animals, may represent a compensatory osteogenic response to load- and/or age-induced degradation of the FC pad and perhaps other connective tissues (Figs. 14.13, 14.15 & 14.17). This interpretation is consistent with the modest ontogenetic increases in symphyseal biomineralization observed in rabbits (Table 14.8), which differ markedly from the initial findings for the aging TMJ (Table 14.9).

Our analyses of a suite of similar TMJ parameters in rabbits and mice represent the first case where adaptive plasticity in experimental organisms can be more readily compared between species. In the over-loaded TMJ, rabbits and mice exhibit larger joint proportions and higher bone-density levels. However, rabbits and mice differ in terms of the response of articular cartilage to joint over-loading. Proteoglycan and type II collagen expression is increased in myostatin-deficient versus control mice, whereas over-use diet rabbits develop lower proteoglycan and type II collagen content as well as greater chondrocyte apoptosis than under-use rabbits. In this regard, the mouse comparisons suggest that articular cartilage exhibits an adaptive response to joint over-loading, much as is the case for TMJ proportions and biomineralization levels (and as in prior shorter-term research). Again, it is important to stress that some proportion of the adaptive response of mouse TMJ articular cartilage, both the altered ECM composition and especially the increased cartilage thickness along the buccal articular surface, is presumably associated with degenerative resorption of subchondral bone along the buccal aspect of the TMJ.

The rabbit findings parallel analyses of myostatin-deficient mice, which document increased differentiation of symphyseal parameters in response to elevated physiological loads (Ravosa et al., 2007a, b). This is suggestive of greater (Tables 14.2–14.4), and perhaps prolonged (Tables 14.8 vs. 14.9), tissue plasticity or norms of reaction for the symphysis versus elsewhere in the masticatory system.

It is thus interesting that the symphysis experiences relatively higher bone-strain levels during biting and chewing, and is characterized by strong positive allometry of joint proportions (Hylander, 1984, 1985; Ravosa, 1991a, b, 1992, 1996, 1998, 2000; Ravosa & Hylander, 1994; Vinyard & Ravosa, 1998; Hogue & Ravosa, 2001; Hogue, 2004; Ravosa & Hogue, 2004; Ravosa et al., 2000), factors that contribute to the potential for greater symphyseal plasticity.

As alluded to above, our study uniquely suggests that the short-term duration of earlier analyses of cranial joint tissues may offer a limited notion of the complex process of developmental plasticity, especially as it relates to the effects of long-term alterations in mechanical loads, when a joint is increasingly characterized by adaptive *and* degradative changes in tissue structure, composition, and function. Perhaps not surprisingly, we also sound a cautionary note that the assessment of masticatory plasticity based solely on external joint proportions can under-represent the amount of change in individual tissues. For instance, the magnitude of the plasticity response differs between loading cohorts according to the level of analysis, e.g., external joint proportions vary less between groups (Tables 14.2 & 14.5) than in comparisons of skeletal biomineralization (Tables 14.3 & 14.4) or internal proportions (Tables 14.2 & 14.5).

Though it is well known that in vivo information is best for detailing how an animal functions during normal behaviors such as biting and chewing (Bock & von Walhert, 1965; Hylander, 1979a, b; Wake, 1982; Wainwright & Reilly, 1994; Lauder, 1995), there is perhaps one shortcoming of the evidence for symphyseal fusion based on the studies of craniomandibular bone strain and jaw-adductor muscle activity. Apart from sound theoretical arguments, the best in vivo support for a functional relationship between symphyseal stress and symphyseal fusion is essentially correlational in linking character–state variation to the way an adult organism loads, or is posited to load, a masticatory structure (Ravosa & Hogue, 2004). While this does not invalidate or diminish the unique and important role of in vivo data for testing hypotheses regarding the biological role and performance of cranial elements, it does imply that when evaluating masticatory function during growth or across a clade, presently one must assume that variation in symphyseal fusion corresponds to specific differences in jaw-loading and jaw-adductor muscle patterns. Indeed, this gap in our knowledge abetted arguments that variation in symphyseal fusion is unrelated to variation in symphyseal loading levels during mastication, with an unfused joint being sufficiently strong to routinely counter significant stresses (Dessem, 1989; Lieberman & Crompton, 2000). It follows from such an interpretation that the tissues of an unfused symphysis would be unresponsive to postnatal variation in long-term, repetitive loads.

This controversy exists because an integrative biomechanical, cellular, and biochemical analysis of adaptive plasticity heretofore had been applied only to cranial synovial joints (TMJ – Bouvier & Hylander, 1982, 1984; Huang et al., 2002, 2003) and syndesmoses (sutures – Byron et al., 2004). To this end, data on tissue plasticity in a cranial arthrosis (rabbit and mouse symphysis) offer a novel perspective on the dynamic inter-relationships among symphyseal fusion, joint performance, and feeding behaviors. In support of prior research (Hylander et al., 1998, 2000, 2005; Ravosa & Hylander, 1994), our analyses where both rabbit cohorts used their incisors similarly, but differed largely in diet-related forces experienced during postcanine chewing and biting, highlights the significant role of stress during *mastication* on postnatal and phylogenetic variation in symphyseal anatomy across diverse mammal clades (e.g., Tables 14.2 & 14.3). In addition, evidence regarding functional adaptation in symphyseal proportions, morphology, and bony properties supports the hypothesis that dynamic alterations in masticatory loads positively influence developmental variation in symphyseal joint strength and integrity. These and other findings are inconsistent with claims that fusion occurs to stiffen, rather than strengthen, the symphyseal joint during mastication (Hogue & Ravosa, 2001; Ravosa & Hogue, 2004).

In sum, by choosing similar joint section/site samples for microCT, histology, and immunohistochemistry, our experimental analyses facilitated a characterization of the integrated suite of dynamic responses (both adaptive and degradative) of skeletal and connective tissues to altered loads. Although soft-tissue responses are similar between older/over-loaded joints on one hand and younger/under-loaded joints on the other hand, it remains to be determined if age-related changes in cartilage composition and properties result from the cumulative effects of repetitive loading and/or other aspects of the aging process. It is likely that a combination of both factors affects the anatomy and function of aging joints. Viewed from an evolutionary perspective, this research suggests that variation in symphysis and TMJ morphology and performance among sister taxa is, in part, an epiphenomenon of interspecific differences in (diet-induced) jaw-loading patterns, characterizing the individual ontogenies of the members of a species (Vinyard & Ravosa, 1998; Ravosa & Hogue, 2004). However, this interspecific behavioral signal may be increasingly mitigated among aging individuals by the (potentially species-specific) interplay between adaptive and degradative tissue responses.

Acknowledgments For the senior author, Bill Hylander long has been a source of scientific inspiration and heartfelt friendship. He also is responsible for instigating a very serious martini fetish in one of the co-authors (MSS). Funding for this research was provided by the Department of Cell and Molecular Biology and Summer Research Opportunity Program, Northwestern University (to MJR), and NIH (AR049717-01A2 to MWH). Brian Shea generously provided the microtome used for histological and immunohistochemical analyses. Barth Wright is thanked for performing the analyses of rabbit food material properties. Al Telser is thanked for kind assistance with, and access to, equipment employed for image analysis of tissue sections. Lidiany González assisted with early phases of the histological and immunohistochemical analyses of mouse TMJs. Ravi Kunwar assisted with the microanatomical analyses of the mice and rabbits. Emily Klopp and Jessie Pinchoff helped with the mouse microCT analyses. Chris Vinyard, Chris Wall, and two anonymous reviewers provided helpful comments. Finally, we acknowledge use of the Scanco MicroCT-40 of Northwestern University's MicroCT facility.

References

Agarwal S, Long P, Gassner R, Piesco NP & Buckley MJ (2001) Cyclic tensile strain suppresses catabolic effects of interleukin-1β in fibrochondrocytes from the temporomandibular joint. Arthritis Rheumatism 44:608–617.

Agrawal AA (2001) Phenotypic plasticity in the interactions and evolution of species. Science 294:321–326.

Bae YC, Park KP, Park MJ & Ihn HJ (1998) Development of vimentin filaments in the cells of the articular disc of the rat squamosomandibular joint with age. Archives Oral Biology 43:579–583.

Bayliss MT, Venn M, Maroudas A & Ali SY (1983) Structure of proteoglycans from different layers of human articular cartilage. Biochem J 209:387–400.

Beecher RM (1977) Function and fusion at the mandibular symphysis. American J Physical Anthropology 47:325–336.

Beecher RM (1979) Functional significance of the mandibular symphysis. J Morphology 159:117–130.

Beecher RM (1983) Evolution of the mandibular symphysis in Notharctinae (Adapidae, Primates). International J Primatology 4:99–112.

Beecher RM & Corruccini RS (1981) Effects of dietary consistency on craniofacial and occlusal development in the rat. Angle Orthodontist 51:61–69.

Beecher RM, Corruccini RS & Freeman M (1983) Craniofacial correlates of dietary consistency in a nonhuman primate. J Craniofacial Genetics Developmental Biology 3:193–202.

Benjamin M & Ralphs JR (1998) Fibrocartilage in tendons and ligaments – An adaptation to compressive load. J Anatomy 193:481–494.

Biewener AA (1993) Safety factors in bone strength. Calcified Tissue International 53:568–574.

Biewener AA & Bertram JEA (1993) Skeletal strain patterns in relation to exercise training during growth. J Experimental Biology 185:51–69.

Biewener AA, Swartz SM & Bertram JEA (1986) Bone modeling during growth: Dynamic strain equilibrium in the chick tibiotarsus. Calcified Tissue International 39:390–395.

Biknevicius AR & Leigh SR (1997) Patterns of growth of the mandibular corpus in spotted hyenas (*Crocuta crocuta*) and cougars (*Puma concolor*). Zoological J Linnean Society 120:139–161.

Biknevicius AR & Ruff CB (1992) The structure of the mandibular corpus and its relationship to feeding behaviours in extant carnivorans. J Zoology, London 228:479–507.

Biknevicius AR & Van Valkenburgh B (1996) Design for killing: Craniodental adaptations of predators. Gittleman JL (Ed): Carnivore Behavior, Ecology, and Evolution. Volume 2. Ithaca: Cornell University Press, pp 393–428.

Block MS, Unhold G & Bouvier M (1988) The effect of diet texture on healing following temporomandibular joint discectomy in rabbits. J Oral Maxillofacial Surgery 46:580–588.

Bock WJ & von Walhert G (1965) Adaptation and the form-function complex. Evolution 19:269–299.

Bouvier M (1986) A biomechanical analysis of mandibular scaling in Old World monkeys. American J Physical Anthropology 69:473–482.

Bouvier M (1987) Variation in alkaline-phosphatase activity with changing load on the mandibular condylar cartilage in the rat. Archives Oral Biology 32:671–675.

Bouvier M (1988) Effects of age on the ability of the rat temporomandibular joint to respond to changing functional demands. J Dental Research 67:1206–1212.

Bouvier M & Hylander WL (1981) Effect of bone strain on cortical bone structure in macaques (*Macaca mulatta*). J Morphology 167:1–12.

Bouvier M & Hylander WL (1982) The effect of dietary consistency on morphology of the mandibular condylar cartilage in young macaques (*Macaca mulatta*). AD Dixon & BG Sarnat (Eds): Factors and Mechanisms Influencing Bone Growth. New York: AR Liss, pp 569–579.

Bouvier M & Hylander WL (1984) The effect of dietary consistency on gross and histologic morphology in the craniofacial region of young rats. American J Anatomy 170:117–126.

Bouvier M & Hylander WL (1996a) The mechanical or metabolic function of secondary osteonal bone in the monkey *Macaca fascicularis*. Archives Oral Biology 41:941–950.

Bouvier M & Hylander WL (1996b) Strain gradients, age, and levels of modeling and remodeling in the facial bones of *Macaca fascicularis*. Z Davidovitch & LA Norton (Eds): The Biological Mechanisms of Tooth Movement and Craniofacial Adaptation. Boston: Harvard Society for the Advancement of Orthodontics, pp 407–412.

Bouvier M & Zimny ML (1987) Effects of mechanical loads on surface morphology of the condylar cartilage of the mandible of rats. Acta Anatomica 129:293–300.

Byron CD, Borke J, Yu J, Pashley D, Wingard CJ & Hamrick M (2004) Effects of increased muscle mass and mouse sagittal suture morphology and mechanics. Anatomical Record 279A:676–684.

Byron CD, Hamrick MW & Wingard CJ (2006) Alterations of temporalis muscle contractile force and histological content from the myostatin and *Mdx* deficient mouse. Archives Oral Biology 51:396–405.

Carrier DR (1996) Ontogenetic limits on locomotor performance. Physiological Zoology 69:467–488.

Carvalho RS, Yen EH & Suga DM (1995) Glycosaminoglycan synthesis in the rat articular disc in response to mechanical stress. American J Orthodontics Dentofacial Orthopaedics 107:401–410.

Cole TM (1992) Postnatal heterochrony of the masticatory apparatus in *Cebus apella* and *Cebus albifrons*. J Human Evolution 23:253–282.

Copray JCVM, Jansen HWB & Duterloo HS (1985) Effects of compressive forces on proliferation and matrix synthesis of mandibular condylar cartilage of the rat *in vitro*. Archives Oral Biology 30:299–304.

Currey JD (2002) Bones: Structure and Mechanics. Princeton: Princeton University Press.

Daegling DJ (1989) Biomechanics of cross-sectional size and shape in the hominoid mandibular corpus. American J Physical Anthropology 80:91–106.

Daegling DJ (1992) Mandibular morphology and diet in the genus *Cebus*. Int J Primatology 13:545–570.

Darvell BW, Lee PKD, Yuen TDB & Lucas PW (1996) A portable fracture toughness tester for biological materials. Meas Sci Technol 7:954–962.

Dessem D (1989) Interactions between jaw-muscle recruitment and jaw-joint forces in *Canis familiaris*. J Anatomy 164:101–121.

Ferguson VL, Ayers RA, Bateman TA & Simske SJ (2003) Bone development and age-related bone loss in male C57BL/6J mice. Bone 33:387–398.

Freeman PW (1979) Specialized insectivory: Beetle-eating and moth-eating molossid bats. J Mammalogy 60:467–479.

Freeman PW (1981) Correspondence of food habits and morphology in insectivorous bats. J Mammalogy 62:166–173.

Freeman PW (1988) Frugivorous and animalivorous bats (Microchiroptera): Dental and cranial adaptations. Biological J Linnean Society 33:249–272.

Fujimura K, Kobayashi S, Suzuki T & Segami N (2005) Histologic evaluation of temporomandibular arthritis induced by mild mechanical loading in rabbits. J Oral Pathol Med 34:157–163.

Gans C, Gorniak GC & Morgan WK (1990) Bite-to-bite variation of muscular activity in cats. J Experimental Biology 151:1–19.

Gerber H-P, Vu TH, Ryan AM, Kowalski J, Werb Z & Ferrara N (1999) VEGF couples hypertrophic cartilage remodeling, ossification and angiogenesis during endochondral bone formation. Nature Medicine 5:623–628.

Goldring MB (2004a) Human chondrocyte cultures as models of cartilage-specific gene regulation. J Picot (Ed): Methods in Molecular Medicine. Human Cell Culture Protocols, 2nd Ed. Totowa, NJ: Humana Press, pp 69–96.

Goldring MB (2004b) Immortalization of human articular chondrocytes for generation of stable, differentiated cell lines. M Sabatini, P Pastoureau & F de Ceuninck (Eds): Methods in Molecular Medicine. Cartilage and Osteoarthritis, Volume 1: Cellular and Molecular Tools. Totowa, NJ: Humana Press, pp 23–36.

Gotthard K & Nylin S (1995) Adaptive plasticity and plasticity as an adaptation: A selective review of plasticity in animal morphology and life history. Oikos 74:3–17.

Grant BR & Grant PR (1989) Evolutionary Dynamics of a Natural Population. Chicago: University of Chicago Press.

Grieser B (1992) Infant development and parental care in two species of sifakas. Primates 33:305–314.

Grodzinsky AJ, Levenston ME, Jin M & Frank EH (2000) Cartilage tissue remodeling in response to mechanical forces. Annual Review Biomedical Engineering 2:691–713.

Guerne PA, Sublet A & Lotz M (1994) Growth factor responsiveness of human articular chondrocytes distinct profiles in primary chondrocytes, subcultured chondrocytes, and fibroblasts. J Cell Physiology 158:476–484.

Guerne PA, Blanco F, Kaelin A, Desgeorges A & Lotz M (1995) Growth factor responsiveness of human articular chondrocytes in aging and development. Arthritis Rheumatism 38:960–968.

Hamrick MW (1999) A chondral modeling theory revisited. J Theoretical Biology 201:201–208.

Hamrick MW (2003) Increased bone mineral density in the femora of GDF8 knockout mice. Anatomical Record 272A:388–391.

Hamrick MW, Pennington C & Byron C (2003) Bone modeling and disc degeneration in the lumbar spine of mice lacking GDF8 (myostatin). J Orthopaedic Research 21:1025–1032.

Haskin CL, Milam SB & Cameron IL (1995) Pathogenesis of degenerative joint disease in the human temporomandibular joint. Critical Reviews Oral Biology Medicine 6:248–277.

Herring SW, Anapol FC & Wineski LE (1991) Motor-unit territories in the masseter muscle of infant pigs. Archives Oral Biology 36:867–873.

Herring SW & Scapino RP (1973) Physiology of feeding in miniature pigs. J Morphology 141:427–460.

Herring SW & Wineski LE (1986) Development of the masseter muscle and oral behavior in the pig. J Experimental Zoology 237:191–207.

Hiiemäe KM & Ardran GM (1968) A cinefluorographic study of mandibular movement during feeding in the rat (*Rattus norvegicus*). J Zoology (London) 154:139–154.

Hirschfeld Z, Michaeli Y & Weinreb MM (1977) Symphysis menti of the rabbit: Anatomy, histology, and postnatal development. J Dental Research 56:850–857.

Hogue AS (2004) On the Relation between Craniodental Form and Diet in Mammals: Marsupials as a Natural Experiment. PhD Thesis. Northwestern University.

Hogue AS & Ravosa MJ (2001) Transverse masticatory movements, occlusal orientation and symphyseal fusion in selenodont artiodactyls. J Morphology 249:221–241.

Holden C & Vogel G (2002) Plasticity: Time for a reappraisal? Science 296:2126–2129.

Holmvall K, Camper L, Johansson S, Kimura JH & Lundgren-Akerlund E (1995) Chondrocyte and chondrosarcoma cell integrins with affinity for collagen type II and their response to mechanical stress. Experimental Cell Research 221:496–503.

Honda K, Ohno S, Taniomoto K, Ijuin C, Tanaka N, Doi T, Kato Y & Tanne K (2000) The effects of high magnitude cyclic tensile load on cartilage matrix metabolism in cultured chondrocytes. European J Cell Biology 79:601–609.

Huang Q, Opstelten D, Samman N & Tideman H (2002) Experimentally induced unilateral tooth loss: Histochemical studies of the temporomandibular joint. J Dental Research 81: 209–213.

Huang Q, Opstelten D, Samman N & Tideman H (2003) Experimentally induced unilateral tooth loss: Expression of type II collagen in temporomandibular joint cartilage. J Oral Maxillof. Surg. 61:1054–1060.

Huang X, Zhang G & Herring SW (1994) Age changes in mastication in the pig. Comparative Biochemistry Physiology 107A:647–654.

Hylander WL (1979a) Mandibular function in *Galago crassicaudatus* and *Macaca fascicularis*: An in vivo approach to stress analysis of the mandible. J Morphology 159:253–296.

Hylander WL (1979b) The functional significance of primate mandibular form. J Morphology 160:223–240.

Hylander WL (1979c) An experimental analysis of temporomandibular joint reaction forces in macaques. American J Physical Anthropology 51:433–456.

Hylander WL (1984) Stress and strain in the mandibular symphysis of primates: A test of competing hypotheses. American J Physical Anthropology 64:1–46.

Hylander WL (1985) Mandibular function and biomechanical stress and scaling. American Zoologist 25:315–330.

Hylander WL (1992) Functional anatomy. BG Sarnat & DM Laskin (Eds): The Temporomandibular Joint. A Biological Basis for Clinical Practice. Philadelphia: Saunders, pp 60–92.

Hylander WL, Johnson KR & Crompton AW (1992) Muscle force recruitment and biomechanical modeling: An analysis of masseter muscle function during mastication in *Macaca fascicularis*. American J Physical Anthropology 88:365–387.

Hylander WL, Ravosa MJ, Ross CF & Johnson KR (1998) Mandibular corpus strain in primates: Further evidence for a functional link between symphyseal fusion and jaw-adductor muscle force. American J Physical Anthropology 107:257–271.

Hylander WL, Ravosa MJ, Ross CF, Wall CE & Johnson KR (2000) Symphyseal fusion and jaw-adductor muscle force: An EMG study. American J Physical Anthropology 112:469–492.

Hylander WL, Wall CE, Vinyard CJ, Ross CF, Ravosa MJ, Williams SH & Johnson KR (2005) Temporalis function in anthropoids and strepsirrhines: An EMG Study. American J Physical Anthropology 128:35–56.

Iinuma M, Yoshida S & Funakoshi M (1991) Development of masticatory muscles and oral behavior from suckling to chewing in dogs. Comparative Biochemistry Physiology 100A:789–794.

Ishibashi H, Takenoshita Y, Ishibashi K & Oka M (1996) Expression of extracellular matrix in human mandibular condyle. Oral Surgery Oral Medicine Oral Pathology Oral Radiology Endodontics 81:402–414.

Ji S, Losinski R, Cornelius S, Frank G, Willis G, Gerrard D, Depreux F & Spurlock M (1998) Myostatin expression in porcine tissues: Tissue specificity and developmental and postnatal regulation. American J Physiology 275:R1265–1273.

Kamelchuk LS & Major PW (1995) Degenerative disease of the temporomandibular joint. J Orofacial Pain 9:168–180.

Kiernan JA (1999) Histological and Histochemical Methods. 3rd Edition. Boston: Butterworh-Heinemann.

Kiliardis S, Engström C & Thilander B (1985) The relationship between masticatory function and craniofacial morphology. I. A cephalometric longitudinal analysis in the growing rat fed a soft diet. European J Orthodontics 7:273–283.

Kim SG, Park JC, Kang DW, Kim BO, Yoon JH, Cho SI, Choe HC & Bae CS (2003) Correlation of immunohistochemical characteristics of the craniomandibular joint with the degree of mandibular lengthening in rabbits. J Oral Maxillofacial Surgery 61:1189–1197.

Kiyoshima T, Kido MA, Nishimura Y, Himeno M, Tsukuba T, Tashiro H & Yamamoto K (1994) Immunocytochemical localization of cathepsin L in the synovial lining cells of the rat temporomandibular joint. Archives Oral Biology 39:1049–1056.

Kiyoshima T, Tsukaba T, Kido MA, Tashiro H, Yamamoto K & Tanaka T (1993) Immunocytochemical localization of cathepsins B and D in the synovial lining cells of the rat temporomandibular joint. Archives Oral Biology 38:357–359.

Kobayashi M, Masuda Y, Fujimoto Y, Matsuya T, Yamamura K, Yamada Y, Maeda N & Morimoto T (2002) Electrophysiological analysis of rhythmic jaw movements in the freely moving mouse. Physiology Behavior 75:377–385.

Langenbach GE, Brugman P & Weijs WA (1992) Preweaning feeding mechanisms in the rabbit. J Developmental Physiology 18:253–261.

Langenbach GEJ & van Eijden TMGJ (2001) Mammalian feeding motor patterns. American Zoologist 41:1338–1351.

Langenbach GEJ, Weijs WA, Brugman P & van Eijden TMGJ (2001) A longitudinal electromyographic study of the postnatal maturation of mastication in the rabbit. Archives Oral Biology 46:811–820.

Langenbach GE, Weijs WA & Koolstra JH (1991) Biomechanical changes in the rabbit masticatory system during postnatal development. Anatomical Record 230:406–416.

Lanyon LE & Rubin CT (1985) Functional adaptation in skeletal structures. M Hildebrand, DM Bramble, KF Liem & DB Wake (Eds): Functional Vertebrate Morphology. Cambridge: Harvard University Press, pp 1–25.

Lauder GV (1995) On the inference of function from structure. JJ Thomason (Ed): Functional Morphology in Vertebrate Paleontology. Cambridge: Cambridge University Press, pp 1–18.

Lee HS, Millward-Sadler SJ, Wright MO, Nuki G & Salter DM (2000) Integrin and mechanosensitive ion channel-dependent tyrosine phosphorylation of focal adhesion proteins and β-catenin in human articular chondrocytes after mechanical stimulation. J Bone Mineral Research 15:1501–1509.

Lemare F, Steimberg N, Le Griel C, Demignot S & Adolphe M (1998) Dedifferentiated chondrocytes cultured in alginate beads: Restoration of the differentiated phenotype and of the metabolic responses to interleukin-1β. J Cellular Physiology 176:303–313.

Lieberman DE & Crompton AW (2000) Why fuse the mandibular symphysis? A comparative analysis. American J Physical Anthropology 112:517–540.

Losos JB (1990) Ecomorphology, performance capability, and scaling of West Indian *Anolis* lizards: An evolutionary analysis. Ecological Monographs 60:369–388.

Lucas PW (1994) Categorization of food items relevant to oral processing. DJ Chivers & P Langer (Eds): The Digestive System in Mammals: Food, Form and Function. Cambridge: Cambridge University Press, pp 197–218.

Lucas PW, Beta T, Darvell BW, Dominy NJ, Essackjee HC, Lee PKD, Osorio D, Ramsden L, Yamashita N & Yuen TDB (2001) Field kit to characterize physical, chemical and spatial aspect of potential primate foods. Folia Primatologica 72:11–25.

Luschei ES & Goodwin GM (1974) Patterns of mandibular movement and jaw muscle activity during mastication in monkeys. J Neurophysiology 37:954–966.

Mankin HJ, Dorfman H, Lippiello L & Zarins A (1971) Biochemical and metabolic abnormalities in articular cartilage from osteo-arthritic human hips. II. Correlation of morphology with biochemical and metabolic data. J Bone Joint Surgery 53:523–537.

Mao JJ, Rahemtulla F & Scott PG (1998) Proteoglycan expression in the rat temporomandibular joint in response to unilateral bite raise. J Dental Research 77:1520–1528.

Marchetti C, Cornaglia I, Casasco A, Bernasconi G, Baciliero U & Stedler-Stevenson WG (1999) Immunolocalization of gelatinase-A (matrix metalloproteinase-2) in damaged human temporomandibular joint discs. Archives Oral Biology 44:297–304.

McPherron AC, Lawler AM & Lee S-J (1997) Regulation of skeletal muscle mass in mice by a new TGF-ß superfamily member. Nature 387:83–90.

McPherron AC & Lee S-J (2002) Suppression of body fat accumulation in myostatin-deficient mice. J Clinical Investigation 109:595–601.

Mizoguchi I, Takahashi I, Nakamura M, Sasano Y, Sato S, Kagayama M & Mitani H (1996) An immunohistochemical study of regional differences in the distribution of type I and type II collagens in rat mandibular condylar cartilage. Archives Oral Biology 41:863–869.

Morenko BJ, Bove SE, Chen L, Guzman RE, Juneau P, Bocan TMA, Peter GK, Arora R & Kilgore KS (2004) In vivo micro-computed tomography of subchondral bone in the rat after intra-articular administration of monosodium iodoacetate. Contemporary Topics Lab Animal Science 43:39–43.

Mow VC, Fithian DC & Keely MA (1990) Fundamentals of articular cartilage and meniscus biomechanics. JW Ewing (Ed): Articular Cartilage and Knee Joint Function. New York: Raven Press, pp 1–18.

Nakano T & Scott PG (1989) A quantitative chemical study of glycosaminoglycans in the articular disc of bovine temporomandibular joint. Archives Oral Biology 34:749–757.

Newton CD & Nunamaker DM (1985) Textbook of Small Animal Orthopaedics. JB Lippincott Co.

Nicholson EK, Stock SR, Hamrick MW & Ravosa MJ (2006) Biomineralization and adaptive plasticity of the temporomandibular joint in myostatin knockout mice. Archives Oral Biology 51:37–49.

Nuzzo S, Lafage-Proust MH, Martin-Badosa E, Boivin G, Thomas T, Alexandre C & Peyrin F (2002) Synchrotron radiation microtomography allows analysis of three-dimensional microarchitecture and degree of mineralization of human iliac crest biopsy specimens: Effect of etidronate treatment. J Bone Mineral Research 17:1372–1382.

Ostergaard K, Andersen CB, Petersen J, Bendtzen K & Salter DM (1999) Validity of histopathological grading of articular cartilage from osteoarthritic knee joints. Annals Rheum Dis 58:208–213.

Patel V, Issever AS, Burghardt A, Laib A, Ries M & Majumdar S (2003) MicroCT evaluation of normal and osteoarthritic bone structure in human knee specimens. J Orthopaedic Research 21:6–13.

Pirttiniemi P, Kantomaa T, Salo L & Tuominen M (1996) Effect of reduced articular function on deposition of type I and type II collagens in the mandibular condylar cartilage of the rat. Archives Oral Biology 41:127–131.

Puzas JE, Landeau JM, Tallents R, Albright J, Schwarz EM & Landesberg R (2001) Degradative pathways in tissues of the temporomandibular joint. Use of in vitro and in vivo models to characterize matrix metalloproteinase and cytokine activity. Cells Tissues Organs 169:248–256.

Ravosa MJ (1991a) The ontogeny of cranial sexual dimorphism in two Old World monkeys: *Macaca fascicularis* (Cercopithecinae) and *Nasalis larvatus* (Colobinae). International J Primatology 12:403–426.

Ravosa MJ (1991b) Structural allometry of the mandibular corpus and symphysis in prosimian primates. J Human Evolution 20:3–20.

Ravosa MJ (1992) Allometry and heterochrony in extant and extinct Malagasy primates. J Human Evolution 23:197–217.

Ravosa MJ (1996) Mandibular form and function in North American and European Adapidae and Omomyidae. J Morphology 229:171–190.

Ravosa MJ (1998) Cranial allometry and geographic variation in slow lorises (*Nycticebus*). American J Primatology 45:225–243.

Ravosa MJ (1999) Anthropoid origins and the modern symphysis. Folia Primatologica 70:65–78.

Ravosa MJ (2000) Size and scaling in the mandible of living and extinct apes. Folia Primatologica 71:305–322.

Ravosa MJ (2007) Cranial ontogeny, diet and ecogeographic variation in African lorises. American J Primatology 68:1–15.

Ravosa MJ & Hogue AS (2004) Function and fusion of the mandibular symphysis in mammals: A comparative and experimental perspective. CF Ross & RF Kay (Eds): Anthropoid Evolution. New Visions. New York: Springer/Kluwer Publishers, pp 413–462.

Ravosa MJ & Hylander WL (1994) Function and fusion of the mandibular symphysis in primates: Stiffness or strength? JG Fleagle & RF Kay (Eds): Anthropoid Origins. New York: Plenum Press, pp 447–468.

Ravosa MJ, Johnson KR & Hylander WL (2000) Strain in the galago facial skull. J Morphology 245:51–66.

Ravosa MJ, Klopp EB, Pinchoff J, Stock SR & Hamrick MW (2007b) Plasticity of mandibular biomineralization in myostatin-deficient mice. J Morphology 268:275–282.

Ravosa MJ, Kunwar R., Stock SR & Stack MS (2007a) Pushing the limit: Masticatory stress and adaptive plasticity in mammalian craniomandibular joints. J Experimental Biology 210:628–641.

Ravosa MJ & Simons EL (1994) Mandibular growth and function in *Archaeolemur*. American J Physical Anthropology 95:63–76.

Ravosa MJ, Stock SR, Simons EL & Kunwar R (2007c) MicroCT analysis of symphyseal ontogeny in *Archaeolemur*. International J Primatology 28:1385–1396.

Rigler L & Mlinsek B (1968) Die Symphyse der Mandibula beim Rinde. Ein Beitrag zur Kenntnis ihrer Struktur und Funktion. Anatomischer Anzeiger 122:293–314.

Scapino RP (1981) Morphological investigation into functions of the jaw symphysis in carnivorans. J Morphology 167:339–375.

Shuto T, Sarkar G, Bronk J, Matsui N & Bolander M (1997) Osteoblasts express types I and II activin receptors during early intramembranous and endochondral bone formation. J Bone Mineral Research 12:403–411.

Sorensen MF, Rogers JP & Baskett TS (1968) Reproduction and development in confined swamp rabbits. J Wildlife Management 32:520–531.

Spencer LM (1995) Morphological correlates of dietary resource partitioning in the African Bovidae. J Mammalogy 76:448–471.

Srinivas R, Sorsa T, Tjaderhane L, Niemi E, Raustia A, Pernu H, Teronen O & Salo T (2001) Matrix metalloproteinases in mild and severe temporomandibular joint internal derangement synovial fluid. Oral Surgery Oral Medicine Oral Pathology Oral Radiology Endodontics 91:517–525.

Stock SR, Nagaraja S, Barss J, Dahl T & Veis A (2003) X-Ray microCT study of pyramids of the sea urchin *Lytechinus variegatus*. J Structural Biology 141:9–21.

Tanaka A, Kawashiri S, Kumagai S, Takatsuka S, Narinobou M, Nakagawa K & Tanaka S (2000) Expression of matrix metalloproteinase-2 in osteoarthritic fibrocartilage from human mandibular condyle. J Oral Pathology Medicine 29:314–320.

Tarnaud L (2004) Ontogeny of feeding behavior of *Eulemur fulvus* in the dry forest of Mayotte. International J Primatology 25:803–824.

Taylor AB, Jones KE, Kunwar R & Ravosa MJ (2006) Dietary consistency and plasticity of masseter fiber architecture in postweaning rabbits. Anatomical Record 288A:1105–1111.

Thexton AJ, Hiiemäe KM & Crompton AW (1980) Food consistency and bite size as regulators of jaw movement during feeding in the cat. J Neurophysiology 44:456–474.

Trevisan RA & Scapino RP (1976a) Secondary cartilages in growth and development of the symphysis menti in the hamster. Acta Anatomica 94:40–58.

Trevisan RA & Scapino RP (1976b) The symphyseal cartilage and growth of the symphysis menti in the hamster. Acta Anatomica 96:335–355.

Vincent JFV (1992) Biomechanics – Materials. A Practical Approach. Oxford: IRL Press.

Vinyard CJ & Ravosa MJ (1998) Ontogeny, function, and scaling of the mandibular symphysis in papionin primates. J Morphology 235:157–175.

Wainwright PC & Reilly SM, Editors (1994) Ecological Morphology. Chicago: University of Chicago Press.

Wainwright SA, Biggs WD, Currey JD & Gosline JM (1976) Mechanical Design in Organisms. Princeton: Princeton University Press.

Wake MH (1982) Morphology, the study of form and function, in modern evolutionary biology, Oxford Surveys in Evolutionary Biology. Volume 8. Oxford. Oxford University Press, pp 289–346.

Watts DP (1985) Observations on the ontogeny of feeding behavior in mountain gorillas (*Gorilla gorilla beringei*). American J Primatology 8:1–10.

Weijs WA (1975) Mandibular movements of the albino rat during feeding. J Morphology 145:107–124.

Weijs WA, Brugman P & Klok EM (1987) The growth of the skull and jaw muscles and its functional consequences in the New Zealand rabbit (*Oryctolagus cuniculus*). J Morphology 194:143–161.

Weijs WA, Brugman P & Grimbergen CA (1989) Jaw movements and muscle activity during mastication in growing rabbits. Anatomical Record 224:407–416.

Weijs WA & Dantuma R (1975) Electromyography and mechanics of mastication in the albino rat. J Morphology 146:1–34.

Weijs WA & de Jongh HJ (1977) Strain in mandibular alveolar bone during mastication in the rabbit. Archives Oral Biology 22:667–675.

West-Eberhard MJ (2003) Developmental Plasticity and Evolution. Oxford: Oxford University Press.

Westneat MW & Hall WG (1992) Ontogeny of feeding motor patterns in infant rats: An electromyographic analysis of suckling and chewing. Behavioral Neurology 106:539–554.

Wong FSL, Elliott JC, Anderson P & Davis GR (1995) Mineral concentration gradients in rat femoral diaphyses measured by x-ray microtomography. Calcified Tissue International 56:62–70.

Wong M & Carter DR (2003) Articular cartilage functional histomorphology and mechanobiology: A research perspective. Bone 33:1–13.

Yardin M (1974) Sur l'ontogénèse de la mastication chez le lapin (*Oryctolagus cuniculus*). Mammalia 38:737–747.

Chapter 15
Mandibular Corpus Form and Its Functional Significance: Evidence from Marsupials

Aaron S. Hogue

Contents

15.1 Introduction 329
    15.1.1 Overview 329
    15.1.2 Corpus Form and Diet 330
15.2 Materials and Methods 336
    15.2.1 Sample 336
    15.2.2 Data 340
    15.2.3 Controlling for Marsupial Phylogeny 341
    15.2.4 Statistical Analyses 342
15.3 Results 344
    15.3.1 Corpus I_x Results 344
    15.3.2 Corpus K Results 346
15.4 Discussion 348
    References 352

15.1 Introduction

15.1.1 Overview

The pioneering work of Hylander on primate craniofacial form and function greatly influenced many investigators, working on a wide array of mammals. This broader taxonomic perspective must be considered when honoring his legacy of technical and theoretical innovation. This chapter seeks to broaden the perspective with respect to one component of the craniofacial complex, the mandibular corpus (i.e., the molar bearing portion of the dentary).

    The connection between mandibular corpus form and loading during biting and chewing in extant primates has been considered by numerous investigators (Hylander, 1979a, b, 1984, 1985; Bouvier and Hylander, 1981, 1996; Demes et al.,

---

A.S. Hogue
Department of Biological Sciences, Salisbury University, 1101 Camden Ave., Salisbury, MD 21801
e-mail: ashogue@salisbury.edu

1984; Bouvier, 1986a, b; Daegling, 1989, 1992; Daegling and Grine, 1991; Ravosa, 1991, 1996, 2000; Anapol and Lee, 1994; Takahashi and Pan, 1994; Pan et al., 1995; Anton, 1996). This interest in corpus shape and function is not restricted to primates. Other mammalian clades in which this topic has been explored include Artiodactyla (Spencer, 1995; Hogue and Ravosa, 2001), Chiroptera (Freeman, 1979, 1981, 1984, 1988, 2000), and Carnivora (Radinsky, 1981a, b; Biknevicius and Ruff, 1992a, b). Given the intimate interaction between the mammalian feeding apparatus and the foods it processes, by examining the degree to which form follows function, we have a unique opportunity to study adaptation and potentially obtain valuable tools for the reconstruction of feeding behaviors in extinct species. The focus of the present chapter is to describe what is known about the relationship between diet and mandibular corpus form in placental mammals, and to examine this putative relationship in an independently evolved group (marsupials), with the hope of clarifying and expanding on previous work.

15.1.2 Corpus Form and Diet

The primary function of the mammalian mandible is to convey the forces generated by the masticatory muscles to the food via the teeth. As such, the shape of the mandibular corpus is primarily important for ensuring that adductor forces are transmitted without being dissipated or causing the mandible to fail structurally (Hylander, 1979b). Mandibular form is related to diet through the magnitude and frequency of adductor muscle forces recruited to bite and chew food objects. The greater the forces needed to fracture foods (or their protective structures), and the more often such forces must be generated (e.g., through repetitive chewing), the stronger the mandible must be to maintain its structural integrity. Foods that impose particularly high demands in this regard fall into roughly two potentially overlapping categories: (a) those that are especially hard, strong, or tough, and (b) those that must be consumed in large quantities and require extensive repetitive cyclical chewing bouts in order to finely subdivide them (all material properties terms used here follow the definitions of Strait, 1997).

Mastication of foods that are especially strong, hard, or tough often results in greater recruitment of working-side (WS) and especially balancing-side (BS) jaw-adductor forces (Hylander and Crompton, 1986; Hylander et al., 1992, 1998, 2000; Hylander and Johnson, 1997). This, in turn, may result in elevated parasagittal bending of the BS corpus, and elevated axial torsion of the WS corpus (Hylander, 1979a, b, 1985). All else held equal (such as incisor size and shape), incisal biting of such foods poses similar, though more extreme, demands on the mandible. In vivo bone strain data in primates indicate that the mandibular corpus is loaded bilaterally in axial torsion and parasagittal bending during incision (Hylander, 1979a). The magnitude of these loads, however, is greater during incision (vis-à-vis *unilateral mastication* of similar items) due to a longer load-arm and elevated activity of the masseter and anterior temporalis muscles (Hylander and Johnson, 1985).

Consequently, incision of highly resistant foods (or their supporting structures – as in tree gouging) exposes the mandible to extensive parasagittal bending and axial torsion loads, presumably requiring corpora better able to resist these loads (though not all tree-gouging primates show relatively larger external corpus dimensions than non-gouging relatives – Vinyard et al., 2003).

The overall significance of biting and chewing resistant foods for mandibular form can be seen most clearly in the studies of mandibular development in animals fed different diets. Specifically, research on primates and marsupials has shown that durophagy (i.e., a hard, strong, or tough diet) elicits greater cortical bone modeling and remodeling in the corpora of growing animals when compared to control subjects fed a non-resistant diet (Bouvier and Hylander, 1981, 1996; Lieberman and Crompton, 1998). Foods that fall into this category include nuts, seeds, bone, certain fruits, and wood (Currey, 1984; Vogel, 1988;Lucas, 1989; Lucas and Corlett, 1991; Lucas et al., 1997). Therefore, species that regularly bite or chew these items are predicted to possess corpora that are better able to resist torsion and parasagittal bending.

The second category of foods thought to require stronger mandibles may generally be described as low-nutrient foods. Resources that are extremely low in available nutrients constrain the animal to masticate more often to meet daily energy needs. This is especially true of foliage. Although only moderately tough (Atkins and Vincent, 1984; Lucas and Pereira, 1990; Vincent, 1990), leaves consist mostly of water (63–92%), and most of the remaining dry matter (30–60%) is in the form of cellulose, which is difficult to digest (Casimir, 1975; Milton, 1979, 1980). Moreover, leaves also contain secondary compounds that decrease nutrient and energy return by limiting digestibility or requiring their mammalian consumers to expend additional energy on detoxification (Williams, 1969; Feeny, 1970; Freeland and Janzen, 1974; Ryan and Green, 1974). These characteristics of foliage require folivores to consume large quantities in order to meet energy requirements. When one considers that they must also finely subdivide these foods through repeated, precise chewing events, folivory appears to demand a vast number of masticatory loading cycles per day (Hylander, 1979b; Ravosa and Hylander, 1994). While this has yet to be demonstrated empirically, the fact that folivores such as cows are known to engage in as many as 51,000 chewing strokes in a given day is consistent with this contention (Stobbs and Cowper, 1972). To the extent that the chewing of apple skins, endive lettuce, or celery leaves is similar to the mastication of leaves consumed by mammalian folivores, in vivo bone strain experiments on primates fed these foods suggest foliage consumption should also be associated with relatively high peak corpus bone strains during occlusion (Hylander, 1979a).

Frequent chewing is considered important for mandibular form because repetitive cyclical loading causes bones to fail at stress magnitudes well below that of a single, comparable loading event (based largely on the data from wet bones loaded in bending – Evans and Lebow, 1957; Morris and Blickenstaff, 1967; Lafferty et al., 1977), and is thought to be a major factor undermining the structural integrity of the mandible (Hylander, 1979b). Given the link between folivory and recurrent parasagittal bending and axial torsion of the corpora, as well as high-peak corpus strains during occlusion, leaves are thought to impose greater mechanical demands on the mandible than high-nutrient resources of limited strength, hardness, and/or toughness (Hylander, 1979b; Bouvier, 1986a; Ravosa, 1996). As such, species feeding on large proportions of foliage should possess corpora that are relatively strong in axial torsion and parasagittal bending.

A third category of "resistant" food not mentioned above is mobile vertebrate prey. In contrast to folivores, carnivores (used here to refer to all eutherian and metatherian mammals that eat vertebrates) do not chew or bite with great frequency. Carnivores go long periods without feeding, a situation made possible by the high concentration of nutrients available from each kill. This results in a low average number of mandibular loading events per day. However, there is reason to believe that the act of killing vertebrates nonetheless imposes extreme corpus torsional loads. Specifically, carnivores typically employ deep, piercing canine bites to restrain, shake, and kill their prey (Biknevicius and Van Valkenburgh, 1996). As a result, large, laterally directed forces are applied to the tall lower canines. Since most carnivores typically have limited mobility of the TMJ, this lateral twisting motion is resisted by the masseter and TMJ, presumably leading to very high torsional loads (Biknevicius and Van Valkenburgh, 1996). If so, highly carnivorous taxa should have relatively strong corpora against axial torsion.

Evidence for the proposed links between diet and corpus form in mammals is decidedly mixed. The ability to predict form from function by these biomechanical models is variable, both within and between clades. Predictive power also varies by diet and the specific measure of corpus strength used. However, the type of data used to evaluate these relationships varies considerably among studies and this alone may account for at least some of the differing results. Therefore, a consideration of both the methods and results of previous studies is in order.

One factor that may account for some of this variation is the specific measure used to assess corpus strength. The two variables thought to most accurately reflect the parasagittal bending and axial torsional strength of a corpus are the second moment of inertia about a horizontal axis (I_x) and Bredt's formula (K), respectively (Demes et al., 1984; Daegling and Grine, 1991; Biknevicius and Ruff, 1992a, b). Put simply, I_x is a measure of the amount of cortical bone in a given section and its distribution about the neutral axis (Hylander, 1984, 1985; Biknevicius and Ruff, 1992a, b). Since parasagittal bending occurs largely in a single plane, I_x gives greater weight to mandibular dimensions (e.g., total diameter, cortical bone thickness) in the plane of bending compared to out-of-plane dimensions (Fig. 15.1a). In contrast to bending, torsional strength is best enhanced by distributing cortical bone in a more uniform, spherical manner. Consequently, maximization of K is achieved by: (a) increasing the circularity of the section, (b) increasing the radius of the circle, (c) increasing the thickness of the cortical bone (particularly where it is thinnest), and (d) distributing the cortical bone uniformly throughout (Fig. 15.1b; Demes et al., 1984; Daegling and Grine, 1991; Daegling and Hylander, 1998).

Although I_x and K appear to be the most appropriate estimates of corpus strength under their respective loading regimes, the necessity of having data on cortical bone distribution to compute these has forced many investigators to use more accessible

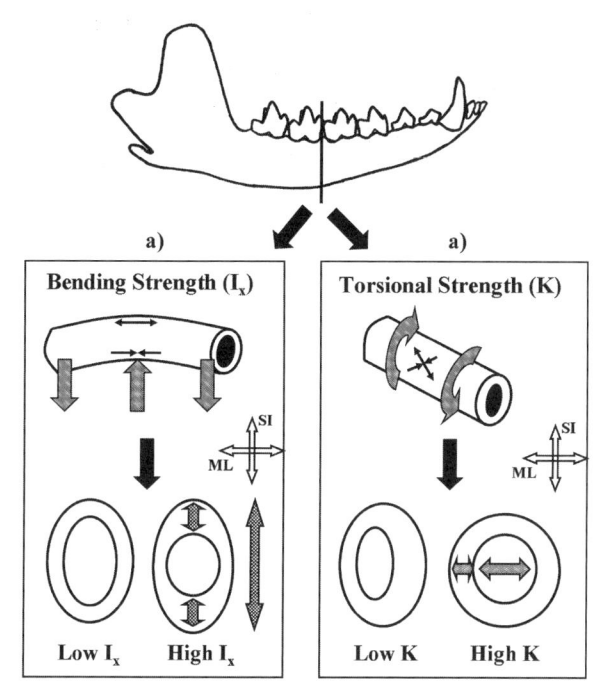

15 Mandibular Corpus Form and Its Functional Significance

Fig. 15.1 Schematic depiction of the principle loading regimes experienced by the mandibular corpus. Each box depicts the corpus as a hollow beam. Below each beam is a pair of simplified corpus cross-sections (e.g., from the M_3/M_4 interdental gap of the mandible above). (**a**) Generally speaking, the strength of a hollow beam against bending is reflected in its second moment of inertia (I). For a given amount of bone, "I" is increased by expanding both the total diameter and the thickness of a section's walls in the plane of bending (checkered arrows). Strength against parasagittal bending (i.e., bending about a horizontal (x) axis) is represented as I_x. (**b**) The strength of the corpus against torsion is calculated using Bredt's formula (K). "K" can be increased in three ways: by increasing the thickness of bone in the thinnest part of the corpus wall, by increasing the diameter of the corpus where its diameter is smallest (wavy-line arrows), and by achieving a more uniform circular cross-section. Key: The vertical line drawn through the corpus represents the location where cross-sections were measured in this study. Stick arrows indicate the location of tension (arrows pointing away from each other) and compression (arrows pointing toward one another) in the walls of the tube or corpus. Diagonally dashed arrows represent generic bending (Fig. 15.1a) and torsional (Fig. 15.1b) loads applied to the tube or corpus. Open arrows indicate the superoinferior (SI) and mediolateral (ML) directions of corpus cross-sections

external proxy measures. In the case of parasagittal bending strength, the proxy most often used is corpus depth (e.g., Hylander, 1979b; Bouvier, 1986a; Ravosa, 1991, 1996; Anapol and Lee, 1994; Anton, 1996). The rationale for this is that the single biggest factor determining the strength of a beam against bending in a given plane is its dimensions in that plane (Hylander, 1979b). Since corpus depth is a direct measure of the maximum external diameter in the plane of bending, differences in this variable are assumed to closely track differences in I_x.

The proxy used for estimating torsional strength follows a similar logic. Specifically, one of the major factors determining the value of K is how closely the cross-section of a mandible approximates a circle (Demes et al., 1984). Given that nearly all mammalian mandibles are at least partly elliptical, and given that the major axis of this ellipse is oriented vertically, one way of increasing the circularity of the corpus is to increase its width. As such, corpus width is generally assumed to be positively correlated with K, and ultimately torsional strength (Hylander, 1979b; Bouvier, 1986a; Ravosa, 1991, 1996; Anapol and Lee, 1994; Anton, 1996).

While external corpus dimensions may reveal important details of jaw function, complete cross-sectional geometry is considered ideal in assessing mechanical design of the corpus (Hylander, 1979b; Daegling, 1989). Thus, where analyses of corpus form using external dimensions fail to uncover the predicted relationship, it is difficult to know whether this is due to an absence of the proposed relationship or the inability of the proxy to uncover it. For example, research on cercopethecids found that the more folivorous colobines have deeper (but not significantly wider) corpora (at a common mandibular length) than the more frugivorous cercopithecines, as expected (Bouvier, 1986a; Ravosa, 1996). Yet similar work on platyrrhines failed to uncover this relationship (Bouvier 1986b; Anapol and Lee, 1994).

The question then arises, is the absence of the proposed relationship in platyrrhines due to an actual lack of such a relationship, a lack of sufficiently detailed morphological data, or some other factor? The fact that platyrrhine seed predators typically do display relatively deeper and wider corpora, as predicted, would seem to suggest that these proxy measures are sufficient to uncover certain ecomorphological associations (Anapol and Lee, 1994). That hard-object feeding is detectable from external corpus form is further confirmed by work on macaques, where the more durophagous *Macaca fuscata* (Japanese macaque) has relatively wider (but not deeper) corpora than the congeners examined (Anton, 1996). Similarly, in a comparison of two sympatric colobines, Daegling and McGraw (2001) found that the apparently more durophagous *Colobus polykomos* (western black-and-white colobus) displays absolutely and relatively deeper corpora than *Procolobus badius* (western red colobus). More durophagous Bornean orangutans (*Pongo pygmaeus*) also have relatively deeper corpora than less durophagous Sumatran orangutans (*Pongo abelii*) (Taylor, 2006). This raises the possibility that it is merely folivory that does not require relatively stronger corpora, and the finding of stronger corpora in folivorous cercopithecids is related to some other clade-specific factor.

Such a conclusion seems unlikely in light of the fact that even within cercopithecine genera, more folivorous species typically display relatively deeper and wider corpora than their less folivorous congeners (Takahashi and Pan, 1994; Ravosa, 1996). Similarly, among gorillas, the more folivorous *Gorilla gorilla beringei* (mountain gorilla) displays relatively wider (but not deeper) corpora than its congener *G. g. gorilla* (western lowland gorilla) (Taylor, 2002). In fact, apes as a whole show positive allometry of corpus dimensions, which is argued to be linked to increases in foliage and other resistant foods in the diets of larger bodied taxa (Ravosa, 2000). Work on bovids even suggests that differences in the types of foliage consumed are reflected in corpus dimensions. Specifically, grazers feed on larger quantities of more fibrous, lower-quality foliage than dicot folivores, presumably placing greater demands on the mandible (Spencer, 1995). Consistent with their more demanding diet, grazers also possess relatively deeper corpora than other folivores (Spencer, 1995). Thus, not only does folivory appear to require more robust mandibles but external corpus measures also appear to be generally capable of detecting this.

It should be noted, however, that the above analyses using external corpus dimensions did not universally reveal the predicted findings. In many cases, only corpus width differed predictably between diet classes (e.g., Anton, 1996; Taylor, 2002), and even in cases where differences in corpus depth were detected, not all species conformed to expectations (Takahashi and Pan, 1994). The lack of a predicted association between external corpus form and diet is even more marked in tree-gouging primates. Specifically, in comparisons of gouging versus non-gouging cheirogaleids, galagids, and callitrichids, only gouging galagids showed relatively deeper and wider corpora than largely non-gouging members of the same family (Williams et al., 2002; Vinyard et al., 2003; similar results were obtained from cross-sectional data – Vinyard, pers. comm.). Optimism about the value of external measures is further diminished when one compares the results of such analyses with those using complete cross-sectional data. For example, *Cebus apella* (the brown capuchin) differs from its congener *C. capucinus* (the white-throated capuchin) in that it includes hard objects in the diet (Daegling, 1992). Consistent with this, analyses of CT scans of the corpora in these two species uncovered significantly larger I_x and K values in the former (Daegling, 1992). However, when only external dimensions are examined, the corpora of *C. apella* are not deeper (absolutely or relative to mandibular length) and appear only slightly relatively wider (Bouvier, 1986b). Taken together, these findings suggest the lack of full cross-sectional data in comparative analyses of corpus shape may leave out critical information for detecting functional patterns, leading to inconsistent results.

Another factor underlying such inconsistent findings is the variable employed for size adjustment. Using body mass in this capacity, Smith (1983) found no functionally significant variation in corpus depth or width in anthropoids. Similarly striking, in a comparison of bats, primates, and marsupials, Dumont (1997) found tree gougers to actually have corpora that were relatively *less* deep than nectarivores (and frugivores). Here again, corpus dimensions were size-adjusted using a geometric mean of skull dimensions. Such measures fail to control for the moment arm acting about a given corpus section. In bending, the stress present at a given corpus section is a product of the moment arm length and magnitude of the applied force. Since longer mandibles have longer moment arms, the corpora in species with long mandibles experience higher bending moments for a given adductor force than corpora in species with short mandibles (Hylander, 1985; Bouvier, 1986a, b; Daegling, 1989; Ravosa, 1991; Biknevicius and Ruff, 1992a). Consequently, when attempting to identify relative differences in mandibular strength, it is essential to control for moment arm lengths.

A third factor that may account for some of the variability in comparative functional analyses of corpus dimensions is the quality of available diet data. Until recently, the diets of few species had been evaluated quantitatively. This has often forced morphologist to compare species allocated to various diet categories based on inconsistent, subjective criteria. Hence, it is not always clear how much of a species actual diet is made up of foods constituting its designated diet category. Variation in food material properties, in addition to percentages of different foods in the diet, could explain why, for example, some folivorous strepsirrhines possess relatively robust corpora, while others do not (Ravosa, 1991). With more detailed diet information, it should be possible to greatly reduce this source of error.

One final confounding input, not typically controlled for in previous studies, is phylogeny. A common assumption of standard statistical tests is that individual data points are independent of one another. Clearly, a sample of multiple related species, each of which shares varying amounts of its evolutionary history with others in the sample, violates this assumption. The net effect is that variation in a sample will likely be biased by the particular pattern of shared ancestry. When this lack of independence and its associated bias are ignored, the degrees of freedom will be overestimated, thereby increasing the risk of Type I error. Thus, while studies of bats indicate hard-object insectivores have relatively wider corpora than soft-object insectivores and carnivores (Freeman, 1981, 1984), and studies of carnivorans (i.e., members of the order Carnivora) have linked an emphasis on powerful killing bites to corpora that are deeper, wider, and have larger second moments of inertia (Radinsky, 1981a; Biknevicius and Ruff, 1992), such findings would carry greater force if obtained in a phylogenetically controlled framework.

Although the weight of the evidence at present suggests mammalian corpus shape is at least partly related to diet, the above noted problems raise questions about the strength of this relationship. Moreover, as this relationship is incompletely explored in non-primates (especially marsupials), it is unclear whether the findings for primates are typical of other mammals. This investigation aims to address these outstanding issues by testing these predictions in an ecologically diverse group of marsupials with well-known diets while controlling for phylogeny.

15.2 Materials and Methods

15.2.1 Sample

The sample consists of 560 adult individuals from 65 species and 44 genera of marsupials (Table 15.1). Where possible, each species is represented by an equal sex sample of 10 individuals (median and mean numbers per species are 10 and 8.6, respectively). All seven extant marsupial orders are represented, as are 15 of 16 nonmacropodoid families (Table 15.1). Macropodoids (kangaroos and their closest relatives) were excluded because they possess masseters that extend into the mandibular corpora (personal observation), making a direct comparison with other marsupials unsound. Species were included in this study based solely on the availability of adequate diet information (i.e., the proportions of major foods had been

15 Mandibular Corpus Form and Its Functional Significance

Table 15.1 Marsupial sample, body mass, dietary classification, and number of specimens

Sample	Body mass (g)[1]	Diet % (by category)[2]	Number of specimens[3]
DIDELPHIMORPHIA			
Caluromyidae			
Caluromys philander	189	Fr/E=90; I=10	M=5, F=5
Didelphidae			
Chironectes minimus	665	V=90; I=10	M=5, F=5
Didelphis aurita	1050	I=50; Fr=40; V=10	M=2, F=2
Didelphis marsupialis	1433	V=45; I=15; Fr=15; P=10	M=5, F=5
Didelphis virginiana	2208	V=35; I=35; Fr=20	M=5, F=5
Lutreolina crassicaudata	537	V=70; I=20; Fr=10	M=2, F=4, U=2
Marmosa murina	45	I=66; Fr=33	M=5, F=5
Marmosa robinsoni	60	I=66; Fr=33	M=5, F=5
Metachirus nudicaudatus	259	I=80; V=15; Fr=5	M=5, F=5
Micoureus demerarae	105	I=80; Fr=20	M=5, F=5
Monodelphis dimidiata	58	I=80; V=20	M=4, F=4
Philander opossum	45	I=85; Fr=10; V=5	M=5, F=5
Thylamys elegans	29	I=90; Fr/S≥5	M=5, F=5
PAUCITUBERCULATA			
Caenolestidae			
Caenolestes fuliginosus	24	I=86; Fr=12	M=4, F=3
Rhyncholestes raphanurus	28	I=55; L=30; Fr/S/M=15	M=5, F=5
MICROBIOTHERIA			
Microbiotheriidae			
Dromiciops gliroides	22	I=70; L=20; Fr=5	M=5, F=5
PERAMELINA			
Peramelidae			
Isoodon macrourus	1438	I=36; Fr/R/S/M=33; P=21	M=5, F=5
Isoodon obesulus	775	I=45; Fr/R/S/M=33; P=22	M=5, F=5
Perameles gunnii	717	I=73; Fr/R/S/M=27	M=5, F=5
Perameles nasuta	1025	I=62; P~15; Fr/R/S/M~13; V=10	M=5, F=5
Thylacomyidae			
Macrotis lagotis	1277	I=57; Fr/S/R=32; P=11	M=2, F=3, U=3
DASYUROMORPHIA			
Dasyuridae			
Antechinus agilis	24	I=95	M=5, F=5
Antechinus flavipes	45	I=95	M=5, F=5
Antechinus swainsonii	53	I=92; Fr=2	M=5, F=5
Dasycercus cristicauda	100	I=75; V=25	M=4, F=3
Dasyurus geoffroii	1100	I=44; V=38	M=2, F=2, U=2
Dasyurus maculatus	5500	V=98; I=1; P=1	M=5, F=2, U=1
Dasyurus viverrinus	1,090	V=51; I=26; Fr/S/L=23	M=5, F=5
Parantechinus apicalis	70	I≥95	F=3
Phascogale tapoatafa	193	I=88; V=10; E=2	M=5, F=5

Table 15.1 (continued)

Sample	Body mass (g)[1]	Diet % (by category)[2]	Number of specimens[3]
Sarcophilus harrisii	8000	V=98	M=5, F=5
Sminthopsis crassicaudata	15	I≥95	M=5, F=5
Sminthopsis leucopsis	28	I=99	M=1, F=1, U=1
Sminthopsis murina	17	I≥95	M=3, F=4, U=1
Myrmecobiidae			
Myrmecobius fasciatus	472	A=99; P=1	M=3, F=4, U=1
Thylacinidae			
Thylacinus cynocephalus	25,000	V≥95	M=3, F=2, U=4
NOTORYCTEMORPHIA			
Notoryctidae			
Notoryctes typhlops	54	A=94; S=6	M=2, F=3
DIPROTODONTIA			
Acrobatidae			
Acrobates pygmaeus	14	M=50; E=40; P=6; I=4	M=4, F=5
Burramyidae			
Burramys parvus	45	I=63; Fr/Fl/S/N=19; P=11	M=1, F=3, U=3
Cercartetus caudatus	20	I=71; Fr/S=27; L/Fl=2	M=5, F=5
Cercartetus nanus	26	N=40; I=40; Fl/S/Fr=20	M=5, F=5
Petauridae			
Dactylopsila trivirgata	423	Ig=78; Eg=20	M=5, F=5
Gymnobelideus leadbeateri	112	Eg=95; I=5	M=2, F=2, U=2
Petaurus australis	555	Eg=84; I=14	M=3, F=2, U=1
Petaurus breviceps	132	Eg=97; I=3	M=5, F=5
Petaurus norfocensis	230	Eg=70; I_l=26; Fr=3	M=5, F=5
Phalangeridae			
Ailurops ursinus	8500	L=85; Fr=5	M=3, F=4
Phalanger carmelitae	2148	L=82; Fr/Fl=8	M=5, F=5
Phalanger gymnotis	3063	Fr~55; L~45	M=5, F=5
Phalanger sericeus	2143	L=96.5; Fr=3.5	M=3, F=5
Phalanger vestitus	1926	L~75; Fr~25	M=3, F=2, U=3
Trichosurus vulpecula	2313	L=63; Fl/E=20; Fr/S=16	M=5, F=5
Phascolarctidae			
Phascolarctos cinereus	10,250	L≥95	M=5, F=5
Pseudocheiridae			
Hemibelideus lemuroides	953	L=94; Fr/Fl=4	M=5, F=5
Petauroides volans	1103	L=100	M=5, F=5
Pseudocheirus peregrinus	612	L=98; Fl=1	M=5, F=5
Pseudochirops archeri	1192	L=91; Fr=7	M=3, F=1, U=2
Pseudochirops cupreus	1707	L=85; Fr=5	M=5, F=5
Pseudochirulus canescens	300	L=95	M=2, F=2
Pseudochirulus herbertensis	1115	L=94; Fr/Fl=5	M=3, F=3
Pseudochirulus mayeri	152	L~80; Fr/M/N~20	M=3, F=3
Pseudochirulus forbesi	596	L=99; I=1	M=5, F=5

Table 15.1 (continued)

Sample	Body mass (g)[1]	Diet % (by category)[2]	Number of specimens[3]
Tarsipedidae			
Tarsipes rostratus	9	N≥95	M=1, F=2, U=1
Vombatidae			
Lasiorhinus latifrons	25,000	L≥95	M=5, F=5
Vombatus ursinus	26,000	L≥95	M=5, F=5

1. See Hogue (2004) for body mass references.
2. Diet category abbreviations key: **A** = **A**nts and/or Termites; **E** = **E**xudates (includes one or more of the following: sap, gum, honeydew, nectar, pollen); **Fl** = **Fl**ower; **Fr** = **Fr**uit; **I** = **I**nsects; **L** = Foliage (**L**eaves); **M** = Fungi (**M**yco); **N** = **N**ectar and/or Pollen; **P** = Miscellaneous **P**lant Material (typically fruit, flower parts, and/or plant vascular tissue); **R** = **R**oots, Tuber, and/or Bulbs; **S** = **S**eeds and/or Nuts. **V** = **V**ertebrate Flesh. Foods obtained by gouging are indicated by the subscript "**g**." Where percentages for two or more food types were lumped together in a study, they are listed together here, separated by a "/." Note: not all species have diet percentages adding to 100%. This is typically due to the fact that some items consumed by a given species could not be adequately identified by the investigator. See Hogue (2004) for diet references.
3. Sex abbreviations key: F = female, M = male, U = unknown sex.

quantified or could be inferred from an analysis of stomach contents, fecal contents, or systematic behavioral observations).

Only wild caught specimens with largely intact skulls and dentitions were included. Individuals were included in the sample only if M_4 was halfway to fully erupted with little wear and M_3 was fully erupted with light wear. Due to the rapid wear of the thin occlusal enamel in *Lasiorhinus latifrons* (southern hairy-nosed wombat) and *Vombatus ursinus* (common wombat), the latter criterion was not employed for these taxa.

Marsupials were chosen to study for two primary reasons. First, marsupials are an inherently interesting clade about which little is known regarding the relationship between corpus form and feeding habits. As predictions about this relationship have been largely explored in placental mammals, it is possible to gain a better understanding of the strength of this association by testing predictions in this ecologically diverse, independent radiation of therian mammals. Second, for understanding the interface between morphological evolution and feeding ecology in primates, no clade has proven to be a more valuable independent test than marsupials (Walker, 1967; Cartmill, 1972, 1974a, b, 1992; Kay and Cartmill, 1977; Kay and Hylander, 1978; Rasmussen, 1990; Strait, 1993; Dumont, 1997; Lemelin, 1999). Their utility in this regard stems largely from the numerous remarkable parallels in diet, foraging habits, and morphology between the two groups. Not only do marsupials overlap primates in most general aspects of habitat use, including the extensive exploitation of arboreal environments, but they also include among their ranks a variety of food resource specialists that are either rare or absent outside Marsupialia and Primates. Examples include terminal branch feeding insectivores and frugivores, arboreal folivores, tree-gouging and opportunistic exudativores, and even

Daubentonia analogs (Charles-Dominique, 1983; Nash, 1986; Rasmussen, 1990; Cartmill, 1972, 1974a, b). These striking similarities in resource utilization are also accompanied by numerous instances of craniodental and postcranial morphological convergence (Walker, 1967; Cartmill, 1974b; Kay and Cartmill, 1977; Kay and Hylander, 1978; Strait et al., 1990; Strait, 1993; Dumont, 1997; Lemelin, 1999). Thus, given these extraordinary parallels in diet and morphology, marsupials provide a unique opportunity to test ecomorphological predictions of the sort examined here and ultimately to help refine our understanding of the functional significance of craniodental variation in primates and other mammals.

15.2.2 Data

Biplanar radiography was used to reconstruct corpus cross-sectional shape (Biknevicius and Ruff, 1992a, b). This approach was chosen because it provides a good estimate of the actual cross-sectional shape of corpora (Biknevicius and Ruff, 1992a, b) and permits the quantification of bone cross-sections in small mammals (Runestad et al., 1993).

External corpus dimensions (used in calculating corpus cross-sectional shape) were corpus depth (maximum height of the corpus at the M_2/M_3 interdental gap) and corpus width (maximum width of the corpus, orthogonal to corpus depth, at M_2/M_3). Both were obtained with digital calipers accurate to 0.03 mm. The M_2/M_3 interdental gap was chosen because it is a representative corpus section in the posterior molar region (where most previous functional analyses of corpus form were focused), and it consistently displays a smooth cylindrical shape in the taxa considered (necessary to model the corpus as hollow elliptical beam for biomechanical analyses). Similar results were obtained in analyses of the M_3/M_4 interdental gap. However, results for the M_2/M_3 interdental gap are presented here, because some taxa were observed to have masseter muscle scars and other modifications of the corpus in the M_3/M_4 region that caused it to deviate more from a cylindrical shape (which is assumed for the biomechanical models used in these analyses).

Mandibular length, which served as the size measure for regression analyses, was measured with digital calipers from infradentale to the midpoint of the mandibular condyle. Mandibular length was selected as the size measure, both because it serves as a rough control for the length of the moment arm in mandibular bending and because most previous comparative studies of mandibular corpus strength used this variable, ensuring comparability with this earlier work. Alternative moment arm estimates (e.g., infradentale to M_2/M_3 and M_2/M_3 to the mandibular condyle) have also been used (e.g., Daegling, 1992), but their applicability depends on where along the toothrow bite forces are being applied, among other things. While less than perfect, mandibular length provides a useful compromise as a size measure when evaluating corpus strength under a wide variety of feeding behaviors.

For internal corpus measurements, lateral and ventral radiographs of the right dentary of each specimen were taken from 70 cm with a MinXray X750G portable X-ray unit. Kodak X-OMAT AR Ready Pack film was exposed at 75 kV and

10 mA for 2–4 s (depending on specimen size) and developed in a Kodak X-OMAT automatic wet developer. Cortical bone thickness was measured along the buccal, lingual, and ventral margins at the M_2/M_3 interdental gap from illuminated radiographs using a 7X minicomparater accurate to 0.05 mm. As trabecular bone obscured the borders of the alveolar cortex, cortical thickness in this region was estimated as a constant fraction (2.5%) of mandibular length.

The second moment of inertia about a labiolingual axis (I_x) and Bredt's formula (K) were calculated from cortical bone thickness, corpus depth, and corpus width using formulas derived from the hollow asymmetrical model of Biknevicius and Ruff, (1992a, b) (see Hogue, 2004 for precise formulas used in calculating these variables). This model provides a good estimate of corpus cross-sectional properties and is among the most accurate in predicting in vitro strain gradients (Daegling and Hylander, 1998).

Data on the feeding habits of each species were obtained from the literature, and are summarized in Table 15.1 (see Hogue, 2004 for further details). In order to test predictions linking corpus I_x and K to diet, it was necessary to compute (for each species) the percentage of the diet composed of foods thought to exhibit specific material properties. Based on limited available experimental work on actual food material properties, foods considered hard, tough, and/or strong include seeds, nuts, bone, and wood (i.e., foods obtained by tree gouging) (Currey, 1984; Vogel, 1988; Lucas, 1989; Lucas and Corlett, 1991; Lucas et al., 1997). Based on the arguments outlined in the introduction, the class of foods thought to require extensive amounts of daily chewing, and repetitive cyclical loading of the mandible, is foliage. For corpus I_x analyses, the percentages of these two food types in each species diet were summed, yielding the total percentage of the diet made up of foods linked to significant parasagittal bending of the corpus. For corpus K analyses, the percentage of vertebrates in the diet was added to this amount, giving the total percentage of foods in the diet presumably related to significant axial torsion of the corpus. While this approach to quantifying dietary material properties is clearly inferior to actually measuring the material properties of foods consumed by each species in the wild, the latter was deemed impractical in a large comparative study such as this.

15.2.3 Controlling for Marsupial Phylogeny

As stated previously, related species cannot be viewed as completely independent data points (Felsenstein, 1985; Cheverud et al., 1985). Numerous phylogenetic control methods (PCMs) are now available to address the issue of non-independence (Cheverud et al., 1985; Felsenstein, 1985; Huey and Bennett, 1987; Grafen, 1989; Lynch, 1991; Smith, 1994; Martins and Hansen, 1997; Diniz-Filho et al., 1998). Each confronts the problem differently and alters Type I error to varying degrees (Martins and Garland, 1991; Martins, 1996; Martins et al., 2002). The two techniques applied here are Independent Contrasts (Felsenstein, 1985) and Phylogenetic Autocorrelation (Cheverud et al., 1985). These differ markedly in how non-independence is addressed, yet perform well at estimating statistical parameters under different models of evolution (Martins and Garland, 1991; Garland et al., 1993; Martins, 1996; Martins et al., 2002).

Phylogenetic Autocorrelation (PA) accounts for phylogeny by decomposing trait values into phylogenetic and phylogeny-free components (Cheverud and Dow, 1985; Cheverud et al., 1985; Gittleman and Kot, 1990; Gittleman and Luh, 1994). Trait decomposition is achieved using a network autocorrelation procedure that expresses a trait's value as a partial function of the species' phylogenetic history. By extracting the portion of a trait's value explained by the phylogeny, the remaining phylogeny-free residual error component can be viewed as statistically independent between species and therefore used to test comparative hypotheses.

The Independent Contrasts (IC) technique (Felsenstein, 1985) differs from the previous approach in that it takes as its starting values the species means for the trait of interest and simulates character evolution down the phylogeny (from the tips to the ancestral node) using Brownian motion. It then computes "contrasts" along each branch (between nodes) to generate a total of $N-1$ contrasts for each trait that are independent of the particular pattern of phylogenetic relationships. These contrasts can then be evaluated using standard statistical procedures (Felsenstein, 1985).

The phylogenetic tree used here (Fig. 15.2) was taken from Hogue (2004). This topology is based on the DNA hybridization data of Kirsch et al. (1997) with interordinal relationships constrained according to the 6.4 kb nuclear gene Bayesian phylogram of Amrine-Madsen et al. (2003). Taxa not included in the analyses of Kirsch et al. (1997) were added as described by Hogue (2004). As relative branch lengths could not be determined empirically for the composite phylogeny, they were set according to the modified speciational model of Pagel (1992). Specifically, all branches were initially set to one, and then modified slightly to ensure all branch tips terminated at the same height in the phylogenetic tree (to recognize that all species, or branch tips, in the phylogeny are extant and therefore contemporaneous).

15.2.4 Statistical Analyses

For Phylogenetic Autocorrelation analyses, phylogeny-free ("autocorrelation-adjusted" or ACA) values were computed as described above for ln corpus I_x and K, ln mandibular length, and diet proportions using COMPARE 4.4 (Martins, 2001). ACA values for ln I_x and K were regressed against ACA ln mandibular length, and Least Squares (LS) residuals were computed (Reduced Major Axis – or RMA – residuals were also computed, but are not included here since analyses of these residuals yielded virtually identical results). The level of support for each prediction was then assessed in two ways. First, the sample was divided into two groups: those with at least 50% of the indicated foods in the diet and those consuming lesser amounts. The use of 50% as the dividing point was somewhat arbitrary, but given its position in the middle of the range of dietary percentages, and the fact that it defines when a food type transitions from being a minority of the diet to a majority, it seemed a reasonable choice. The two groups were then compared with a Kruskal-Wallis One-Way ANOVA to test the prediction that species in the high-diet category

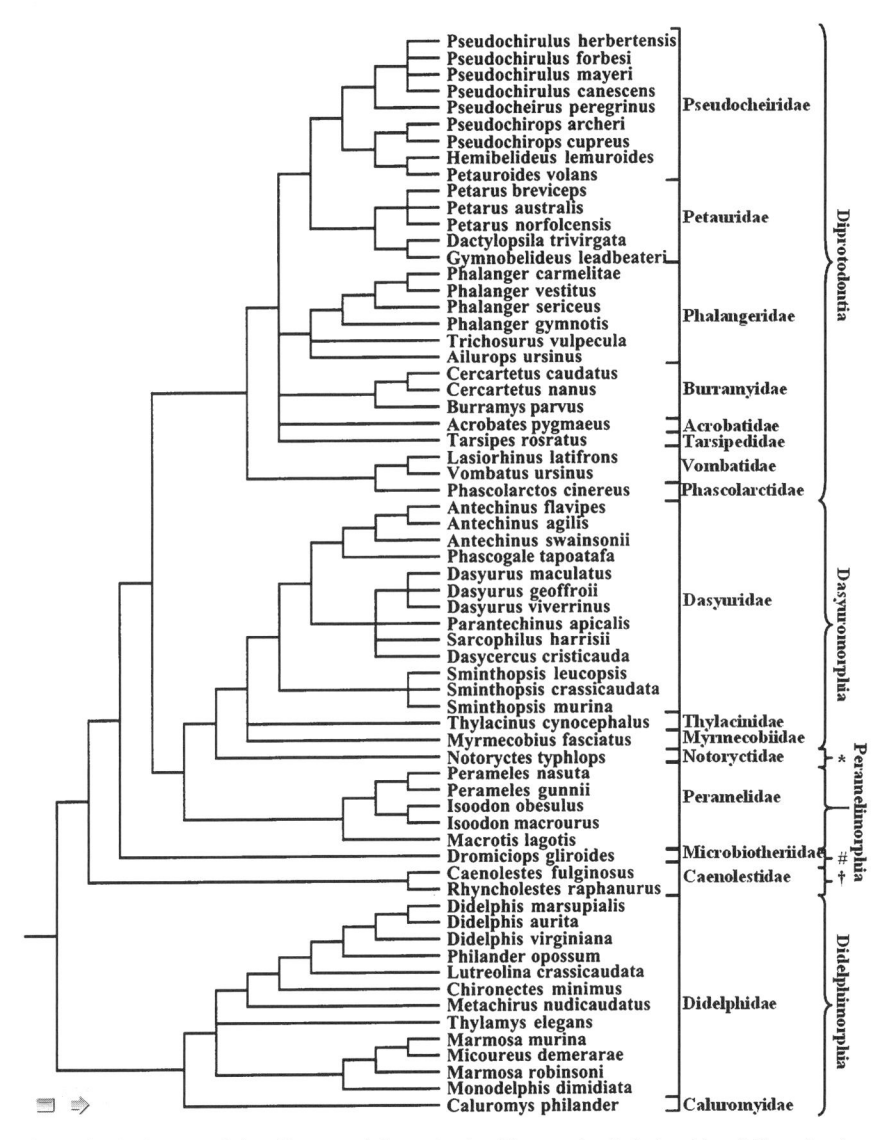

Fig. 15.2 Phylogeny of the 65 marsupial species in this sample. Relationships follow Amrine-Madsen et al. (2003) and Kirsch et al. (1997) with modifications as described in Hogue (2004). Key: * = Notoryctemorphia; # = Microbiotheria; † = Paucituberculata

would display significantly larger I_x and K residuals. This non-parametric approach was chosen over the parametric One-Way ANOVA because it provides a more conservative test of predictions. Second, Pearson's Product–Moment Correlations were computed of ACA corpus I_x and K residuals versus ACA diet proportions (i.e., the percentage of foliage, nuts, seeds, bone, hard fruits, and the products of gouging in corpus I_x analyses, and the percentage of all these plus vertebrates in corpus K analyses; $\alpha = 0.05$). Note that Spearman's Rank Order Correlations were also computed, but are not included here as they yielded similar results. The prediction is upheld if these variables exhibit the predicted significant positive correlation.

For Independent Contrast analyses, corpus I_x and K residuals were calculated relative to LS ln–ln bivariate regression lines versus mandibular length (as above, RMA residuals were also computed, but are not presented here due to the similarity of findings). Subsequently, contrasts of these residuals and dietary proportions were computed using the PDTREE module of PDAP (Garland et al., 1993, 1999; Garland and Ives, 2000). To permit the use of standard probability tables, contrasts were standardized by dividing them by their standard deviation (Garland et al., 1992). Given that the sign of a contrast depends on the (arbitrary) direction in which it was computed, standardized contrasts of I_x and K residuals were "positivized" prior to analysis (Gittleman and Luh, 1994; Garland et al., 1999; Garland and Ives, 2000). Following standard approaches to evaluating independent contrasts (Garland et al., 1992), predictions were tested by calculating Pearson's Product–Moment Correlations and LS regression lines through the origin ($\alpha = 0.05$) between contrasts of corpus residuals and diet percentages (Spearman's Rank Order Correlations, though not presented, yielded similar results in all analyses). As changes in I_x and K residuals along a branch are predicted to be positively correlated with changes in diet, a significant positive correlation and regression slope indicates the prediction was upheld.

15.3 Results

15.3.1 Corpus I_x Results

ACA ln M_2/M_3 corpus I_x is significantly positively correlated with ACA ln mandibular length ($r = 0.956$; $p < 0.001$; Table 15.2). The LS regression slope for ACA corpus I_x versus ACA ln mandibular length shows no significant allometry (slope = 0.947, 95% CI = 0.873–1.021; Table 15.2; Fig. 15.3). As predicted, a comparison of ACA corpus I_x residuals versus ACA diet proportions found these two variables to be significantly positively correlated ($r = 0.681$; $p < 0.001$; Table 15.3; Fig. 15.4). This indicates increases in the proportion of foods in the diet thought to impose large and/or repetitive loads on the mandible are indeed linked to increases in relative

Table 15.2 Regression equations for ACA *ln* corpus I_x and K versus ACA *ln* mandibular length

Dependent variable (vs. ACA ln mandibular length)	N	LS slope (95% CI)	LS intercept (95% CI)	'r'[1]
ACA ln Corpus I_x	64	0.947 (0.873–1.021)	−0.002 (−0.010, 0.006)	0.956 ***
ACA ln Corpus K	64	0.947 (0.871–1.023)	0.000 (−0.010, 0.010)	0.953 ***

[1]. Key: *** = $p \leq 0.001$.

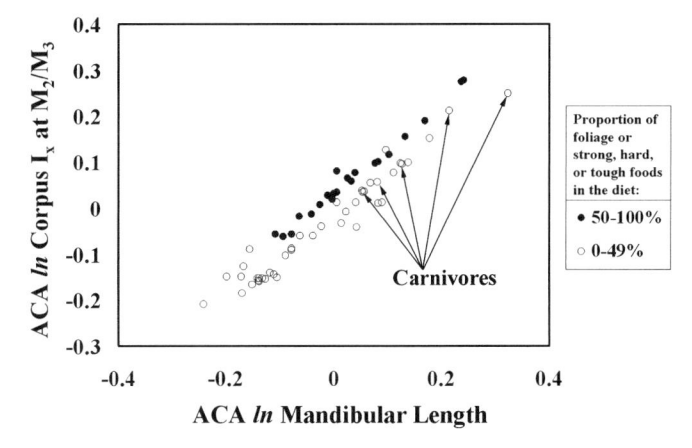

15 Mandibular Corpus Form and Its Functional Significance

Fig. 15.3 Plot of ACA *ln* M_2/M_3 corpus I_x versus ACA *ln* mandibular length. As predicted, folivores and durophages have significantly larger LS residuals than all other taxa (Table 15.2). Carnivorous species are indicated with arrows for comparison with results in Fig. 15.6

corpus vertical bending strength. This is further supported by the fact that species feeding on at least 50% foliage, gouging products, or tough, strong, hard foods consistently have significantly higher residuals than other taxa (Kruskal-Wallis One-Way ANOVA, $p < 0.001$; Table 15.4; Fig. 15.3). Thus, as predicted, the former have corpora that are stronger against vertical bending loads than the latter. This supports the proposition that gougers, folivores, and/or durophages require stronger corpora for resisting parasagittal bending because they experience higher magnitudes or frequencies of these loads during biting and chewing.

Based on the predictions outlined above, relative corpus I_x dimensions should increase along branches of the phylogeny where the proportions of foliage and resistant foods in the diet also increase (and decrease where these foods decrease). If so, contrasts of corpus I_x residuals should display significant positive correlation coefficients versus contrasts of dietary proportions. As predicted, the Pearson's correlation coefficient and LS regression slope relating contrasts of corpus I_x residuals to contrasts of diet proportions are significantly positive (slope $= 0.007$; $r = 0.310$; $p < 0.01$; Table 15.5; Fig. 15.5). This indicates that lineages in which the proportion of foliage or resistant foods in the diet increases or decreases do indeed show a corresponding increase or decrease, respectively, in relative corpus I_x.

Table 15.3 Correlation of ACA corpus I_x and K residuals with ACA diet proportions

LS Residuals (vs. Diet Percentage)	N	r	P-values
ACA Corpus I_x Residuals	64	0.681	**p< 0.001**
ACA Corpus K Residuals	64	0.478	**p< 0.001**

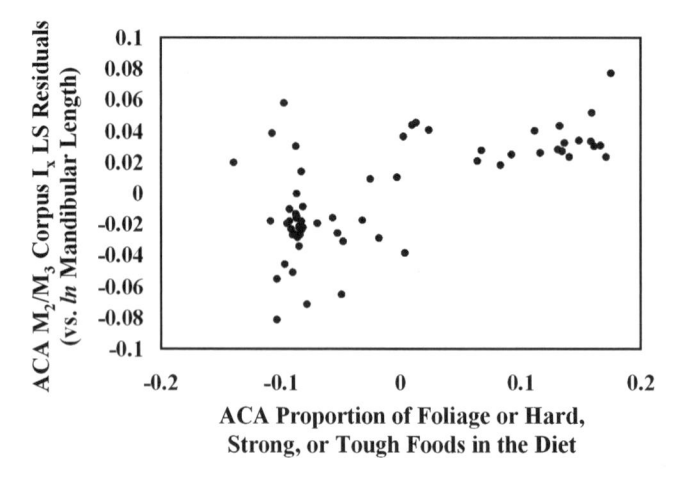

Fig. 15.4 Plot of ACA M$_2$/M$_3$ corpus I_x LS residuals (covariate = ACA ln mandibular length) versus the proportion of foliage or hard, strong, or tough foods in the diet. As predicted, these two variables are significantly positively correlated (Table 15.3)

Table 15.4 Kruskal-Wallis One-Way ANOVA results contrasting ACA corpus I_x and K residuals between diet categories

LS Residuals	Resistant foods > 50% of Diet N Mean (SD)	Resistant foods < 50% of Diet N Mean (SD)	P values
ACA Corpus I_x Residuals	22 0.035 (0.013)	42 −0.018 (0.028)	*p* < 0.001
ACA Corpus K Residuals	28 0.024 (0.029)	36 −0.019 (0.032)	*p* < 0.001

Table 15.5 Slopes of LS regression lines (through the origin) and correlation coefficients for standardized contrasts of corpus I_x and K residuals versus standardized contrasts of diet proportions

LS Residual Contrasts (vs. Diet Percentage Contrasts)	N	LS Slope[1]	r[1]
Corpus I_x Residuals	63	0.007**	0.310**
Corpus K Residuals	63	0.002 NS	−0.047NS

[1]. Key: ** = $p \leq 0.01$, NS = Not Significant.

15.3.2 Corpus K Results

ACA ln corpus K is significantly positively correlated with ACA ln mandibular length ($r = 0.953$; $p < 0.001$; Table 15.2; Fig. 15.6). As with corpus I_x, the LS regression slope for ACA ln corpus K versus ACA ln mandibular length does not depart significantly from isometry (slope = 0.947; 95% CI = 0.871–1.023, Table 15.2; Fig. 15.6). As predicted, the correlation between corpus K residuals and diet proportions is highly significant ($r = 0.478$; $p < 0.001$, Table 15.3; Fig. 15.7). Similarly, a comparison of corpus K residuals between the two diet groupings found that residuals are significantly higher in species with diets of at

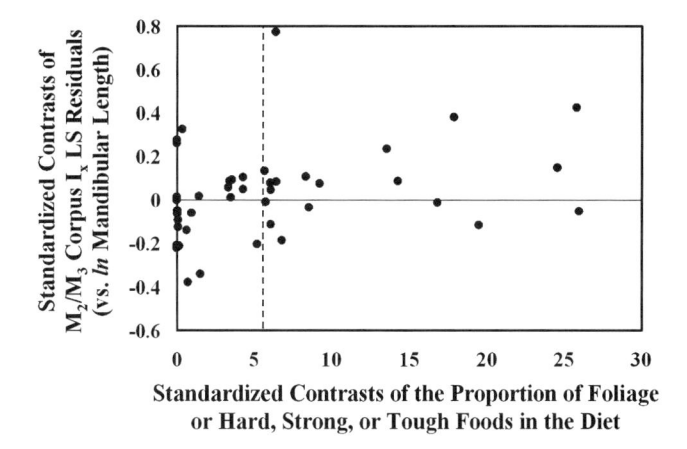

15 Mandibular Corpus Form and Its Functional Significance

Fig. 15.5 Plot of standardized contrasts of M_2/M_3 corpus I_x LS residuals (covariate $= ln$ mandibular length) versus standardized contrasts of the proportion of foliage or hard, strong, or tough foods in the diet. As predicted, increases or decreases in diet proportions are significantly positively correlated with increases or decreases, respectively, in corpus I_x residuals (Table 15.5). Note that, as expected, of the 19 contrasts where the proportion of foliage or hard, strong, or tough foods in the diet increased by more than five points (*vertical dashed line*), 12 also exhibited a corresponding increase in M_2/M_3 corpus I_x residuals

least 50% vertebrates, foliage, and/or other resistant foods than in those consuming these items in proportions less than 50% (Kruskal-Wallis One-Way ANOVA, $p < 0.001$; Table 15.4; Fig. 15.6). These findings bear out the prediction that folivores, carnivores, and gougers should have mandibles that are relatively better at resisting torsion than other taxa.

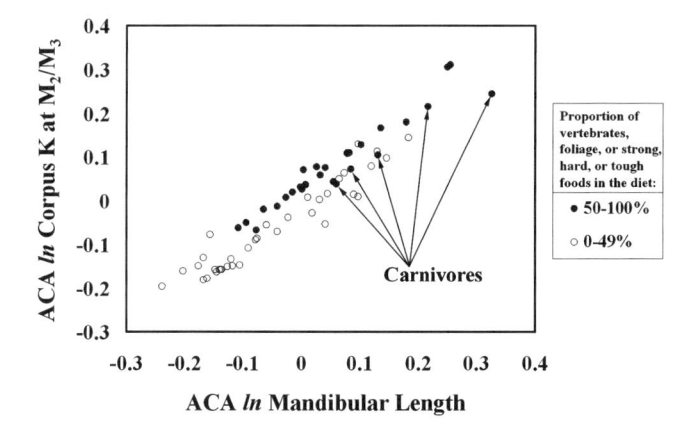

Fig. 15.6 Plot of ACA ln M_2/M_3 corpus K versus ACA ln mandibular length. As predicted, species with diets high in the indicated items have significantly larger LS residuals than those with diets low in these items (Table 15.4). Carnivorous species, which deviate from predicted findings, are indicated with arrows

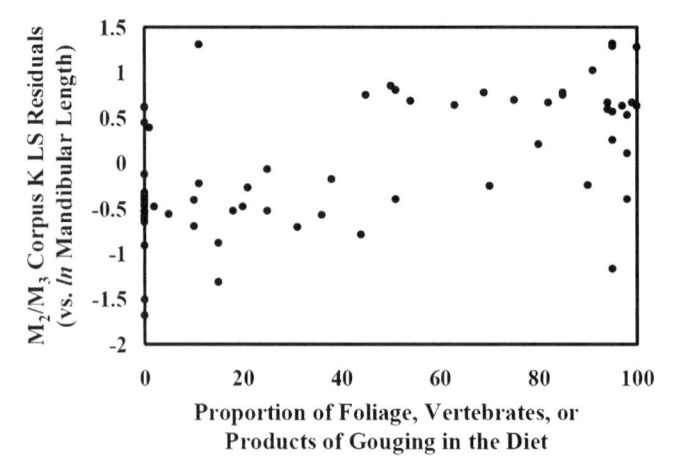

Fig. 15.7 Plot of ACA M_2/M_3 corpus K LS residuals (covariate = ACA ln mandibular length) versus the proportion of foliage, vertebrates, or products of tree gouging in the diet. As predicted, these two variables were significantly positively correlated (Table 15.3)

Given the positive relationship posited for relative mandibular torsional strength and the proportion of vertebrates, foliage, and/or hard, tough, or strong foods in the diet, contrasts of torsional strength and diet proportions should be related to each other through significantly positive slopes and correlation coefficients. Unlike the autocorrelation findings, this prediction receives no support from independent contrast analyses. That is, the Pearson's correlation coefficient and LS regression slope relating corpus torsional strength and diet proportions are not significantly positive (slope = 0.002; $r = -0.047$; $p > 0.05$, Table 15.5; Fig. 15.8).

15.4 Discussion

These results reveal strong evidence for a link between morphometric estimates of corpus vertical bending strength and a durophagous and/or folivorous diet in marsupials (Tables 15.3–15.5). This indicates corpus form does indeed reflect details of a species' feeding habits, and may therefore be useful for elucidating the feeding habits of extinct species. Support for the functional significance of corpus torsional strength was somewhat less, with only the autocorrelation analyses uncovering significant relationships (Tables 15.3–15.5). These findings differ from some previous studies, which found that corpus torsional strength (inferred from corpus width) tracked diet more consistently than vertical bending strength (inferred from corpus height) (e.g., Anton, 1996; Taylor, 2002). The reason for this difference appears related to the classification of carnivores.

Carnivorous mammals are thought to incur large corpus torsional loads during the act of killing, and are therefore expected to display relatively large corpus K values (Section 1.2; Biknevicius and Van Valkenburgh, 1996). However, as Fig. 15.6

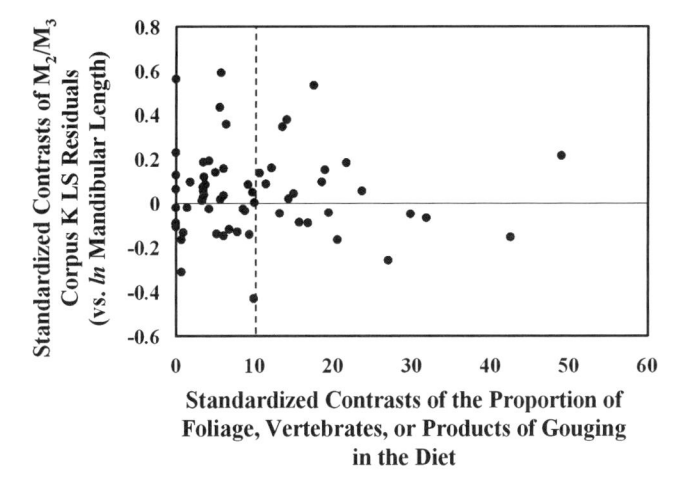

15 Mandibular Corpus Form and Its Functional Significance

Fig. 15.8 Plot of standardized contrasts of M_2/M_3 corpus K LS residuals (covariate = ln mandibular length) versus standardized contrasts of the proportion of foliage, vertebrates, or products of gouging in the diet. Contrary to the prediction, changes in the proportions of these foods in the diet along the phylogeny are not significantly positively correlated with changes in corpus K residuals (Table 15.5)

reveals, this is not the case in carnivorous marsupials. In fact, plots of corpus K versus mandibular length (Fig. 15.6) are strikingly similar to plots of corpus I_x versus mandibular length (Fig. 15.3), with most species remaining in roughly similar positions in both figures (including the marsupial carnivores, which consistently display weak mandibles relative to folivores, tree gougers, and other durophages). In fact, a subsequent correlation analysis of corpus I_x versus corpus K revealed an extremely high significant positive correlation between them ($r = 0.998$). The principle difference between the corpus K and I_x analyses was the categorization of carnivory as a "resistant" food for torsional loading, but not vertical bending. Given that it is the marsupial carnivores that violate the prediction relating to torsion, this likely accounts for the lower overall support for corpus K prediction. This conclusion is born out by the fact that when carnivores are removed, the correlation between ACA corpus K residuals and ACA diet increases substantially (from 0.478 to 0.671). These results suggest three possible explanations: the large mandibular torsional loads presumed to occur in carnivores do not actually occur (at least in marsupial carnivores), these loads do not occur with sufficient frequency to require robust corpora, or jaws are not consistently adapted to loads in the specific ways assessed here, if at all. Determining which of these explanations holds true will require in vivo mandibular bone strain work on marsupial and placental carnivores.

The similarity between plots of corpus K and I_x versus mandibular length also has implications for the relative importance of torsion and vertical bending across diet classes. The fact that species with relatively higher K values also have higher I_x values, and vice versa, suggests that either durophagy and folivory yield increases in both mandibular torsion and bending or strengthening the corpus to resist high levels of one loading regime results in corpora that are also stronger against the other. Given the complexities of jaw organization, and the fact that both loading regimes appear to be elevated during chewing and biting of resistant foods (Hylander, 1979a, b, 1985), it seems likely that the former is at least partly true. Still, it is difficult to rule out the possibility that variation in torsion or bending alone underlies differences in corpus form between species. This study provides stronger support for the role of vertical bending, but many previous studies provide greater support for the importance of axial torsion. Given these conflicting results, it is difficult to say which loading regime is more important. Perhaps, all we can say at present is that durophages and folivores typically have more robust (stronger) mandibles than other taxa, presumably related to relatively larger or more frequent loads during biting and/or chewing.

Not all species conformed to expectations, even in corpus I_x analyses, (Figs. 15.3 and 15.6). In the case of the *Burramys parvus* (mountain pygmy-possum), the relatively robust corpora likely relate to its winter feeding habits. These animals are known to cache seeds and nuts, which likely serve as their primary food source during the winter (Mansergh et al., 1990; Smith and Broome, 1992). Since the diet percentages reported in Table 15.1 for *B. parvus* are based on feeding habits during non-winter months, the figure of 19% seeds and nuts is almost certainly an underestimate. Even if seeds and nuts are less than 50% of the annual diet, the fact that these animals are likely to feed principally on this hard, strong food resource during part of the year may be sufficient to require relatively strong corpora.

The unexpectedly strong corpora of a second species, *Phalanger gymnotis* (the ground cuscus), are likely related to the consumption of foliage. Based on limited diet data, *P. gymnotis* is estimated to consume roughly 45% foliage (Hume et al., 1993, 1997; Hogue, 2004). Though not quite half the diet, this high consumption of foliage may account for this species' elevated corpus dimensions. Also, the diet data used to obtain this value were of lower quality than for most species. Consequently, it remains entirely possible that foliage actually constitutes more than 50% of the diet, thereby explaining its seemingly atypical corpus dimensions.

The robust mandibles of the remaining aberrant taxa (i.e., *Cercartetus nanus* – the eastern pygmy-possum, *C. caudatus* – the long-tailed pygmy-possum, and *Notoryctes typhlops* – the marsupial mole) are more difficult to explain. Seeds appear to be the only resistant foods consumed by these species, yet they make up well under 25% of the diet in all cases (Smith, 1986; Winkel and Humphery-Smith, 1988; van Tets and Whelan, 1997). Moreover, it is not clear whether these foods are masticated or simply incidentally ingested with associated fruits. As no other paramasticatory behaviors have been identified in these animals that would account for their robust corpora, other factors, such as phylogenetic inertia not removed by the PCMs used here, may underlie these findings.

The fact that all species found to display anomalous corpus I_x values consistently have relatively large dimensions raises an interesting question: Do species with moderately soft, weak, and/or fragile diets *need* to have relatively small corpus I_x values? From an optimal design standpoint, one would assume that making, using, and supporting an excessively robust mandible would be an unnecessary energetic drain on an animal. If so, results for several species studied here would seem to suggest such excesses may be tolerable in some cases. Given the consistency with which species deviate from expectations in the direction of overbuilt corpora (except carnivores), and given that no explanation for these unexpected findings could be identified in most cases, the costs of an overbuilt mandible may simply be too low to result in strong selection against it. Consequently, relatively large corpora may not consistently indicate a folivorous or durophagous diet.

Aside from the individual species noted above to deviate from the predictions, it is important to keep in mind only a moderate amount of the variance in corpus form was accounted for by diet (46% in the case of corpus I_x, and only 23% for corpus K). This suggests a variety of other factors may be influencing corpus form, or complicating its relationship to diet. Perhaps, the most likely factors undermining the association are the quality and nature of the diet data. First of all, the quality of the diet data varied considerably between species. In some cases, data may have come from a small sample of individuals studied in a single season of a single year. Given that, for many species, diet varies from year to year, and season to season, such results would like fail to fully characterize all functionally significant details of the species' diet. Furthermore, it is important to keep in mind that the analyses performed here relied on the assumption that corpus form is connected principally to the total percentage of foods consumed by a species (ideally over the course of a year or more). While this may be true, it is plausible that some species may go through periodic dietary bottlenecks, wherein they are heavily dependent on certain mechanically demanding foods that they otherwise rarely eat. If so, corpus form could be adapted to confronting these short-term challenges, which may not factor significantly into diet percentages computed over the course of a year or more.

Another drawback of the diet data used here is the lack of material properties data for the foods actually consumed by each of the species in the sample. Variability in the precise hardness, toughness, strength, and other characteristics of foods alone could account for much of the unexplained variance. Future work quantifying the material properties of foods consumed by a smaller sample of related species that vary in corpus form could go a long way toward assessing the degree and strength of the relationship between jaw form and diet.

One final potentially confounding input to consider in these analyses is tooth form. Different molar shapes and sizes could have profoundly different effects on the transmission of forces to foods, regardless of the masticatory muscle forces generated. For example, a small number of tall, sharp cusps (as in carnivorous mammals) could deliver very high, localized pressure to certain resistant foods, thereby breaking these items down without powerful contractions of the masticatory muscles. Hence, by considering the role of the teeth, which intervene between the corpus and ingested food items, it may be possible to more precisely identify the connection between the latter two.

In conclusion, this study has added greatly to our understanding of the covariation between diet and corpus form in marsupials. This work has shown that with the use of appropriate size measures, sufficiently detailed dietary data, full corpus cross-sectional data, and phylogenetic control, corpus form does reflect diet in this ecologically diverse, yet poorly understood clade. More broadly, the confirmation, in this independently evolved group, of predictions previously confirmed in placental mammals (at least in part) provides strong evidence for the putative link between overall corpus robusticity and diet. As such, this morphological system would likely prove very useful for reconstructing feeding habits in fossil mammals.

Acknowledgments I am indebted to M. Ravosa for his unfailing advice, encouragement, and helpful comments on the original manuscript. Thanks to C. Vinyard and two anonymous reviewers for their assistance in improving the clarity of this manuscript. Thanks are also due to M. Dagosto, B. Shea, R. Kay, J. Flynn, and A. Yoder for their many contributions in the development and completion of this research. I thank L. Salas, S. Stephens, and M. Springer for making available data prior to publication. Finally, I am grateful to the following individuals for providing access to museum specimens: L. Gordon and R. Thorington (Smithsonian Institution), L. Frigo (Museum of Victoria), P. Jenkins (British Museum of Natural History), M. Archer, T. Ennis, and S. Ingleby (Australian Museum), T. Flannery, D. Stemmer, and C. Kemper (South Australian Museum), L. Abraczinskas and B. Lundrigan (Michigan State University Museum), R. Voss, C. Norris, and B. Randall (American Museum of Natural History), L. Heaney, B. Patterson, and W. Stanley (Field Museum of Natural History). This project was funded by grants from the American Society of Mammalogists, the National Science Foundation (Grant # NSF BCS-0127915), and Sigma Xi.

References

Amrine-Madsen, H., Scally, M., Westerman, M., Stanhope, M. J., Krajewski, C., and Springer, M. S. (2003). Nuclear gene sequences provide evidence for the monophyly of australidelphian marsupials. *Mol Phylogenetics Evol* 28:186–196.

Anapol, F., and Lee, S. (1994). Morphological adaptation to diet in platyrrhine primates. *Am J Phys Anthropol* 94:239–261.

Anton, S. C. (1996). Cranial adaptation to a high attrition diet in Japanese Macaques. *Int J Primatol* 17:401–427.

Atkins, A. G., and Vincent, J. F. V. (1984). An instrumented microtome for improved histological sections and the measurement of fracture toughness. *J Mater Sci* 3:310–312.

Biknevicius, A. R., and Ruff, C. B. (1992a). The structure of the mandibular corpus and its relationship to feeding behaviors in extant carnivorans. *J Zool* 228:479–507.

Biknevicius, A. R., and Ruff, C. B. (1992b). Use of biplanar radiographs for estimating cross-sectional geometric properties of mandibles. *Anat Rec* 232:157–163.

Biknevicius, A. R., and Van Valkenburgh, B. (1996). Design for killing: Craniodental adaptations of predators. In: Gittleman, J. L. (ed.), *Carnivore Behavior, Ecology, and Evolution*. Cornell University Press, New York, pp. 393–428.

Bouvier, M. (1986a). A biomechanical analysis of mandibular scaling in old world monkeys. *Am J Phys Anthropol* 69:473–482.

Bouvier, M. (1986b). Biomechanical scaling of mandibular dimension in new world monkeys. *Int J Primatol* 7:551–567.

Bouvier, M., and Hylander, W. L. (1981). Effect of bone strain on cortical bone structure in macaques (*Macaca mulatta*). *J Morphol* 167:1–12.

Bouvier, M., and Hylander, W. L. (1996). The mechanical or metabolic function of secondary osteonal bone in the monkey *Macaca fascicularis*. *Arch Oral Biol* 41:941–950.

Cartmill, M. (1972). Arboreal adaptations and the origin of the order Primates. In: Tuttle, R. H. (ed.), *The Functional and Evolutionary Biology of Primates*. Aldine-Atheton, Chicago, pp. 3–35.

Cartmill, M. (1974a). Rethinking primate origins. *Science* 184:436–443.

Cartmill, M. (1974b). *Daubentonia, Dactylopsila*, woodpeckers and klinorhynchy. In: Doyle, G. A., Martin, R. D., and Walker, A. (eds.), *Prosimian Biology*. Duckworth, London, pp. 655–670.

Cartmill, M. (1992). New views on primate origins. *Evol Anthropol* 1:105–111.

Casimir, N. J. (1975). Feeding ecology and nutrition of an eastern gorilla group in the Mt. Kahuzi region (Republic of Zaire). *Folia Primatol* 24:81–136.

Charles-Dominique, P. (1983). Ecological and social adaptations in didelphid marsupials: comparison with eutherians of similar ecology. In: Eisenberg, J. F., and Kleiman, D. G. (eds.), *Advances in the Study of Mammalian Behavior. Special publication no. 7.* American Society of Mammalogists, Shippensburg, PA, pp. 395–422.

Cheverud, J. M., and Dow, M. M. (1985). An autocorrelation analysis of the effect of lineal fission on genetic variation among social groups. *Am J Phys Anthropol* 67:113–121.

Cheverud, J. M., Dow, M. M., and Leutenegger, W. (1985). The quantitative assessment of phylogenetic constraints in comparative analyses: Sexual dimorphism in body weight among primates. *Evolution* 39:1335–1351.

Currey, J. D. (1984). *The Mechanical Adaptations of Bone*. Princeton University Press, Princeton, NJ.

Daegling, D. J. 1989. Biomechanics of cross-sectional size and shape in the hominoid mandibular corpus. *Am J Phys Anthropol* 80:91–106.

Daegling, D. J. 1992. Mandibular morphology and diet in the genus *Cebus*. *Int J Primatol* 13:545–570.

Daegling, D. J., and Grine, F. E. (1991). Compact bone distribution and biomechanics of early hominid mandibles. *Am J Phys Anthropol* 86:321–339.

Daegling, D. J., and Hylander, W. L. (1998). Biomechanics of torsion in the human mandible. *Am J Phys Anthropol* 105:73–87.

Daegling, D. J., and McGraw, W. S. (2001). Feeding, diet and jaw form in West African Colobus and Procolobus. *Int J Primatol* 22:1033–1055.

Demes, B., Preuschoft, H., and Wolff, J. E. A. (1984). Stress-strength relationships in the mandibles of hominoids. In: Bilsborough, A. (ed.), *Food Acquisition and Processing in Primates*. Plenum Press, New York, pp. 369–390.

Diniz-Filho, J. A. F., Sant'Ana, C. E. R., and Bini, L. M. (1998). An eigenvector method for estimating phylogenetic inertia. *Evolution* 52:1247–1262.

Dumont, E. R. (1997). Cranial shape in fruit, nectar, and exudates feeders: Implications for interpreting the fossil record. *Am J Phys Anthropol* 102:187–202.

Evans, F. G., and Lebow, M. (1957). Strength of human compact bone under repetitive loading. *J Appl Physiol* 10:127–130.

Feeny, P. (1970). Seasonal changes in oak leaf tannins and nutrients as a cause of spring feeding by winter moth Caterpillars. *Ecology* 51:565–580.

Felsenstein, J. (1985). Phylogenies and the comparative method. *Am Nat* 125:1–15.

Freeland, W. J., and Janzen, D. H. (1974). Strategies of herbivory in mammals; The role of plant secondary compounds. *Am Nat* 108:269–289.

Freeman, P. W. (1979). Specialized insectivory: Beetle-eating and moth-eating molossid bats. *J Mammal* 60:467–479.

Freeman, P. W. (1981). Correspondence of food habits and morphology in insectivorous bats. *J Mammal* 62:164–166.

Freeman, P. W. (1984). Functional cranial analysis of large animalivorous bats (Microchiroptera). *Biol J Linn Soc* 21:387–408.

Freeman, P. W. (1988). Frugivorous and animalivorous bats (Microchiroptera): Dental and cranial adaptations. *Biol J Linn Soc* 33:249–272.

Freeman, P. W. (2000). Macroevolution in Microchiroptera: Recoupling morphology and ecology with phylogeny. *Evol Ecol Res* 2:317–335.

Garland, T., Jr., and Ives, A. R. (2000). Using the past to predict the present: Confidence intervals for regression equations in phylogenetic comparative methods. *Am Nat* 155:346–364.

Garland, T., Jr., Harvey, P. H., and Ives, A. R. (1992). Procedures for the analysis of comparative data using phylogenetically independent contrasts. *Syst Biol* 41:18–32.

Garland, T., Jr., Dickerman, A. W., Janis, C. M., and Jones, J. A. (1993). Phylogenetic analysis of covariance by computer simulation. *Syst Biol* 42:265–292.

Garland, T., Jr., Midford, P. E., and Ives, A. R. (1999). An introduction to phylogenetically based statistical methods, with a new method for confidence intervals on ancestral values. *Am Zool* 39:374–388.

Gittleman, J. L., and Kot, M. (1990). Adaptation: Statistics and a null model for estimating phylogenetic effects. *Syst Zool* 39:227–241.

Gittleman, J. L., and Luh, H.-K. (1994). Phylogeny, evolutionary models and comparative methods: A simulation study. In: Vane-Wright, D. (ed.), *Phylogenetics and Ecology*. Academic Press, London, pp. 103–122.

Grafen, A. (1989). The phylogenetic regression. *Philosophical Transactions of the Royal Society of London Series B-Biological Sciences* 326:119–157.

Hogue, A. S. (2004). On the relation between craniodental form and diet in mammals: Marsupials as a natural experiment. Ph.D. dissertation, Northwestern University, Evanston, IL.

Hogue, A. S., and Ravosa, M. J. (2001). Transverse masticatory movements, occlusal orientation, and symphyseal fusion in selenodont artiodactyls. *J Morphol* 249:221–241.

Huey, R. B., and Bennett, A. F. (1987). Phylogenetic studies of coadaptation: Preferred temperatures versus optimal performance temperatures of lizards. *Evolution* 41:1098–1115.

Hume, I. D., Jazwinski, E., and Flannery, T. F. (1993). Morphology and function of the digestive tract in New Guinea possums. *Aust J Zool* 41:85–100.

Hume, I. D., Runcie, M. J., and Caton, J. M. (1997). Digestive physiology of the ground cuscus (*Phalanger gymnotis*), a New Guinean phalangerid marsupial. *Aust J Zool* 41:85–100.

Hylander, W. L. (1979a). Mandibular function in *Galago crassicaudatus* and *Macaca fascicularis* – *In vivo* approach to stress analysis of the mandible. *J Morphol* 159:253–296.

Hylander, W. L. (1979b). The functional significance of primate mandibular form. *J Morphol* 160:223–239.

Hylander, W. L. (1984). Stress and strain in the mandibular symphysis of primates – A test of competing hypotheses. *Am J Phys Anthropol* 64:1–46.

Hylander, W. L. (1985). Mandibular function and biomechanical stress and scaling. *Am Zool* 25:315–330.

Hylander, W. L., and Crompton, A. W. (1986). Jaw movements and patterns of mandibular bone strain during mastication in the monkey *Macaca fascicularis*. *Arch Oral Biol* 31:841–848.

Hylander, W. L., and Johnson, K. R. (1985). Temporalis and masseter muscle function during incision in macaques and humans. *Int J Primatol* 6:289–322.

Hylander, W. L., and Johnson, K. R. (1997). *In vivo* bone strain patterns in the zygomatic arch of macaques and the significance of these patterns for functional interpretations of craniofacial form. *Am J Phys Anthropol* 102:203–232.

Hylander, W. L., Johnson, K. R., and Crompton, A. W. (1992). Muscle force recruitment and biomechanical modeling: An analysis of masseter muscle function during mastication in *Macaca fascicularis*. *Am J Phys Anthropol* 88:365–387.

Hylander, W. L., Ravosa, M. J., Ross, C. F., and Johnson, K. R. (1998). Mandibular corpus strain in primates: Further evidence for a functional link between symphyseal fusion and jaw-adductor muscle force. *Am J Phys Anthropol* 107:257–71.

Hylander, W. L., Ravosa, M. J., Ross, C. F., Wall, C. E., and Johnson, K. R. (2000). Symphyseal fusion and jaw-adductor muscle force: An EMG study. *Am J Phys Anthropol* 112:469–492.

Kay, R. F., and Cartmill, M. (1977). Cranial morphology and adaptations of *Palaechthon nacimienti* and other Paromomyidae (Plesiadapoidea, ?Primates), with a description of a new genus and species. *J Hum Evol* 6:19–53.

Kay, R. F., and Hylander, W. L. (1978). The dental structure of mammalian folivores with special reference to primates and Phalangeroidea (Marsupalia). In: Montgomery, G. G. (ed.), *The Ecology of Arboreal Folivores*. Smithsonian Institution Press, Washington D. C., pp. 173–191.

Kirsch, J. A. W., Lapointe, F. J., and Springer, M. S. (1997). DNA-hybridisation studies of marsupials and their implications for metatherian classification. *Aust J Zool* 45:211–280.

Lafferty, J. F., Winter, W. G., and Gambaro, S. A. (1977). Fatigue characteristics of posterior elements of vertebrae. *J Bone Joint Surg* 59A:154–158.

Lemelin, P. (1999). Morphological correlates of substrate use in didelphid marsupials: implications for primate origins. *J Zool (London)* 247:165–175.

Lieberman, D. E., and Crompton, A. W. (1998). Response of bone to stress: Constraints on symmorphosis. In: Bolis, L. (ed.), *Principles of Animal Design: The Optimization and Symmorphosis Debate*. Cambridge University Press, Cambridge, pp. 78–86.

Lucas, P. W. (1989). Significance of *Mezzettia leptopoda* fruits eaten by orang-utans for dental microwear analysis. *Folia Primatol* 52:185–190.

Lucas, P. W., and Corlett, R. T. (1991). Quantitative aspects of the relationship between dentitions and diets. In: Vincent, J. F. V. (ed.), *Feeding and the Texture of Food*. Cambridge University Press, New York, pp. 93–121.

Lucas, P. W., and Pereira, B. (1990). Estimation of the fracture-toughness of leaves. *Funct Ecol* 4:819–822.

Lucas, P. W., Tan, H. T. W., and Cheng, P. Y. (1997). The toughness of secondary cell wall and woody tissue. *Philos T Roy Soc B* 352:341–352.

Lynch, M. (1991). Methods for the analysis of comparative data in evolutionary ecology. *Evolution* 45:1065–1080.

Mansergh, I., Baxter, B., Scotts, D., Brady, T., and Jolley, D. (1990). Diet of the mountain pygmy possum *Burramys parvus* (Marsupalia: Burramyidae) and other small mammals in the alpine environment at Mt. Higginbotham Victoria, Australia. *Aust Mammal* 13:167–178.

Martins, E. P. (1996). Phylogenies, spatial autoregression, and the comparative method: A computer simulation test. *Evolution* 50:1750–1765.

Martins, E. P. (2001). COMPARE, version 4.4. Computer programs for the statistical analysis of comparative data. Distributed by the author at http://compare.bio.indiana.edu/. Department of Biology, Indiana University, Bloomington IN.

Martins, E. P., and Garland, T. (1991). Phylogenetic analyses of the correlated evolution of continuous characters: A simulation study. *Evolution* 45:534–557.

Martins, E. P., and Hansen, T. F. (1997). Phylogenies and the comparative method: A general approach to incorporating phylogenetic information into analysis of interspecific data. *Am Nat* 149:646–667.

Martins, E. P., Diniz-Filho, J. A. F., and Housworth, E. A. (2002). Adaptive constraints and the phylogenetic comparative method: A computer simulation test. *Evolution* 56:1–13.

Milton, K. (1979). Factors influencing leaf choice by howler monkeys: A test of some hypotheses of food selection by generalist herbivores. *Am Nat* 114:362–378.

Milton, K. (1980). *The Foraging Strategy of Howler Monkeys: A Study in Primate Economics*. Columbia University Press, New York.

Morris, J. M., and Blickenstaff, L. D. (1967). *Fatigue Fractures: A Clinical Study*. Charles C. Thomas, Springfield, IL.

Nash, L. T. (1986). Dietary, behavioral, and morphological aspects of gumnivory in primates. *Yearb Phys Anthropol* 29:113–137.

Pagel, M. D. (1992). A method for the analysis of comparative data. *J Theor Biol* 156:431–442.

Pan, R., Peng, Y., Ye, Z., Wang, H., and Yu, F. (1995). Comparison of masticatory morphology between *Rhinopithecus bieti* and *R. roxellana*. *Am J Primatol* 35:271–281.

Radinsky, L. B. (1981a). Evolution of skull shape in carnivores. 1. Representative modern carnivores. *Biol J Linn Soc* 15:369–388.

Radinsky, L. B. (1981b). Evolution of skull shape in carnivores. 2. Additional carnivores. *Biol J Linn Soc* 16:337–355.

Rasmussen, D. T. (1990). Primate origins: lessons from a neotropical marsupial. *Am J Primatol* 22:263–277.

Ravosa, M. J. (1991). Structural allometry of the prosimian mandibular corpus and symphysis. *J Hum Evol* 20:3–20.

Ravosa, M. J. (1996). Jaw morphology and function in living and fossil old world monkeys. *Int J Primatol* 17:909–932.
Ravosa, M. J. (2000). Size and scaling in the mandible of living and extinct apes. *Folia Primatol* 71:305–322.
Ravosa, M. J., and Hogue, A. S. (2004). Function and fusion of the mandibular symphysis in mammals: A comparative and experimental perspective. In: Kay, R. F., and Ross, C. (eds.), *Anthropoid Origins: New Visions*. Plenum Publishing Corporation, New York, pp. 413–462.
Ravosa, M. J., and Hylander, W. L. (1994). Function and fusion of the mandibular symphysis in primates: Stiffness or strength? In: Kay, R. F. (ed.), *Anthropoid Origins*. Plenum Press, New York, pp. 447–468.
Runestad, J. A., Ruff, C. B., Nieh, J. C., Thorington, R. W., and Teaford, M. F. (1993). Radiographic estimation of long-bone cross-sectional geometric properties. *Am J Phys Anthropol* 90:207–213.
Ryan, C. A., and Green, T. R. (1974). Proteinase inhibitors in natural plant protection. In: Conn, E. E. (ed.), *Metabolism and Regulation of Secondary Plant Products*. Academic Press, New York, pp. 123–140.
Smith, R. J. (1983). The mandibular corpus of female primates: Taxonomic, dietary, and allometric correlates of insterspecific variations in size and shape. *Am J Phys Anthropol* 61:315–330.
Smith, A. P. (1986). Stomach contents of the long-tailed pygmy-possum *Cercartetus caudatus* (Marsupalia: Burramyidae). *Aust Mammal* 9:135–137.
Smith, R. J. (1994). Degrees of freedom in interspecific allometry: An adjustment for the effects of phylogenetic constraint. *Am J Phys Anthropol* 93:95–107.
Smith, A. P., and Broome, L. (1992). The effects of season, sex and habitat on the diet of the mountain pygmy-possum (*Burramys parvus*). *Wildlife Res* 19:755–768.
Spencer, L. M. (1995). Morphological correlates of dietary resource partitioning in the African Bovidae. *J Mammal* 76:448–471.
Stobbs, T. H., and Cowper, L. J. (1972). Automatic measurement of the jaw movements of dairy cows during grazing and rumination. *Trop Grasslands* 6:107–112.
Strait, S. G. (1993). Molar morphology and food texture among small-bodied insectivorous mammals. *J Mammal* 74:391–402.
Strait, S. G. (1997). Tooth use and the physical properties of food. *Evol Anthropol* 5:199–211.
Strait, S. G., Fleagle, J. G., Bown, T. M., and Dumont, E. R. (1990). Diversity in body size and dietary habits of fossil caenolestid marsupials from the Miocene of Argentina. *J Vert Paleontol* 10:44A.
Takahashi, L. K., and Pan, R. (1994). Mandibular morphometrics among macaques: The case of *Macaca thibetana*. *Int J Primatol* 15:597–621.
Taylor, A. B. (2002). Masticatory form and function in the African apes. *Am J Phys Anthropol* 117:133–156.
Taylor, A. B. (2006). Feeding behavior, diet, and the functional consequences of jaw form in orangutans, with implications for the evolution of *Pongo*. *J Hum Evol* 50:377–393.
van Tets, I. G., and Whelan, R. J. (1997). *Banksia* pollen in the diet of Australian mammals. *Ecography* 20:499–505.
Vincent, J. F. V. (1990). Fracture properties of plants. *Advances in Botanical Research* 17:235–282.
Vinyard, C. J., Wall, C. E., Williams, S. H., and Hylander, W. L. (2003). Comparative functional analysis of skull morphology of tree-gouging primates. *Am J Phys Anthropol* 120:153–170.
Vogel, S. (1988). *Life's Devices*. Princeton University Press, Princeton, NJ.
Walker, A. (1967). Locomotor adaptations in recent and fossil Madagascan lemurs. Ph.D. thesis, University of London, London
Williams, R. T. (1969). *Detoxification Mechanisms*. Wiley, New York.
Williams, S. H., Wall, C. E., Vinyard, C. J., and Hylander, W. L. (2002). A biomechanical analysis of skull form in gum-harvesting galagids. *Folia Primatol* 73:197–209.
Winkel, K., and Humphery-Smith. I. (1988). Diet of the marsupial mole, *Notoryctes typhlops* (Stirling 1889) (Marsupialia: Notoryctidae). *Aust Mammal* 11:159–162.

Chapter 16
Putting Shape to Work: Making Functional Interpretations of Masticatory Apparatus Shapes in Primates

Christopher J. Vinyard

Contents

16.1　Introduction 357
　　　16.1.1　What Question Are You Asking? 358
16.2　Choice of Size-Adjustment Technique: Statistically Controlling for Size Versus
　　　Relative Size 358
　　　16.2.1　"Controlling" for Size Via Variance Partitioning 359
　　　16.2.2　Ratios and Relative Size 361
16.3　Choice of Size-Adjustment Variable: Overall Size Versus Functional Estimates 366
16.4　How Does a Shape Variable Denominator Affect a Hypothesis Test? A Case Study ... 370
16.5　What Can We Reliably Interpret from Comparative Analyses of Shapes? 378
　　　References 380

16.1 Introduction

While publishing a functional analysis of skull shapes in tree-gouging primates (Vinyard et al., 2003), we entered into an important discussion with our reviewers regarding our approach to size adjustment (see also Vinyard et al., 2003). In particular, reviewers expressed concern over our decision to apply what I will define below as a "biomechanical standard" for comparing skull morphologies across a range of different-sized animals. I realized from this fruitful, but unpublished, discussion that (1) biological anthropologists studying the functional morphology of the primate masticatory apparatus have not adequately explained some of our decisions regarding size adjustment, (2) biological anthropologists have neglected certain theoretical concerns in applying various size estimates in functional analyses, and (3) we need to carefully consider what can be reliably concluded from comparative morphometric analyses of skull shapes. Rather than attempt to summarize this debate without the input from these reviewers, I focus on addressing these issues and in doing so

---

C.J. Vinyard
Department of Anatomy, NEOUCOM, 4209 St. Rt. 44, Box 95, Rootstown, OH 44272
e-mail: cvinyard@neoucom.edu

discuss reasons for using shape ratios relative to a biomechanical standard when studying primate masticatory apparatus function.

16.1.1 What Question Are You Asking?

Before starting down the path to size adjustment, it is necessary to specify the question one is addressing. While certainly a trite reminder, size-adjustment techniques will vary with and be dictated by the goals of a study. A significant amount of debate concerning size-adjustment methods likely could have been avoided with a clearer explication of the goals from various studies. In this chapter, I am specifically interested in comparing the functional consequences (i.e., the ability to do a given task) of changing masticatory apparatus form among groups of primates. Even though I focus on primate jaws, many of the points raised here pertain to functional studies of shape in any organ system. While it is unmistakably clear that for many questions variation in size is a confounding factor warranting the statistical removal of its effects, I am focusing on how the functional abilities of the masticatory apparatus change with size. By concentrating on this question, I explicitly sacrifice the ability to partition the potential factors driving change in form during primate evolution in favor of understanding how relative functional abilities vary across organisms. Depending on one's view of comparative analyses, this approach can be characterized as non-evolutionary in focus. As such, examining the functional consequences of size-related variation in form would represent one part of a larger study integrating these findings with physiological, ecological, ontogenetic, phylogenetic, genetic, and/or evolutionary allometric results in a holistic attempt to describe the evolutionary morphology of some feature.

16.2 Choice of Size-Adjustment Technique: Statistically Controlling for Size Versus Relative Size

One of the first decisions in size adjustment must be choosing a technique for rendering different-sized animals comparable. Recently, Smith (2005) has eloquently contrasted two philosophies of size adjustment as those techniques attempting to statistically control for size versus methods focusing on relative size, or proportional adjustment. The merits of "residuals" versus "ratios", as these methods are often respectively labeled, has a lengthy history of discussion in morphometric analyses (Mosimann, 1970, 1975; Sprent, 1972; Mosimann and James, 1979; Lemen, 1983; Reist, 1985; Corruccini, 1987, 1995; Bookstein, 1989; James and McCulloch, 1990; Albrecht et al., 1993, 1995; Falsetti et al., 1993; Jungers et al., 1995; Shea, 1995; Ravosa, 1996a; Smith, 2005). It is not my aim to describe these two approaches in detail here nor develop them in a historical perspective for two reasons. First, the papers cited above, particularly Corruccini (1987), Jungers et al. (1995), and Smith (2005), have already done this for us. Second, I am primarily interested in discussing how these methods relate to studying masticatory apparatus functional morphology. In fact, it is worth emphasizing that the following critique of the merits and disadvantages of these two approaches is only appropriate in the context of studying the functional consequences of form among different-sized organisms.

16.2.1 *"Controlling" for Size Via Variance Partitioning*

Residuals taken from the regression of a variable(s) of interest on a size estimate are often examined as size-adjusted data in comparative studies of primate masticatory apparatus form (e.g., Kay, 1975, 1978; Smith, 1983, 1984a; Smith et al., 1983; Bouvier, 1986a,b; Martin and MacLarnon, 1988; Ravosa, 1991a,b, 1996b; Anthony and Kay, 1993; Jablonski, 1993; Takahashi and Pan, 1994; Pan et al., 1995; Gauld, 1996; Godfrey et al., 2001; Bastir and Rosas, 2004; Vinyard and Hanna, 2005). Although examining residuals from a best-fit regression slope is a procedure commonly used by morphometric researchers, this method represents one of a larger set of approaches all aimed at statistically controlling for a given variable through variance partitioning (Smith, 2005). These additional procedures include multiple regression, analysis of covariance, and partial correlations.[1] The statistical outcome of these approaches is the extraction of a component from the variable being studied. This component is uncorrelated with the variable(s) being controlled for, such as a size estimate. For residuals from allometric regression slopes, the extracted portion of the variable of interest estimates non-allometric shape, or shape that is uncorrelated with the size estimate. Proponents of residuals as size-adjusted data (e.g., Gould, 1975; Reist, 1985) argue that the statistical removal of a confounding size variable beneficially allows the comparison of shapes for a variable of interest across organisms that differ in both size and this variable.

While analyses of residuals have proven extremely useful in several contexts, four specific issues arise when making functional interpretations of residuals as size-adjusted data. As shown by Jungers et al. (1995) and noted by earlier researchers (Lemen, 1983; Corruccini, 1987), residuals often fail to demonstrate organisms of the same proportion relative to a size criterion. Because the allometric regression slope used in creating a residual captures size and size-correlated variation (i.e., regression slopes are not restricted to being isometric), residuals usually do not capture proportionality across size. Thus, two different-sized organisms with different mandibular proportions might have the same residual value of zero (as an example), because the regression slope is allometric and passes through both organisms (Smith, 1980). Interpreting the relative functional abilities of forms across different-sized organisms typically involves a comparative assessment of proportionality – for a given control variable, is morphology "a" or "b" better suited to accomplish a task? The absence of an inherent proportionality framework makes functional interpretations of residuals from empirically fit regression slopes more problematic.

---

[1] Throughout this section I will focus on residuals, but many of the same comments apply to these other techniques.

Deriving residuals from an empirically fit regression slope creates a second concern for the functional interpretation of size-adjusted forms. We all recognize that adding or removing organisms from a sample will likely change the regression slope as it is computed based on that sample. (All of the techniques aimed at variance partitioning will behave similarly.) Despite these potential changes in the slope and residual values, the intrinsic shapes of the organisms remaining in the sample have neither changed nor have the associated functional consequences of their shapes. Thus, residuals provide estimates of shape that are defined relative to the clade, or more-specifically the sample group, being studied (Corruccini, 1987, 1995; Jungers et al., 1995). In many cases, these clade-based estimates of shape provide important information for addressing evolutionary questions. However, these residual-based shape estimates may differ considerably from a feature's functional capabilities.

Studying the functional consequences of residuals as size-adjusted data is also hindered by the often ambiguous functional interpretation of a regression slope arrived at through a statistical line-fitting technique (Smith, 1980).[2] This uncertainty regarding the biological meaning of the regression slope, including its functional interpretation, prompted Jungers et al. (1998:309) to state that "allometry ...explains nothing" (see also Smith, 1980, 1984b, 2005; Shea, 1983, 1985a,b, 1995; Fleagle, 1985; Shea and Gomez, 1988; Falsetti et al., 1993; Jungers et al., 1995). Without delineating the functional significance of the slope, its use as a criterion of subtraction for creating residuals as size-adjusted data undermines our ability to make meaningful functional interpretations (Gould, 1966; Jerison, 1973; Smith, 1980, 1984b, 2005; Shea, 1985a, 1995; Falsetti et al., 1993; Jungers et al., 1995, 1998; Ravosa and Vinyard, 2002). In direct contrast, taking residuals from theoretically derived lines with explicit functional interpretations (McMahon, 1973; Smith, 1980, 1984b; Shea, 1981, 1983, 1995; Vinyard and Ravosa, 1998) provides adjusted data that facilitate comparing the functional consequences of morphological variation across organisms.

Finally, it is worth remembering the difference between the statistical control achieved in approaches involving variance partitioning and the biological control aimed at by the researcher (Smith et al., 1986; Shea, 1995). This distinction is essential to all size-adjustment approaches, but it is particularly relevant for the methods employing variance partitioning because of the ease of conflating statistical and biological control. The statistical control achieved in variance partitioning, as in other size-adjustment techniques, is contingent on both the sample and the measurements used to estimate the biological parameters of interest. How closely the estimate of size approximates size (or the specific variable that the researcher wants to control for) will have a significant effect on the biological interpretation of the results (Wright, 1932; Smith, 1980). Unfortunately, size is very difficult to measure, and the existence of widespread pleiotropy (Wright, 1968; Atchley and Hall, 1991) suggests that single variables, even those composed of multiple measurements, are

---

[2] In addition to discussions of the functional interpretation of a statistically fit line, there has been considerable debate over the best statistical line-fitting technique for estimating allometric relationships (see e.g., Kermack and Haldane 1950; Hills and Woods, 1984; Ricker, 1973; Rayner, 1985; Jungers 1988; McArdle, 1988; Martin and Barbour, 1989; Riska, 1991; Smith, 1994).

unlikely to completely capture size. Thus, statistical control via variance partitioning is unlikely to provide the biological control that researchers set out to achieve when trying to remove the effects of another variable.

16.2.2 Ratios and Relative Size

To facilitate comparing organisms of varying magnitudes, morphometricians also create ratios by dividing a variable of interest by a second, typically size-descriptive, variable. When the numerator and denominator have the same scale, the resulting ratio is dimensionless (i.e., scale free) and hence comparable across organisms of different magnitudes (Mosimann, 1970; Corruccini, 1987; Jungers et al., 1995). This ratio provides an estimate of shape, or relative size, based on the proportionality constant and geometric similarity. Proportionality facilitates the identification of organisms with the same shape as defined relative to a geometric criterion (Mosimann, 1970; Corruccini, 1987; Jungers et al., 1995). For morphologists, preservation of shape across organisms of different sizes indicates isometry and would be manifest as a zero correlation between the shape ratio and the size estimate across these organisms (Mosimann, 1970; Mosimann and James, 1979; Jungers et al., 1995) Alternatively, a shape ratio that is correlated with a size estimate is allometric or changing shape with size.

As noted by Smith (2005), ratios are new variables that provide information not measured by either the numerator or the denominator variable. A relevant biomechanical example is the mechanical advantage of a lever system, such as the idealized jaw in biting (Fig. 16.1). A lever has a moment arm (in-lever) and a load arm (out-lever), which when combined in a ratio describes its mechanical advantage. This derived ratio has clear mechanical relevance, but cannot be known from either constituent dimension by itself. Placed in an allometric context, this means that examining the scaling pattern of only one of these component variables relative to size might lead to an incorrect assessment of how the mechanics of that system change with size (see e.g., Vinyard and Ravosa, 1998). For example, both the moment and the load arm could be increasing with strong positive allometry relative to a size estimate, but when combined to examine the system's leverage, this ratio might indicate no change in mechanical advantage with size (Fig. 16.2).

The proportionality inherent in ratios is beneficial for understanding the functional consequences of masticatory apparatus form when using these ratios as shape estimates (see also, Oxnard, 1978; Kidd and Oxnard, 2005). Many of the functional questions we ask about the morphology of the primate masticatory apparatus are explicitly, or at least implicitly, stated in a proportional framework. For example, several researchers have compared the relative load-resisting ability of the jaws among primate species (e.g., Hylander, 1979, 1985, 1988; Demes et al., 1984; Bouvier, 1986a,b; Daegling, 1989, 1992, 1993, 2001; Ravosa, 1990, 1991c, 1996b,c, 2000; Cole, 1992; Takahashi and Pan, 1994; Anton, 1996; Vinyard and Ravosa, 1998; Ravosa et al., 2000; Daegling and McGraw, 2001; Williams et al., 2002; Vinyard et al., 2003, Taylor, 2002, 2005). Hylander (1979, 1985) pioneered an argument that load-resistance ability should be compared relative to the

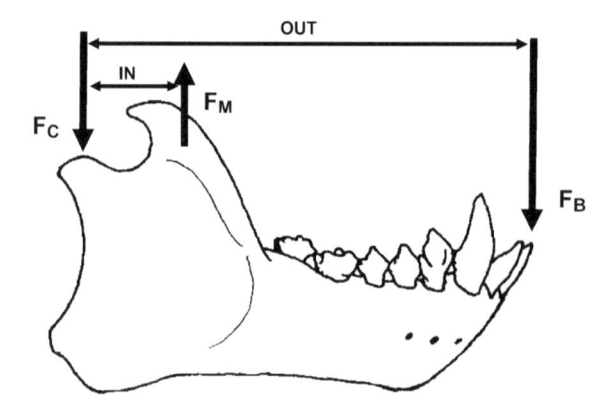

F_B = IN * F_M / OUT

Mechanical Advantage for Jaw Muscles in generating F_B = IN / OUT

Fig. 16.1 A diagrammatic representation of the mechanical advantage of the jaw-closing muscles for producing vertical bite force at the incisors. The mechanical advantage of these jaw muscles is estimated as the ratio of the perpendicular distance from the line of action of the jaw-closing muscle (arrow labeled F_m) to the temporomandibular joint (in-lever) divided by the distance from the joint to the incisal bite point (out-lever). In this idealized example, mechanical advantage (IN/OUT) represents a shape ratio with biomechanical significance for biting performance. This derived ratio has clear mechanical relevance, but cannot be known from either moment arm by itself. (Note that the arrows are not vectors and their lengths do not represent force values. Furthermore, the location of the jaw closing muscle line of action is arbitrarily placed for the purpose of illustration.) Abbreviations: F_m = vertical force generated by the jaw-closing muscles, F_b = vertical bite force, F_c = vertical reaction force at the condyle, IN = the in-lever or moment arm for the jaw-closing muscles and OUT = the out-lever or load arm for biting at the incisors

external forces placed upon the jaw during chewing or biting if we are going to accurately describe the mechanical implications of morphological variation in primate jaws. This line of reasoning lends itself to using ratios of estimates of load-resistance ability relative to estimates of the external forces placed on the jaw during chewing or biting.

From a functional perspective, it does not matter whether the numerator or denominator is responsible for size-correlated changes in a shape ratio (Fig. 16.3). The functional consequence of this change, that is, the increase or decrease in ability to accomplish a task, occurs regardless of which part of the ratio changes. Where the change occurs is less relevant than its impact on the feature's ability to function in a given task. For instance, observed interspecific differences in load-resisting ability relative to an external force estimate may be related solely to interspecific variation in that external force estimate. Alternatively, if a proportional amount of morphological change had occurred only in the load-resisting feature, then the functional consequences for relative load resistance would be the same in both conditions. This ability of shape ratios to describe proportionality with an explicit functional interpretation offers researchers the capacity to describe the functional consequences of changing masticatory apparatus form across primate groups of varying sizes and shapes.

16 Putting Shape to Work

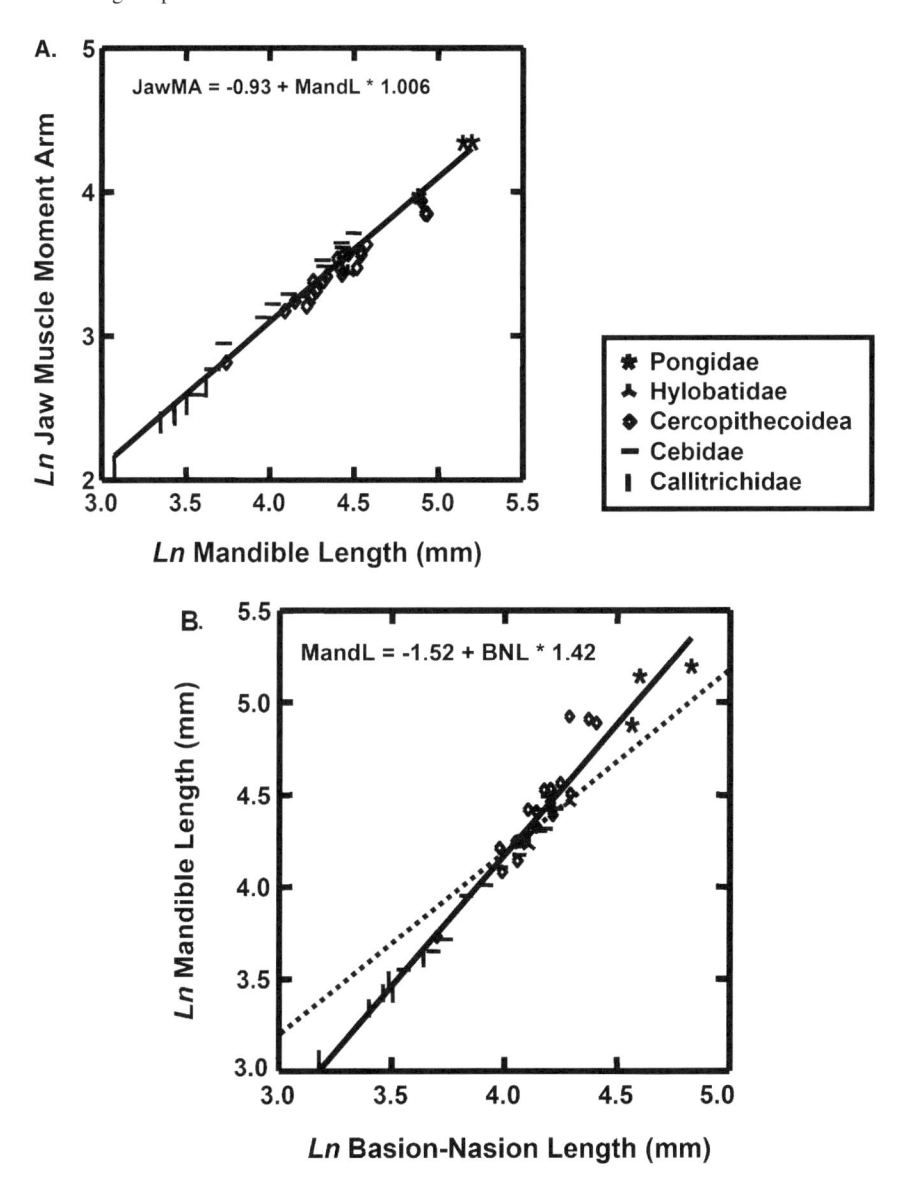

Fig. 16.2 Examination of the functional consequences and scaling of mechanical advantage for the jaw-closing muscles across 45 haplorhine primates. In this graphical example, mechanical advantage is a shape ratio with a biomechanical relevance for biting performance. (**a**) The jaw-muscle moment arm scales nearly isometrically relative to mandible length (Least-squares regression slope = 1.006). These two measures represent the in-lever and out-lever of the jaw-muscle mechanical advantage for biting, respectively. The jaw-closing muscle moment arm is estimated as the average of the moment arms of the masseter, temporalis, and medial pterygoid. Mandible length is taken as the distance from the back of the condyle to the central incisor. (**b**) Mandible length scales with positive allometry relative to basion-nasion length (BNL) as an estimate of skull size (LS slope = 1.42). (**c**) Similarly, jaw-muscle moment arm scales with positive allometry relative to BNL (LS slope = 1.45). In both (**b**) and (**c**), the dotted line indicates isometry.

C.

JawMA = -2.54 + BNL * 1.45

Ln Jaw Muscle Moment Arm (mm) vs *Ln* Basion-Nasion Length (mm)

D.

JawMA/MandL= -1.02 + BNL * 0.03 (p=0.4)

Ln Jaw Muscle Moment Arm / Mandible Length vs *Ln* Basion-Nasion Length (mm)

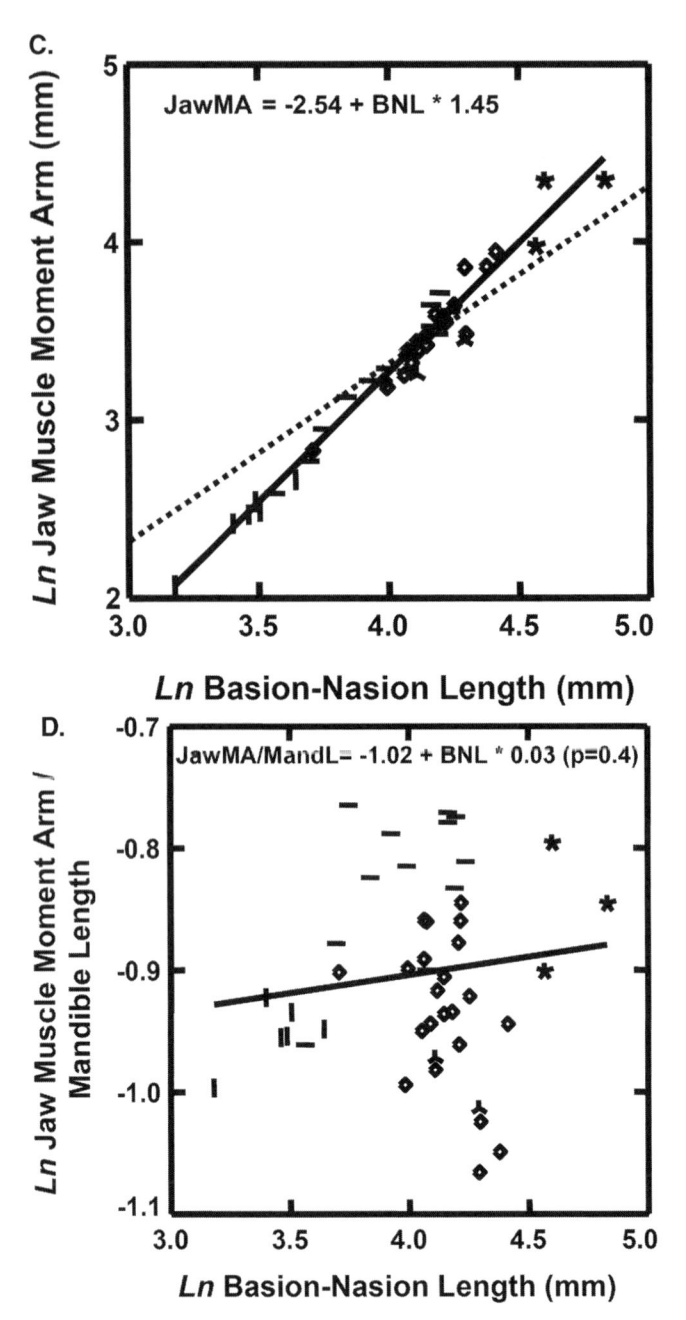

Fig. 16.2 (*Continued*) (**d**) Mechanical advantage does not change significantly with BNL, indicating that on average larger and smaller haplorhines have similar incisor biting leverage. This conclusion would not be apparent from examining the positively allometric slope of either component of the leverage ratio or from only comparing these components to a particular size estimate

16 Putting Shape to Work 365

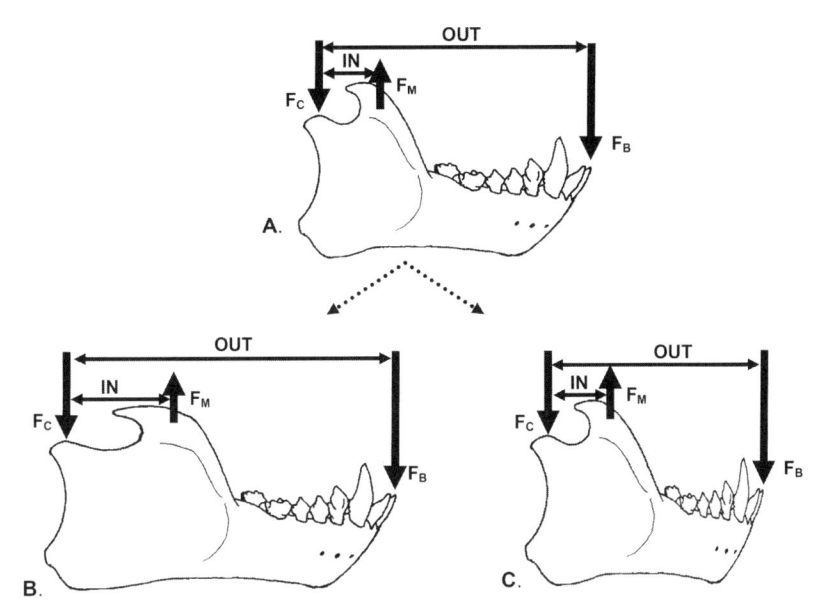

Fig. 16.3 A cartoon example of changing jaw shapes with similar functional consequences. (**a**) The initial jaw morphology, as seen in Fig. 16.1, where the mechanical advantage is approximately 1–4 (IN/OUT). One can change either the in-lever or out-lever to improve the mechanical advantage. (**b**) The ramus is elongated while corpus length is held constant to improve mechanical advantage (approximately 1–3). (**c**) The corpus is shortened while the ramus is held constant to yield a similar improvement in mechanical advantage (approximately 1–3). Thus, the biomechanical shape ratio, here mechanical advantage, is similar in (**b**) and (**c**) despite changes in different components of the ratio and concomitantly very different looking jaws

One of the drawbacks of studying shape ratios created using two functionally related variables is that it will be impossible to describe how either constituent variable is changing relative to overall size (Cole, 1992; Albrecht et al., 1993; Shea, 1995). Thus, a functional analysis based on shape ratios will not describe the allometric changes occurring in these variables. It will also be difficult to describe how either of these functional variables might be evolving across species. To describe these patterns, additional allometric analyses of these constituent variables relative to size will be required. Combining these allometric descriptions with a functional analysis of shape ratios can provide a more robust understanding of both the size-related changes in a feature and the functional consequences of this change (Darroch and Mosimann, 1985).

A final issue with using ratios in functional analyses relates to the identification of an appropriate denominator for a study.[3] The importance of this choice seems

---

[3] I have not presented an exhaustive list of the problems raised against using ratios in analyzing biological data. Additional concerns include the non-normal statistical distributions of ratios (e.g., Atchley et al., 1976), lack of statistical power (Packard and Boardman, 1987), and generation of spurious correlations (Weil, 1962; Atchley et al., 1976; Jackson et al., 1990). Each of these issues is worth considering when using ratios in an analysis, but beyond the limited scope of this chapter.

to have been neglected by many biological anthropologists and is the subject of the following two sections. In contrast to residuals and variance partitioning procedures where adequate sampling among individuals or species will help dictate the accuracy of shape estimates, shape ratios are intrinsic to the organism. They suffer a similar concern, however, in that the relevance of a shape ratio for a given research question will depend on how accurately the denominator and numerator fit the biological requirements of that question.

In summarizing this section, I argue that carefully constructed shape ratios present a number of appealing qualities for studying the functional consequences of size-related variation in primate masticatory apparatus form. This argument should be placed in the broader, and obvious, context that no one size-adjustment approach will satisfy the needs of all morphometric research. From a functional perspective, mechanically relevant shape ratios extend previous approaches aimed at examining shape variation relative to predetermined estimates of how functional equivalence changes with size (e.g., Gould, 1966; McMahon, 1973, 1975a,b; Smith, 1980; Shea, 1981, 1984, 1995; Ravosa and Vinyard, 2002; Lucas, 2004). The a priori determination of functional equivalence has proven extremely difficult and frankly biological anthropologists rarely employ this approach. Shape ratios created relative to a mechanical criterion provide an alternative method that allows us to examine the functional consequences of size-related change in form based on proportionality.

16.3 Choice of Size-Adjustment Variable: Overall Size Versus Functional Estimates

After deciding on a size-adjustment technique, the next step is to choose an appropriate variable to use in adjusting for size or other relevant dimension that will facilitate comparisons among animals of different magnitudes. I argue that biological anthropologists can do a better job in developing appropriate adjustments tailored to their questions involving primate morphology. Following the argument above in favor of using shape ratios in functional analyses, this section focuses on these ratios and the variable used as the denominator in relative adjustment. Much of the discussion, however, is also relevant to size adjustment techniques involving variance partitioning.

Emphasizing the importance of the denominator in a shape ratio is not new. Mosimann and colleagues stressed the significance of choosing an appropriate denominator in the analysis of morphological shapes over 30 years ago (Mosimann, 1970, 1975; Mosimann and James, 1979; Mosimann and James, 1979; Darroch and Mosimann, 1985). They provided a theoretical geometric argument demonstrating that a shape can change isometrically with only one size variable (Mosimann, 1970, 1975; Mosimann and James, 1979). Of course with statistical criteria for establishing isometry and added variation due to sampling, a shape variable can appear to be isometrically related to multiple size variables. Their key theoretical point, however, persists that "there is an infinite choice of size variables, . . . and one size variable is not generally recoverable from another" (Darroch and Mosimann, 1985:242). Thus, "there is no reason to expect similar size–shape associations for the different choices of a size variable" (Mosimann and James, 1979: 453).

Throughout the series of publications discussing shape ratios and their utility for studying morphology, Mosimann and colleagues frequently estimate size as a geometric mean of the variables included in their analysis (Mosimann, 1970; Mosimann and James, 1979; Darroch and Mosimann, 1985). They offer three reasons for using this geometric mean as a size estimate. First, it provides scale invariance with unequal changes of scale (Mosimann, 1970). Second, the geometric mean of all variables provides independence of shape and size for random data in a lognormal population (Mosimann, 1970). Finally, it corresponds to the multivariate isometry test of Jolicoeur (1963).

Despite these beneficial mathematical characteristics, Mosimann (1970:939) argues that the geometric mean of all variables "may be a desirable choice, if only because of a lack of precision in the definition of the biological problem." He goes on to state that "the choice of size variable is most contingent on a careful consideration of the biological purposes of the study, and on a precise definition of the size-shape problem involved" (Mosimann, 1970:939). Mosimann is arguing for a plurality of potential size variables stemming directly from the range of biological questions that can be addressed using shape variables. He holds the researcher responsible for developing the appropriate denominator that meets the needs of the biological question being considered (Mosimann, 1970).

Over the past 15 years, many biological anthropologists have come to rely on the geometric mean of several variables, usually all of those included in a study, as a size estimate in analyzing form and function (e.g., Jungers and Cole, 1992; Grine et al., 1993; Churchill, 1996; Hamrick, 1996a,b, 1998; Lague and Jungers, 1996; Dumont, 1997; Masterson, 1997; McCollum, 1997; Johnson and Shapiro, 1998; Richmond et al., 1998; Hamrick et al., 1999; Spencer, 1999; Vinyard and Smith, 2001; Jantz and Jantz, 2000; Williams et al., 2002; Lague, 2003; Vinyard et al., 2003; Ackermann, 2005; Lycett and Collard, 2005; O'Connor et al., 2005; Shapiro et al., 2005). In my opinion, this variable has been adopted uncritically as a size estimate in many cases and does not necessarily benefit the questions being addressed. Despite its ease of use, the perfunctory utilization of the geometric mean of all variables as a shape ratio denominator contradicts Mosimann's argument that the denominator be chosen with the biological question in mind.

The choice of denominator in a shape ratio is crucial for analyses involving comparative assessment of functional abilities across organisms (see also Smith, 1980, 1993). Even though ratios capture both shape intrinsic to an individual and proportionality across organisms, these estimates must be considered relative to the denominator used in that shape ratio (Mosimann, 1970; Mosimann and James, 1979; Jungers, 1985; Rohlf, 1990). As indicated above, a variable of interest may appear to have different shapes when assessed relative to different denominators, and hence potentially altered functional interpretations in comparisons with other individuals.

Placed in the context of analyzing primate form, it is difficult to provide a functional interpretation of shapes based on the geometric mean of all variables when the functional relevance of this denominator is unclear. This issue is exacerbated when a study lacks explanation of the functional significance of measurements and/or why various measurements were included in the study. As indicated above, the geometric mean of all variables is often interpreted as a measure of overall or regional size in the analyses of primate form. However, as pointed by Hylander (1985) for the masticatory apparatus, it is unclear how body or even head size directly relates to chewing function and performance. Thus, while overall size can have a clear functional interpretation for the locomotor system relating to the resistance of gravitational force, overall size estimates typically do not have an analogous functional conception with respect to the masticatory apparatus. This lack of a direct functional link, however, does not preclude size-correlated influences on masticatory apparatus form due to changes in body size.

When comparing the functional morphology of the masticatory apparatus in primates of different sizes, Hylander (1979, 1985) advocated examining jaw form relative to an appropriate mechanical parameter, such as a load arm estimate for chewing or biting. This approach allows the researcher to examine variation in form relative to a specifically defined functional criterion. As noted above, several researchers studying masticatory apparatus functional morphology in primates have applied this strategy. With respect to shape variables, I have referred to this functionally relevant denominator as a "biomechanical standard" (Vinyard et al., 2003:157). A biomechanical standard essentially aims to hold constant specific functional aspects of the system being studied. The resulting biomechanical shape variable capitalizes on proportionality to offer a morphometric estimate of the relative ability to perform a given task. Thus, this approach allows researchers to examine the functional consequences of changing form across primates that differ in size and masticatory apparatus form.

Using a biomechanical standard to create a shape variable departs from the more common approach of dividing a variable of interest by an overall size estimate (e.g., Falsetti et al., 1993; Jungers et al., 1995). Even though applying a biomechanical standard results in a shape variable that is mathematically equivalent to these other shape variables, the goals of these two approaches differ considerably. A biomechanical shape variable facilitates the comparison of functional abilities based on form among organisms, but it may not reliably indicate how shape changes relative to a size estimate. Thus, biomechanical shape variables may not be appropriate for examining size-correlated changes in shape as is typical of allometric studies (e.g., Smith, 1993; Jungers et al., 1995; Shea, 1995; Ravosa and Vinyard, 2002; Vinyard et al., 2003). Smith (1993) has previously distinguished between two categories of allometry based on whether overall size versus a biomechanical size variable are used as an independent variable. As an extension of the terminology offered by Smith (1993) for allometric descriptions, the approach discussed here can be described as a "biomechanical" shape analysis.

As an illustrative example of how these two approaches can yield different biological insights, we can re-visit the comparison of jaw shapes between cercopithecine and colobine monkeys. This comparison provided the empirical backdrop for a productive debate regarding overall size-related vs. biomechanical approaches to studying primate jaw form (Hylander, 1979, 1985; Smith, 1983, 1993;

Bouvier, 1986a; Ravosa, 1991c, 1996a). If we first compare jaw depth divided by the cube root of body mass, we see that cercopithecines have relatively deeper jaws than colobine primates (Fig. 16.4a) (see Smith, 1983). Alternatively, examining jaw depth relative to a biomechanical standard, in this case jaw length, indicates

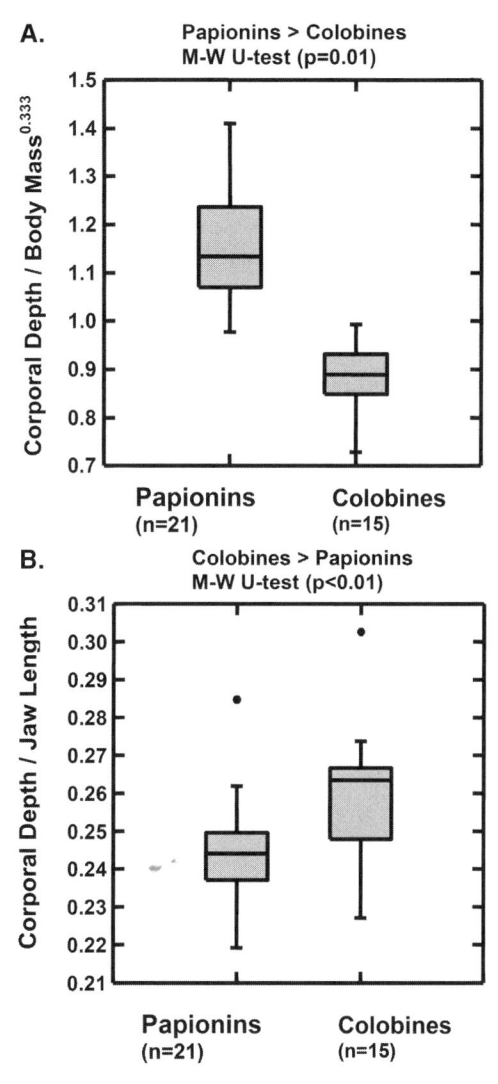

Fig. 16.4 Box plots comparing corporal depth of the mandible at M_1 in papionins and colobines relative to body mass$^{1/3}$ (**a**) and jaw length (**b**). Papionins have significantly deeper corpora than colobines when compared relative to body mass$^{1/3}$ (Mann-Whitney U-test, $p = 0.01$). Conversely, when compared relative to jaw length as a biomechanical estimate of the load arm during parasaggital bending, colobines have significantly deeper jaws than papionins (Mann-Whitney U-test, $p < 0.01$). This reversal in jaw depth shape between colobines and papionins emphasizes the importance of the denominator in shape variable construction. See text for further discussion. Data are from species means of papionin ($n = 21$) and colobine ($n = 15$) taxa

that colobines have significantly deeper jaws than cercopithecines (Fig. 16.4b) (see Hylander, 1979; Bouvier, 1986a). This reversal of relative jaw depth between groups likely relates to a combination of the positive of allometry of jaw length relative to body size (Hylander, 1985; Vinyard and Ravosa, 1998), as well as differences in relative body mass between these groups.

The choice of denominator in this example significantly alters the interpretation of relative jaw depth between these groups. As stated above, denominator choice should be dictated by the goals of a study. If a researcher aims to describe size-related differences in jaw shape between cercopithecines and colobines, then calculating shape variables relative to body mass is appropriate. Alternatively, the functional relevance of jaw length in masticatory mechanics makes it an appropriate biomechanical standard for assessing the functional consequences of differences in cercopithecine and colobine jaw shapes. Given that load arms lengths affect chewing and incisal external forces and internal loads, we can more readily interpret the functional consequences of variation in jaw depth between these groups using a biomechanical shape variable to hold these factors constant. After adjusting for this important biomechanical component, the relative deeper colobine jaw suggests that these primates are able to resist increased parasagittally directed bending of their jaws (Hylander, 1979; Bouvier, 1986a). This finding fits the behavioral data showing that colobines eat more leaves than cercopithecines (traditionally thought to be a tough food requiring numerous chewing cycles to break down). From a functional perspective, it is the proportional relationship between the two mechanically relevant components of the shape variable that is important. Identifying which component(s) of the shape variable changes with size is secondary. Without accounting for the positive allometry of cercopithecoid jaw length in the shape variable, we might misinterpret the functional relationship between diet and jaw shape between these two groups of Old World monkeys.

The argument in favor of examining functional abilities relative to a biomechanical standard in the previous example should not be mistaken for an argument against using geometric means as denominators for other questions relating to primate masticatory apparatus form. I am arguing for a pluralistic approach, particularly for functional comparisons where shapes based on geometric means are difficult to interpret, emphasizing multiple shape variables tailored to the biological questions being considered (Mosimann, 1970; Sprent, 1972; Darroch and Mosimann, 1985; Smith, 1993; Shea, 1995). A pluralistic approach that considers both functional and evolutionary allometric perspectives will benefit our understanding of primate masticatory apparatus form and function in the long run.

16.4 How Does a Shape Variable Denominator Affect a Hypothesis Test? A Case Study

The partiality that biological anthropologists show toward employing the geometric mean of all dimensions as a shape variable denominator coupled with the frequent lack of functional explanation for this geometric mean and its constituent dimensions suggests that we have not fully explored how the denominator of a shape variable impacts a hypothesis test. In this section, I use a case study focusing on jaw functional morphology in tree-gouging primates to examine how the results of hypothesis tests are altered by changing the denominator of a shape variable. Furthermore, these comparisons highlight how the functional interpretation of a jaw shape can hinge on the shape variable denominator.

To investigate these patterns, I re-analyze the morphometric data examined in Vinyard et al. (2003). This paper explored the functional morphology of the masticatory apparatus in habitual tree-gouging primates. We compared masticatory apparatus shapes in three tree-gouging species, *Phaner furcifer*, *Euoticus elegantulus*, and *Callithrix jacchus*, to closely related non-gouging species within their respective families (Vinyard et al., 2003; Table 16.1). Thirteen predicted morphologies were compared between gouging and non-gouging species. For most predictions, we calculated shape variables relative to jaw length as a biomechanical standard. This dimension captures both the load arm for biting and a key component of maximum jaw gape.[4] We performed Mann-Whitney U-tests on pairwise shape comparisons of a gouging species to each non-gouging species within its respective clade to assess these predictions.

I repeat this analysis here using shape variables created from six different denominators (Table 16.1). Four of these denominators are geometric means that estimate overall craniofacial size ($GM_{overall}$), jaw load-resistance ability (GM_{load}), the combination of overall size and load-resistance ability ($GM_{combined}$), and the

Table 16.1 The six denominators used in shape variable comparisons[1]

$GM_{combined}$ = (Mandibular Length * Corporal Depth * Corporal Breadth * Symphyseal Width * Symphyseal Length * M^1 Width * Nasion-Prosthion Height * Cranial Length * Palate Breadth at M^1 * Bi-Glenoid Breadth * Postcanine Toothrow Length)$^{1/11}$

GM_{load} = (Corporal Depth * Corporal Breadth * Symphyseal Width * Symphyseal Length * Condylar Length * Condylar Width)$^{1/6}$

$GM_{overall}$ = (Mandibular Length * M^1 Width * Nasion-Prosthion Height * Cranial Length * Palate Breadth at M^1 * Bi-Glenoid Breadth * Postcanine Toothrow Length)$^{1/7}$

GM_{all} = (Mandibular Length * Corporal Depth * Corporal Breadth * (Condylar Area$^{1/2}$) * Symphyseal Length *AP Mandibular Length * Condylar Height above Toothrow * Condylar Length * Glenoid Length * Minimum Skull Width * Masseter Moment Arm)$^{1/11}$

VarSum = (Mandibular Length + Corporal Depth + Corporal Breadth + (Condylar Area$^{1/2}$) + Symphyseal Length + AP Mandibular Length + Condylar Height above Toothrow + Condylar Length + Glenoid Length + Minimum Skull Width + Masseter Moment Arm)

Biomechanical Standard = Mandible length

---

[1] Landmark definitions of these variables are available in Vinyard (1999) or from the author.

---

[4] For predictions involving mandible length, we examined variation in this dimension relative to a skull size geometric mean. One other prediction involved a functional ratio of two scalars, and as such was already dimensionless.

geometric mean of all variables (GM$_{all}$), respectively (Table 16.1). The fifth denominator is the sum of all variables (VarSum) (Mosimann, 1970; Mosimann and James, 1979), while the final denominator is the biomechanical standard applied in Vinyard et al. (2003). It is reasonable to ask why these six denominators were chosen from the long list of possible denominators. Their importance rests in that, with the exception of the VarSum, denominators analogous to the GM$_{overall}$ (e.g., O'Connor et al., 2005), GM$_{load}$ (i.e., strictly functional) (e.g., Hamrick et al., 1999), GM$_{combined}$ (e.g., Vinyard et al., 2003), GM$_{all}$ (e.g., Dumont, 1997), and biomechanical standard (e.g., Hylander, 1985) can be identified in earlier studies. Thus, the results of these comparisons may highlight an important analytical issue that deserves further consideration by biological anthropologists.

For each denominator, 11 shape ratios are created based on the 13 predictions made by Vinyard et al., (2003). (The difference in the number of shape ratios and original predictions reflect the facts that the same variable was used previously in two distinct predictions, and the prediction involving a ratio of two scalars was not repeated here.) The predictions including the variables of interest are listed in Table 16.2. Vinyard et al. (2003) provides a more detailed discussion of these predictions.

Following Vinyard et al. (2003), compared shape variables using Mann-Whitney U-tests between each gouging species and non-gougers within their respective families for these 11 predictions. In this case, the same set of hypothesis tests were repeated six times for the shape variables created from each of the six different denominators. To quantify how these outcomes differed across these six denominators, I developed a ranking scale based on the interpretations of the Mann-Whitney U-test results. A five-level rank scale was developed, which focuses on statistical test results to differentiate outcomes based on varying the shape variable denominator (Table 16.3). Admittedly, there are a number of interpretative scales that could be developed to evaluate the influence of the shape variable denominator. In this case, I focused on the statistical "bottom-line," emphasizing the role that the outcomes of statistical tests play in our biological interpretations of morphological patterns. After assigning each Mann-Whitney U-test result to one of the five ranked categories, I calculated the percentage of similar ranks across the 11 comparisons for all possible pairs of shape variable denominators.

The results of these repeated hypothesis tests indicate a wide range of variation in the percentage similarity of ranks among shape variable denominators (Table 16.4). In some cases, two shape variable denominators yielded very similar ranks throughout hypothesis tests involving a particular gouging species. For example, the biomechanical standard showed similar ranks in 91% of the comparisons relative to both the variable sum in cheirogaleids and the GM$_{all}$ in galagids (Table 16.4). This can be contrasted with the 9% similarity between the GM$_{overall}$ and the GM$_{load}$ for hypothesis tests involving galagids. Across all comparisons, the six shape variable denominators had similar ranks for corresponding hypothesis tests in 60% of the cases. These results indicate that the shape variable denominator can have a significant effect on hypothesis testing and the subsequent interpretation of morphological shapes (see also Mosimann and James, 1979).

Table 16.2 Morphological predictions used to examine how shape variable denominators influence functional hypothesis tests. All predictions come from Vinyard et al. (2003) and relate to predicted differences in the skull form of tree-gouging primates relative to non-gouging forms

Prediction[1]	Variable of interest (numerator)	Mechanical prediction
1.	Mandibular corpus depth at M_1	Relatively larger in gouging primates to provide increased load-resistance ability.
2.	Mandibular corpus width at M_1	Relatively larger in gouging primates to provide increased load-resistance ability.
3.	Symphyseal length	Relatively larger in gouging primates to provide increased load-resistance ability.
4.	Condylar area	Relatively larger in gouging primates to provide increased load-resistance ability.
5.	Minimum cranial width	Relatively larger in gouging primates to provide increased load-resistance ability.
6.	Mandibular length	Relatively shorter in gouging primates to increase biting mechanical advantage.
7.	Condylar height	Predicted to be either relatively taller to increase masticatory muscle size and force or relatively shorter to increase maximum jaw gape in gouging primates.
8.	Masseter moment arm	Relatively larger in gouging primates to increase biting mechanical advantage.
9.	AP mandibular length	Relatively longer in gouging primates to provide increased maximum jaw gapes.
10.	AP condylar length	Relatively longer in gouging primates to provide increased maximum jaw gapes.
11.	Temporal articular surface length	Relatively longer in gouging primates to provide increased maximum jaw gapes.

[1] See Vinyard et al. (2003) for further description of each prediction including references.

Table 16.3 The five-level ranking scale for assessing similarity among hypothesis test results. Each Mann-Whitney U-test result was assigned to one of these levels to facilitate comparison among tests involving different shape-variable denominators

1. Gouging species significantly larger than non-gouging species after Bonferroni correction.[1]
2. Gouging species tends to be larger than non-gouging species ($p < 0.05$), but not after Bonferroni correction.
3. No statistical difference between gouging and non-gouging species ($p > 0.05$).
4. Non-gouging species tends to be larger than gouging species ($p < 0.05$), but not after Bonferroni correction.
5. Non-gouging species significantly larger than gouging species after Bonferroni correction.

[1] Because of the multiple (11) statistical comparisons for each pair of species, a sequential Bonferroni technique was applied to set a pairwise significance level of $\alpha = 0.05$ for all statistical comparisons involving a gouging and non-gouging species (see Rice, 1989).

We see a comparable range of variation in rank similarity for the corresponding tests focusing on each of the three gouging species (Table 16.4). The highest percentage of similarity in outcomes (74%) is seen in comparisons of *Callithrix jacchus* to non-gouging callitrichids, while the least similarity (54%) is apparent for hypothesis tests involving *Euoticus elegantulus*. *Phaner furcifer* and cheirogaleids are intermediate, exhibiting 62% similarity in ranks across these six denominators. In sum, the relative impact of the shape variable denominator will vary across species of interest, but based on these results it appears that this denominator will affect the statistical outcomes in most species.

The differences in percentage similarity of ranks among species might be attributed to numerous, non-mutually exclusive effects. This analysis was not designed to parcel out these potential effects. However, one effect worth mentioning relates to the magnitude of morphological differences between the species of interest and its comparative taxa. For example, *C. jacchus* might exhibit greater morphological divergence relative to non-gouging callitrichids when compared to the other gouging species and their respective comparative taxa. If true, then comparisons involving marmosets might have a greater morphological buffer to shape changes linked to varying the shape variable denominator. Unfortunately, it will be difficult to determine a priori which species of interest might exhibit greater divergence.

Table 16.4 Percentage similarity of ranks based on statistical outcomes from hypothesis tests using various shape variable denominators.

A. *P. furcifer* vs. cheirogaleids[1]

Shape variable denominator	GM combined	GM load	GM overall	GM all	Variable Sum	Biomechanical standard
GM Combined	—					
GM Load	40% (22/55)[2]	—				
GM Overall	67% (37/55)	29% (16/55)	—			
GM Relative Size	76% (42/55)	35% (19/55)	73% (40/55)	—		
Variable Sum	71% (39/55)	44% (24/55)	65% (36/55)	75% (41/55)	—	
Biomechanical Standard	60% (27/45)[3]	47% (21/45)	73% (33/45)	89% (40/45)	91% (41/45)	—

Mean Similarity Among Shape Variable Denominators = 62%

B. *E. elegantulus* vs. galagids[4]

GM Combined	—					
GM Load	15% (10/66)	—				
GM Overall	70% (46/66)	9% (6/66)	—			
GM Relative Size	79% (52/66)	17% (11/66)	64% (42/66)	—		
Variable Sum	65% (43/66)	33% (22/66)	56% (37/66)	74% (49/66)	—	
Biomechanical Standard	72% (39/54)	30% (16/54)	63% (34/54)	91% (49/54)	81% (44/54)	—

Mean Similarity Among Shape Variable Denominators = 54%

Table 16.4 (continued)

C. *C. jacchus* vs. callitrichids[5]

Shape variable denominator	GM combined	GM load	GM overall	GM all	Variable sum	Biomechanical standard
GM Combined	—					
GM Load	77% (17/22)	—				
GM Overall	82% (18/22)	82% (18/22)	—			
GM Relative Size	59% (13/22)	73% (16/22)	64% (14/22)	—		
Variable Sum	68% (15/22)	86% (19/22)	68% (15/22)	68% (15/22)	—	
Biomechanical Standard	72% (13/18)	78% (14/18)	72% (13/18)	72% (13/18)	89% (16/18)	—

Mean Similarity Among Shape Variable Denominators = 74%

[1] *Phaner furcifer* was compared to five non-gouging cheirogaleids (i.e., *Cheirogaleus major, C. medius, Microcebus murinus, M. rufus,* and *Mirza coquereli*).
[2] Percentage of results from hypothesis tests sharing the same rank. The numbers in parentheses indicate the number of tests yielding similar ranks relative to the total number of statistical tests.
[3] Hypothesis tests involving jaw length as the variable of interest were omitted for shape variables created using the biomechanical standard (i.e., jaw length) as the denominator. Therefore, pairwise comparisons involving the biomechanical standard are based on nine rather than eleven hypothesis tests per species pair.
[4] *Euoticus elegantulus* was compared to six non-gouging galagids (i.e., *Galagoides alleni, Galago moholi, G. senegalensis, G. zanzibaricus, G. demidoff,* and *G. gallarum*).
[5] *Callithrix jacchus* was compared to two non-gouging callitrichids (i.e., *Leontopithecus rosalia* and *Saguinus fuscicollis*).

Lack of concordance in outcome ranks for a significant percentage of the statistical tests is an important observation helping us understand how the denominator of a shape variable affects our morphological hypothesis tests. However, we also need to consider how the resulting functional interpretations might vary by altering the shape variable denominator. Admittedly, a high percentage of dissimilarity in outcome ranks might be mollified if it can be demonstrated that the bottom-line functional interpretation does not change appreciably.

As a graphical example of how the functional interpretation can differ, Fig. 16.5 illustrates changes in relative jaw length in *E. elegantulus* compared to non-gouging galagids when varying five of these shape variable denominators. The first three plots (Fig. 16.5a–c) similarly show *E. elegantulus* with a jaw length shape that is within the range of shapes for non-gouging galagids. This pattern markedly changes in the final two plots (Fig. 16.5d, e). Relative to the GM_{load}, *E. elegantulus* has the shortest jaw among these galagids (Fig. 16.5d). This pattern is reversed for GM_{over} with *E. elegantulus* having the relatively longest jaw among these species (Fig. 16.2e). The functional importance of jaw length as both the load arm during biting and a key component of maximum jaw gape has been noted and is emphasized by using this dimension as the biomechanical standard in Vinyard et al. (2003).

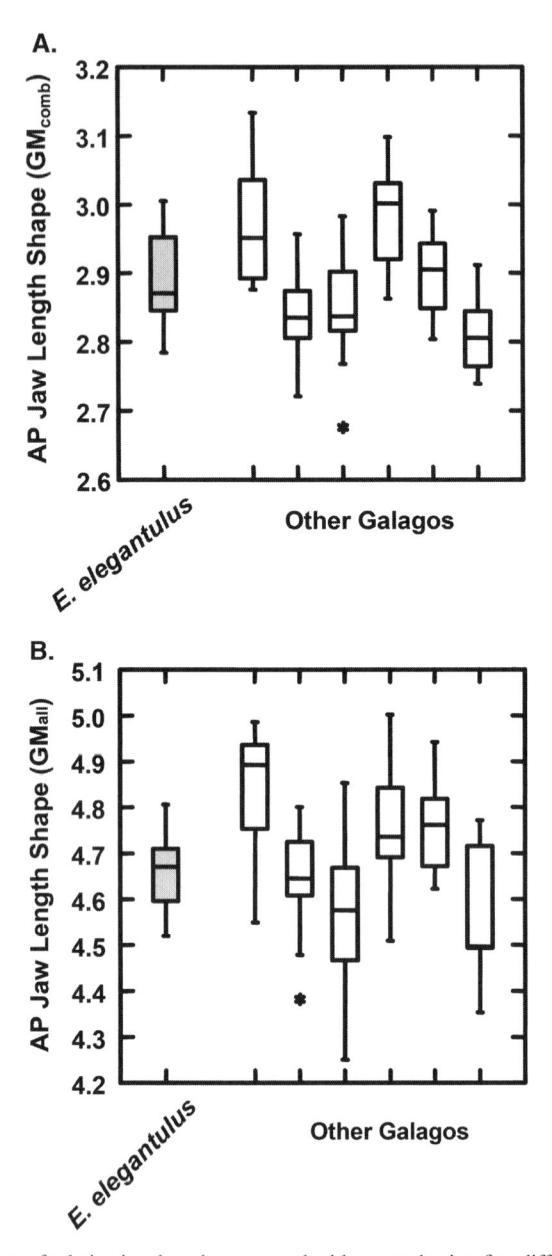

Fig. 16.5 Box plots of relative jaw length among galagids created using five different shape ratios. Each shape ratio was developed by dividing jaw length by a different denominator (as indicated in parentheses). The functional interpretation of relative jaw length in the tree-gouging *Euoticus elegantulus* differs in comparison to non-gouging galagids depending on the shape variable denominator. Data from Vinyard et al., (2003)

16 Putting Shape to Work

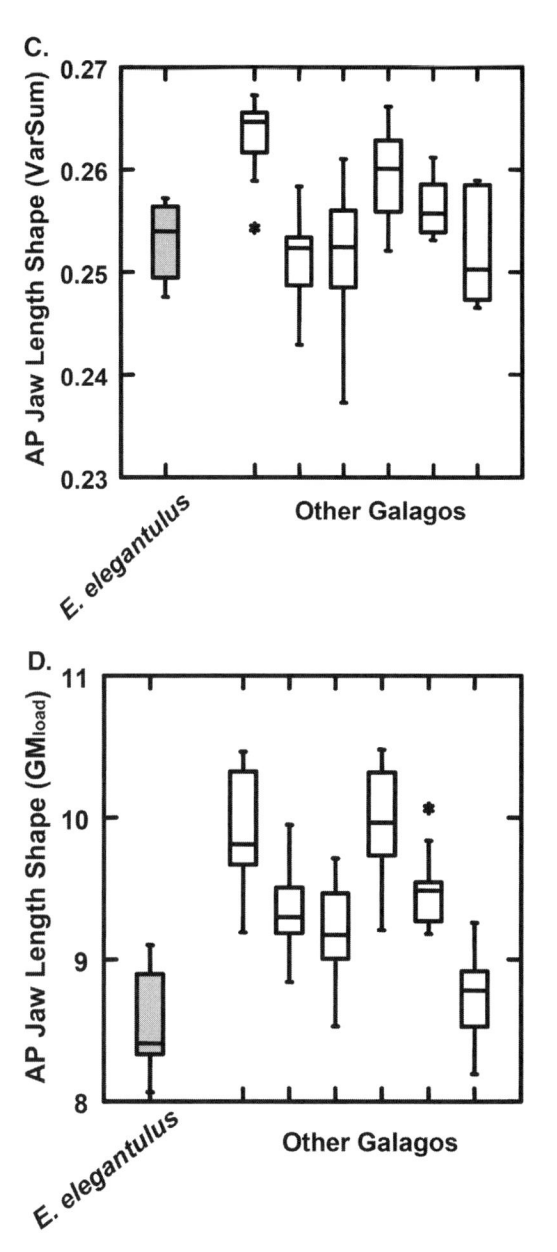

Fig. 16.5 (Continued)

By choosing among these five denominators, it is possible to make practically any functional interpretation of jaw length shape in *E. elegantulus* relative to nongouging galagids. Given the range of possible functional interpretations based on varying the shape variable denominator, this example stresses the significance of

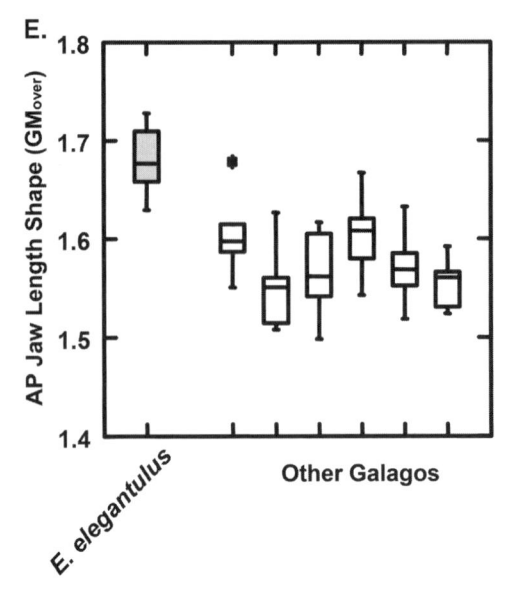

Fig. 16.5 (Continued)

determining an appropriate denominator for examining the functional consequences of proportional change in a variable of interest.

The analysis and example presented in this section emphasize the importance of shape variable denominators for functional analyses of primate morphology. This denominator can have a significant impact on both the statistical outcome and the functional interpretation of comparative analyses. Based on this potential influence, denominator choice deserves a thorough consideration in studies of primate jaw functional morphology, as well as other studies of primate form. Even though some denominators yielded the same bottom-line results in this analysis, it was not possible to identify this similarity a priori. Therefore, determining how the denominator might influence an analysis must be done on a case-by-case basis. This lack of predictability coupled with its significant potential influence on our biological interpretations strongly suggest that we pay close attention to the choice of shape variable denominators.

16.5 What Can We Reliably Interpret from Comparative Analyses of Shapes?

The methodological and interpretative challenges for shape variables demonstrated in the previous section raise the important question, "What can we reasonably expect to learn about function and adaptation by comparing jaw shapes?" We can examine this question using the previous example involving tree-gouging primates. Recent comparative morphometric analyses of jaw shapes in tree-gouging primates have variably concluded that these primates *do* (Dumont, 1997), *may* (Williams et al., 2002, Viguier, 2004), or *do not* (Vinyard et al., 2003) exhibit relatively robust jaws as compared to non-gouging species. Given these conflicting interpretations, it is likely that morphometric analyses by themselves will be insufficient to fully understand the functional and evolutionary morphology of the jaws in tree-gouging primates (Vinyard et al., 2003). I would argue that a main reason for this shortcoming of comparative morphometric analyses relates to the explicit assumptions, such as those highlighted in the previous sections, that must be made in these analyses (see also Corruccini, 1987). If one or more of these assumptions turns out to be incorrect, then the resulting functional interpretations are invalidated.

It is admittedly tempting to compare jaw shapes among organisms and then interpret the observed differences in morphology as indicative of distinct jaw functions and/or adaptations. In my opinion, it is this attraction that led to the contradictory interpretations listed above for the jaw's functional role in resisting loads during tree gouging. Unfortunately, attempts to discern function and adaptation from form via comparative analyses are fraught with assumptions and difficulties that severely limit our ability to succeed in this endeavor (Bock, 1977, 1989; Brandon, 1990; Leroi et al., 1994; Lauder, 1995). The assumptions and limitations to discerning adaptation from comparative morphometric analyses are well known and do not require detailed explanation here. In brief, these issues highlight that comparative morphometrics often fail to indicate 1) the ancestral morphology, 2) the link between morphological variation and relative performance, 3) what the biological role(s) of a variable of interest is in its natural environment, and/or 4) whether this biological role has been influenced by natural selection (e.g., what are the fitness consequences of variation in form). All of these factors combine to offer serious challenges to the interpretations of function and adaptation from the comparisons of form.

Given these limitations, what can we reasonably expect to learn by comparing jaw forms among groups? I argue that the basic insight gained in this effort involves an improved understanding of the *functional consequences* of a given shape. If we hold all other variables constant, we can evaluate the relative functional capabilities of different forms in a given task. This argument assumes that we have used reasonable proxies of functional abilities in our measurements (e.g., Daegling, 2007). For shape analyses, this concern relates specifically to the identification of an appropriate biomechanical standard for the shape variable denominator. The interpretation of variation in functional consequences is not contingent on a given performance, natural environment, or character polarity. The functional consequence of a difference in form simply reports its relative ability to execute an assumed task. For tree gouging, the functional consequence of corporal shape might be considered in the context of its ability to resist parasagittal bending loads that are hypothesized to occur when an animal bites a tree.

By focusing on the functional consequences of a given form, we can reduce some, but not all, of the assumptions inherent in interpreting comparative morphometric analyses. Through this approach, we relegate the determination of whether a task is functionally relevant or actually takes place in an animal's environment to in vivo laboratory and field work, respectively (Bock and von Wahlert, 1965). While advocating for a more restricted interpretation of morphometric comparative analyses, I am not arguing that we refrain from generating adaptive hypotheses based on findings from morphometric analyses. By recognizing and focusing on what comparative morphometric analyses can tell us about the primate masticatory apparatus, we can use these analyses to more effectively stimulate future studies that integrate morphometric, laboratory, and field research for improving our understanding of primate cranial form.

Acknowledgments William Hylander has made an unparalleled contribution to our understanding of primate craniofacial function. As someone interested in this topic, I thank Bill for his many years of hard work and dedication. Any success that my generation has in advancing our understanding of primate craniofacial function will be made on the foundations that Bill built. More personally, Bill has been a terrific mentor, great colleague, and wonderful friend. I am grateful for all that you have done for me and I am indebted to you.

I thank W. Hylander, M. Ravosa, B. Shea, R. Smith, A. Taylor, P. Vinyard, and C. Wall for numerous helpful discussions and critical comments that greatly improved this chapter. Additionally, I thank W. Jungers and B. Shea for bringing the work of J. Mosimann to my attention. Finally, I am grateful to the following institutions and individuals for providing access to the primate skulls analyzed here: Field Museum of Natural History – L. Heaney, B. Patterson, W. Stanley; National Museum of Natural History – L. Gordon, R. Thorington; American Museum of Natural History – R. MacPhee; Natural History Museum, London- P. Jenkins; Muséum National d'Histoire Naturelle – J. Cuisin; Naturhistorisches Museum Basel – M. Sutermeister, F. Weidenmayer. Funding for the morphometric data collection was provided by Sigma-Xi, NSF (SBR-9701425, BCS-0094666), and the L.S.B. Leakey Foundation.

References

Ackermann, R.R. (2005). Ontogenetic integration of the hominoid face. *J. Hum. Evol.* 48:175–197.
Albrecht, G.H., Gelvin, B.R., and Hartman, S.E. (1993). Ratios as a size adjustment in morphometrics. *Am. J. Phys. Anthropol.* 91:441–468.
Albrecht, G.H., Gelvin, B.R., and Hartman, S.E. (1995). Ratio adjustments in morphometrics: a reply to Dr. Corruccini. *Am. J. Phys. Anthropol.* 96:193–197.
Anthony, M.R.L., and Kay, R.F. (1993). Tooth form and diet in Ateline and Alouattine primates: reflections on the comparative method. *Am. J. Sci.* 293-A:356–382.
Anton, S.C. (1996). Cranial adaptation to a high attrition diet in Japanese macaques. *Int. J. Primatol.* 17:401–427.
Atchley, W.R., Gaskins, C.T., and Anderson, D. (1976). Statistical properties of ratios. I. Empirical results. *Syst. Zool.* 25:137–148.
Atchley, W.R., and Hall, B.K. (1991). A model for development and evolution of complex morphological structures. *Biol. Rev. Camb. Phil. Soc.* 66:101–158.
Bastir, M., and Rosas, A. (2004). Facial heights: evolutionary relevance of postnatal ontogeny for facial orientation and skull morphology in humans and chimpanzees. *J. Hum. Evol.* 47:359–381.
Bock, W.J. (1977). Adaptation and the comparative method. In: Hecht, M.K., Goody, P.C., and Hecht, B.M. (eds.), *Major Patterns in Vertebrate Evolution*, Volume 14. Plenum Press, New York, pp. 57–82.
Bock, W.J. (1989). Principles of biological comparison. *Acta Morphol. Neerl.-Scand.* 27:17–32.

Bock, W.J., and von Wahlert, G. (1965). Adaptation and the form-function complex. *Evolution* 19:269–299.
Bookstein, F.L. (1989). "Size and shape": a comment on semantics. *Syst. Zool.* 38:173–180.
Bouvier, M. (1986a). A biomechanical analysis of mandibular scaling in Old World monkeys. *Am. J. Phys. Anthropol.* 69:473–482.
Bouvier, M. (1986b). Biomechanical scaling of mandibular dimension in New World Monkeys. *Int. J. Primatol.* 7:551–567.
Brandon, R.N. (1990). *Adaptation and Environment*. Princeton University Press, Princeton.
Churchill, S.E. (1996). Particulate versus integrated evolution of the upper body in late pleistocene humans: a test of two models. *Am. J. Phys. Anthropol.* 100:559–583.
Cole, T.M. (1992). Postnatal heterochrony of the masticatory apparatus in *Cebus apella* and *Cebus albifrons*. *J. Hum. Evol.* 23:253–282.
Corruccini, R.S. (1987). Shape in morphometrics: comparative analyses. *Am. J. Phys. Anthropol.* 73:289–303.
Corruccini, R.S. (1995). Of ratios and rationality. *Am. J. Phys. Anthropol.* 96:189–191.
Daegling, D.J. (1989). Biomechanics of cross-sectional size and shape in the hominoid mandibular corpus. *Am. J. Phys. Anthropol.* 80:91–106.
Daegling, D.J. (1992). Mandibular morphology and diet in the genus *Cebus*. *Int. J. Primatol.* 13:545–570.
Daegling, D.J. (1993). The relationship of in vivo bone strain to mandibular corpus morphology in (*Macaca fascicularis*). *J. Hum. Evol.* 25:247–269.
Daegling, D.J. (2001). Biomechanical scaling of the hominoid mandibular symphysis. *J. Morphol.* 250:12–23.
Daegling, D.J. (2007). Morphometric estimation of torsional stiffness and strength in primate mandibles. *Am. J. Phys. Anthropol.* 132:261–266.
Daegling, D.J., and McGraw, W.S. (2001). Feeding, diet, and jaw form in West African *Colobus* and *Procolobus*. *Int. J. Primatol.* 22:1033–1055.
Darroch, J.N., and Mosimann, J.E. (1985). Canonical and principal components of shape. Biometrika 72:241–252.
Demes, B., Preuschoft, H., and Wolff, J.E.A. (1984). Stress-strength relationships in the mandibles of hominoids. In: Chivers, D.J., Wood, B.A., and Bilsborough, A. (eds.), *Food Acquisition and Processing in Primates*. Plenum Press, New York, pp. 369–390.
Dumont, E.R. (1997). Cranial shape in fruit, nectar, and exudate feeders: implications for interpreting the fossil record. *Am. J. Phys. Anthropol.* 102:187–202.
Falsetti, A.B., Jungers, W.L., and Cole, T.M. (1993). Morphometrics of the callitrichid forelimb: a case study in size and shape. *Int. J. Primatol.* 14:551–572.
Fleagle, J.G. (1985). Size and adaptation in primates. In: Jungers, W.L. (ed.), *Size and Scaling in Primate Biology*. Plenum Press, New York, pp. 1–19.
Gauld, S.C. (1996). Allometric patterns of cranial bone thickness in fossil hominids. *Am. J. Phys. Anthropol.* 100:411–426.
Godfrey, L.R., Samonds, K.E., Jungers, W.L., and Sutherland, M.R. (2001). Teeth, brains, and primate life histories. *Am. J. Phys. Anthropol.* 114:192–214.
Gould, S.J. (1966). Allometry and size in ontogeny and phylogeny. *Biol. Rev.* 41:587–640.
Gould, S.J. (1975). Allometry in primates, with emphasis on scaling and the evolution of the brain. In: Szalay, F.S. (ed.), *Approaches to Primate Paleobiology*, Karger, Basel, pp. 244–292.
Grine, F.E., Demes, B., Jungers, W.L., and Cole, T.M. (1993). Taxonomic affinity of the early *Homo* cranium from Swartkrans, South Africa. *Am. J. Phys. Anthropol.* 92:411–426.
Hamrick, M.W. (1996a). Functional morphology of the lemuriform wrist joints and the relationship between wrist morphology and positional behavior in arboreal primates. *Am. J. Phys. Anthropol.* 99:319–344.
Hamrick, M.W. (1996b). Locomotor adaptations reflected in the wrist joints of early Tertiary primates (Adapiformes). *Am. J. Phys. Anthropol.* 100:585–604.
Hamrick, M.W. (1998). Functional and adaptive significance of primate pads and claws: evidence from New World anthropoids. *Am. J. Phys. Anthropol.* 106:113–127.

Hamrick, M.W., Rosenman, B.A., and Brush, J.A. (1999). Phalangeal morphology of the Paromoyidae (?Primates, Plesiadapiformes): the evidence for gliding behavior reconsidered. *Am. J. Phys. Anthropol.* 109:397–413.

Hills, M., and Woods, B.A., (1984). Regression lines, size and allometry. In: Chivers, D.J., Wood, B.A., and Bilsborough, A. (eds.), *Food Acquisition and Processing in Primates.* Plenum Press, New York, pp. 557–567.

Hylander, W.L. (1979). The functional significance of primate mandibular form. *J. Morphol.* 160:223–240.

Hylander, W.L. (1985). Mandibular function and biomechanical stress and scaling. *Am. Zool.* 25:315–330.

Hylander, W.L. (1988). Implications of in vivo experiments for interpreting the functional significance of "robust" australopithecine jaws. In: Grine, F.E. (ed.), *Evolutionary History of the "Robust" Australopithecines,* Aldine de Gruyter, New York, pp. 55–83.

Jablonski, N.G. (1993). Evolution of the masticatory apparatus in *Theropithecus*. In: Jablonski, N.G., (ed.), *Theropithecus: The Rise and Fall of a Primate Genus*, Cambridge University Press, Cambridge, pp. 299–329.

Jackson, D.A., Harvey, H.H., and Somers, K.M. (1990). Ratios in aquatic sciences: statistical shortcomings with mean depth and the morphoedaphic index. *Can. J. Fish. Aquat. Sci.* 47:1788–1795.

James, F.C., and McCulloch, C.E. (1990). Multivariate analysis in ecology and systematics: panacea or Pandora's box? *Ann. Rev. Ecol. Syst.* 21:129–166.

Jantz, R.L., and Jantz, L.M. (2000). Secular changes in craniofacial morphology. *Am. J. Hum. Biol.* 12:327–338.

Jerison, H.J. (1973). *Evolution of the Brain and Intelligence,* Academic Press, New York.

Johnson, S.E., and Shapiro, L.J. (1998). Positional behavior and vertebral morphology in atelines and cebines. *Am. J. Phys. Anthropol.* 105:333–354.

Jolicoeur, P. (1963). The multivariate generalization of the allometry equation. *Biometrics* 19:497–499.

Jungers, W.L. (1985). Body size and scaling of limb proportions in primates. In: Jungers, W.L. (ed.), *Size and Scaling in Primate Biology,* Plenum Press, New York, pp. 345–382.

Jungers, W.L. (1988). Relative joint size and hominoid locomotor adaptations with implications for the evolution of hominid bipedalism. *J. Hum. Evol.* 17:247–265.

Jungers, W.L., Burr, D.B., and Cole, M.S. (1998). Body size and scaling of long bone geometry, bone strength, and positional behavior in cercopithecoid primates. In: Strasser, E., Fleagle, J., Rosenberger, A., and McHenry, H. (eds.), *Primate Locomotion: Recent Advances,* Plenum Press, New York, pp. 309–330.

Jungers, W.L., and Cole, M.S. (1992). Relative growth and shape of the locomotor skeleton in lesser apes. *J. Hum. Evol.* 23:93–105.

Jungers, W.L., Falsetti, A.B., and Wall, C.E. (1995). Shape, relative size, and size-adjustments in morphometrics. *Yrbk. Phys. Anthropol.* 38:137–162.

Kay, R.F. (1975). The functional significance of primate molar teeth. *Am. J. Phys. Anthropol.* 43:195–215.

Kay, R.F. (1978). Molar structure and diet in extant Cercopithecidae. In: Butler, P.M., and Joysey, K., (eds.), *Development, Function and Evolution of Teeth,* Academic Press, London, pp. 309–339.

Kermack, K.A., and Haldane, J.B.S. (1950). Organic correlation and allometry. *Biometrika* 37:30–41.

Kidd, R., and Oxnard, C. (2005). Little Foot and big thoughts – a re-evaluation of the Stw573 foot from Sterkfontein, South Africa. *Homo* 55:189–212.

Lague, M.R. (2003). Patterns of joint size dimorphism in the elbow and knee of catarrhine primates. *Am. J. Phys. Anthropol.* 120:278–297.

Lague, M.R., and Jungers, W.L. (1996). Morphometric variation in Plio-Pleistocene hominid distal humeri. *Am. J. Phys. Anthropol.* 101:401–427.

Lauder, G.V. (1995). On the inference of function from structure. In: Thomason, J. (ed.), *Functional Morphology in Vertebrate Paleontology*, Cambridge University Press, Cambridge, pp. 1–18.

Lemen, C.A. (1983). The effectiveness of methods of shape analysis. *Field. Zool.* 15:1–17.

Leroi, A.M., Rose, M.R., and Lauder, G.V. (1994). What does the comparative method reveal about adaptation? *Am. Nat.* 143:381–403.

Lucas, P.W. (2004). *Dental Functional Morphology: How Teeth Work*, Cambridge University Press, Cambridge.

Lycett, S.J., and Collard, M. (2005). Do homoiologies impede phylogenetic analyses of the fossil hominids? An assessment based on extant papionin craniodental morphology. *J. Hum. Evol.* 49:618–642.

Martin, R.D., and Barbour, A.D. (1989). Aspects of line-fitting in bivariate allometric analyses. *Folia Primatol.* 53:65–81.

Martin, R.D., and MacLarnon, A.M. (1988). Quantitative comparisons of the skull and teeth in guenons. In: Gautier-Hion, A., Bourliere, F., Gautier, J.P., and Kingdon, J., (eds.), *A Primate Radiation: Evolutionary Biology of the African Guenons*, Cambridge University Press, Cambridge, pp. 160–183.

Masterson, T.J. (1997). Sexual dimorphism and interspecific cranial form in two capuchin species: *Cebus albifrons* and *C. apella*. *Am. J. Phys. Anthropol.* 104:487–511.

McArdle, B.H. (1988). The structural relationship: regression in biology. *Can. J. Zool.* 66:2329–2339.

McCollum, M. (1997). Palatal thickening and facial form in *Paranthropus* examination of alternative developmental models. *Am. J. Phys. Anthropol.* 103:375–392.

McMahon, T.A. (1973). Size and shape biology. *Science* 179:1201–1204.

McMahon, T.A. (1975a). Allometry and biomechanics: limb bones in adult ungulates. *Am. Nat.* 109:547–563.

McMahon, T.A. (1975b). Using body size to understand the structural design of animals: quadrupedal locomotion. *J. Appl. Physiol.* 39:619–627.

Mosimann, J.E. (1970). Size allometry: size and shape variables with characterizations of the log-normal and generalized gamma distributions. *J. Am. Stat. Assoc.* 65:930–945.

Mosimann, J.E. (1975). Statistical problems of size and shape. I. Biological applications and basic theorems. In: Patil, G.P., Kotz, S., Ord, J.K., (eds.), *Statistical Distributions in Scientific Work, Vol. 2: Model Building and Model Selection*, D. Reidel Publishing Company, Dordrecht, pp. 187–217.

Mosimann, J.E., and James, F.C. (1979). New statistical methods for allometry with application to Florida Red-winged blackbirds. *Evolution*. 33:444–459.

Mosimann, J.E., and Malley, J.D. (1979). Size and shape variables. In: Orloci, L., Rao, C.R., and Stiteler, W.M., (eds.), *Multivariate Methods in Ecological Work*, International Co-operative Publishing House, Fairland MD, pp. 175–189.

O'Connor, C.F., Franciscus, R.G., and Holton, N.E. (2005). Bite force production capability and efficiency in Neandertals and modern humans. *Am. J. Phys. Anthropol.* 127:129–151.

Oxnard, C.E. (1978). One biologist's view of morphometrics. *Annu. Rev. Ecol. Syst.* 9:219–241.

Packard, G.C., and Boardman, T.J. (1987). The misuse of ratios to scale physiological data that vary allometrically with body size. In: Feder, M.E., Bennett, A.F., Burggren, W.W., and Huey, R.B., (eds.), *New Directions in Ecological Physiology*, Cambridge University Press, Cambridge, pp. 216–236.

Pan, R., Yanzhang, P., Zhizhang, Y., Hong, W., and Yu, F. (1995). Comparison of masticatory morphology between *Rhinopithecus bieti* and *R. roxellana*. *Am. J. Primatol.* 35:271–281.

Ravosa, M.J. (1990). Functional assessment of subfamily variation in maxillomandibular morphology among Old World monkeys. *Am. J. Phys. Anthropol.* 82:199–212.

Ravosa, M.J. (1991a). Ontogenetic perspective of mechanical and nonmechanical models of primate circumorbital morphology. *Am. J. Phys. Anthropol.* 85:95–112.

Ravosa, M.J. (1991b). Interspecific perspective on mechanical and nonmechanical models of primate circumorbital morphology. *Am. J. Phys. Anthropol.* 86:369–396.

Ravosa, M.J. (1991c). Structural allometry of the prosimian mandibular corpus and symphysis. *J. Hum. Evol.* 20:3–20.

Ravosa, M.J. (1996a). Jaw scaling and biomechanics in fossil taxa. *J. Hum. Evol.* 30:159–160.

Ravosa, M.J. (1996b). Jaw morphology and function in living and fossil Old World monkeys. *Int. J. Primatol.* 17:909–932.

Ravosa, M.J. (1996c). Mandibular form and function in North American and European Adapidae and Omomyidae. *J. Morphol.* 229:171–190.

Ravosa, M.J. (2000). Size and scaling in the mandible of living and extinct apes. *Folia Primatol.* 71:305–322.

Ravosa, M.J., and Vinyard, C.J. (2002). On the interface between ontogeny and function. In: Plavcan, J.M., Kay, R.F., Jungers, W.L., and van Schaik, C.P., (eds.), *Reconstructing Behavior in the Primate Fossil Record*, Kluwer Academic/Plenum Publishers, New York, pp. 73–111.

Ravosa, M.J., Vinyard, C.J., Gagnon, M., and Islam, S.A. (2000). Evolution of anthropoid jaw loading and kinematic patterns. *Am. J. Phys. Anthropol.* 112:493–516.

Rayner, J.M.V. (1985). Linear relations in biomechanics: the statistics of scaling functions. *J. Zool. Lond.* 206:415–439.

Reist, J.D. (1985). An empirical evaluation of several univariate methods that adjust for size variation in morphometric data. *Can. J. Zool.* 63:1429–1439.

Rice, W.R. (1989). Analyzing tables of statistical tests. *Evolution* 43:223–225.

Richmond, B.G., Fleagle, J.G., Kappelman, J., and Swisher, C.C. (1998). First hominoid from the Miocene of Ethiopia and the evolution of the catarrhine elbow. *Am. J. Phys. Anthropol.* 105:257–277.

Ricker, W.E. (1973). Linear regressions in fishery research. *J. Fish Res. Board Can.* 30:409–434.

Riska, B. (1991). Regression models in evolutionary allometry. *Am. Nat.* 138:283–299.

Rohlf, F.J. (1990). Morphometrics. *Annu. Rev. Ecol. Syst.* 21:299–316.

Shapiro, L.J., Seiffert, C.V.M., Godfrey, L.R., Jungers, W.L., Simons, E.L., and Randria, G.F.N. (2005). Morphometric analysis of lumbar vertebrae in extinct Malagasy strepsirrhines. *Am. J. Phys. Anthropol.* 128:823–839.

Shea, B.T. (1981). Relative growth of the limbs and trunk of the African apes. *Am. J. Phys. Anthropol.* 56:179–202.

Shea, B.T. (1983). Size and diet in the evolution of African ape craniodental form. *Folia Primatol.* 40:32–68.

Shea, B.T. (1984). An allometric perspective on the morphological and evolutionary relationships between pygmy (*Pan paniscus*) and common (*Pan troglodytes*) chimpanzees. In: Susman, R.L. (ed.), *The Pygmy Chimpanzee*, Plenum Press, New York, pp. 89–130.

Shea, B.T. (1985a). Bivariate and multivariate growth allometry: statistical and biological considerations. *J. Zool. Lond.* 206:367–390.

Shea, B.T. (1985b). Ontogenetic allometry and scaling: a discussion based on the growth and form of the skull in African apes, In: Jungers, W.L., (ed.), *Size and Scaling in Primate Biology*, Plenum Press, New York, pp. 175–205.

Shea, B.T. (1995). Ontogenetic scaling and size correction in the comparative study of primate adaptations. *Anthropology* 33:1–16.

Shea, B.T., and Gomez, A.M. (1988). Tooth scaling and evolutionary dwarfism: an investigation of allometry in human pygmies. *Am. J. Phys. Anthropol.* 77:117–132.

Smith, R.J. (1980). Rethinking allometry. *J. Theor. Biol.* 87:97–111.

Smith, R.J. (1983). The mandibular corpus of female primates: taxonomic, dietary, and allometric correlates of interspecific variations in size and shape. *Am. J. Phys. Anthropol.* 61:315–330.

Smith, R.J. (1984a). Comparative functional morphology of maximum mandibular opening (gape) in primates. In: Chivers, D.J., Wood, B.A., and Bilsborough, A., (eds.), *Food Acquisition and Processing in Primates*, Plenum Press, New York, pp. 231–255.

Smith, R.J. (1984b). Determination of relative size: the "criterion of subtraction" problem in allometry. *J. Theor. Biol.* 108:131–142.

Smith, R.J. (1993). Categories of allometry: body size versus biomechanics. *J. Hum. Evol.* 24:173–182.
Smith, R.J. (1994). Regression models for prediction equations. *J. Hum. Evol.* 26:239–244.
Smith, R.J. (2005). Relative size versus controlling for size. *Curr. Anthropol.* 46:249–273.
Smith, R.J., German, R.Z., and Jungers, W.L. (1986). Variability of biological similarity criteria. *J. Theor. Biol.* 118:287–293.
Smith, R.J., Petersen, C.E., and Gipe, D.P. (1983). Size and shape of the mandibular condyle in primates. *J. Morphol.* 177:59–68.
Spencer, M.A. (1999). Constraints on masticatory system evolution in anthropoid primates. *Am. J. Phys. Anthropol.* 108:483–506.
Sprent, P. (1972). The mathematics of size and shape. *Biometrics* 28:23–37.
Takahashi, L.K., and Pan, R. (1994). Mandibular morphometrics among macaques: the case of *Macaca thibetana*. *Int. J. Primatol.* 15:596–621.
Taylor, A.B. (2002). Masticatory form and function in African apes. *Am. J. Phys. Anthropol.* 117:133–156.
Taylor, A.B. (2005). A comparative analysis of temporomandibular joint morphology in the African apes. *J. Hum. Evol.* 48:555–574.
Viguier, B. (2004). Functional adaptations in the craniofacial morphology of Malagasy primates: shape variations associated with gummivory in the family Cheirogaleidae. *Ann. Anat.* 186:495–501.
Vinyard, C.J. (1999), *Temporomandibular Joint Morphology and Function in Strepsirhine and Eocene Primates*. Ph.D. Thesis. Northwestern University.
Vinyard, C.J. (2003). Functional interpretations of jaw shapes: beware of morphometricians bearing geometric means. *Am. J. Phys. Anthropol. Suppl.* 36:216.
Vinyard, C.J., and Hanna, J. (2005). Molar scaling in strepsirrhine primates. *J. Hum. Evol.* 49:241–269.
Vinyard, C.J., and Ravosa, M.J. (1998). Ontogeny, function, and scaling of the mandibular symphysis in papionin primates. *J. Morphol.* 235:157–175.
Vinyard, C.J., and Smith, F.H. (2001). Morphometric testing of structural hypotheses of the supraorbital region in modern humans. *Z. Morphol. Anthropol.* 83:23–41.
Vinyard, C.J., Wall, C.E., Williams, S.H., and Hylander, W.L. (2003). Comparative functional analysis of skull morphology of tree-gouging primates. *Am. J. Phys. Anthropol.* 120:153–170.
Weil, W.B. (1962). Adjustment for size – a possible misuse of ratios. *Am. J. Clin. Nutr.* 11:249–252.
Williams, S.H., Wall, C.E., Vinyard, C.J., and Hylander, W.L. (2002). A biomechanical analysis of skull form in gum-harvesting galagids. *Folia Primatol.* 73:197–209.
Wright, S. (1932). General, group and special size factors. *Genetics* 17:603–619.
Wright, S. (1968). *Evolution and the Genetics of Populations 1. Genetics and Biometric Foundations*. University of Chicago Press, Chicago.

Chapter 17
Food Physical Properties and Their Relationship to Morphology: The Curious Case of *kily*

Nayuta Yamashita

Contents

17.1	Background	387
17.2	Materials and Methods	390
	17.2.1 Study Site and Species	390
	17.2.2 Plant Collection and Mechanical Tests	390
	17.2.3 Tooth Morphometrics	392
	17.2.4 Analyses	393
17.3	Results	394
	17.3.1 Kily Contribution to Lemur Diets	394
	17.3.2 Mechanical Properties of Kily Plant Parts	395
	17.3.3 Molar Morphometrics	398
17.4	Discussion	400
	17.4.1 Comparisons of Tooth Features	400
	17.4.2 Comparisons with Larger Diet	402
17.5	Conclusions	403
	References	404

17.1 Background

When diet is related to masticatory design, it is with reference to physical parameters of food items (e.g., Hylander, 1975; Kay, 1975; Rosenberger and Kinzey, 1976; Seligsohn, 1977; Lucas, 1979; Hiiemae and Crompton, 1985; Happel, 1988; Kinzey and Norconk, 1990; Strait, 1993; Yamashita, 1998a; Wright, 2003). Functional dental morphology assumes a direct relationship between the external environment and the form that interacts with it. Physical interactions with foods, however, are just one parameter of the much broader issue of food choice. Herbivores must weigh nutritional benefits against the costs of eating specific parts. At the same time, the plant protects itself from being eaten (with the obvious exception of fruits) by

---

N. Yamashita
Department of Anthropology, University of Southern California, Grace Ford Salvatori Building, 3601 Watt Way, Suite 120, Los Angeles, CA 90089-1692, 213.740-1746
e-mail: nayutaya@usc.edu

manufacturing or possessing chemical or mechanical defenses to prevent herbivory. In terms of morphological interactions, the key defenses of concern are mechanical, and the ability of herbivorous species to overcome these defenses efficiently, by possessing appropriate morphological tools, determine in part the criterion for acceptance or rejection of particular plant parts.

Physical properties of foods are described in terms of external factors, such as size and shape, and internal properties that are related to the material composition of the particular food (see Strait, 1997; Lucas, 2004 for discussions). Food reduction depends on crack formation and propagation. The mechanical properties of the food in question describe its construction and determine how it resists crack propagation. *Toughness* is the work of fracture and is represented as the area underneath the curve in a force–displacement graph. Tough foods are good at stopping cracks once they start and are often able to undergo large deformations before they fail. *Brittleness* is its converse and is a relative lack of toughness. *Elastic* or *Young's modulus* is a measure of stiffness or resistance to bending. *Strength* at fracture is the force at which unrecoverable breakage occurs. *Hardness* is the resistance to indentation.

The relationship between two properties, toughness (R), and elastic modulus (E) describes how foods fragment and how they are used by plants to mount mechanical defenses against herbivory (Agrawal et al., 1998; Lucas et al., 2000; Lucas, 2004). The square root of the product of elastic modulus and toughness ($\sqrt{ER}$) describes stress-limited defenses. Stress-limited foods are brittle and shatter when sufficient stress levels are reached. A plant that invests in this type of defense relies on an herbivore not being able to generate sufficient force to induce breakage. Displacement-limited defenses are represented by the square root of toughness divided by elastic modulus ($\sqrt{R/E}$). These defenses rely on the herbivore not being able to strain the plant part to failure in order to fragment it (Agrawal et al., 1998). Thin sheets of material, such as leaves, tend to rely solely on toughness as a defense.

Mechanical plant defenses require appropriate tools to overcome them. Tooth features most suited for fragmenting foods with certain mechanical properties can be predicted based on engineering principles (Lucas, 2004). In earlier work, molar features were quantified and correlated with food properties for the lemurids (excluding *Hapalemur* spp.) and indriids (Yamashita, 1998a, b). Several functional complexes were identified: hard food items were correlated with short cusps in lemurids, tight occlusal fit, small trigon and large talonid areas, and deep, acute basins. Large, shallow trigons; shallow, unrestricted talonids; and large upper molar basins were indicative of a diet with high shear strength. The hardest and strongest foods eaten had higher correlations with tooth features than the most frequently eaten foods.

Tough foods (high in displacement-limited defenses) should require longer molar crests or blades with edges to cut them. The edges are necessary to continue and direct crack propagation to fragment the food. However, Yamashita (1998a, b) found that neither hardness nor shear strength demonstrated strong correlations with crest lengths, which were correlated with eating foods with a flat shape, such as leaves. Food toughness may have a more significant relationship.

Hard and/or brittle foods (low in displacement-limited defenses) should emphasize blunt cusps to propagate and drive cracks. The fit of occluding cusp/basin pairs should be loose for tough foods to allow for greater excursions of the cusp and its adjacent crest to cut the food. Basins should be correspondingly unrestricted. Cusp height may increase as a means to increase crest length or overall tooth height (hypsodonty) as a defense against wear from abrasive diets.

Ecomorphological studies integrate ecology and functional morphology (Wainwright and Reilly, 1994). In this regard, identifying which dietary elements have the strongest correlations with the masticatory apparatus necessitates conducting mechanical tests in the field. In this study, I examine mechanical properties of a key resource in the diets of two Malagasy lemurs as an example of how diet and a particular aspect of the masticatory apparatus, tooth morphology, interrelate. *Lemur catta* (ringtailed lemur) and *Propithecus v. verreauxi* (white sifaka) differ markedly in behavior and morphology (Tattersall, 1982; Gebo, 1987; Richard et al., 1991; Sussman, 1991; Demes et al., 1996; Yamashita, 1998b; Richard et al., 2002; Gould et al., 2003).

At the site of Beza Mahafaly in southwestern Madagascar, the two species are sympatric and occur in high densities. The ringtailed lemur is a generalist herbivore that is frequently terrestrial (Sauther et al., 1999), and the sifaka is a folivore with complementary specialized morphology (e.g., spiral colon, sacculated cecum for housing bacterial symbionts for breaking down cellulose, highly crested molars) (Tattersall, 1982; Campbell et al., 2000).

In terms of masticatory morphology, lemurids as a family are more variable than indriids in molar crest length, tooth area, cusp radius and height, and basin areas (Yamashita, 1998a,b). Compared to lemurids, indriids are more dentally uniform, have notable molar crest development on quadritubercular teeth, fewer teeth, accelerated dental development, deep and robust jaws, and a partially fused symphysis (Beecher, 1977; Tattersall, 1982; Schwartz and Tattersall, 1985; Yamashita, 1998b; Godfrey et al., 2001, 2004).

Though there is a certain amount of dietary overlap, the lemur species eat plant parts from different phenophases (developmental stages) of a resource common to both their diets. *Tamarindus indica* (*kily*) is one of the most common trees in the shared forest habitat of the two lemur species (Sussman and Rakotozafy, 1994). At present, it is unknown whether *T. indica* is a native or an introduced species on the island (Binggeli, 2003). Regardless of its origins, it is well established in the dry forests of Madagascar. Its current distribution is pantropical (Missouri Botanical Garden, 2006). *T. indica* is a dietary staple for both lemur species throughout the year (Sauther, 1998; Yamashita, 2002). Mature and immature leaves, unripe and ripe fruits, seeds, and flowers are eaten as they become available. I concentrate on a single food resource in this chapter because of its prevalence in the diets of the lemur species. Furthermore, the partitioning of its various parts by the two lemurs appears to reflect larger patterns of food selection, which may indicate mechanical segregation. Tooth morphologies are predicted to be congruent with differences in the mechanical properties of the plant parts eaten. Specifically, in comparisons between the lemur species, longer tooth crests are expected to be correlated with tougher foods, blunt cusps with hard/brittle foods, looser occluding cusp/basin pairs with tougher foods, and unrestricted basins and higher cusps with tough foods.

17.2 Materials and Methods

17.2.1 Study Site and Species

Observations of *Propithecus v. verreauxi* and *Lemur catta* were conducted in the deciduous tropical dry forest of Beza Mahafaly special reserve in southwestern Madagascar from February 1999 to February 2000. This region is characterized by distinct wet and dry seasons. The majority of rainfall occurs between the rainy season months of November to March with almost no rain during the dry season approximately from April to October. The primary study site, Parcel 1, is a small (80 ha) area with a diversity of microhabitats ranging from a riverine gallery forest in the east to a xeric habitat to the west (Sussman and Rakotozafy, 1994). Parcel 1 contains dense populations of the two diurnal lemur species studied (Richard et al., 2002; Yamashita, 2002; Gould et al., 2003).

Five ringtailed lemur and six sifaka groups were observed. Ringtailed lemur groups contained from ten to more than fourteen individuals. Sifaka group sizes ranged from four to seven individuals. Identifying collars and pendants on individual animals facilitated focal observations. Continuous bout observations were conducted on focal animals that were changed every 10 min (Altmann, 1974). Time spent on basic behaviors of feeding, movement, resting, and social activities was recorded. Feeding behaviors were further detailed by noting the plant species eaten, the exact part eaten (e.g., young or mature leaves, fruit pulp, etc.), food preparation techniques employed, and ingestive behaviors.

Ringtailed lemurs and sifakas ate different developmental phases of specific kily parts throughout the year. Ripe kily fruits are pods with a brittle exocarp surrounding a jelly-like mesocarp. While both lemurs eat the ripe fruits, sifakas restrict ripe fruit eating to a few months at the end of the dry season. Ringtailed lemurs eat ripe fruit pulp throughout the year. Seeds from ripe fruit are not eaten by either species; sifakas, however, preferentially eat seeds from unripe fruits. The pinnate leaves flush bright pink and gradually expand and green as they mature. Ringtailed lemurs eat immature leaves, whereas sifakas concentrate on the more mature phase.

17.2.2 Plant Collection and Mechanical Tests

Food trees were flagged during observations for later plant part collection. In some cases, animals dropped foods that were collected at the time of observation. Foods were usually collected and tested on the same day as observations, or at least within 24 h. Care was taken to collect the exact plant part from the tree that the animals were observed feeding on. Many of the foods tested were either chewed and dropped by the animals or had adjacent bite marks. Specific *T. indica* parts that were eaten and collected included young and mature leaves, unripe fruit, exocarp of ripe fruit, seeds from unripe and ripe fruits, and flowers (Fig. 17.1).

17 Food Physical Properties and Their Relationship to Morphology

Fig. 17.1 Unripe kily fruit and seeds dropped by sifakas. Coin diameter = 21 mm

Feeding observations focused on the specific point where foods were bitten off, and the toughness (in J/m^2) of these plant parts was tested with a portable mechanical tester (Darvell et al., 1996; Lucas et al., 2000). The instrument has a number of interchangeable pieces that can be used to perform a variety of mechanical tests.

The toughness (R) (J/m^2) for most individual food parts was determined by a scissors cutting test. In the test, a platform with an attached load cell (either 10 or 100 N) is lowered onto the scissor handles. A controlled crack forms as the scissor blades close down on the food item. After the food is cut to a preset length, a second, empty pass of the scissor blades alone subtracts out the work of friction between the scissor blades. The output to a computer is a force–displacement diagram, such as the one in Fig. 17.2. Figure 17.2 depicts graphs of ripe and unripe *Tamarindus indica* fruit, where force is on the Y-axis and displacement is on the X-axis. The computer program returns toughness values in J m^{-2}, which is the area under the force–displacement curve (shaded area), or the work of fracture.

Fruits were tested according to the first part encountered, in most cases the fruit shell or fleshy mesocarp with attached exocarp. Scissors tests of ripe kily fruit shells were conducted by cutting small test pieces out of the shell. Bipinnate kily leaves were tested on the rachis or pinnae that support individual leaflets and through individual pinnules. In some cases where the pinnules were too delicate, rows of pinnules were stacked and cut together. Flowers were most often tested at the individual pedicel that supports the flower, and toughness of either individual flower parts (petals, sepals) or through the nectary was taken.

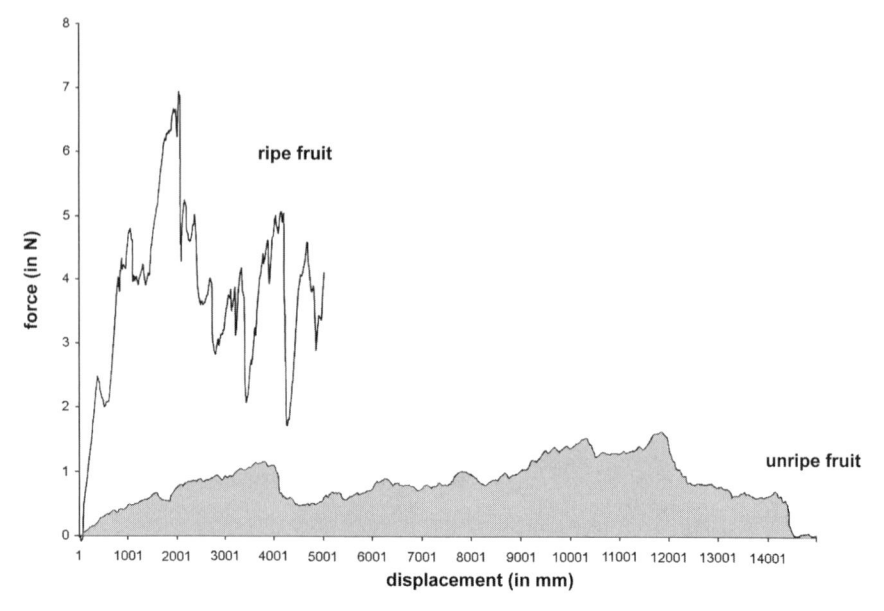

Fig. 17.2 Force–displacement graph of scissors cutting test of ripe kily fruit shell and unripe kily fruit. Peaks in ripe fruit trace represent crack formation. Average toughness of all fruits tested: unripe fruit ($R = 1199\,\mathrm{J\,m^{-2}}$, $n = 12$), ripe fruit ($R = 3112\,\mathrm{J\,m^{-2}}$, $n = 24$). Figure adapted from Yamashita (2002; Fig. 17.2); used with the permission of Springer

Toughness was also measured with a wedge test, in which a wedge is lowered into a prepared block of food. Elastic modulus (in MPa) of kily fruit shell was tested in three-point bending. Hardness (in MPa) was tested with an indentation test in which a 1-mm ball bearing attached to the upper tester platform was pressed onto the food. The hardness sample comprised seeds and fruit casings (exocarp).

The numbers of times plants were tested are consistent with the numbers of observations of lemurs eating a specific plant part. Three to four individual parts were tested for each plant species each time the plant species was tested.

17.2.3 Tooth Morphometrics

Occlusal tooth areas, basin areas, crest lengths, radii of curvature of cusps, cusp heights, and ratios of occluding cusps and basins were measured from second upper and lower molars made from epoxy tooth casts of field and museum specimens (see Yamashita, 1998a for details). Second molars were chosen for study because of their intermediate position in the molar row.

Tooth features were measured with JAVA (Jandel Video Analysis software) and a Reflex microscope. Procedures and protocols are reported in Yamashita (1998a).

17 Food Physical Properties and Their Relationship to Morphology

JAVA measures video images in two dimensions via either edge-tracking software that follows contours based on the contrast of the image or by tracing non-linear features manually from the keyboard. Teeth were oriented so that the plane of the occlusal surface of the second molar was parallel with the videocamera lens. Molar crown areas and basin areas were measured in millimeter square with JAVA.

Indriids and lemurids differ in the presence of some basins. Both families have an upper molar trigon and lower molar trigonid and talonid; however, the trigonid in lemurids does not occlude with an upper molar cusp. The indriid quadritubercular molar has both a talon and a trigonid basin. Areas of the occlusal surfaces of upper and lower second molars were digitized in the same orientation as the basins. Two occlusal cusp/basin pairs, the protocone/talonid and hypoconid/trigon, were measured in both lemur taxa.

Measurements in three dimensions were obtained with a Reflex microscope. A high-intensity LED point acts as the measuring mark at the center of the microscope view. The microscope digitizes 3D coordinate data of the object of interest. Crest lengths, cusp heights, and cusp radii were measured with the Reflex scope. (See Yamashita, 1998a for details on specific measurements).

17.2.4 Analyses

Tooth features were compared using residuals from least squares regression analyses of individual features regressed separately against lower molar area within each family (Lemuridae and Indriidae; Table 17.1). Analyses were confined to residuals

Table 17.1 Lemur species included in regressions

Taxa	Sample size[a]
Lemuridae	
Lemur catta	24
Eulemur coronatus	10
E. fulvus albocollaris	3
E. fulvus albifrons	10
E. fulvus collaris	8
E. fulvus fulvus	10
E. fulvus mayottensis	10
E. fulvus rufus	15 [2]
E. fulvus sanfordi	3
E. macaco flavifrons	2
E. macaco macaco	7
E. mongoz	10
E. rubriventer	24 [4]
Varecia variegata rubra	4
V. variegata variegata	6

(*Continued*)

Table 17.1 (Continued)

Taxa	Sample size[a]
Indriidae	
Avahi laniger laniger	10
Indri indri	10
Propithecus diadema candidus	7
P. diadema diadema	8 [2]
P. diadema edwardsi	10 [9]
P. diadema holomelas	3
P. verreauxi coquereli	10
P. verreauxi coronatus	8
P. verreauxi deckeni	10
P. verreauxi verreauxi	25 [1]

[a] Numbers in brackets are field specimens.

since earlier analyses of covariance showed that, though the slopes of the two families do not differ significantly from isometry, the indriid intercept is significantly higher than that of the lemurids (Yamashita, 1998b). The dental samples, from museum and field collections, often did not have associated body weights and so required size surrogates for all regression analyses. Lower molar area was found to be a more appropriate size surrogate for comparisons than skull length in Yamashita (1998b).

Raw data values were compared with non-parametric Mann-Whitney U and Kruskal-Wallis tests for all data sets comparing food properties. Tooth features were compared with non-parametric Mann-Whitney U tests on residuals derived from least squares regressions of individual tooth features against lower molar area within each family. However, raw data values were used for comparisons of cusp/basin ratios. Due to small sample sizes in some comparisons, results must be interpreted cautiously.

17.3 Results

17.3.1 Kily Contribution to Lemur Diets

T. indica fruit parts were eaten year-round and contributed to the diets of both lemur species, but especially to the diets of the ringtailed lemurs (Fig. 17.3). Ringtailed lemurs spent 29% of total annual feeding time on various kily plant parts (ripe and unripe fruits, flowers, and young leaves). Sifakas spent 11% of total feeding time on flowers, unripe fruit seeds, ripe fruits, and mature leaves.

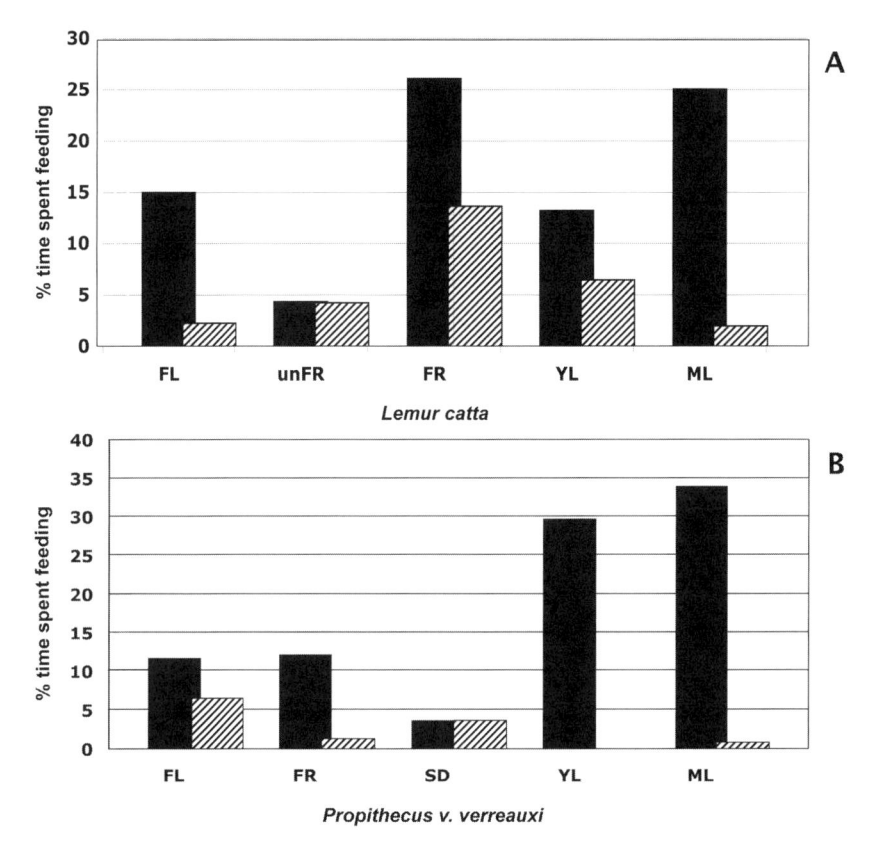

17 Food Physical Properties and Their Relationship to Morphology 395

Fig. 17.3 Time spent feeding on kily parts throughout year compared to total annual feeding time on same food parts. Food categories of total diet confined to match categories of kily plant parts eaten. (**A**) *Lemur catta*; (**B**) *Propithecus v. verreauxi*. Solid bars, total annual diet; cross-hatched bars, kily diet expressed as percentage of total diet. FL = flowers; unrp FR, unripe fruit; FR pulp, ripe fruit pulp; SD, unripe fruit seed; YL, young leaves; ML, mature leaves. No kily YL parts for sifakas

17.3.2 Mechanical Properties of Kily Plant Parts

Toughnesses of different kily plant parts were mechanically distinct from one another (Kruskal-Wallis $\chi^2 = 42.547$, $P < 0.0001$; Figs. 17.2, 17.4). Average toughness values for all fruits tested in Fig. 17.2 were: unripe fruit ($R = 1199 \, \text{J m}^{-2}$, $n = 12$) and ripe fruit ($R = 3112 \, \text{J m}^{-2}$, $n = 24$).

The toughness of kily parts eaten by ringtailed lemurs was significantly higher than that eaten by sifakas (Mann-Whitney U, $Z = -2.148$, $P = 0.032$, $n = 45$ for ringtailed lemurs and 26 for sifakas; Fig. 17.5). However, in comparisons of individual plant parts, there were no significant differences between the two species (Mann-Whitney U; see Fig. 17.6, Table 17.2 for comparisons of specific plant parts).

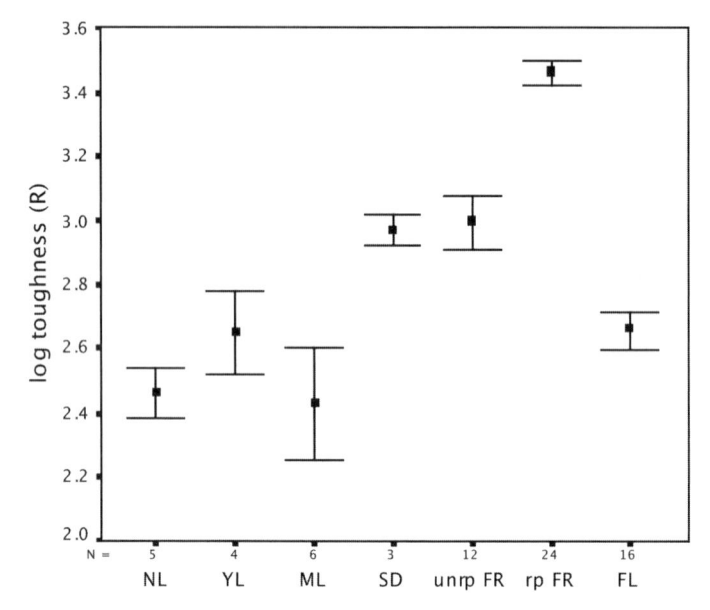

Fig. 17.4 Logged toughness (*R*) values of phenophases of *Tamarindus indica*. NL, new leaves; YL, young leaves; ML, mature leaves; SD, unripe fruit seed; unrp FR, unripe fruit; rp FR, ripe fruit; FL, flower. Error bars represent one standard error of the mean. Toughnesses of plant parts are significantly different from one another at $P < 0.0001$ (Kruskal-Wallis on raw values; figure is logged for clarity)

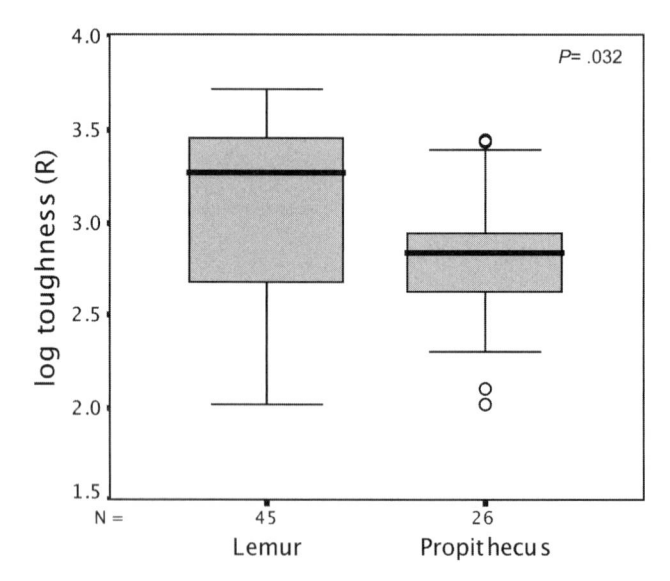

Fig. 17.5 Toughness (*R*) comparisons of entire *kily* diet of two lemur species. Boxes represent central half of data divided by median line; whiskers are data points that lie from the top of the box to 1.5 times the data range of the box; circles indicate outliers that lie between 1.5 and 3 times the data range of the box; and asterisks are data points that lie beyond 3 times the data range of the box. $Lc\ R = 1900.22\,\mathrm{J\,m^{-2}}$ (SE, 230.82); $Pvv\ R = 892.19\,\mathrm{J\,m^{-2}}$ (SE, 151.24). Comparison is significant (P = 0.032, Mann-Whitney U on raw values; figure is logged for clarity)

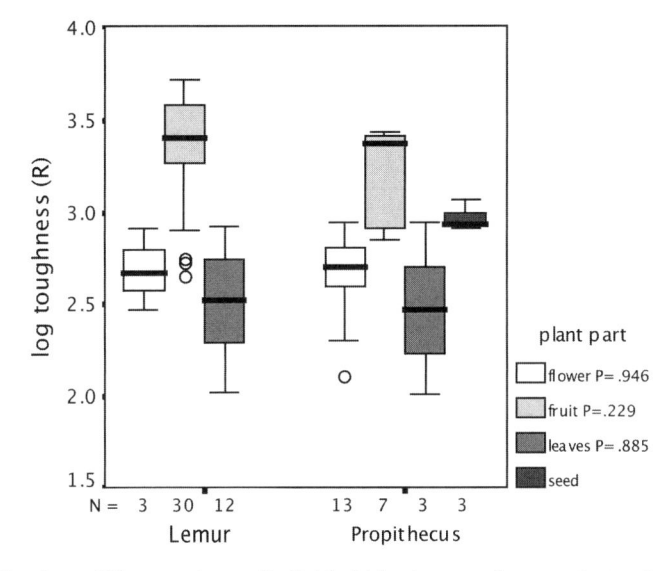

17 Food Physical Properties and Their Relationship to Morphology

Fig. 17.6 Toughness (R) comparisons of individual *kily* plant parts between the two lemur species. Comparisons of individual plant parts between lemur species are not significant (Table 17.1; non-parametric comparisons on raw values; figure is logged for clarity.). No comparisons of seeds since *L. catta* was not observed eating them. See Fig. 17.5 caption for explanation of symbols

Table 17.2 Comparisons between lemur species of toughness of kily parts eaten[a]

Plant part	n[b]	Z	P-value
Flower	3, 13	−0.067	0.946
Fruit	30, 7	−1.202	0.229
Leaves	12, 3	−0.144	0.885

[a] Mann-Whitney U-test.
[b] Sample sizes for *L. catta* and *P. verreauxi*, respectively.

Indentation hardness of unripe and ripe fruits and seeds eaten by the lemurs were also compared, and the hardness values of the lemur species were not significantly different from one another (Mann-Whitney U, $Z = -0.682$, $P = 0.495$; Lc $H = 6.75$ MPa, SE $= 2.16$, $n = 8$; Pvv $H = 22.10$ MPa, SE $= 11.05$, $n = 4$). Hardness values are calculated using a model that may change by a constant in future analyses. Relative differences in indentation hardness of foods between species are unaffected.

Ripe kily fruit shells were subjected to bending tests, and the two fragmentation criteria, $\sqrt{ER}$ and $\sqrt{R/E}$, were calculated ($\sqrt{ER}$: mean $= 1459.29$, SE $= 238.45$, $n = 5$; $\sqrt{R/E}$: mean $= 2.161$, SE $= 0.207$, $n = 5$). Values for stress-limited defenses were several orders of magnitude higher than those for displacement-limited, suggesting that the ripe fruit shell is quite brittle.

17.3.3 Molar Morphometrics

Ringtailed lemurs had relatively longer crests than sifakas when these features were compared using within-family residuals (Mann-Whitney U, $Z = -3.430$, $P = 0.001$; Fig. 17.7, Table 17.3). In addition, the fit of the protocone/talonid occlusal pair was significantly tighter in the ringtailed lemurs (Mann-Whitney U, $Z = -3.867$, $P < 0.001$; Fig. 17.8, Table 17.3). All other comparisons were not statistically significant (Figs. 17.9, 17.10, Table 17.3).

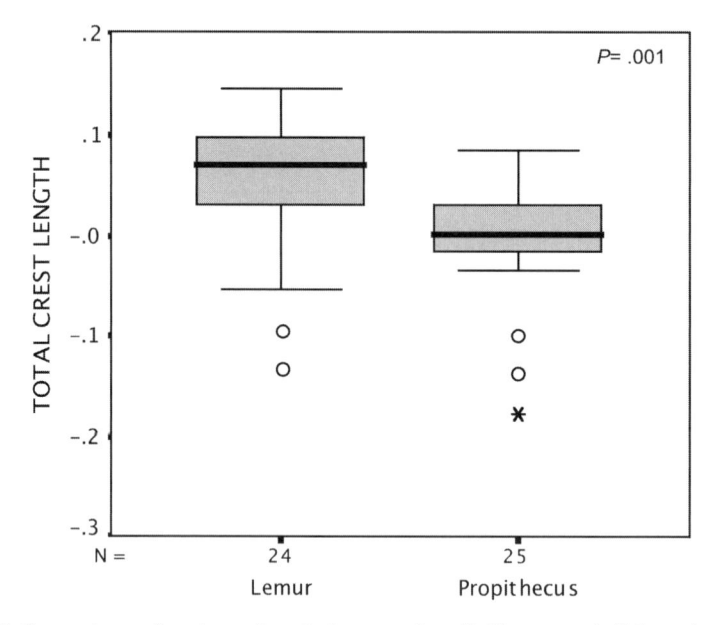

Fig. 17.7 Comparisons of total crest lengths between ringtailed lemurs and sifakas using within-family residuals. Comparison is significant (P = 0.001). See Fig. 17.5 caption for explanation of symbols

Table 17.3 Comparisons of tooth features between the two lemur species[a]

Tooth feature	n[b]	Z	P-value
Talonid	24, 26	−0.272	0.786
Total crest length[c]	24, 25	−3.430	0.001[g]
Average cusp radius[d]	17, 21	−1.864	0.062
Average cusp height[e]	24, 26	−0.194	0.846
Protocone/talonid radii[f]	15, 20	−3.867	< 0.001[h]
Hypoconid/trigon radii[f]	11, 18	−0.584	0.559

[a] Mann-Whitney U-tests on lower molar residuals calculated within families.
[b] Sample sizes for *L. catta* and *P. verreauxi*, respectively.
[c] Sum of all upper and lower molar crests.
[d] Mean of all upper and lower molar cusp radii.
[e] Mean of all upper and lower molar cusp heights.
[f] Raw data values used instead of residuals.
[g] Longer crests in *L. catta*.
[h] Smaller ratio (looser fit) in *P. verreauxi*.

17 Food Physical Properties and Their Relationship to Morphology 399

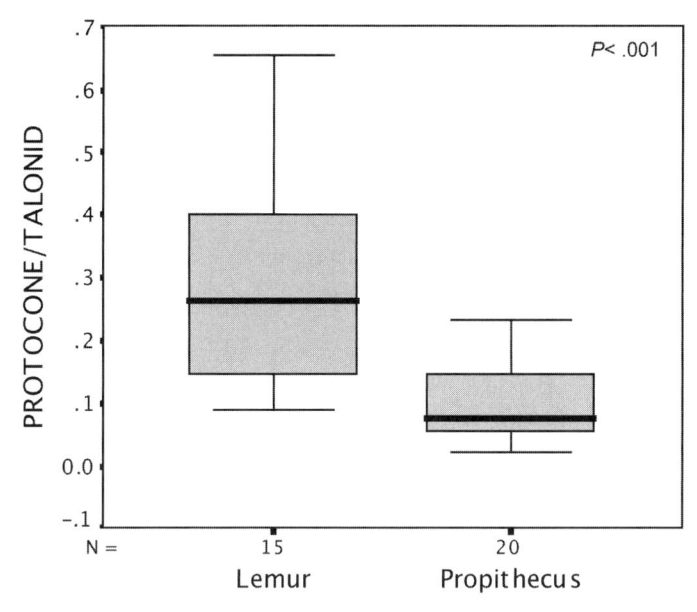

Fig. 17.8 Comparisons of protocone/talonid radii between ringtailed lemurs and sifakas using raw data values. Comparison is significant at P < 0.001. See Fig. 17.5 caption for explanation of symbols

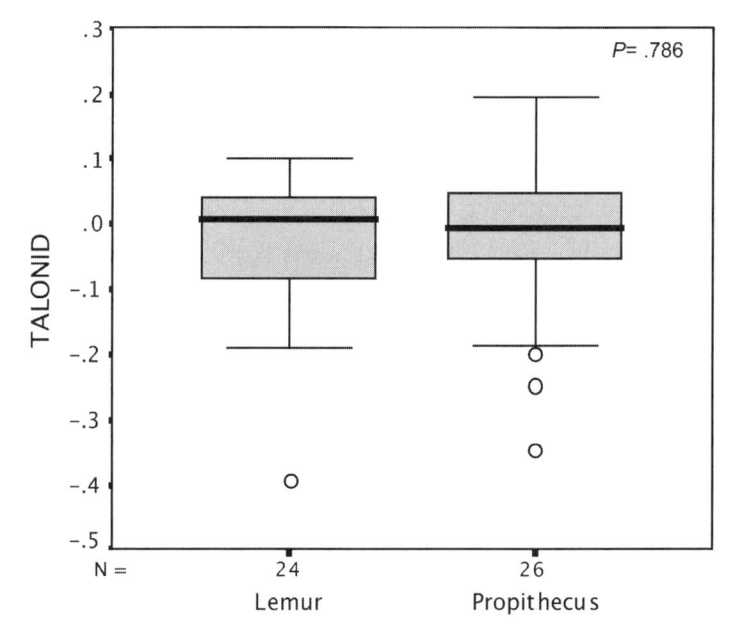

Fig. 17.9 Comparisons of talonid basin area between ringtailed lemurs and sifakas using within-family residuals. Comparison is not significant (P = 0.786). See Fig. 17.5 caption for explanation of symbols

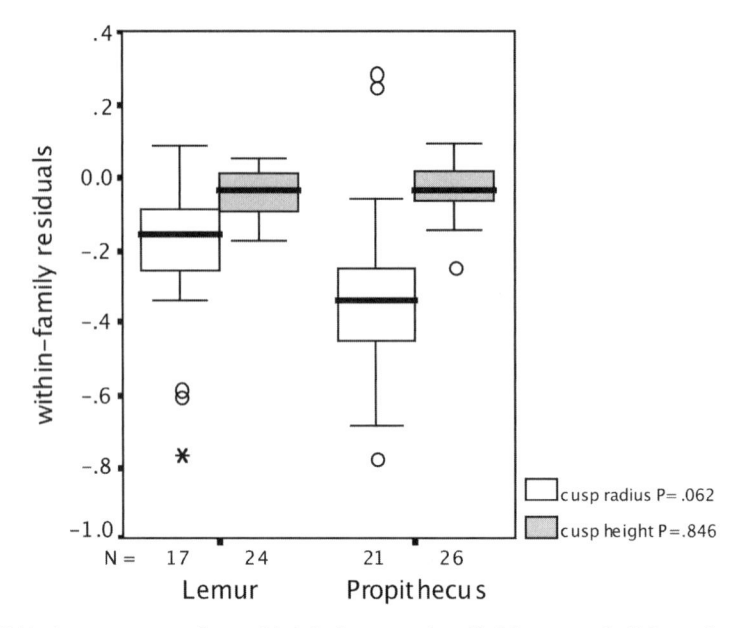

Fig. 17.10 Average cusp radius and height between ringtailed lemurs and sifakas using within-family residuals. Comparisons are not significant (P = 0.062 and 0.846, respectively). See Fig. 17.5 caption for explanation of symbols

17.4 Discussion

17.4.1 Comparisons of Tooth Features

The predicted relationships between mechanical food properties and specific tooth features fared equivocally in light of the lack of separation between the lemur species in food toughness and hardness. As predicted, ringtailed lemurs with their relatively longer crests ate tougher foods, though this may be overestimated (see below). Though the species with the blunter cusps did not eat the hardest food (the difference was not significant), the slightly blunter cusps of ringtailed lemurs may help them crack brittle kily fruit shells. Sifakas had a looser occlusal fit of the protocone to the talonid, though it had a less-tough diet than ringtailed lemurs.

However, the high toughness values for the ringtailed lemur diet are largely due to the ripe kily fruit shell. Having a loose occlusal fit, as was predicted for tough foods, would not have helped fragment kily shell since, being a brittle, stress-limited material, it must be cracked not cut (Fig. 17.2). Most of the toughness differences between the lemur species are attributable to their selection of kily fruits at different developmental stages. The force–displacement graph in Fig. 17.2 illustrates the differences in toughness between unripe and ripe fruit stages. Though the ripe kily shell is tougher, the dropoffs from the multiple peaks show where side cracks and runaway cracks appear. Because the scissors were controlling the crack direction, runaway propagation was limited, and as a result, in all likelihood, the test overestimated the actual toughness of the shell. When ringtailed lemurs break open kily fruit, they do not guide the formation of the crack. They instead apply multiple bites with their canines and premolars to start a crack (Sauther et al., 2001), then tear off the parts that break. In contrast, opening an unripe fruit requires the application of constant stress to initiate and continue crack formation, since runaway propagation does not occur. Unripe fruit ($R = 1199 \, \text{J m}^{-2}$) may not be as tough externally as ripe fruit ($R = 3112 \, \text{J m}^{-2}$), though again this may be overestimated, but the pulp and exocarp has to be continuously worked. The exterior of unripe kily fruit is also quite hard (8.964 MPa, unripe fruit; 4.568 MPa, ripe fruit). In contrast to unripe fruit, once the brittle casing of ripe fruit has been cracked, the jelly-like pulp is licked off. Unripe fruit (and leaf material) may impose cyclical loads on the masticatory apparatus that is countered by the greater robusticity of the sifaka mandible. Ringtailed lemurs exploit unripe fruit pulp when no ripe fruit is available. The combined effects of exploiting ripe and unripe kily fruits may contribute to the greater postcanine tooth loss and wear seen in ringtailed lemurs compared to sifakas (Cuozzo and Sauther, 2006).

Furthermore, Yamashita (1998a) showed that a loose occlusal fit of cusp to basin was related to eating large quantities of leaf material and not necessarily to food toughness. Crest length was another feature that was correlated with flat leaf geometry. Although sifakas ate less kily leaf material than ringtailed lemurs (Fig. 17.3), young and mature leaves form the bulk of their larger diet (Yamashita, 2002; in prep.).

Perhaps somewhat surprisingly, some tooth features, such as crest length, are relatively greater in size in ringtailed lemurs. In fact, Yamashita (1998b) found that *Lemur catta* and *Indri indri* had the longest crests within their respective families (Table 17.1). Of course, the results presented here are predicated on the *relative* differences within families being the functionally significant comparison. If tooth features are compared to lower molar area without regard to family (Yamashita, 1998b), then sifakas have larger and longer features than ringtailed lemurs, with the sole exception of the ratio of the protocone to the talonid. Food size, especially of a shared resource such as kily, would presumably be invariant, which supports the idea that "absolute" size of tooth features is more critical. However, analyses of covariance between the lemurids and the indriids demonstrated that slope differences were not and intercept differences were significantly different for the tooth features examined here. Taxonomic affiliation constrains tooth features along parallel slopes between families. While larger and longer features may be preferable, the species have inherited morphologies imposed by their respective families. That indriid features are generally larger at the same tooth size speaks to the homogeneity and potential stabilizing selection on these features within the Indriidae. The relative elongation of crests in ringtailed lemurs is indicative of an adaptive shift within the family. Ringtailed lemurs are in a sense approaching the sifaka condition. The reasons for this are purely speculative, but may be related to commonalities of diet. Yamashita (1996) found that the diets of ringtailed lemurs, in terms of hardness, shear strength, and food type, had more in common with sympatric sifakas than with their rainforest confamilials, *Eulemur fulvus rufus*, and *Eulemur rubriventer*.

17.4.2 Comparisons with Larger Diet

Kily food parts are not necessarily representative of the larger diet in terms of the actual time spent on specific parts or their mechanical properties. One of the cautionary tales learned here is that one has to be careful about extrapolating to the larger diet even from a prominent, though highly seasonal, component. Somewhat contrary to their expected dietary classifications, the ringtailed lemurs focused on kily fruit pulp and young leaves and sifakas spent the most time on flowers. This deviates somewhat from the larger diet, in which ringtailed lemurs spend equal amounts of time on fruits and leaf material (31% and 38%, respectively), while sifakas overwhelmingly spend the majority of feeding time on leaf material (12% on fruits and 64% on leaves; Yamashita, 2002). The differences found between the whole diet and the kily subset is most likely related to the relative importance of kily to the two lemurs. Kily food parts, and especially ripe fruits, are major components of the ringtailed lemur diet (29% of total time spent feeding), and, while important, they are less prevalent in the diet of sifakas (11%).

Earlier work found that the two species' diets did not differ significantly in hardness and toughness, though sifaka diets had higher seed hardness thresholds (Yamashita, 2000, 2002). Up to around 12 MPa, seeds were masticated so that nothing identifiable appeared in the feces. Above this level, the seed was discarded, as ripe kily seeds were. Ring-tailed lemurs had a lower threshold at 4 MPa. Seeds generally passed through the digestive tract intact and were defecated, and the defecated seeds were not as hard as those rejected by sifakas.

In the current study, the toughness of the overall kily diet was significantly higher in ringtailed lemurs than in sifakas. As discussed above, this is most likely due to the toughness of kily fruit shells. Kily plant parts represent extremes on toughness and hardness scales with respect to the larger diet for both lemur species. Comparing the kily diet to the entire diet, average R was $1900.22\,\mathrm{J\,m^{-2}}$ and $810.87\,\mathrm{J\,m^{-2}}$, respectively, for ringtailed lemurs, and $892.19\,\mathrm{J\,m^{-2}}$ and $585.63\,\mathrm{J\,m^{-2}}$ for sifakas. Hardness values also represented extremes in the kily diet: 6.75 MPa (kily parts) and 4.91 MPa (whole diet) for ringtailed lemurs and 22.10 MPa and 12.65 MPa for sifakas, respectively.

Because kily plant parts lie at the mechanical extremes of the diets of the two lemur species, they may represent fallback foods, eaten when preferred foods are not available (Wrangham et al., 1998; Furuichi et al., 2001; Lambert et al., 2004). However, ripe kily fruit, though the toughest food eaten, is an important component of the ringtailed lemur diet throughout the year (Sauther, 1998; Yamashita, 2002). Seeds eaten from unripe kily fruit by sifakas may represent a fallback food, since they are eaten at restricted times of the year and the unripe fruit exterior is harder than the overall diet. Indriids, however, are seed predators in addition to being folivores (Hemingway, 1996; Yamashita, 1998a; Powzyk and Mowry, 2003), so interpreting unripe kily fruit seeds as "fallback foods" may not be strictly accurate.

Kily food parts appear to have a greater significance to ringtailed lemurs in terms of total dietary contribution and the demands they place on the masticatory apparatus. The specific molar features of ringtailed lemurs, though in some cases relatively larger, are absolutely smaller than in sifakas, which makes breaking down foods with similar mechanical properties a comparatively greater challenge. Cuozzo and Sauther (2004) report the most severe tooth wear and loss on the P3-M1, where ripe kily fruit pods are often initially inserted. Ripe kily fruit has a high sugar content that proves irresistible to these lemurs despite the mechanical challenges (Yamashita, 2008). Since ringtailed lemur molar features are elaborated in the same directions as those of sifakas, kily food parts may be exerting some sort of selection on the ringtailed lemur masticatory apparatus toward a sifaka-like morphology.

Ripe kily fruit relies on stress-limited defenses since the lemur has to exert enough force to crack the shell and initiate crack formation. The shell is quite brittle. For unripe fruit and the seeds contained within, toughness appears to be the primary mechanical defense. This also applies to leaves at different phenophases. The fit of the protocone to the talonid basin in sifakas may also be an indication of displacement-limited defenses in leaf material since the looser fit allows greater excursion of the cusp and crest within the basin for cutting leaves. However, other properties may be more appropriate for offering defenses against predation.

Chemical properties of kily plant parts support their mechanical distinctiveness and offer further insights as to why the two species differ in selection of kily plant parts that are related to metabolic requirements and tolerances for secondary plant compounds (Yamashita, 2008). The ringtailed lemurs' preference for ripe kily fruit is probably related to its high sugar content. Young kily leaves provide the highest amounts of protein to a species that is relatively lacking in other protein sources. In a complementary fashion, sifakas take in more protein and less sugar. The seeds and flowers eaten by sifakas are high in protein, and mature kily leaves contain higher amounts of sugar than earlier phenophases. At the same time, the ripe fruit largely consumed by ringtailed lemurs contains relatively low amounts of secondary plant compounds. Sifakas take in much higher levels in combination in mature leaves, seeds, and flowers.

17.5 Conclusions

The "curious" nature of kily lies in how thoroughly the two sympatric lemur species exploit its edible parts while minimizing overlap on them. Though there is some small degree of overlap in some of the plant parts, actual feeding times reveal biases for key parts that are only partially related to their mechanical properties. The distinctive preferences for specific kily parts mirror chemical preferences and aversions in their larger diet.

The expected relationships between molar features and mechanical properties were generally upheld. Ringtailed lemurs had relatively longer crests and ate tougher kily parts than sifakas, though the toughness of the primary diet (ripe kily fruit) may have been overestimated. Though food hardness was not significantly different between the lemur species, the slightly blunter cusps of ringtailed lemurs may help them crack brittle kily fruit shells. The looser protocone/talonid occlusal fit in sifakas is probably related to reducing tough leaf material with a uniform geometry.

As important as kily is to the diets of both species, it is not representative in terms of the mechanical properties of its parts. Kily food parts lie at the extremes of toughness and hardness ranges representing the entire diets of both lemur species. Though the individual parts are mechanically distinctive, the kily diets of the two lemurs did not segregate mechanically, except for ripe kily fruit shells that were tougher than other foods.

Since kily plant parts represent the mechanical extremes of the overall diets for both species, they could potentially help clarify the debate concerning the relative importance of the properties of the most frequently eaten or the most stressful foods on morphology. The data presented here and from other fieldwork on mechanical food properties, however, have shown that the debate cannot readily be resolved. In the present case, the geometry of the most frequently eaten food, leaf material, relates to the unconstricted basins of sifakas, whereas the extreme toughness of kily fruit shell is related to the long crests in ringtailed lemurs.

Kily parts form a greater proportion of the total diet for ringtailed lemurs, and, because these parts represent mechanical extremes compared to the rest of the diet, it is probably more stressful for ringtailed lemurs than sifakas when eating them. Interestingly, ringtailed lemurs have relatively larger molar features than sifakas (though sifaka features are "absolutely" larger), which indicates that they are approaching a sifaka-like morphology for fracturing similar diets.

The integration of fieldwork with morphology yields insights into how animals use their particular morphologies and, conversely, how morphologies constrain and/or direct behavior and food choice. In the case of the two lemurs investigated here, food properties appear to be directing morphology in the ringtailed lemurs, while, in sifakas, their morphology constrains them to foods within a range of properties.

Acknowledgments I am indebted to Dr. Hylander for his 1975 paper (among others) on incisor size and diet, which piqued my growing interest in tooth morphometrics. In one paper, he included associations between tooth form and food type, which offered insights into physical processes interacting with functional morphology.

I also thank Peter Lucas for his support and Mary Blanchard for her field assistance. The staff at Beza Mahafaly special reserve, ANGAP, the Departement des Eaux et Fôret, and the Université d'Antananarivo in Madagascar greatly facilitated research. My thanks to the curatorial staffs at the Museum of Comparative Zoology-Harvard, American Museum of Natural History, Field Museum of Natural History, Museum National d'Histoire Naturelle, Rijksmuseum of Natural History, Natural History Museum-London) for access to their collections. Alison Richard kindly provided field casts of sifaka dentitions. Data collection for this paper was funded by the Royal Grants Council to Peter Lucas and NSF dissertation improvement grant SBR-9302279, a Northwestern University dissertation grant, Sigma Xi, and an AMNH collection study grant.

References

Agrawal, K.R., Lucas, P.W., Bruce, I.C., and Prinz, J.F. (1998). Food properties that influence neuromuscular activity during human mastication. *J. Dent. Res.* 77:1931–1938.
Altmann, J. (1974). Observational study of behavior: sampling methods. *Behaviour* 69:227–267.

Beecher, R.M. (1977). Function and fusion at the mandibular symphysis. *Am. J. Phys. Anthropol.* 47:325–336.

Binggeli, P. (2003). Introduced and invasive plants. In: *The Natural History of Madagascar*. Goodman, S.M. and Benstead, J.P. (eds.), University of Chicago Press, Chicago, pp. 257–268.

Campbell, J.L., Eisemann, J.H., Williams, C.V., and Glenn, K.M. (2000). Description of the gastrointestinal tract of five lemur species: *Propithecus tattersalli, Propithecus verreauxi coquereli, Varecia variegata, Hapalemur griseus*, and *Lemur catta*. *Am. J. Primatol.* 52:133–142.

Cuozzo, F.P., and Sauther, M.L. (2004). Tooth loss, survival, and resource use in wild ring-tailed lemurs (*Lemur catta*): implications for inferring conspecific care in fossil hominids. *J. Hum. Evol.* 46:623–631.

Cuozzo, F.P., and Sauther, M.L. (2006). Severe wear and tooth loss in wild ring-tailed lemurs (Lemur catta): a function of feeding ecology, dental structure, and individual life history. *J. Hum. Evol.* 51:490–505.

Darvell, B.W., Lee, P.K.D., Yuen, T.D.B., and Lucas, P.W. (1996). A portable fracture toughness tester for biological materials. *Meas. Sci. Technol.* 7:954–962.

Demes, B., Jungers, W.L., Fleagle, J.G., Wunderlich, R.E., Richmond, B.G., and Lemelin, P. (1996). Body size and leaping kinematics in Malagasy vertical clingers and leapers. *J. Hum. Evol.* 31:367–388.

Furuichi, T., Hashimoto, C., and Tashiro, Y. (2001). Fruit availability and habitat use by chimpanzees in the Kalinzu Forest, Uganda: examination of fallback foods. *Int. J. Primatol.* 22:929–945.

Gebo, D.L. (1987). Locomotor diversity in prosimian primates. *Am. J. Primatol.* 13:271–281.

Godfrey, L.R., Samonds, K.E., Jungers, W.L., and Sutherland, M.R. (2001). Teeth, brains, and primate life histories. *Am. J. Phys. Anthropol.* 114:192–214.

Godfrey, L.R., Samonds, K.E., Jungers, W.L., Sutherland, M.R., and Irwin, J.A. (2004). Ontogenetic correlates of diet in Malagasy lemurs. *Am. J. Phys. Anthropol.* 123:250–276.

Gould, L., Sussman, R.W., and Sauther, M.L. (2003). Demographic and life-history patterns in a population of ring-tailed lemurs (*Lemur catta*) at Beza Mahafaly reserve, Madagascar: a 15-year perspective. *Am. J. Phys. Anthropol.* 120:182–194.

Happel, R. (1988). Seed-eating by West African cercopithecines, with reference to the possible evolution of bilophodont molars. *Am. J. Phys. Anthropol.* 75:303–327.

Hemingway, C.A. (1996). Morphology and phenology of seeds and whole fruit eaten by Milne-Edwards' sifaka, *Propithecus diadema edwardsi*, in Ranomafana National Park, Madagascar. *Int. J. Primatol.* 17:637–659.

Hiiemae, K.M., and Crompton, A.W. (1985). Mastication, food transport and swallowing. In: *Functional Vertebrate Morphology*. Hildebrand, M., Bramble, D.M., Liem, K.F. and Wake, D.B. (eds.), Belknap Press, Cambridge, pp. 262–290.

Hylander, W.L. (1975). Incisor size and diet in Anthropoids with special reference to Cercopithecidae. *Science* 189:1095–1097.

Kay, R.F. (1975). The functional adaptations of primate molar teeth. *Am. J. Phys. Anthropol.* 43:195–216.

Kinzey, W.G., and Norconk, M.A. (1990). Hardness as a basis of fruit choice in two sympatric primates. *Am. J. Phys. Anthropol.* 81:5–15.

Lambert, J.E., Chapman, C.A., Wrangham, R.W., and Conklin-Brittain, N.L. (2004). Hardness of cercopithecine foods: implications for critical function of enamel thickness in exploiting fallback foods. *Am. J. Phys. Anthropol.* 125:363–368.

Lucas, P.W. (1979). The dental-dietary adaptations of mammals. *N. Jb. Palaont. Mh.* 8:486–512.

Lucas, P.W. (2004). *Dental Functional Morphology: How Teeth Work*. Cambridge University Press, Cambridge.

Lucas, P.W., Turner, I.M., Dominy, N.J., and Yamashita, N. (2000). Mechanical defenses to herbivory. *Ann. Bot.* 86:913–920.

Missouri Botanical Garden, VAScular Tropicos (VAST) database, 2006, St. Louis (June 8, 2006); http://mobot.mobot.org/W3T/Search/vast.html.

Powzyk, J.A., and Mowry, C.B. (2003). Dietary and feeding differences between sympatric *Propithecus diadema diadema* and *Indri indri*. *Int. J. Primatol.* 24:1143–1162.
Richard, A.F., Rakotomanga, P., and Schwartz, M. (1991). Demography of *Propithecus verreauxi* at Beza Mahafaly, Madagascar: sex ratio, survival, and fertility, 1984–1988. *Am. J. Phys. Anthropol.* 84:307–322.
Richard, A.F., Dewar, R.E., Schwartz, M., and Ratsirarson, J. (2002). Life in the slow lane? Demography and life histories of male and female sifaka (*Propithecus verreauxi verreauxi*). *J. Zool. Lond.* 256:421–436.
Rosenberger, A.L., and Kinzey, W.G. (1976). Functional patterns of molar occlusion in platyrrhine primates. *Am. J. Phys. Anthropol.* 45:281–298.
Sauther, M.L. (1998). Interplay of phenology and reproduction in ring-tailed lemurs: implications for ring-tailed lemur conservation. *Folia Primatol.* 69 (Suppl 1):309–320.
Sauther, M.L., Sussman, R.W., and Gould, L. (1999). The socioecology of the ringtailed lemur: thirty-five years of research. *Evol. Anthropol.* 8:120–132.
Sauther, M.L., Cuozzo, F.P., and Sussman, R.W. (2001). Analysis of dentition of a living wild population of ring-tailed lemurs (*Lemur catta*) from Beza Mahafaly, Madagascar. *Am. J. Phys. Anthropol.* 114:215–223.
Schwartz, J.H., and Tattersall, I. (1985). Evolutionary relationships of living lemurs and lorises (Mammalia, Primates) and their potential affinities with European Eocene Adapidae. *Anthropol. Papers Am. Mus. Nat. Hist.* 60:1–100.
Seligsohn, D. (1977). Analysis of species-specific molar adaptations in Strepsirhine primates. *Contr. Primatol.* 11:1–116.
Strait, S.G. (1993). Differences in occlusal morphology and molar size in frugivores and faunivores. *J. Hum. Evol.* 25:471–484.
Strait, S.G. (1997). Tooth use and the physical properties of food. *Evol. Anthropol.* 5:199–211.
Sussman, R.W. (1991). Demography and social organization of free-ranging *Lemur catta* in the Beza Mahafaly Reserve, Madagascar. *Am. J. Phys. Anthropol.* 84:43–58.
Sussman, R.W., and Rakotozafy, A. (1994). Plant diversity and structural analysis of a tropical dry forest in southwestern Madagascar. *Biotropica* 26:241–254.
Tattersall, I. (1982). *The Primates of Madagascar.* Columbia University Press, New York.
Wainwright, P.C., and Reilly, S.M., (eds.), (1994). *Ecological Morphology*. University of Chicago Press, Chicago.
Wrangham, R.W., Conklin-Brittain, N.L., and Hunt, K.D. (1998). Dietary response of chimpanzees and cercopithecines to seasonal variation in fruit abundance. I. Antifeedants. *Int. J. Primatol.* 19:949–970.
Wright, B.W. (2003). The critical function of the 'robust' jaws of tufted capuchins. *Am. J. Phys. Anthropol.* Supplement 36:228.
Yamashita, N. (1996). Seasonality and site-specificity of mechanical dietary patterns in two Malagasy lemur families (Lemuridae and Indriidae). *Int. J. Primatol.* 17:355–387.
Yamashita, N. (1998a). Functional dental correlates of food properties in five Malagasy lemur species. *Am. J. Phys. Anthropol.* 106:169–188.
Yamashita, N. (1998b). Molar morphology and variation in two Malagasy lemur families (Lemuridae and Indriidae). *J. Hum. Evol.* 35:137–162.
Yamashita, N. (2000). Mechanical thresholds as a criterion for food selection in two prosimian primate species. *Proc. 3rd Plant Biomech. Conf.*, Freiburg-Badenweiler. Thieme Verlag, pp. 590–595.
Yamashita, N. (2002). Diets of two lemur species in different microhabitats in Beza Mahafaly special reserve, Madagascar. *Int. J. Primatol.* 23:1025–1051.
Yamashita, N. (2008). Chemical properties of the diets of two lemur species in southwestern Madagascar. *Int. J. Primatol.* 29:339–364.

Chapter 18
Convergence and Frontation in Fayum Anthropoid Orbits

Elwyn L. Simons

Contents

18.1	Introduction	407
18.2	Methodology	410
18.3	Results	414
18.4	Discussion	417
	18.4.1 Arboreal Theories	418
	18.4.2 Visual Predation Theory	420
	18.4.3 Angiosperm Radiation Theory	421
18.5	Conclusions	422
	References	426

18.1 Introduction

There have been several theories proposed to explain changes in the arrangement of the axis of the orbits of primates through time. Primitive eutherian mammals are known from numerous Cretaceous and Paleocene crania and these, in general, have orbits that are situated laterally. Hence, none of these early mammals have eyeballs that are located in sockets or that are rimmed by a lateral bar or "postorbital" bar. It is widely supposed that placental mammals were originally terrestrial, small, presumably insectivorous, and nocturnal, and such ideas have certainly been applied to the Plesiadapoidea or Proprimates (see Kay and Cartmill, 1977). None of the fossil mammals known so far from this early period, whether plesiadapoid or generalized placental, have any struts or bars surrounding the lateral margins of their eyeballs. Several Cretaceous mammal skulls are now known (not all placentals) and their orbital anatomy is well illustrated in Kielan-Jaworowska et al. (2004).

E.L. Simons
Duke Primate Center, Division of Fossil Primates, Biological Anthropology and Anatomy, Duke University, 1013 Broad Street, Durham, NC 27705
e-mail: esimons@duke.edu

In contrast to these forms, primates have either bars or sockets surrounding the eyeballs and with them the orbital alignment is typically more forward directed than in most other kinds of mammals. Modern placentals related to Order Primates such as members of the extant tree shrews, or the extinct plesiadapoids, could be used as proxies for probable primitive orbital orientation of the earliest and, so far, unknown primates, or the stages transitional to them. As is well known, both of these groups have frequently in the past been considered primates. The development of these characteristic primate structures: first bars and then sockets of the orbit seem to have appeared successively in time, and these changes are therefore of interest in an evolutionary context. Curiosity has grown as to what groups among the placentals, lacking bars or sockets of the eye, are most closely related to primates that show these structures. Equally, scientists want to know what is the region or continent where (1) euprimates; and (2) higher primates (Anthropoidea) and their visual adaptations arose. These questions have been reviewed by Holroyd and Maas (1994), who suggested several possibilities for time and place of origin (also see Miller et al., 2005). Among their suggestions for the African higher primate radiation, Holroyd and Maas suggest as alternates: (1) early migration from Europe to Africa; (2) similar early migration from Asia to Africa; and (3) origin in Africa. They also proposed a dispersal for New World anthropoids either from Africa or North America. Regrettably, little definite new evidence about this matter has been presented since 1994. Anthropoids could have had an origin in deep time in Africa or even in southern Asia, the exact location is both unknown and immaterial at present.

In addition to ideas about time and place of origin, there are uncertainties as to the adaptive context for the appearance of anthropoidean characters. Theories that postorbital bars arose in association with arboreality are weakened by the fact that there are many terrestrial mammals that exhibit postorbital bars as has been itemized by Heesy (2005). These include horses, some xenarthrans, some hyraxes, many artiodactyls, certain carnivores and, on the arboreal side: tree shrews and bats (flying foxes). There are numerous illustrations showing such bars, but see for instance the Traité de Paléontologie, volume IV edited (1958) by Jean Piveteau.

At present, nothing is known about the time or region when either the primate postorbital bar or postorbital closure arose. In the context of considering place, manner, and time of these orbital changes among primates, it seems useful to quote from Rasmussen and Sussman (2007, in press) who raise the question whether such deductions are even possible:

> Primates originated once, long ago, in an unknown place, without leaving a fossil record of the event. At face value, learning anything about primate origins seems implausible. Can our curiosity about such a singular, unobservable, historical event be investigated within the realm of science?

Actually, the following two processes or stages are involved in these questions, but they may be connected and interdependent. The first is a question. What is the nature of the basic primate lifestyle? The second question is: what caused the ongoing changes in visual acuity seen throughout the history of Order

Primates, but best exemplified among the Anthropoidea? In sum, the two stages are, first, origin of Primates (development of postorbital bar and beginnings of rotation toward stereoscopy) and, second, origin of Anthropoidea (development of a closed orbit and further advances in convergence). Separation of the previous two questions depends on what phylogeny for Order Primates is accepted as proven. In the case that three major divisions: Lemuroidea, Tarsioidea, and Anthropoidea had a near trichotomy, or unresolvable trichotomy in their origins, this kind of family tree co-mingles these two questions and makes them difficult to separate (see Yoder, 2003). In contrast to this is the work summarized by Schmitz and Zischler (2004), who found three out of 118 "alu-SINE markers" shared between tarsiers and anthropoids. SINE markers are complex DNA structures that, it is believed, are unlikely to independently evolve. Hence, these few SINE markers shared in common between Anthropoidea and Tarsioidea are thought to mark a short common stem. In stark contrast to this is the work reported by Murphy et al. (2001), who state that there is a "gulf" between tarsiers and strepsirrhines on one hand and anthropoids on the other hand. What is one to believe? Presumably, the problem is that molecular primate genetics is in its infancy and, so far, results are based on a few selected and different molecules from a particular genome that is comparatively huge.

The most obvious osteological feature associated with the visual system that distinguishes living primates from most other mammals (excepting the groups already mentioned) is the possession of a postorbital bar. Many hypotheses as to the functional reason for development of postorbital, actually periorbital, bars have been proposed. One theory as to the appearance of postorbital bars in primates is that the bar is part of a framework evolved to resist chewing stresses. For instance, DuBrul (1988) was of the opinion that the facial skeleton is adapted for the purpose of resisting masticatory stress. However, research carried out principally by Hylander's group at Duke has shown that at least in *Macaca* and *Galago* the postorbital bar receives relatively weak stresses and hence is overbuilt or over-engineered as far as these stresses are concerned (see Hylander and Johnson, 1997; Noble et al., 2000; Ravosa et al., 2000). The latter authors suggest that a thin postorbital bar—all that is needed for resistance to the chewing stresses—could easily be broken and would limit the fitness of the organism. The more strongly built bar hence would provide greater safety against trauma as suggested by Simons (1962). Heesy (2003) has also reviewed this complex subject. Long ago I remarked (Simons, 1962): A few years later Cartmill (1970) observed that such bars would not protect the eye from attacks with sharp teeth due to the slenderness of the bar, but the dangers I had in mind were of blows from branches being struck from the side, not tooth punctures. No bar, brow ridge, or socket protects mammalian eyes from punctures coming in from the front, or even from behind. Incidentally, a lateral bar around the eyeball might, with further selection, lead to increased strengthening of the orbital framework by developing bony plates surrounding the back of the eyeball. Ravosa et al. (2000) pointed out that the postorbital bar appears to provide rigidity so that chewing and biting activities, carried out as muscles move, do not impair visual acuity. They also suggest that advances in convergence and frontation of the orbit among small-sized primates derive from a combination of increase in brain size and improvements in nocturnal visual predation.

More recently, Heesy (2005:378) has convincingly demonstrated that lacking "a stiff lateral orbit, deformation due to temporalis contraction would displace soft tissues contributing to normal oculomotor function." This implies that during chewing activities visual distortions would occur, which could jeopardize the safety of the animal concerned by interfering with the detection of predators (see also Heesy, 2003). This argument applies equally to arboreal and nonarboreal species. Heesy's work has also summarized in detail numerous theories as to the reasons, or reason for the development of postorbital bars in primates. Of interest is the fact that among many non-primate mammals and with almost all primates, the actual eyeballs are capable of pointing essentially forward-allowing stereoscopic vision, even when the sockets are not fully forward directed (Starck, 1995). One striking exception among primates may be the giant lemur *Megaladapis* where eye sockets are so placed that the two overlapping fields of vision are interrupted by the snout. Many theories have arisen as to why, through time, most primate fossils have shown greater and greater degrees of both convergence and frontation of the eye sockets, of the sort reported here. Most hypotheses are attempts to explain what the consequent stereopsis is for and include: (1) enhanced perception in foraging for fruit and leaves; (2) locating insect prey (visual predation); (3) identifying conspecifics; and (4) making out locomotor routes through aerial substrates, including leaping from branch to branch.

Considering anthropoid origins, most extant higher primates, members of suborder or infraorder Anthropoidea, have fully forward-directed eye sockets (high orbital convergence). In addition, higher primates have greater frontation of the orbits, which grow to become oriented such that the dorsal rim of the orbital opening in the cranium is located vertically directly above the lower orbital rim (when the cranium is oriented in the Frankfurt plane). Hence, two processes of rotation of the orbital sockets have taken place through time.

18.2 Methodology

Scholars have differed as to the methodology for measuring these orientational changes, and most particularly in their hypotheses as to why these changes have happened. Another less significant issue has been concerned with precision in recording these angles. This is important for modern, fresh, undistorted crania where very precise angles can be determined, but since the study of orbital changes through evolutionary time must depend in considerable part on fossils, extreme precision is an elusive concern with them for almost every known fossil has been subjected to some sort of distortion, however slight. Equally, biologists know that every species population shows individual variation in its anatomy, as well as changes during the life of an individual, documented here in the case of *Aegyptopithecus zeuxis*. It is to be expected that angles of convergence and frontation will slightly vary among different individuals of both extinct and living species. One must add to this as well the problems that arise from having to base conclusions on very small samples as is the case for almost every extinct species. Both of these angles reported here can be considered elements in expressing change in orbital function or equally morphological estimates used to capture change in orbital form through time among primates. Their differences, which become evident in the fossils found from successive Tertiary time periods, beg for hypotheses explaining the causation of such changes.

In this chapter, I provide information about the morphology and orientation of the bony orbits in those fossil anthropoids from the Fayum for which there is suitable cranial material available. I then discuss the significance of these results and how they better our understanding of the pattern of orbit evolution among primates and the pattern's implications for evaluating various theories that have been advanced in order to explain the evolution of primate orbital orientation.

I have been able to collect data on some aspects of the orbital orientation for seven species and six genera of early anthropoids from the Fayum region of Egypt of both Eocene and Oligocene age. The systematics of the Fayum primates have been discussed in numerous previous papers and need not be repeated here (e. g. Simons, 1989, 1997; Simons and Rasmussen, 1991; Simons et al. 2001; Seiffert et al., 2004). *Aegyptopithecus* and *Catopithecus* are members of separate subfamilies within the Propliopithecidae—first considered a family by Strauss (1961), but see also Rasmussen (2001) who proposes a distinct family for Oligopithecidae; *Parapithecus* and *Apidium* are members of the Parapithecidae proposed by Schlosser (1911), while *Proteopithecus* and *Arsinoea* are each placed in their own families, the Proteopithecidae (Simons, 1997) and Arsinoeidae (Simons et al., 2001). Hence, these genera and the species they contain represent a considerable taxonomic and apparent phyletic diversity. Where these primates can be compared postcranially, there are many adaptive differences among them, suggesting considerable adaptive diversity and potentially a considerable evolutionary divergence among them prior to the age of the L-41 locality, a site dated to ~34/35 ma (Seiffert, 2006).

Because fossils are rarely complete, measurements of the orbits can be difficult and it is important to carefully define the angles being measured and compared and how these were measured. It should also be understood that the forward directedness of the actual eyeballs in living primates may be different from that suggested by the arrangement of the skull bones involved in the eye socket.

Convergence angle among fossil primates can be measured as was done by Simons and Rasmussen (1996, Figure 6), showing the angle of orbital convergence in three Eocene *Catopithecus* skulls. Like most crania of Paleogene as well as Miocene primates, which document the early stages of cranial transformation during primate evolutionary history, the frontal region is flattened dorsoventrally either naturally or by crushing, or both, and it is not difficult to locate the medial and lateral points of bending of the dorsal orbital margin and draw a line across each, as indicated in Fig. 18.1A. It is also typical of many of these early primates that

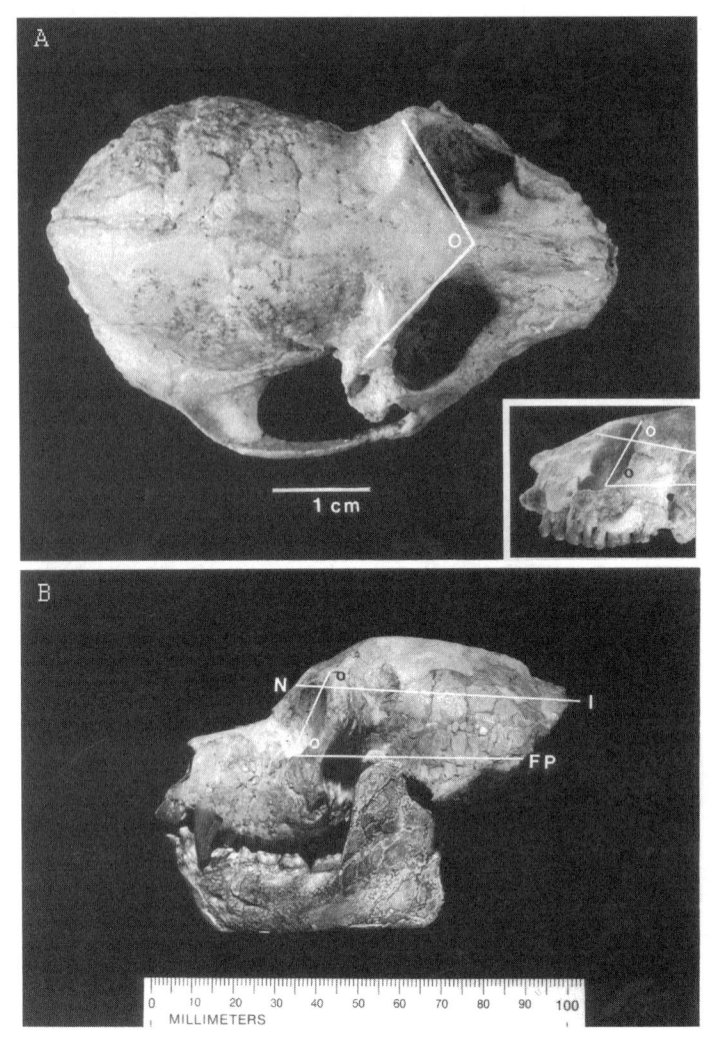

Fig. 18.1 (**A**) Two cuts of a *Parapithecus grangeri* cranium, DPC 18651, housed at the Duke University Primate Center, Durham, NC. The view of the dorsal aspect shows the convergence angle at the degree sign, as discussed. The inset shows how different angles of frontation are obtained at the black and white degree signs. This is because in this particular parapithecid the two horizontals are not parallel; (**B**) *Aegyptopithecus zeuxis* cranium with unassociated mandible, CGM 40237, housed at the Egyptian Geological Museum, Cairo, Egypt. The upper horizontal line runs from nasion to inion and lower line from the top of the auditory opening to the bottom of the orbit FP = Frankfurt plane. Note that the two horizontals are more or less parallel. The frontation angles by both measures, indicated at the black and white degree signs, are almost the same in this propliopithecid

the dorsal orbital margin tends not to curve but to be aligned more as a straight or linear margin than that seen in many extant anthropoideans such as *Homo* or *Pongo,* where the upper orbital rim is rounded. A simple protractor suffices to measure such an angle, which increases with greater orbital forward orientation toward 180° or complete convergence. A second methodology for determining convergence was used by Ross (1995) following Cartmill (1970). In his case, the measurement was taken from a line drawn along the midline or sagittal line and the upper margin of one orbit. Hence angles taken that way can be made comparable to those reported here or by other authors, such as Noble et al. (2000), by doubling. Lanèque (1993) has published a series of convergence angles for Eocene adapid prosimians and various other primates, and this can be referred to as a basis of comparison with anthropoideans.

Frontation angle can be defined by establishing a line along the long axis of the orbital opening as viewed from the side and measuring the angle between this and the line of the Frankfurt horizontal plane, named for the city in which the term was established in the 1880s. This plane is defined as a line running from the most superior point of the external auditory canal to the lowermost point of the orbital rim. Frontation is considered to be the degree of verticality of the orbital apertures as observed from the side. In a fully frontated orbit, this angle would be approximately 90°, while in a less-frontated primate the angle would be lower than 90°. Some prefer to see frontation as being measured as the degree of verticality of the orbital apertures as observed from the side. This angle can be defined by establishing a line along the long axis of the orbital opening as viewed from the side and measuring the angle between this and the line of the Frankfurt horizontal plane. See Fig. 18.1B, for frontation angle.

A somewhat different angle of frontation was determined by Cartmill (1970), who measured this angle using a unique "dihedral goniometer" invented in order to carry out his thesis research. An identical angle, using Cartmill's method, was calculated in Simons (2004, Figure 2) for *Parapithecus grangeri*. By this methodology, the angle is measured from a line drawn as a chord between the nasion and the inion, and from this the angle α is determined as an angle that is measured anteriorly between the latter chord and one drawn between the upper and the lower orbital margins (Fig. 18.1) (orbitale inferius-OI and orbitale superius-OS see Ross, 1995, Figure 3). In Cartmill's methodology, the angle of orbital frontation is then determined as 180°-minus the angle. Since the line of the Frankfurt plane drops from the ear opening toward a somewhat more ventral point (OI) than the nasion, it is likely that frontation angles determined from this chord will be lower than those determined by either Cartmill (1970) or Ross (1995), see Fig. 18.1A, and for both frontation chords and angles see Fig. 18.1B. In this contribution, I prefer the former angle of frontation calculated from the Frankfurt plane as it is easier to determine from any lateral photograph of a skull, Fig. 18.1B. Also, many fossil primates do not have an easily locatable inion as the back of the cranium is rounded. Heesy (2005:368) also illustrates the methods of taking these angles.

18.3 Results

Many primate cranial fossils from the Fayum Eocene and Oligocene deposits, ranging in preservation from isolated frontals to relatively complete crania, are amenable to the determination of the convergence angle as viewed from above (Table 18.1). Frontation angles in fossils are only likely to be correct if the skulls have little or no distortion due to dorsoventral compression. Of all Fayum primate discoveries,

Table 18.1 Convergence angles calculated for Fayum primates who belong in diverse taxonomic groups

Species	Specimen	Convergence angle
Catopithecus browni	DPC #11594	125°
	DPC #11388	132°
	DPC #13604	121°
	CGM #42222	121°
	DPC #12367	134°
Proteopithecus sylviae	DPC #14095	148°
	CGM #42214	122°
Aegyptopithecus zeuxis	CGM #40237	183°
	CGM #85785	192°
	*DPC #8794	190°
Parapithecus grangeri	DPC #18651	106°
	*DPC #6642	99°
Apidium phiomense	AMNH #14556	113°
	DPC #9867	110°
	*DPC #3887	112°
	*DPC #1067	110°
Apidium bowni	DPC #5264	105°
Arsinoea kallimos	*DPC #11434	125°

*Based on frontal orbital margins alone. Specimen identification numbers abbreviated as CGM are held at the Egyptian Geological Museum Cairo, Egypt, while specimen identification numbers abbreviated DPC are held at the Duke University Primate Center, Durham, NC, and specimen identification numbers abbreviated AMNH are held at the American Museum of Natural History, New York.

Table 18.2 Frontation angles for Fayum primates

Species	Specimen	Frontation angle* NP	FP
Aegyptopithecus zeuxis	CGM # 40237	69°	69°
	CGM # 85785	70°	70°
Parapithecus grangeri	DPC # 18651	62°	72°

* Frontation was calculated relative to both the NP (NP = nasion/inion) and FP (FP = Frankfurt plane) planes. There are two known crania of *Aegyptopithecus zeuxis*, both recovered from Fayum Quarry M.

18 Convergence and Frontation in Fayum Anthropoid Orbits 415

so far only three are complete enough to measure with any certainty the angle of frontation (Table 18.2).

As a further presentation of the evidence, Figs. 18.2–18.4 have been prepared in order to show the convergence angles in some of the best preserved specimens of Fayum primates. Figure 18.2 shows how convergence angles of *Catopithecus* and *Proteopithecus* have been determined. These genera represent two different Fayum primate families. In the one species of the former genus and family, this angle ranges

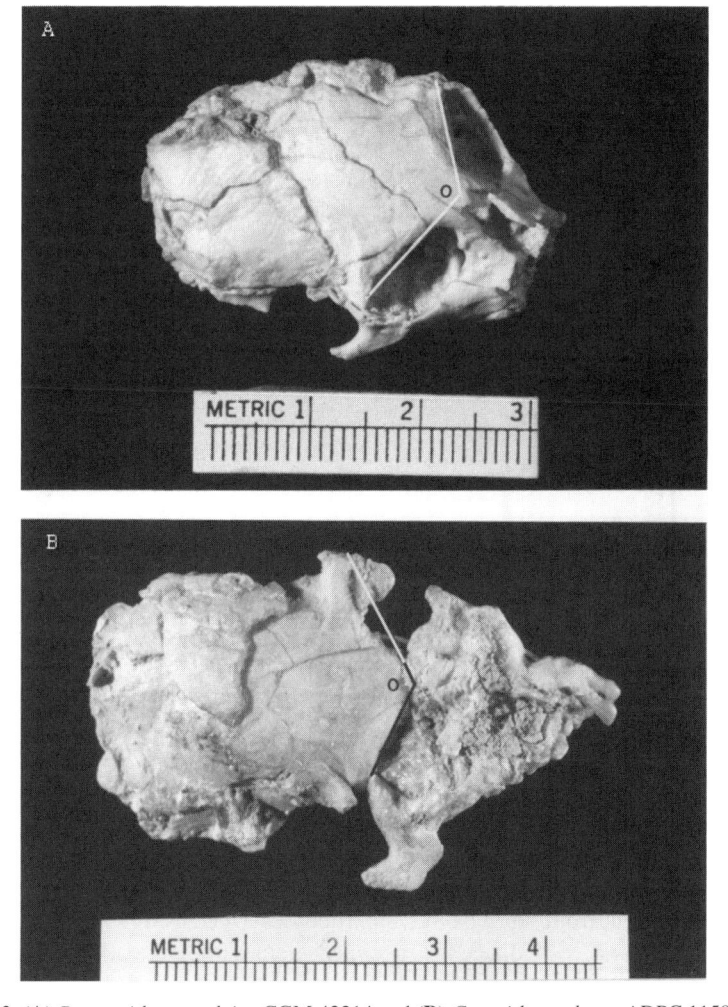

Fig. 18.2 (**A**) *Proteopithecus sylviae* CGM 42214 and (**B**) *Catopithecus browni* DPC 11594 show convergence angles measured from the degree symbols in individuals of these two species

A

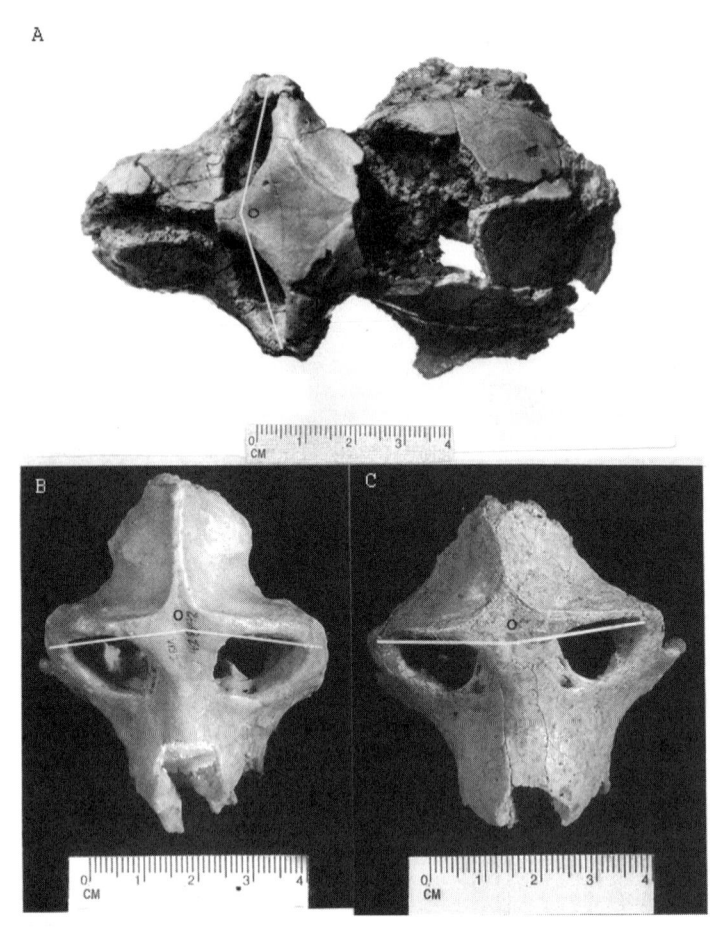

Fig. 18.3 Crania of *Aegyptopithecus zeuxis* showing age-related changes in convergence angles. (**A**) subadult male CGM 40237, angle 150°; (**B**) older male CGM 42842, angle 195°; and C: adult male DPC 8794, angle 171°

from 121° to 134°, while in the latter the angle (two specimens only) is from 122° to 148°. If the DPC specimen #14095 of the latter, *Proteopithecus sylviae*, is the more accurate one, then the convergence angle of proteopithecids is likely to be distinctly higher than in *Catopithecus*, see also Fig. 18.4. Figure 18.3 shows the convergence angle in three specimens of *Aegyptopithecus zeuxis*, the only Fayum Oligocene species where this angle, in mature adults, can exceed 180°, a situation quite different from other Fayum anthropoidean species. Figure 18.4 provides additional comparative evidence for specimens belonging to three different Fayum anthropoid (or anthropoidean) families: Parapithecidae, Arsinoeidae, and Proteopithecidae. The samples are small, but it seems to be the case that parapithecids have distinctly lower convergence angles than do any other Fayum anthropoid groups, see also Table 18.2.

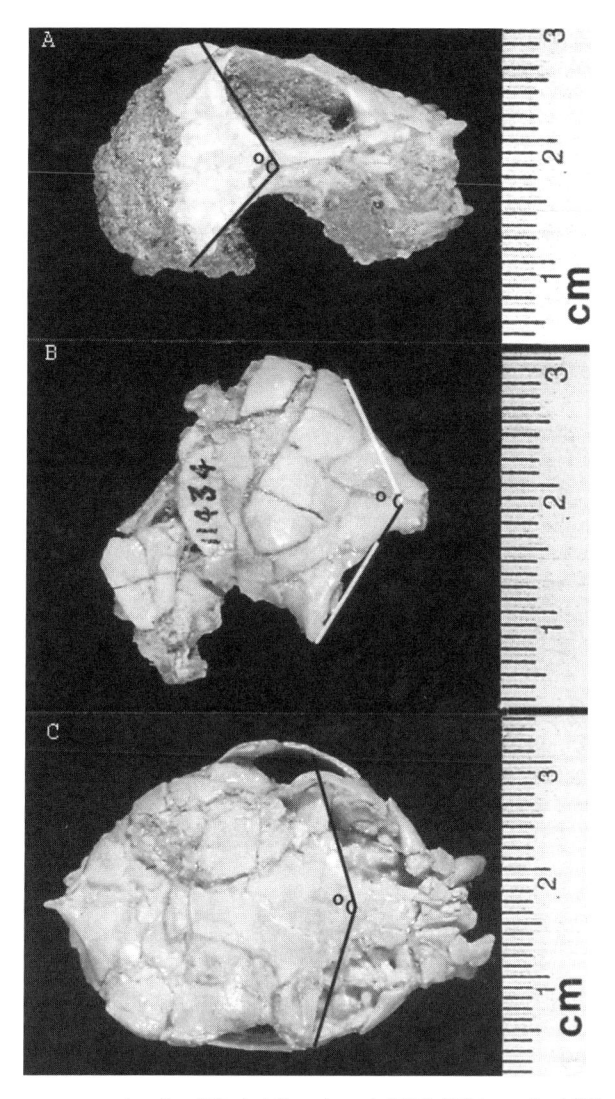

Fig. 18.4 Convergence angles for (**A**) *Apidium bowni* DPC 5264, angle 105°, (**B**) *Arsinoea kallimos* DPC 11434, angle 125°, and (**C**) *Proteopithecus sylviae* DPC 14095, angle 148°

18.4 Discussion

In as much as the actual orientation of the eyeballs of living primates are more forward directed than their socket orientation, this suggests that one could question the significance of even determining the angles of convergence. Ravosa and Savakova (2004) have pointed out that within groups the smaller species are more likely to exhibit "more divergent orbits due to the presence of relatively large eyes." This would seem to be the reason that the convergence angle of *Apidium bowni* is so low. Its cranium happens to be less than half the size of that of *Apidium phiomense*. The reasons for variation in angle of frontation are more complex, but it has often been thought to be due to variations in habitual posture of the head. Many prosimians have been observed in life to lower the head and in effect focus their visual field above the snout. Such orientation is enhanced by a lower degree of frontation than in most modern catarrhines, for instance, see Simons (1972:72). Ravosa and Savakova (2004) conclude that recent fossil evidence concerning the pattern of anthropoid origins, particularly as reviewed by Fleagle (1999), suggests a stepwise appearance of the most relevant specialization that have come to characterize the group. Such a conclusion is supported by the evidence presented here.

18.4.1 Arboreal Theories

Many theories have been proposed to explain the reasons for early primate arboreality, and this is not the place to discuss them in detail. These theories have been put forward both to explain the origin of the primate order or of anthropoids (Anthropoidea) only. However, the evolution of both frontation and convergence in primate orbits is intimately involved in both these discussions. Collins (1921) proposed that grasping hands and feet are highly useful for living in an arboreal world. This is also expressed in the arboreal theory first proposed by Wood-Jones (1916), suggesting that we may be derived from a tarsier-like tree-dwelling primate good at leaping and grasping branches. To this, Grafton Elliot-Smith (1924) in a series of essays on the evolution of man added the concept that stereoscopic vision was important for tree-dwelling life and that this feature was correlated with relative brain enlargement among primates. Later, LeGros Clark (1959) elaborated on this concept that the coordination of hands and feet with stereoscopic vision supported an arboreal theory of primate or of anthropoid origins. Knowing him, I feel that he was primarily thinking about the latter origin. He certainly believed that a number of features such as forward-directed eyes with stereoscopy and depth perception indicated greater dependency on vision than the other senses, and a consequent enlargement of the visual centers of the brain made up a total morphological pattern characterizing higher primates. In general, these views are mainly descriptive and are an attempt to summarize primate adaptations rather than a codified theory about the reason for advanced orbital convergence but, nevertheless, the two are interrelated.

Modern arboreal primates are excellent at movement in the trees although many researchers have pointed out that it is not necessary to have fully forward-directed eye sockets to adapt to tree life, as exemplified by arboreal rodents such as squirrels or by marsupial opossums. It is hard to understand the relevance to the acquisition of primate arboreal agility of the observation that seemingly less well-adapted tree-dwelling rodents also are arboreal. How can the way arboreal rodents are constructed challenge hypotheses about the reasons why primates developed an arboreal way of living? Arboreal rodents must have restricted and different abilities for moving about in the fine branches. Moreover, rodents scurry upon trunks and branches with dependence on their claws and frequently fall out of trees, as primates do also for somewhat different reasons (see Schultz, 1956). I know of no data on relative frequency of primate and rodent accidents during arboreal life. Crompton (1995) has made an important contribution in citing a very wide range of alternative theories and field observations concerning primate arboreality and that of other mammals. In doing this, he has brought attention to the complexity of the adaptation. Although those authors that expressed the arboreal theory early did not explicitly discuss the point, all of them must have believed that primates were up in the trees for a *reason*, that is, to feed there or avoid terrestrial predators.

Primate behaviorists know that it is not necessary to leave an office or the city to understand the extraordinary degree of hand/foot and eye coordination exhibited by living primates. The ubiquity of nature films showing primate movements in the trees have made knowledge of the arboreal abilities of modern primates commonplace. It seems almost self-evident that this is the result of evolutionary development of an adaptation for improved arboreal maneuvering. Such wildlife movies frequently show prosimians, monkeys, and apes, leaping unconcernedly from one tree branch to another. This leaping seems often to be done without pausing to judge the exact goal, but instead individuals behave as if confident holds will be found wherever they land.

While participating in wild-captures of lemurs in Madagascar that were intended for outbreeding at the Duke Primate Center, I had occasion to observe such extraordinary abilities.

When highly drugged lemurs, after darting, had already lost hold of the branches and were falling toward a waiting net below, they sometimes fell past or through lower branches. Such already drugged individuals would instantly seize hold of such branches when falling, for even under extremely impaired conditions their visual/grasping acuity was not lost. Crompton (1995) has discussed stereopsis and its origin, perhaps considering issues such as "breaking crypsis" first identified by Julesz (1971) and in references cited. Crompton (1995:14) remarks:

> Full binocular stereopsis, as opposed to monocular depth perception, does, then, seem to involve comparison of neural disparity between two extensively overlapping fields of vision, and does, therefore, require orbital frontality. However, excessive approximation of the two eyes will *at some point* actually impair the accuracy of reconstruction of the stereo image by excessively reducing the angle between the two axes of vision.

Moreover, studies of the cranial and skeletal parts discovered for the Fayum anthropoid genera *Aegyptopithecus, Catopithecus, Apidium, Parapithecus,* and *Proteopithecus* show that all these primates were both *diurnal* and *arboreal*. Since species of these genera belong to at least four or more taxonomic families, they certainly appear to establish what the life style of anthropoids in Africa was at that time. Regardless of what sort of selection induces the development of higher and higher angles of orbital convergence, these primates differ among themselves and not all achieved the modern condition where convergence angles are at or over 180° (Table 18.1). It is possible that orbital enlargement, resulting from a nocturnal adaptation, produces a higher convergence angle, and high convergence angles resulting from this adaptation may eventually be demonstrated in early anthropoids. The one known Fayum anthropoid suspected of being nocturnal, *Biretia megalopsis*, recently described by Seiffert et al. (2005) belongs to a superfamily, Parapithecoidea, which cannot possibly have given rise to "crown" Anthropoidea (Fleagle and Kay, 1987). The species is atypical, in as much as other members of the parapithecid group are all diurnal. It is also too incompletely preserved to measure either of the angles discussed here.

18.4.2 Visual Predation Theory

This concept originated in an important series of papers by Cartmill (1970, 1972, 1974) and it combines several different concepts associated with the origin of primate arboreality. If anthropoids originated very early in the history of this order, then this system may contribute to understanding both sets of origins. Basically, this theory expresses: (1) the view that the forward direction of the orbits and development of grasping hands and feet coupled with the reduction of claws came about to facilitate nocturnal foraging, mainly for insects, in the finer branches of forest under-story as tarsiers do; (Nevertheless, it is well known that if a mouse lemur, small possum, or bush baby is feeding on nectar or gum and an insect passes, it will be seized and eaten. Few primates, other than tarsiers, show complete exclusivity in diet.) (2) that increased orbital convergence has arisen to assist in estimating the location of prey without the necessity of moving the head, evidently the case with both modern tarsiers and owls; and (3) that reduction in olfaction could have received acceleration because convergence brings together the inner sides of the eyeballs, hence reducing the space available for the sense of olfaction (olfactory bulbs).

Observations on tarsiers at the Duke Primate Center has clearly demonstrated that these primates do identify and stare at insect prey without movement, presumably assessing distance, before leaping to seize live food. They will not take food that is not alive, not "on the hoof", see also Haring and Wright (1989). When first brought into captivity, tarsiers were often presented with fruits, leaves, and berries that they did not eat. Dr. Faust, Former Director of the Frankfurt Zoo, once told me that in order to cause tarsiers to eat meat it was necessary to insert insect wings (from Maikäfer beetles) into tiny pieces of ground meat. It appeared the Frankfurt tarsiers were languishing because of insufficient vitamins and minerals and would not eat any non-living preparation. At the Duke Primate Center, in order to maintain tarsiers, they had to be fed crickets, lizards, mealworms, and other insects, and it was an entirely live diet. Tarsiers are the most predatory and carnivorous of all primates and do not smell their way to food sources. What is the relevance of considerations of tarsier feeding habits in the wild and captivity to the considerations of convergence and frontation angles? It is because, unfortunately, it remains to be established whether the earliest primates or the earliest anthropoideans bore any structural–functional resemblance to tarsiers. With such predators as tarsiers, there is no need for a developed sense of smell, as is believed to be the case for folivorous and frugivorous primates, but the greatly enlarged orbits, enhancing stereopsis, and resultant high angles of convergence do improve nocturnal predation. Equally, the reduction of the size of the rostrum, consequent on diminished dependence on the sense of smell, coupled with the comparatively enormous eye balls, causes tarsiers to have more frontated orbits. Also the loss of the glandular rhinarium (characteristic of strepsirrhine prosimians) among tarsiers may not be a shared-derived advance linking the group with Anthropoidea, even though loss of the rhinarium is not directly connected with decreased olfactory ability. Hofer (1980) who examined a series of tarsiers found a strepsirrhine nostril in a tarsier and thought, "the shape of the nostrils [is] irrelevant for primate taxonomy." Members of genus *Homo* and *Tarsius,* considered overall, are among the most derived of all extant primates, and hence the latter could be of limited significance in reflecting the nature of earliest haplorhines or as an example of an animal whose adaptations suggest reasons for ancient orbital forward rotation. The high orbital convergence and frontation which species of genus *Tarsius* exhibit may be independently acquired resemblances to advanced Anthropoidea and need not necessarily tell us anything about anthropoid evolution.

Nevertheless, some go so far to conclude that almost everything is settled, as do Ross and Kay (2004:725). "The origin of Primates was associated with a shift to nocturnal visual predation in the shrub layer of tropical rainforests." This is a conclusion completely modeled on the lifestyle of the tarsier, and consequently is dependent, first, on whether tarsiers have or do not have a derived behavior and, secondly, it assumes that there is a case for initial insectivory as well as, third, the demonstrated existence of a haplorhine clade. In their defense on this point, however, these latter authors do stress that SINE markers (see Schmitz and Zischler, 2004) are the evidence suggesting a common stem ancestry for tarsiers and anthropoids.

Finally, there are different categories of locomotor style among prosimians and some of them, regardless of whether they eat insects or forage on leaves and fruit, do not jump. There are the slow-climbing living lorisiformes such as *Nycticebus, Potto, Arctocebus,* and *Loris,* as well as the extinct genera, *Adapis* and *Leptadapis,* which are also slow climbing and were presumably folivorous (see Dagosto, 2007). The extinct sloth lemurs and *Megaladapis* are browsers, and with them stereoscopy is less important as they do not leap. I am indebted to Dr. Patricia Wright for pointing out and confirming my own impression that many lemurs such as the indriids, *Hapalemur*, *Lepilemur,* and *Varecia* although all folivorous show rostral reduction, convergent visual fields, and can jump skillfully—almost with abandon. Hence, locomotion must be considered a factor in the primate arboreal adaptation.

18.4.3 Angiosperm Radiation Theory

This concept, proposed by Sussman (1991, 1995, 1999) is that the adaptive radiation, basically the origin, of primates was correlated with the appearance of flowering plants in earth history (during the Cretaceous) or at least came after such plants were available. This can also be described as a terminal branch-feeding hypothesis.

Long ago, Grand (1972) described primate utilization of the terminal branch niche and cited earlier work, but he and others had concentrated on large primates. In Sussman's view, the adaptations of hand and eye evolved early in order to reach terminal tree branches where nutrient-rich flowers, nectars, insects (and elsewhere gums) could be harvested as food sources. The hypothesis is attractive because it provides a causative factor, the spread of angiosperm plants that is believed to have happened during a particular range of time in the past during the Cretaceous Period. Nevertheless, one might question the theory because these plants evidently radiated long before the primates made the adaptation, but the angiosperm flowers' being there does not mean that there need have been an immediate occupancy. In effect, the theory has been questioned for the reason that there are no known fossil euprimates until the end of the Paleocene and Proprimates, such as plesiadapids or carpolestids, for instance, do not show orbital convergence, see Conroy (1990). Plesiadapiformes (=Proprimates), including both plesiadapids and carpolestids among other groups, were defined as an Order of mammals by Gingerich (1989, 1990). However, recent work by Bloch and Boyer (2002) has shown that carpolestids had evolved grasping hands before there was orbital convergence and that this may be pertinent to the question of the origin of primate arboreality, if such proprimates are actually related to euprimates, but the latter has been contested. In addition, Martin (1990) and Miller et al. (2005) have made a case that Order Primates arose in the Cretaceous, even though there are no fossils, the order may have differentiated and made its arboreal adaptation in southern continents or another place where, so far, nothing in the way of fossil evidence has been found. Even so, origins always mean definition. What was it that characterized the first identifiable primate?

Recent research has demonstrated that trichromatic color vision has been spreading, seemingly during the later Tertiary, among arboreal frugivorous/folivorous primates, and has not yet been achieved by many species. Like forwardness in the direction of orbits, this is an ongoing process. It would seem that being a trichromat would also contribute to the process, to perception of depth and increase precision of identification of both foodstuffs and sources of danger, see Lucas et al. (2003) for a current analysis of color vision. Dominy (2004) has recently reviewed evidence of the degree of dependency on color in food selection. His conclusions stress that arboreal primates smell fruits, perhaps to detect ethanol, typically associated with sugar concentrations, and test them with "digital and/or dental evaluation of texture," Dominy (2004:295).

18.5 Conclusions

The most interesting thing about the varying convergence angles, reproduced in Table 18.1 above, for several Fayum primates is that the different ancient families for which this angle can be determined are not the same in degree of convergence, showing that the development of the process of orbital rotation forward of these early anthropoids was still in progress in Eocene/Oligocene times. Also they do demonstrate that although the Propliopithecidae (*Aegyptopithecus*) have prosimian-sized brains (Simons, 1993) when compared to estimated body size, this can be coupled, as shown above, with advanced convergence. Hence, in this case, at least, it was not directly due to brain enlargement and its expansion forward that the orbits have rotated anteriorly. Equally, it is not the case that arboreal/visual adaptations are something that was completed during the obscure and undocumented beginnings of the order. These developments may more likely be linked with the origin of Anthropoidea. Fayum primates reveal these processes in progress. Certainly, these Egyptian higher primates were arboreal before most of them had fully perfected development of stereoscopic vision, at least as is suggested by those families that possessed the lower convergence angles. If full stereoscopy is possible when the axis of the socket is not rotated fully forward, then much of the point of determining such angles is lost.

Convergence angles must somehow relate to the underlying eye socket structure. Among all or all but one (depending on how they are classified) of these Fayum anthropoid families, the jugal bone forms a broad spoon-like plate or extension inward below the frontal to reach the alisphenoid bone. The inferior orbital fissure is very small in these anthropoids and is smaller in some than is typical even of platyrrhines. These relationships of frontal, jugal, and alisphenoid do not resemble the condition here in *Tarsius* species. In contrast, among some, but not all, tarsiers the alisphenoid bone has a narrow toothpick-like lateral extension running out under the frontal bone to make its only contact with the jugal. Moreover, tarsiers have unique thin, but extensive, periorbital bony flanges extending around the orbit and directed outward from the cranium so that broadening or expansion of the jugal is also largely pericranial. If these unique extensions are disregarded, then the jugal is not very broad and, quite unlike the Fayum anthropoids, contributes little inwardly to the postorbital septum. Because of this, the arrangement of jugal, frontal, and alisphenoid bones are not similar between Fayum anthropoids and present-day *Tarsius* species. Nevertheless, when one is scoring this contact—between jugal and alisphenoid—cladistically as a numerical example of character, coding the character code score is the same, while the anatomy is quite different. One can assert that only tarsiers and anthropoids share this feature, but is it (the alisphenoid-jugal contact) a real synapomorphy or just another odd feature about *Tarsius*, perhaps brought about by the eye ball hypertrophy on a tiny skull? This could mean that postorbital closure or development of a postorbital septum in *Tarsius* is not due to relatedness to anthropoids but is co-incidental, and has developed independently for reasons related to the enormous enlargement of the eyeballs. No one has explained why, in some tarsiers, this splinter of bone runs out from the nearly basicranial alisphenoid and touches the jugal, a reverse of the condition in Fayum anthropoid orbits. Contact between alisphenoid and jugal is one of the pillars of the series of assumptions that tarsiers belong with anthropoids in Haplorhini. Cladists represent character coding as being unbiased, but here is a perfect example of how subjectivity creeps in. Often, the decisions which characters to select and how they are weighted are far from being impartial. This process is not unlike politics: if something is proclaimed repeatedly with the air of being the truth, before long it will be believed by many. The question arises what a cladistic character actually is. There are many variable definitions, thus documenting that the entity "character" is changeable and thus biased in itself. Here are but three examples of such definitions:

- A character in systematics may be defined as any feature which may be used to distinguish one taxon from another (Mayr et al., 1953).
- A character is a feature of an organism that can be evaluated as a variable with two or more mutually exclusive and ordered states (Pimentel and Riggins, 1987).
- A character is a theory that two attributes which appear different in some way are nevertheless the same (Platnick, 1979).

The official website of the British Paleontological Association simply states:

The main point to be made in this article is that the selection and coding of characters is the key stage of cladistic analysis. The way we code characters can influence the phylogenetic hypothsis (Sic).

This website concludes by stating that the laboriously assembled "characters" finally will be evaluated by a variety of computer programs and "computerized phylogenetic tree manipulators." Consequently, this documents that it is the computer program, not the researcher, who makes the final decision about how to generate the outcome of the "unbiased" character collection.

These Fayum anthropoids I have reviewed here do not look like tarsiers cranialogically. However, if primate and anthropoid origins date to 35–55 ma before these Fayum anthropoids (at 34/36 ma), there is little reason to suppose that those early animals looked like or acted like either tarsiers or these ancient anthropoids of Egypt. The acceptance of tarsiers as haplorhines and their use as inferential models for earliest anthropoids, however indirectly drawn, remains an uncertain procedure. Starck (1955) has discounted the similarities in placentation between the two groups, but his points are widely disregarded, perhaps because they were expressed in German only, but see Simons (2003) for translation. In any case, the argument has primarily shifted to the molecular evidence. Yoder (2003) has discussed the problems arising from the possibility of there being short internal phyletic branches shared between, either tarsiers and strepsirrhines (lemurs/lorises), or tarsiers and anthropoids. If short enough, these might summate as an unresolvable trichotomy. Some work such as Murphy et al. (2001) favor the former linking, while Ross and Kay (2004) favor the latter grouping. The view that anthropoids had a deep time origin becoming distinct somewhere back in the Cretaceous period as discussed by Miller et al. (2005) also affects how we interpret these relationships. Of course, no creatures that might be primates have been convincingly demonstrated from such an early time.

There are various locomotor differences between Fayum anthropoids, but it is not clear at present how these affect, if they do at all, the visual system. Parapithecids are much more adapted to leaping and springing than are propliopithecids, which appear to have been slow climbers. Compare for instance the locomotor determinations of Fleagle and Simons (1983, 1995) of the parapithecid *Apidium* with papers analyzing the skeleton of *Aegyptopithecus* or *Propliopithecus* such as Fleagle and Simons (1978, 1982) and Ankel-Simons et al. (1998). Proteopithecids are also quite different postcranially and their hind limb bones suggest similarity to those of platyrrhines (Simons and Seiffert, 1999).

Among the Fayum Oligocene primates, only *Aegyptopithecu zeuxis* shows full orbital convergence with angles, depending on age, from 150° to as high as 195° (see Fig. 18.3). In this species, the increase in forward orientation of the orbits appears to be due to the anterior growth and spread of the temporalis muscles with age and is not the result of relative brain enlargement. Although most parapithecids are somewhat smaller animals, the skull of *P. grangeri* (DPC 18651) is about the same size as female *A. zeuxis* (CGM 85785), but convergence and frontation angles among these two species are quite distinct, Tables 18.1 and 18.2. Also these differences are not due to differences in visual adaptation, whether crepuscular, nocturnal, or diurnal, as the comparatively unexpanded orbital dimensions of these primates indicate that all species studied here are diurnal, with the possible exception of one species of *Biretia*. While it can be argued as to whether Fayum primates belong in four of five distinct families, their diversity is definitely considerable and they are not all the same or at the same stage as regards convergence and frontation. Species in each of the several families display different degrees of "modernization". Their diversity makes it unlikely that there was much earlier uniformity in the possession of either feature. Hence, the basal primates or basal anthropoids are unlikely to have resembled in these orbital features any particular modern forms. None of these taxonomic groups occur outside Africa although claims have been made, for instance, see Ross (2000) who erroneously placed *Arsinoea kallimos* in the Asian clade [Family] Pondaungidae + Amphipithecidae of Kay (2005).

Dominy (2004) has also emphasized that papers by Kay and Simons (1980) and Kirk and Simons (2001) give evidence that the same Fayum primates discussed here do suggest the approximate time of appearance of catarrhine folivory, at or somewhat after 34/36 ma. If folivory is a factor in selection that favored increasing the degree of frontation and convergence among primates, this would perhaps help to date the approximate time that development of these features began. Nevertheless, this date would have to be altered if the clearly folivorous, or even seed predating, Asian amphipithecids from the late middle Eocene of Myanmar (Burma), ~37 ma, are actually related to catarrhine anthropoids, see Kay (2005) for recent discussion of amphipithecids. Presumably, no one now thinks that this family could belong with archaic Catarrhini.

Evaluation of the various theories, suggesting reasons for development of full orbital convergence among primates as reviewed, indicates that all of them have merits, whether or not they apply to the time of origin of arboreal primates or of anthropoids alone. As Crompton (1995:25–26) remarks in his conclusion,

> Visual isolation and three-dimensional location of diverse targets (not only insects, but equally small branches, and fruit) using monochromatic, scotopic vision, in this densely-packed, complexly-shaded forest zone, would have been well served by the twin benefits of orbital frontality: steropsis and enhanced scotopic acuity. Since scotopic acuity is a major benefit of orbital frontality, to the extent to which orbital frontality is an element of the adaptive suite of ancestral primates, nocturnality is also implied as a basic primate adaptation, whether it be an ancestral or derived character.

Therefore, in view of all of the above, it seems advisable to propose here a new *Synergy Theory* incorporating all of the best of the above hypotheses. Evolution of higher primates must have been a continuous process of change during the long period before sometime in the Miocene, and a process that is but scantily documented. Analogies can be drawn from the study of living forms, but the relationship of such species to earliest ancestors is not understood. A staged-out sequence of events changing these animals seems implied, but as it concerns the origin of the order we have almost no real evidence, only inference. Proof of the timing of events is yet to come. Climbing abilities, stereoscopy, orbital convergence, and consequent improvements in visual perception and brain enlargement, as well as the development in some cases of trichromacy, all work together to improve both identification of food items whether plant or animal, add to an ability to deal with arboreal substrates and equally enhance capacities for detection of any kind of predator that enters or lurks in the arboreal environment.

Acknowledgments We thank the Egyptian Geological Museum, Cairo Egypt, for allowing us to make measurements on the specimen housed there. Also, we thank the American Museum of Natural History, New York, for allowing us to make measurements on a specimen housed there. Other measurements were from specimen housed at the Division of Fossil Primates, Duke University Primate Center. Photos were modified by Dr. Elwyn L. Simons at the Division of Fossil Primates. The National Science Foundation has provided ongoing support for research in the Fayum. This manuscript is DUPC publication number 1015.

References

Ankel-Simons, F., Fleagle, J. G., and Chatrath, P. S. (1998). Femoral anatomy of *Aegyptopithecus zeuxis*, an early Oligocene anthropoid. *Am. J. Phys. Anthropol.* 106:413–424.
Bloch, J. J. and Boyer, D. M. (2002). Grasping primate origins. *Science* 298:1606–1610.
Cartmill, M. (1970). *The Orbits of Arboreal Mammals a Reassessment of the Arboreal Theory of Primate Evolution*. Ph. D. Dissertation, University of Chicago, 671pp.
Cartmill, M. (1972). Arboreal adaptations and the origin of the order primates. In: Tuttle, R. H. (ed.), *The Functional and Evolutionary Biology of Primates*, Aldine-Atherton, Chicago/ New York, pp. 97–122.
Cartmill, M. (1974). Rethinking primate origins. *Science* 184:436–443.
Clark, W. E. L. (1959). *The Antecedents of Man*. Edinburgh University Press, Edinburgh, 374pp.
Collins, E. T. (1921). Changes in the visual organs associated with the adoption of arboreal life with the assumption of the erect posture. *Trans. Ophthalmol. Soc. U.K.*, 41:10–90.
Conroy, G. C. (1990). *Primate Evolution*. W. W. Norton & Co., New York, 492pp.
Crompton, R. H. (1995). Visual predation, habitat structure, and the ancestral primate niche. In: Alterman, L., Doyle, G. A., and Izard, M. K., (eds.), *Creatures of the Dark: The Nocturnal Prosimians*, Plenum Press, New York, pp. 11–30.
Dagosto, M. (2007). The postcranial morphotype of primates. In: Ravosa, M. J., and Dagosto, M., (eds.), *Primate Origins: Evolution and Adaptation*, Springer Press, New York, pp. 489–534.
Dominy, N. L. (2004). Fruits, fingers, and fermentation: the sensory cues available to foraging primates. *Integr. Comp. Biol.*, 44:295–303.
DuBrul, E. L. (1988). *Sicher and DuBrul's Oral Anatomy*, 8th Ed., St. Louis-Tokyo Ishiyaka EuroAmerica Press, 356pp.
Elliot Smith, G. (1924). *The Evolution of Man: Essays*. Oxford University Press, London, 155pp.

Fleagle, J. G. (1999). *Primate Adaptation and Evolution,* 2nd Ed., Academic Press, New York, 596pp.
Fleagle, J. G. and Kay, R. F. (1987). The phyletic position of the Parapithecidae. *J. Hum. Evol.* 16:483–532.
Fleagle, J. G. and Simons, E. L. (1978). Humeral morphology of earliest apes. *Nature* 267:705–707.
Fleagle, J. G. and Simons, E. L. (1982). Skeletal remains of *Propliopithecus chirobates* from the Egyptian Oligocene. *Folia Primatol.* 39:161–177.
Fleagle, J. G. and Simons, E. L. (1983). The tibio-fibular articulation in *Apidium phiomense,* an Oligocene anthropoid. *Nature* 301:238–239.
Fleagle, J. G. and Simons, E. L. (1995). Limb skeleton and locomotor adaptations of *Apidium phiomense*, an Oligocene anthropoid from Egypt. *Am. J. Phys. Anth.* 97:235–289.
Gingerich, P. D. (1989). New earliest Wasatchian mammalian fauna from the Eocene of northwestern Wyoming: composition and diversity in a rarely sampled high-flood plain assemblage. University Michigan Papers on Paleontology, 28:1–97.
Gingerich, P. D. 1990. Mammalian order Proprimates. *J. Human Evol.* 19:821–822.
Grand, T. I. (1972). A mechanical interpretation of terminal branch feeding. *J . Mamm.* 53:98–201.
Haring, D. M. and Wright, P. C. (1989). Hand-raising a phillipine tarsier. *Tarsius syrichta. Zoo Biol.* 8:265–274.
Heesy, C. P. (2003). *The Evolution of Orbit Orientation in Mammals and the Function of the Primate Postorbital Bar*. Ph. D. dissertation. Stony Brook University, Stony Brook, NY.
Heesy, C. P. (2005). Function of the mammalian postorbital bar. *J. Morphol.* 264:363–380.
Hofer, H. O. (1980). The external anatomy of the oro-nasal region of primates. *Z. Morph. Anthrop.*, Stuttgart 71:233–249.
Holroyd, P. A., and Maas, M. C. (1994). Paleogeography, paleobiogeography, and anthropoid origins. In: Fleagle J. G., and Kay, R. F. (eds.), *Anthropoid Origins*, Plenum Press, New York. pp. 297–334.
Hylander, W. L. and Johnson, K. R. (1997). In vivo bone strain patterns in the zygomatic arch of macaques and the significance of these patterns for functional interpretations of craniofacial form. *Am. J. Phys. Anth.* 102:203–232.
Jones, F. W. (1916). *Arboreal Man*. Edward Arnold & Company, London, 230pp.
Julesz, B. (1971). *Foundations of Cyclopean Perception.* University Chicago Press, Chicago, 406pp.
Kay, R. F. (2005). A synopsis of the phylogeny and paleobiology of Amphipithecidae, South Asian middle and late Eocene Primates. *Anthropol. Sci.,* Nippon, 113:33–42, online at www.jstage.jst.go.jp.
Kay, R. F., and Cartmill, M. (1977). Cranial morphology and adaptations of *Palaechton nacimienti* and other Paromomyidae (Plesiadapoidea, ?Primates) with a description of a new genus and species. *J. Hum. Evol.* 6:19–53.
Kay, R. F., and Simons, E. L. (1980). The ecology of Oligocene African Anthropoidea. *Int. J. Primatol.* 1:21–37.
Kielan-Jaworowska, Z., Cifelli, R. L., and Luo, Z.-X. (2004). *Mammals from the Age of Dinosaurs: Origins, Evolution, and Structure,* Columbia University Press, New York, pp. i–xv, 1–630, 239 figures.
Kirk, E. C., and Simons, E. L. (2001). Diets of fossil primates from the Fayum Depression of Egypt: a quantitative analysis of molar shearing. *J. Hum. Evol.* 40:203–229.
Lanèque, L. (1993). Variation of orbital features in adapine skulls. *J. Hum. Evol.* 25:287–317.
Lucas, P. W., Dominy, N. J., Riba-Hernandez, P., Stoner, K. E., Yamashita N., Loria-Calderon, E., Peterson-Pereira, W., Rojas-Duran, Y., Salas-Pena, R., Solis-Madrigal, S., Osorio, D., and Darvell, B. W. (2003). Evolution and function of routine Trichromatic vision in primates.*Evolution* 57:2636–2643.
Martin, R. D. (1990). *Primate Origins and Evolution*, Princeton University Press, Princeton, 804pp.
Mayr, E., Linsley, E. G., and Usinger, R. L. (1953). *Methods and Principles of Systematic Zoology,* New York: McGraw-Hill.

Miller, E. M., Gunnell, G. F., and Martin, R. D. (2005). Deep time and the search for anthropoid origins. *Yearbk. Phys. Anth.* 48:60–95.

Murphy, W. J., Eizirik, E., Johnson, W. E., Zhang, Y. P., Ryder, O. A., and O'Brian, S. J. (2001). Molecular phylogenetics and the origins of placental mammals. *Nature* 409:614–618.

Noble, V. E., Kowalski, E. M., and Ravosa, M. J. (2000) Orbit orientation and the function of the mammalian postorbital bar. *J. Zool.* 250:405–418.

Pimentel, R. A. and Riggins, R. (1987). The natureof cladistic data. *Cladistics* 3:201–209.

Platnick, N. I (1979). Philosophy and the transformation of cladistics. *Syst. Zool.*, 28: 537546.

Rasmussen, D. T. (2001). Primate origins. In: Hartwing, W. C. (ed.), *The Primate Fossil Record*, Cambridge University Press, London, pp. 5–10.

Rasmussen, D. T. and Sussman, R. W. (2007). Parallelisms among primates and possums. In: Ravosa, M. J. and Dagosto, M., (eds.), *Primate Origins: Evolution and Adaptation*, Springer Press, New York, pp. 775–803.

Ravosa, M. J., Johnson, K. R., and Hylander, W. L. (2000) Strain in the galago facial skull. *J. Morphol.* 245:51–66.

Ravosa, M. J., Noble, V. E., Hylander, W. L, Johnson, K. R., and Kowalski, E. M. (2000). Masticatory stress, orbital orientation and the evolution of the primate postorbital bar. *J . Hum. Evol.* 38:667–693.

Ravosa, M. J., and Savakova. D. G. (2004). Euprimate origins: the eyes have it. *J. Hum. Evol.* 46:357–364.

Ross, C. F. (1995). Allometric and functional influences on primate orbit orientation and the origins of Anthropoidea. *J. Hum. Evol.* 29:201–227.

Ross, C. F. (2000). Into the light: the origin of Anthropoidea. *Ann. Rev. Anthropol.* 29:47–94.

Ross, C. F. and Kay, R. F. (2004). Anthropoid origins: retrospective and prospective. In: Ross, C. F., and Kay, R. F. (eds.), *Anthropoid Origins: New Visions*, Kluwer Academic/Plenum Publishers, New York, pp. 701–737.

Schlosser, M. (1911). *Beiträge zur Kenntnis der Oligozänen Landsäugetiere aus dem Fayum: Ägypten. Beitr. Palaont*, Geo. Österreich-Ungarns 24: 167pp, plates 1–16 Wien und Leipzig.

Schmitz, J., and Zischler, H. (2004). Molecular cladistic markers and the infraordinal phylogenetic relationships of primates. In: Ross, C. F. and Kay, R. F. (eds.), *Anthropoid Origins: New Visions*, Kluwer Academic/Plenum Publishers, New York, pp. 65–77.

Schultz, A. H. (1956). Postembryonic age changes. In: Hofer, H., Schultz, A. H., and Stark, D. (eds.), *Primatologie I*, pp. 687–1014.

Seiffert, E. R. (2006). Revised age estimates for the later Paleogene mammal faunas of Egypt and Oman. *Proc. Natl. Acad. Sci.* USA 103:5000–5005.

Seiffert, E. R., Simons, E. L., and Simons, C. V. M. (2004). Phylogenetic, biogeographic and adaptive implications of new fossil evidence bearing on crown anthropoid origins and early stem catarrhine evolution. In: Ross, C. F. and Kay, R. F. (eds.), *Anthropoid Origins: New Visions*, Kluwer Academic/Plenum Publishers, New York, pp. 157–181.

Seiffert, E. R., Simons, E. L., Clyde, W. C., Rossie, J. B., Attia, Y., Bown, T. M., Chatrath, P., and Mathison, M. E. (2005). Basal anthropoids from Egypt and the antiquity of Africa's higher primate radiation. *Science* 310:300–304.

Simons, E. L. (1962). Fossil evidence relating to the early evolution of primate behavior. *Ann N. Y. Acad. Sci.* 102:282–294.

Simons, E. L. (1972). *Primate Evolution: An Introduction to Man's Place in Nature*. Macmillan Publ. Co., New York, 322pp.

Simons, E. L. (1989). Description of two genera and species of late Eocene Anhropoidea from Egypt. *Proc. Natl. Acad. Sci.* USA 86:9956–9960.

Simons, E. L. (1993). New endocasts of *Aegyptopithecus*: oldest well-preserved record of the brain in Anthropoidea. *Am. J. Sci.* 295-A:385–390.

Simons, E. L. (1997). Preliminary description of the cranium of *Proteopithecus sylviae*, an Egyptian late Eocene anthropoidean primate. *Proc. Natl. Acad. Sci.* USA 94:14970–14975.

Simons, E. L. (2003). The fossil record of tarsier evolution. In: Wright, P. C., Simons, E. L., and Gursky, S., (eds.), *Tarsiers, Past, Present and Future*, Rutgers University Press, New Brunswick, NJ, pp. 9–34.

Simons, E. L. (2004). The cranium and adaptations of *Parapithecus grangeri*, a stem catarrhine. In: Ross, C. F. and Kay, R. F. (eds.), *Anthropoid Origins: New Visions,* Kluwer Academic/Plenum Publishers, New York, pp. 183–204.

Simons, E. L. and Rasmussen, D. T. (1991). The generic classification of Fayum Anthropoidea. *Int. J. Primatol.* 12:163–178.

Simons, E. L. and Rasmussen, D. T. (1996). Skull of *Catopithecus browni*, an early Tertiary catarrhine. *Am. J. Phys. Anth.* 100:261–292.

Simons, E. L. and Seiffert, E. R. (1999). A partial skeleton of *Proteopithecus sylviae* (Primates, Anthropoidea): first associated dental and postcranial remains of an Eocene anthropoidean. *Comptes Rendus De L'Academie Des Sciences Serie II, Fascicule a—Sciences De La Terre Et Des Planetes* 329:921–927.

Simons, E. L., Seiffert, E. R., and Chatrath, P. S. (2001). Earliest record of a parapithecid anthropoid from the Jebel Qatrani Formation, Northern Egypt. *Folia Primatol.* 72:316–331.

Smith, G. E. (1924). *The Evolution of Man: Essays*. Oxford University Press, London, 155pp.

Starck, D. (1955). *Embryologie,* Thieme-Verlag, 688pp.

Starck D. (1995). *Lehrbuch der speziellen Zoologie Säugetiere*, Jena, Gustav Fischer, 2 vols., 1241pp.

Strauss, W. L. Jr. (1961). Primate taxonomy and oreopithecus. *Science* 133:760–761.

Sussman, R. W. (1991). Primate origins and the evolution of angiosperms. *Am. J. Phys. Anth.* 23:209–223.

Sussman, R. W. (1995). How primates invented the rainforest and vice versa. In: Alterman, L., Doyle, G. A., and Izard, M. K., (eds.), *Creatures of the Dark: The Nocturnal Prosimians*, Plenum Press, New York, pp. 1–10.

Sussman, R. W. (1999). *Primate Ecology and Social Structure, Vol. I: Lorises, Lemurs and Tarsiers*, Pearson Custom Publishing, Needham Heights, MA, 284pp.

Yoder, A. D. (2003). The phylogenetic position of Genus *Tarsius*: whose side are you on? In: Wright, P. C., Simons, E. L., and Gursky, S. (eds.), *Tarsiers, Past, Present and Future*, Rutgers University Press, New Brunswick, NJ., pp. 161–175.

Chapter 19
What Else Is the Tall Mandibular Ramus of the Robust Australopiths Good For?

Yoel Rak and William L. Hylander

Contents

19.1 Introduction 431
19.2 Mandibular Ramus Height in the Robust Australopiths: Metrics 431
19.3 The Functional Significance of a Tall Mandibular Ramus 433
19.4 Occlusal Topography and Wear Patterns in the Robust Australopiths 435
19.5 Discussion 437
19.6 Conclusions 440
    References 441

19.1 Introduction

The height of the mandibular ramus in the robust australopiths is hard to ignore, as are its mechanical effects on the masticatory system. One such effect is the increased anterior movement of the lower postcanine occlusal surfaces against the upper surfaces during the last stage of the power stroke of mastication. It is our contention that the increased anterior movement along with the medial movement of the working-side (chewing-side) teeth during mastication creates an arguably unique grinding pattern. We attribute the perfectly flat occlusal surfaces typical of *Australopithecus robustus* and *A. boisei* to this combination of movements.

19.2 Mandibular Ramus Height in the Robust Australopiths: Metrics

Ever since the first discovery of a robust australopith mandible, attention has been drawn to the great height of its mandibular (vertical) ramus. Indeed, as early

---

Y. Rak
Department of Anatomy and Anthropology, Sackler Faculty of Medicine,
Tel Aviv University, Tel Aviv 69978, Israel
e-mail: yoelrak@post.tau.ac.il

as 1952, Broom and Robinson described this height in specimens SK 23 and SK 34[1], which are complete enough to permit observations as well as measurements. They expressed ramal height as the vertical height of the mandibular condyle measured from the inferior (horizontal) margin of the mandibular angle. In SK 34, for example, the height of the ramus—that is, of the condyle—comes to 87 mm. Of course, the deep mandibular body so characteristic of the robust species enhances (and confounds) the appearance and measurement of a high ramus.

Alternatively, a measurement that reflects a different aspect of the mandible's function is the vertical height of the condyle relative to the occlusal plane. On SK 23, the mean vertical height of the left and right condyles is 59 mm, and on SK 34, the height of the right condyle is 60 mm. In specimen SK 12, which has only one (reconstructed) condyle, this height is also substantial at 58 mm. The somewhat shorter height of the right condyle of the Peninj mandible is 52 mm.

In a sample of 62 modern human mandibles, the mean condylar height above the occlusal plane for males and females is about 36 mm. Samples of 10 male and 10 female chimpanzee mandibles yield a mean height of 41 mm and 40 mm, respectively. Similarly, 10 male and 10 female gorilla mandibles yield a mean height of 72 mm and 60 mm, respectively.

Note that all of these values are absolute values, and of course absolute measurements of ramal height are of limited value. In contrast, the height of the condyle relative to the size of the animal is arguably a more significant variable to consider. Thus, we compared condyle height relative to biorbital breadth. (Kimbel et al. [2004] propose using the biorbital breadth to represent the overall size of a primate.) In *Homo sapiens*, the height of the condyle above the occlusal plane constitutes only 38% of the biorbital breadth. In chimpanzees this value is 41% for males and 40% for females, and in gorillas this value is 61% for both males and females. In fossil mandibles, determining this ratio is usually problematic since most fossil mandibles are not associated with crania. This compelled us to adopt a less than ideal method: we determined the mandibular condyle height value as a percentage of the mean of the available skulls' biorbital breadth measurements. Thus, the height of the condyle of SK 23 constitutes 64% of the mean *A. robustus* biorbital breadth (92 mm, derived from specimens SK 48 and TM 1517); in SK 34, the value is 65%, and in SK 12, 63%. In the relatively small mandible of the *A. boisei* Peninj specimen, the height of the condyle constitutes 53% of the mean *A. boisei* biorbital breadth (99 mm, derived from specimens OH 5, KNM-ER 406, KNM-ER 13750, and KNM-ER 732).

We also obtained additional data on condylar height by measuring the distance between the maxillary occlusal plane and the Frankfurt horizontal. This method yields a relative approximation of the data obtained on the height of the condyle

---

[1] The prefixes in the specimens' names indicate their origin or storage location: SK: Swartkrans cave, South Africa; TM: Transvaal Museum, South Africa (these fossils were discovered in Kromdraii cave, South Africa); OH ("Olduvai hominid"): Olduvai Gorge, Tanzania; KNM-ER ("Kenya National Museum-East Rudolph"): housed in the Kenya National Museum and discovered on the east side of Lake Rudolph, today called Lake Turkana, in Kenya; Sts: Sterkfontein cave, South Africa.

above the occlusal plane, although the skull measurements differ from the measurement of the actual condylar height above the occlusal plane by 20–30% (as borne out on specimens in which both sets of measurements are available). The distance between the occlusal plane and the Frankfurt horizontal constitutes 80% of the biorbital distance in OH5, 70% in SK 48, 75% in TM 1517, 78% in KNM ER 406, and only 64% in the small female *A. boisei* specimen KNM ER 732. By comparison, this value is 47% in male and female *Homo sapiens,* 46% in male and female chimpanzees, 70% and 68% in male and female gorillas, respectively. Note that the gorilla values are smaller than those of the available fossil mean, but not by much.

Thus, despite the small fossil sample, its poor taphonomic quality, and what appears to be substantial variation in absolute condylar height, these measurements confirm early investigators' intuitive impressions regarding vertical height of the condyle in robust australopiths. These early observations became an accepted truth although condylar height was never evaluated relative to the occlusal plane.

19.3 The Functional Significance of a Tall Mandibular Ramus

Much has been written about the mechanical advantages of the height of the condyle above the occlusal plane (or above the base of the mandibular body). In 1980, Ward and Molnar published seminal research that experimentally examined the effect of ramal height (and the anterior position of the masseter) on the magnitude of the occlusal load and the nature of the load's distribution along the dental arcade. In their article, the authors review the literature about the advantages of a tall ramus as follows:

> 1. *Leverage Improvement*: A tall ramus with the joint elevated above the tooth rows may favor an increase in the moment arms (about an intercondylar axis) of the anterior fibers of the temporalis muscle and all fibers of the masseter and medial pterygoid (Du Brul, 1974, 1977; Scapino, 1972; Crompton, 1963; Maynard-Smith [sic] and Savage, 1959). 2. *Bite Force Variation*. As the joint is elevated, a bolus located on the occlusal table is subjected to increasing antero-posterior forces (Davis, 1964; **A. de Wolf-Exalto [sic] 1951**; Moss, 1968; Stallard, 1923). 3. *Occlusal Simultaneity*. Elevation of the jaw joint promotes simultaneous cusp contact in the terminal phase of occlusion, ensuring an extended area of high magnitude bite forces over the cheek teeth (Greaves, 1974; Tattersall, 1972) (pp. 390–391).

As for leverage improvement (number 1 in the previous list), a tall ramus (or condyle) would have little or no influence on increasing the moment arm of the anterior temporalis muscle. Nevertheless, an increase in vertical height of the ramus should indeed result in an increase in the vertical orientation of the masseter and medial pterygoid muscles (when seen in a frontal view) (Ravosa et al., 2000), as well as increase in the moment arms of these two muscles (when seen in a lateral view) (Crompton and Hiiemae, 1969). We should note that this "height of the ramus" refers not to the condyle's height above the occlusal plane but, rather, to its height above the inferior margins of the mandibular angle (the insertion site of many of the masseter and medial pterygoid muscle fibers). The vertical increase between the origin of the masseter muscle and its insertion also plays a major role in maximizing the verticality of the robust australopith masseteric orientation, when considered in an anterior view, particularly because of the lateral flare of the zygomatic arches (the origin of the masseter) so characteristic of the robust species (Rak, 1983). Furthermore, it has been suggested that, in some extreme cases, an increase in the lateral flare of the zygomatic arches resulted in an elevation of the zygomatic arches relative to the transverse plane of the orbital floors so as to maximize the masseter's verticality (Rak, 1983).

A tall vertical ramus and a more anterior position of the masseter likely result in a small gape (and therefore relatively small bolus), unless muscle architecture is modified by increasing the length of the masseter muscle fibers, as well as all the additional mechanical advantages that such an altered muscle position affords (Hylander, 1972; Rak, 1983). Parenthetically and as noted elsewhere, small gape and reduced canine height are likely functionally linked (Hylander and Vinyard, 2006).

Along with occlusal simultaneity (number 3 in the previous list), Ward and Molnar nicely demonstrated experimentally (summarized in their Fig. 9) that a gradual increase in ramal height is, indeed, accompanied by "an obvious trend in the reduction of differences in absolute load between the premolars and molars...A gradient of occlusal forces appears in low positions of the ramus....[The magnitude of forces is greater] at the third molar than at the third premolar. In the tallest position this disparity essentially disappears" (p. 393).

The well-known phenomenon of premolar molarization in the robust australopiths is in full accordance with Ward and Molnar's results and a logical outcome of what those results imply. Nevertheless, the effect of the condylar height above the occlusal plane on the anterior movement of the lower occlusal surfaces against the upper ones in the last stage of occlusion (number 2 in Ward and Molnar's list) has been largely overlooked, especially in discussions of the robust australopiths' masticatory system. Even Ward and Molnar refer to this effect only cursorily: "...the assertion that tall rami favor increased mesio-distal force application to resistant boli is found to be true only if the mandible acts as a pure hinge" (p. 393). (The hinge-like function of the mandible is worthy of note and is treated later.)

Not only does the height of the ramus play a role here, but the length of the mandible is also of major importance. For example and by way of clarification, based on the work dealing with mandibular movements in both humans and non-human primates, during the terminal portion of the power stroke the working-side mandible in the lateral projection rotates about a mediolateral axis that passes through or very close to the working-side condyle. [Note that this instantaneous axis of rotation does not pass through the balancing-side condyle, and therefore this axis is not perpendicular to the sagittal plane. Instead, this axis passes through or behind the balancing-side ramus well below the balancing-side condyle (Obrez and Gallo, 2006).] Thus, the rotary movement of the working-side mandible (in the lateral projection) during the terminal portion of the power stroke is very similar to the closing of a hinge. Parenthetically, the working-side mandible also rotates about a vertical axis passing through or near the working-side condyle, and this rotary movement contributes to medial movements of the working-side lower teeth against the uppers. Since in lateral view the working-side mandible closes in a near hinge-like movement without any significant posterior translation, the larger the ratio between the condylar height above the occlusal plane relative to the horizontal distance (parallel to the occlusal plane) from the projection of the condyle to the central incisors, the larger the anterior component of movement of the lower incisors relative to the upper incisors (Hylander, 1972). Similarly, the larger the ratio between the condylar height above the occlusal plane relative to the horizontal distance from the condyle to the postcanine teeth (e.g., the molars), the larger the anterior component of movement of the working-side lower molars relative to the upper molars (Hylander, 1972). For our purposes, however, we will analyze the height of the condyle above the tooth row relative to the total horizontal length of the mandible.

SK 23 yields the highest ratio of condylar height to mandibular length in our entire sample, at 50%. The value for the Peninj mandible stands at 47%; for male and female *H. sapiens* it is 38%; for male and female chimpanzees it is 32% and 38%, respectively. For male and female gorillas it is 46% and 43%, respectively. Here, too, as with the relative condylar height values, the gorilla ratios are smaller than in the fossil mandibles, though not by much.

19.4 Occlusal Topography and Wear Patterns in the Robust Australopiths

That researchers concentrating on the robust australopith clade have failed to comment on the effect of ramal height on the last stage of occlusion is surprising. Nevertheless, we suspect that this effect is arguably linked to wear patterns of the postcanine occlusal surfaces that are uniquely identified with the robust australopith masticatory system. Robinson (1956) seems to have been the first to discuss this wear pattern but accords it little space, noting only that "in advanced wear the [*Australopithecus africanus*] postcanine teeth...wear down most strongly buccally, whereas [in *Paranthropus robustus*] wear is most heavy lingually" (p. 19). Subsequently, descriptions of the robust australopith pattern, which is basically the wearing down of the occlusal plane to a completely flat, polished surface, are much more accurate (Wallace, 1972, 1975, 1978; Grine, 1981), and the distinction was made clearer between the robust pattern and the more generalized pattern typical of *A. africanus*. The subsequent increase in the South African sample and the discovery of *A. boisei* in East Africa added even more detail and consequently a greater refinement to these descriptions (Tobias, 1967; Grine, 1981; Grine and Martin, 1988).

Currently, the topography of worn teeth in the robust australopith clade is generally described as consisting of large, near perfectly flat occlusal surfaces that differ from those of worn teeth in the more generalized masticatory system as in *A. africanus*. The more generalized lower postcanine teeth typically exhibit a flaring of the sharp lingual margins along with more advanced wear and rounding of the

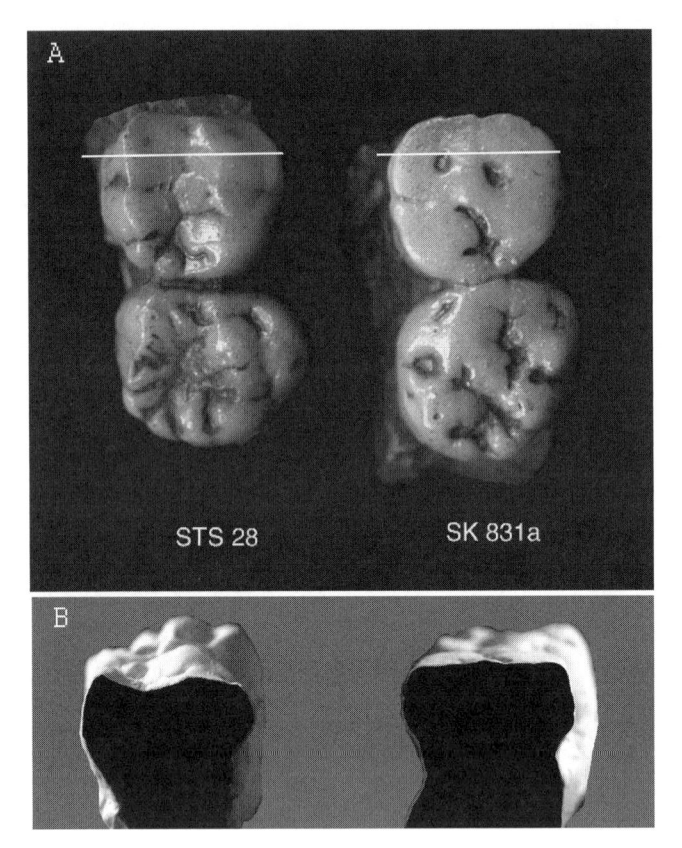

Fig. 19.1 Occlusal topography of *Australopithecus africanus* (*left*) and *A. robustus* (*right*). **A.** the M^2 teeth. Note the differences between the wear patterns of the two species, particularly the sharp buccal margins in *A. africanus* and the completely flat occlusal plane in *A. robustus*. A more advanced stage of wear would have produced even greater differences. **B.** A buccal-lingual cross section of the M^2 teeth, demonstrating the differences between their occlusal topography. The two white lines in the upper photographs represent the approximate plane of the incision for the cross-section

buccal margins (Fig. 19.1), with the pattern reversed in the upper teeth. On the other hand, the wear pattern in the robust australopiths is expressed as a horizontally worn surface of enamel or dentine, the latter of which is delineated by a ring of enamel (the walls of the tooth). In a more advanced stage of wear, the dentine in the robust australopiths is scooped out, lending the entire occlusal surface a basin-like appearance. In specimens that bear the more generalized masticatory system and exhibit advanced wear of the lower postcanine teeth—for example, the Sterkfontein specimens Sts 7 and Sts 36—the dentine is exposed and excavated mesiodistally as a gutter all along the buccal margins.

In his important paper on tooth wear in australopiths, Grine (1981), besides presenting his own observations, summarizes and elaborates on previous observations regarding the *process* of tooth wear throughout an individual's life for the robust and more generalized hominins. For example, there are differences in the sequence of molar facet appearance, the orientation of the facets' surfaces, and the sequence and speed of dentine exposure. Enamel thickness, the topography and height of the cusps, the depth of the crevices surrounding them, and the sub-enamel topography of the dentine cusps all play a role in producing processual differences in the nature and sequence of postcanine tooth wear (Grine, 1981, and references therein; Wallace, 1972, 1975, 1978).

19.5 Discussion

The flatness of the wear pattern of the postcanine teeth in the more specialized system—that of the robust australopiths—suggests that the chewing motion might be more horizontal and that the pattern might not be simply the outcome of increased mediolateral directed chewing. We propose that increased mediolateral movements during chewing *combine with* the anterior movement of the lower teeth relative to the uppers during the final stage of occlusion to produce a rotatory motion. The anterior movement of the lower teeth, during which "a bolus located on an occlusal table is subjected to increasing antero-posterior forces" (Ward and Molnar, 1980), stems from the height of the ramus (Fig. 19.2) and the relatively short horizontal mandibular length. The increased rotatory chewing motion gains further credence in light of the obtuse angle between phase I and phase II facets of mediolateral chewing, as suggested by Wallace: "...because of their low cusps and almost horizontal incline planes, robust australopithecines probably had a more horizontal glide path into centric occlusion than gracile australopithecines" (Wallace, 1975, p. 214).

Robinson (1972), as well, refers to an anteroposterior motion of the mandible:

> The anatomical characteristics of this region suggest that, besides the powerful, mainly vertical component of the muscle, there was also a more than usually important and powerful posterior component that pulled in a more nearly horizontal direction. This suggests that perhaps chewing was not merely powerful but involved also considerable fore and aft, as well as rotatory, movements of the mandible (p. 226).

Note an important distinction here in that Robinson suggests that the horizontal movement of the mandibular teeth (chewing-side teeth?) is in a posterior direction [attributing this direction, strangely enough, to the relatively decreased posterior pull of the more vertically aligned posterior fibers of the robust temporalis muscle (Robinson, 1958; Rak, 1978; Kimbel and Rak, 1985; Kimbel et al., 2004)], whereas we suggest that the horizontal movement of the working-side lower teeth as a function of ramal height can be only in an anterior, not posterior, direction.

Two factors in addition to increased ramal height appear to play an important role in causing the anterior movement of the lower occlusal surfaces during the last stages of occlusion. The first is the extent of anterior condylar translation as the mouth opens; the more the condyle advances anteriorly, the farther the instantaneous axis of rotation descends (cf. Hylander, 2006). According to this logic, animals in

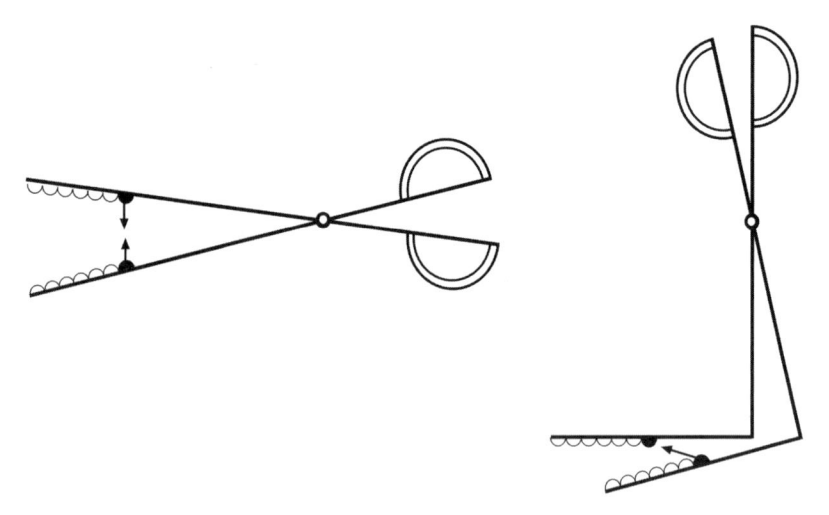

Fig. 19.2 A comparison of the effect of a short mandibular ramus (left) and a tall mandibular ramus (right) on the movement of the lower teeth relative to the upper teeth in the final stages of occlusion. The black arrow on the right represents the AP movement. Note how the tall ramus produces a more anterior movement of the lower teeth

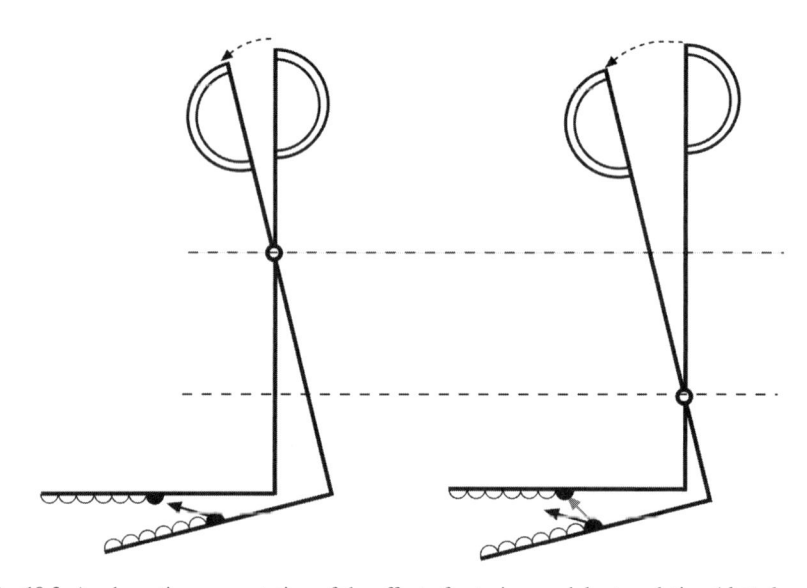

Fig. 19.3 A schematic representation of the effect of anterior condylar translation (*dotted curved arrows*) on the height of the axis of rotation and the magnitude of the anterior movement of the lower teeth. Left: Limited condylar translation. Right: Extensive condylar translation. The distance between the dotted lines represents the difference in the height of the axes of rotation. The black arrows represent the AP movement of the lower teeth when the axis of rotation is high, and the gray arrow represents the shorter AP movement when the axis is low

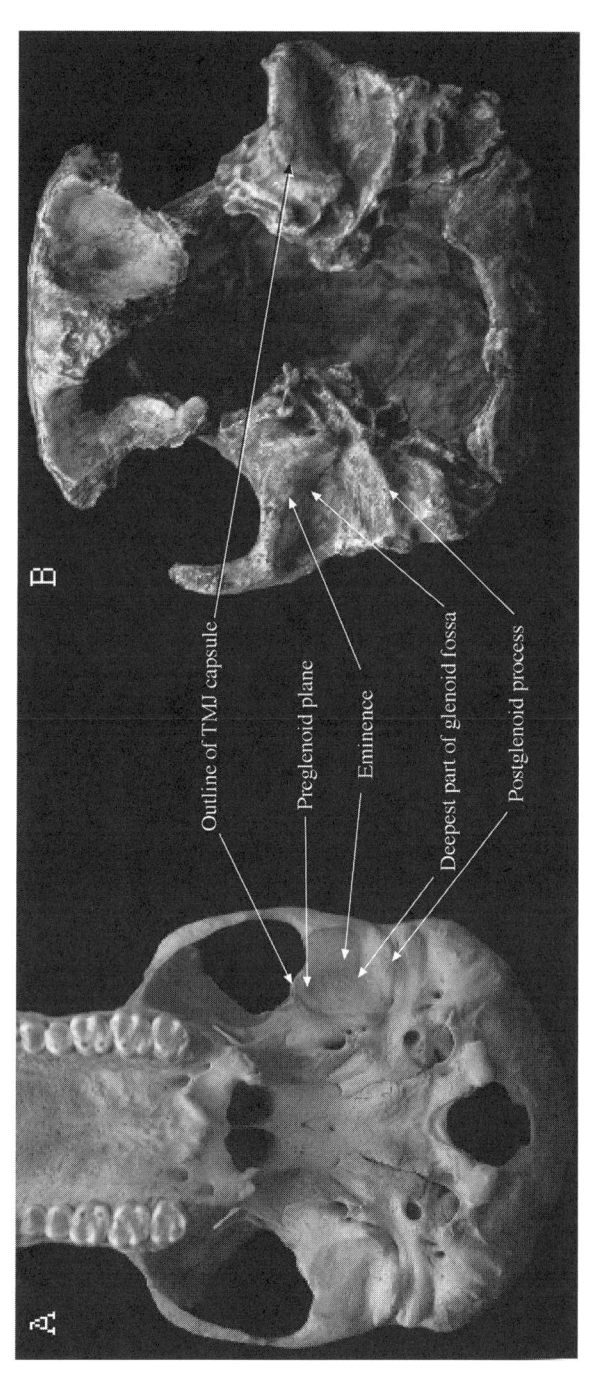

Fig. 19.4 The glenoid fossa and articular eminence in a chimpanzee (**A**) and *A. boisei* specimen KNM-ER 23000 (**B**). The chimpanzee represents the generalized (common) configuration, and *A. boisei* represents the specialized configuration. Note, in the chimpanzee, the weakly developed articular eminence, shallow glenoid fossa, and extensive preglenoid plane anterior to the eminence. In *A. boisei*, the articular eminence is extremely steep, its articular surface terminates in a sharp crest, and the preglenoid plane is absent

which ramal height is crucial for the functions listed earlier will exhibit the mechanisms that minimize the fore and aft (AP) translation of the (working-side) condyle during the power stroke of mastication. And, indeed, much of the topography of the glenoid fossa and the articular eminence in the robust australopiths, particularly in *A. boisei*, restricts the extent of AP translation (for both condyles). The steep articular surface of the *A. boisei* articular eminence, which terminates as a crest, does not extend anteriorly as a preglenoid plane and thus minimizes or restricts the translation of the robust australopiths' mandibular condyle, in contrast to the more generalized masticatory system (Figs. 19.3 and 19.4). This topography of the glenoid fossa and articular eminence in the robust australopiths is what produces the near hinge-like rotation of the mandible about a mediolateral axis during chewing and/or biting, referred to earlier. (Presumably, the mediolateral translation of the condyles is not restricted.)

The second factor is the topography of the occlusal surface of both the individually unworn teeth and the dental arcade as a whole. This dental topography permits the anterior movement of the lower teeth against the uppers during the terminal phase of the power stroke; thus, only a flat occlusal surface, as in the robust australopiths, can accommodate an AP movement during the occlusal portion of chewing. In the more generalized masticatory system, the interlocking of the tooth cusps (and of the projecting canines) restricts this anterior movement, and thus mandibular motion is more or less confined to mediolateral movements during tooth–tooth contacts. Because of the absence of such interlocking in gorillas, their wear pattern of the postcanine teeth does not resemble that of the robust australopiths, despite the considerable vertical height of the gorilla ramus. Furthermore, the extended preglenoid plane in gorillas, unlike that of the robust australopiths, allows a considerable amount of AP condylar translation during chewing. Finally, in those situations where the working-side condyle is allowed to translate posteriorly throughout the power stroke of mastication, this has the effect of reducing or eliminating the anterior component of movement of the working-side lower molars relative to the upper molars.

19.6 Conclusions

It is, indeed, feasible that three factors—the considerable height of the mandibular ramus in the robust australopiths, their more rostrally positioned jaw muscles, and the resulting small gape—when combined with the topography of the glenoid fossa, the unique morphology of the articular eminence, the bulbous, low-relief topography of the unworn cusps, and the tiny, non-projecting canines, are importantly linked to the postulated "rotatory" motion of the mandible during chewing. This motion in turn must be the explanation for the typically flat, worn surfaces of the postcanine teeth. Furthermore, these derived masticatory movements and forces are also likely linked to the robust australopiths' unusually thick (mediolateral) mandibular corpora, presumably because of forceful twisting of the corpora during the power stroke of mastication (Hylander, 1979, 1988).

References

Arendsen de Wolf-Exalto, E. (1951). On the differences in the lower jaw of animalivorous and herbivorous mammals. *II. Proc. Kon. Ned. Ak. v. Wet. Ser. C* 54:405–410.

Broom, R., and Robinson, J. T. (1952). Swartkrans ape-man *Paranthropus crassidens*. *Mem. Transvaal mus.* 6:1–123

Crompton, A. W. (1963). On the lower jaw of *Diarthrognathus* and the origin of the mammalian lower jaw. *Proc. Zool. Soc. Lond.* 140:697–753.

Crompton, A. W., and Hiiemae, K. (1969). How mammalian teeth work. *Discovery.* 5:23–34.

Davis, D. D. (1964). The giant panda: A morphological study of evolutionary mechanisms. *Fieldiana: Zoology Memoirs* 3:1–339.

Du Brul, E. L. (1974). Origin and evolution of the oral apparatus. In: Kawamura, Y. (ed.), *Frontiers of Oral Physiology.* Karger Publishers, Basel, pp. 1–30.

Du Brul, E. L. (1977). Early hominid feeding mechanisms. *Am. J. Phys. Anthropol.* 47:305–320.

Greaves, W. S. (1974). Functional implications of mammalian jaw joint position. *Forma et Functio* 7:363–376.

Grine, F. E. (1981). Trophic differences between 'gracile' and 'robust' australopithecines: A scanning electron microscope analysis of occlusal events. *S. Afr. J. Sci.* 77:203–230.

Grine, F. E., and Martin, L. B. (1988). Enamel thickness and development in *Australopithecus* and *Paranthropus*. In: Grine, F. E. (ed.), *Evolutionary History of the "Robust" Australopithecines.* Aldine de Gruyter, New York, pp. 3–42.

Hylander, W. L. (1972). The adaptive significance of Eskimo craniofacial morphology, Ph.D. thesis, University of Chicago.

Hylander, W. L. (1979). The functional significance of primate mandibular form. *J. Morphol.* 160:223–240.

Hylander, W. L. (1988). Implications of *in vivo* experiments for interpreting the functional significance of "robust" australopithecine jaws. In: Grine, F. E. (ed.), *Evolutionary History of the "Robust" Australopithecines.* Aldine de Gruyter, New York, pp. 55–80.

Hylander, W. L. (2006). Functional anatomy. In: Laskin, D. M., Greene, C., and Hylander, W. L. (eds.), *Temporomandibular Disorders: An Evidenced-Based Approach to Diagnosis and Treatment.* Quintessence Publishing, Chicago, pp. 3–34.

Hylander, W. L. and Vinyard, C. J. (2006). The evolutionary significance of canine reduction in hominins: Functional links between jaw mechanics and canine size. *Am. J. Phys. Anthrop. Suppl.* 42:107.

Kimbel, W. H., and Rak, Y. (1985). Functional morphology of the asterionic region in extant hominoids and fossil hominids. *Am. J. Phys. Anthropol.* 66:31–54.

Kimbel, W. H., Rak, Y., and Johanson, D. C. (2004). The Skull of *Australopithecus afarensis*. Oxford University Press, Oxford.

Maynard Smith, J., and Savage, R. J. G. (1959). The mechanics of mammalian jaws. *School Sci. Rev.* 141:289–301.

Moss, M. L. (1968). Functional cranial analysis of mammalian mandibular ramal morphology. *Acta Anat.* 71:423–447.

Obrez, A., and Gallo, L. M. (2006). Anatomy and function of the TMJ. In: Laskin, D. M., Greene, C., and Hylander, W. L. (eds.), *Temporomandibular Disorders: An Evidenced-Based Approach to Diagnosis and Treatment.* Quintessence Publishing, Chicago, pp. 35–52.

Rak, Y. (1978). The functional significance of the squamosal suture in *Australopithecus boisei*. *Am. J. Phys. Anthropol.* 49:71–78.

Rak, Y. (1983). *The Australopithecine Face.* Academic Press, New York.

Ravosa, M. J., Vinyard, C. J., Gagnon, M., and Islam, S. A. (2000). Evolution of anthropoid jaw loading and kinematic patterns. *Am. J. Phys. Anthropol.* 112:493–516.

Robinson, J. T. (1956). The dentition of the australopithecine. *Mem. Transvaal mus.* 9:1–179.

Robinson, J. T. (1958). Cranial cresting patterns and their significance in the Hominoidea. *Am. J. Phys. Anthropol.* 16:397–428.

Robinson, J. T. (1972). *Early Hominid Posture and Locomotion*. University of Chicago Press, Chicago.

Scapino, R. P. (1972). Adaptive radiation of mammalian jaws. In: Schumacher, G. (ed.), *Morphology of the Maxillo-Mandibular Apparatus*. Thieme, Leipzig, pp. 33–39.

Stallard, H. (1923). The anterior component of the force of mastication and its significance to the dental apparatus. *Dental Cosmos* 65:457–479.

Tattersall, I. (1972). The functional significance of airorhynchy in *Megaladapis*. *Folia Primatol.* 18:20–26.

Tobias, P. V. (1967). *The cranium and maxillary dentition of Australopithecus (Zinjanthropus) boisei. Olduvai Gorge*, Volume 2. Cambridge University Press, Cambridge.

Wallace, J. A. (1972). The dentition of the South African early hominids: A study of form and function, Ph.D. thesis, University of the Witwatersrand, Johannesburg (unpublished).

Wallace, J. A. (1975). Dietary adaptations of *Australopithecus* and early *Homo*. In: Tuttle, R. (ed.), *Paleoanthropology, Morphology and Paleoecology*. Mouton, The Hague, pp. 203–223.

Wallace, J. A. (1978). Evolutionary trends in the early hominid dentition. In: Jolly, C. (ed.), *Early Hominids of Africa*. Duckworth, London, pp. 285–310.

Ward, S. V., and Molnar, S. (1980). Experimental stress analysis of topographic diversity in early hominid gnathic morphology. *Am. J. Phys. Anthropol.* 53:383–395.

Chapter 20
Framing the Question: Diet and Evolution in Early *Homo*

Susan C. Antón

Contents

20.1 Introduction 443
    20.1.1 Foraging Shifts in the Fossil Record 445
20.2 Materials and Methods 448
    20.2.1 Fossil Individuals and Chimera 448
    20.2.2 Variables and Scaling Analyses 450
20.3 Results 455
    20.3.1 Is *Homo erectus* Large-Bodied and Large-Brained? 455
    20.3.2 Do *Homo erectus* Jaws and Teeth Suggest a Foraging Shift? 456
20.4 Discussion 464
    20.4.1 Evidence for a Foraging Shift 464
    20.4.2 Suggestions About Scaling 470
    20.4.3 Trends with Time 471
    20.4.4 Implications for Understanding Early *Homo* 472
    References 477

20.1 Introduction

The morphology of the primate skull is strongly influenced by feeding adaptations and paramasticatory activity. In the 1970s, Bill Hylander began his career by considering the morphology of Inuit cranial remains with reference to biomechanical principles related to masticatory loading (Hylander, 1972, 1977). Over the course of that career, this tradition has brought together morphologists and experimentalists to interpret skeletal morphology in light of in vivo data. Occasionally, paleoanthropologists have used this method to interpret fossils functionally. The mandible, relatively abundant in fossil assemblages, has been the most frequently analyzed fossil element, and in hominid paleontology this has resulted in insights into australopithecine evolution (e.g., DuBrul, 1977; Hylander, 1988). Early genus *Homo* has been less amenable to this biomechanical approach, given a preference

---
S.C. Antón
Department of Anthropology, 25 Waverly Place, New York University, New York, NY 10003
e-mail: susan.anton@nyu.edu

for cranial remains in analyses and perhaps less interspecific variation in mandibular form. More frequently, the dentognathic complex has been used in phylogenetic rather than functional discussions (e.g. Wood, 1991; but see Wood and Aiello, 1998; Dobson and Trinkaus, 2002).

Due to the incompleteness of the hominin fossil record, many of the advances in Paleoanthropology come as the result of the discovery of new fossils that tell us something unexpected about our forebears. Because fossil sample sizes adequate for testing most biological hypotheses accumulate only slowly, if at all, the questions we ask are often driven by perceived differences between fossil bits rather than by framing an inherently interesting question and going to the fossil sample to address it. That is, we work with what we have not with what we would like to have. Yet over time, we develop hypotheses about behaviors and lifeways and implications of the same. Here, I attempt to frame some of the questions related to diet, which we might address with the current fossil assemblage of early *Homo*.

Testing these hypotheses is frequently fraught with methodological issues including the lack of associated remains, the fragmentary nature of those that exist, and, equally importantly, the temporal range of fossils of a single taxon. Thus, tests are skewed to the parts that are most represented, and the more complete remains tend to be overemphasized in these tests. In addition, not only do we often have small, anatomically unrepresentative samples but we are also often forced to aggregate together specimens of vastly different ages and sometimes different locations to even begin to attain a statistically reasonable sample size. Thus, there is an issue of scale between comparisons made on extant primate skeletal samples that aggregate conspecifics separated by at most hundreds of years (or in the case of humans thousands of years) and those aggregated for fossil hominins that routinely span hundreds of thousands of years and thousands if not tens of thousands of miles (cf. Wood and Xu, 1991). The idea that we might be dealing with true biological populations is thus infinitely relaxed in the case of the fossil samples. Despite this limitation, we often can tease apart fossil samples into more appropriate subsamples, and we can increase sample size by considering more fragmentary bits. I will attempt to do some of that here.

More common than the logistical issues of sample composition is that a hypothesis all too frequently becomes so integrated into the conventional wisdom that we fail to realize that, because we never really framed the initial question, we also never gathered sufficient data to support or refute it. Or we have not acknowledged that perhaps we do not really have the data to do either one. Such, I would argue, is the case with ideas regarding foraging shifts between members of early genus *Homo*. Here I attempt to lay out some of the arguments and use some of the bits to test these. Although commonly accepted, the suggestion of a foraging shift between species of early *Homo* has not been rigorously tested. I will begin that process here although much more substantial fossil samples are necessary to definitively test these hypotheses.

20.1.1 Foraging Shifts in the Fossil Record

Foraging shifts are said to occur early in genus *Homo* and to account for morphological changes from body size to cranial shape and, at least in part, for the dispersal of *Homo* from Africa (e.g. Shipman and Walker, 1989; Antón et al., 2002). Yet the fossil bits that can be used as evidence of this shift are few, and arguably not all of these have been used systematically. At least two foraging shifts are argued to have occurred between *Australopithecus* and early *Homo*: one based on an *Australopithecus* to early *Homo erectus* shift, the other on a specific-level shift between *Homo habilis* and *H. erectus*.

The 'Expensive Tissue Hypothesis' argues that shifts in body size and shape between *Australopithecus* and early *H. erectus* are evidence of a change in energy requirements, which could be addressed only by a foraging shift to a higher-quality diet (Leonard and Robertson, 1992, 1994, 1997; Aiello and Wheeler, 1995). The hypothesis posits a trade-off between brain size and gut size- two expensive tissues- and suggests that in genus *Homo* a shift to a high-quality diet allowed the reduction of gut size and the increase in brain size. Although the shift is presumed to have occurred at the base of the *Homo* lineage, the question of earliest *Homo* is in large part unaddressable, due to the paucity of the fossil record for *H. habilis sensu lato*. This is because, although body size may be somewhat retrievable from isolated postcranial remains (e.g., McHenry, 1992, 1994), body shape (particularly of the thorax) and proportions are hard to reconstruct, given the dearth of associated remains for *H. habilis sensu lato* (*s.l.*; e.g., Richmond et al., 2002; Haeusler and McHenry, 2004). Yet it is these shape differences that are critical evidence of gut reduction. Body size and shape differences exist between the well-preserved, associated skeletons of *A. afarensis* and early *H. erectus* (e.g., McHenry and Coffing, 2000), including elongation of the lower limb and loss of a funnel-shaped thorax in the latter. Changes in thorax shape suggest that the gut reduction predicted by the hypothesis was in place by the time of *H. erectus*. And the enlarged body size offers indirect support for increased diet quality or at least enhanced nutrition, since both are correlated with larger size (e.g., Frisancho, 1978; Fogel and Costa, 1997; Cole, 2000). These differences, coupled with differences in jaw and tooth size (e.g., Wood and Abbott, 1983; Wood, 1991), strongly support a foraging shift between the two genera.

The intensification of this foraging shift within genus *Homo*, including a greater dependence on animal protein, is argued to be one of a number of factors involved in the dispersal of *Homo ergaster*, rather than any earlier hominin, from Africa (Shipman and Walker, 1989; Antón et al., 2001, 2002); still others have argued for a foraging shift that involved the cooking of underground tubers (Wrangham et al., 1999; O'Connell et al., 1999). Arguments for a shift to a higher-quality diet in *H. erectus* are based largely on postcranial evidence that indicates an increase in body size as early as 1.95 Ma (based on an innominate from Koobi Fora, Kenya; KNM-ER 3228). Although *H. habilis* is often divided into larger-bodied and smaller-bodied morphs[1] (e.g., Wood and Collard, 1999; Gabunia et al., 2000a), the average body size of *H. erectus* is larger than both, although some regional samples of *H. erectus* overlap the range of *Homo rudolfensis* (Table 20.1). In mammals, larger body sizes are correlated with larger home-range sizes (Harestad and Bunnell, 1979; Nunn and Barton, 2000), which in turn would facilitate dispersal in *H. erectus* (Antón et al., 2002). Intensified dependency on higher-quality food stuffs is suggested to drive subsequent brain size increases, as well (Shipman and Walker, 1989; Foley and Lee, 1991; Antón et al., 2002; Leonard et al., 2003). Thus, changes in body size, brain size, home range size, and ultimately dispersal are predicated on this increased reliance on this higher-quality food, whatever it may be.

Both meat and cooked-tubers should offer the same mechanical consequences relative to previous hominin food stuffs; a higher-quality diet that is also less difficult to masticate. The fallback foods for *Australopithecus*, that is, those which are likely to be reflected in their anatomical adaptations (cf. Robinson and Wilson, 1998), were likely some form of raw vegetation (Kay, 1985; Teaford and Ungar, 2000). These may include the types of foods eaten by living great apes, although Wrangham and colleagues (1999) argue on ecological grounds that fallback foods similar to those of extant chimpanzees were unlikely. Raw tubers, widely available in arid landscapes, have been argued as one possible source of *Australopithecus* fallback foods (Hatley and Kappelman, 1980; Wrangham et al., 1999; Laden and Wrangham, 2005; Sponheimer et al., 2005). However, dental topography supports the idea that *A. afarensis*, with relatively low relief on its molar crowns, was adapted to brittle, less deformable fallback foods, than those of extant chimpanzees, gorillas, and early *Homo* (Ungar, 2004). These data are more supportive of seeds, roots, and rhizomes (cf. Ryan and Johanson, 1989) than underground tubers as fallback

Table 20.1 Body size by geographic region in *H. habilis* and *H. erectus*

Taxon	Africa *H. habilis s.s.*	Africa *H. rudolfensis*	Africa *H. erectus*	Georgia*** *H. erectus*	China *H. erectus*	Indonesia *H. erectus*
Femur length (mm)						
– mean		397.5	459	–	395	445
– range (n)	280–395 (3)	395–400 (2)	430–500 (5)*	386 (1)	378–413 (2)	433–455 (3)
Stature (cm)						
– mean	128	149	174	148	148	164
– range (n)	106–148 (3)	148–150 (2)	161–186 (6)**	134–154 (1)	141–154 (2)	158–170 (4)

*excludes KNM WT 15000.
**includes estimated adult value for KNM WT 15000.
***Georgian data includes estimated values from Dmanisi D4167 (Lordkipanidze et al., 2007). These values were added in proof and were not available for these analyses. However, they do not differ from the metatarsal D2021 derived values used in this study.

---

[1] The smaller-bodied morph has been removed by some from the genus on the argument that its body size and shape more closely approximates *Australopithecus* (Wood and Collard, 1999). In my opinion, without more abundant data on proportions, this is premature.

foods for *Australopithecus*. In either case, the mechanical properties of these fallback foods are likely to be different than those composing the diet of early *Homo*. It should be noted that although the work of Ungar (2004) on dental topography supports the idea of tough, more elastic food properties for early *Homo*, perhaps also supporting the idea of more meat in their diet, his study used a mixed *H. habilis* and *H. erectus* sample and thus cannot address the issue of a dietary shift between the two.

The work of Hylander and associates on extant primates suggests that adaptation to foods requiring greater occlusal forces to process results in structural changes to the skull and jaws, changes that we might expect to see from earlier to later *Homo* if a softer diet is present in *H. erectus*. We also might see such changes between geographically separated members of *H. erectus* if significant differences in dietary composition existed across regions or times. Structural consequences related to dietary adaptation include anteroposteriorly shorter, vertically deeper faces, more anteriorly placed masseter attachment areas, and broader, taller mandibular corpora (Kinzey, 1974; Hylander, 1977, 1979, 1988; Ravosa, 1991, 1992; Daegling, 1992). For example, the more frugivorous cercopithecines tend to be relatively prognathic and have thinner and shallower mandibular corpora (relative to length) than do folivorous colobines (Hylander, 1979; Bouvier, 1986). Cercopithecines, such as Japanese macaques (*M. fuscata*), adapted to tougher fallback foods (Suzuki, 1965; Koganezawa, 1975), are less prognathic, and have more robust mandibular corpora than their congeners (Antón, 1996). Similarly, great apes with greater amounts of tough, herbaceous vegetation in their diets, especially gorilla subspecies, exhibit labiolingually deeper symphyses, more robust corpora, and taller mandibular rami, than those with less-tough diets (Taylor, 2002).[2] By extension, mandibular robusticity at the corpus and symphysis should be expected to decrease from earliest *Homo* to *H. erectus*. Molar occlusal area scales with body size (possibly positively allometrically, Gould, 1977, but see Vinyard and Hanna, 2005) and probably isometrically within dietary group (Kay, 1975; Gingerich et al., 1982; Vinyard and Hanna, 2005). A shift to a mechanically softer diet should transpose *H. erectus* below *H. habilis*, and should suggest that molar area will decrease regardless of body size.

A shift in the mechanical properties of the hominin diet seems well supported between *Australopithecus* and early *H. erectus*. However, a foraging shift between *H. habilis* and *H. erectus* seems less secure. If a shift to a higher-quality, softer diet occurred or intensified between *H. habilis* and *H. erectus*, then we should see the consequences in mandibular and dental morphology. To test the claims about foraging shifts in *Homo*, one might look for the above-mentioned structural and scaling changes in the fossil record.

---

[2] It should be noted that in most of the above-mentioned studies the mechanical properties of the food items have not themselves been tested but have been inferred. Similarly, studies of internal bone architecture do not always support the same conclusions as do external measurements (e.g., Daegling, 1989). Thus, while it is generally considered that these external dimensions provide insight into masticatory loading and diet, the correlations between the two at an intra- and interspecific level may still be questioned.

20.2 Materials and Methods

If there was a shift in diet quality (and mechanical properties of that diet) from earliest *Homo* to *H. erectus*, we might anticipate seeing in *H. erectus*: (1) absolutely larger body size relative to *H. habilis*; (2) absolutely larger brain size, but not necessarily relatively larger brain size than in *H. habilis*; and (3) absolutely and relatively smaller jaws and teeth.

To evaluate the proposed foraging shifts in early genus *Homo*, I accumulate or review three sets of data: postcranial/body size data, cranial capacity/brainsize data, and jaw and tooth size data. The first two of these have been extensively marshaled in previous arguments related to foraging shifts (e.g., McHenry, 1992; Leonard and Robertson, 1994; Leonard et al., 2003), but are reconsidered here because new samples have been found and because previous analyses have often relied on species averages rather than ranges of variation. Jaw and tooth data have also been considered previously (e.g., Wood and Abbott, 1983, Wood et al., 1983; Wood and Aiello, 1998; Kaifu et al., 2005), but not from the biomechanical and scaling perspective adopted here. Additionally, I focus on comparisons nor to this extent between species of early *Homo* and between Asian and African *H. erectus*, and place greater emphasis on individual data points.

20.2.1 Fossil Individuals and Chimera

The fossil record poses particular challenges to scaling studies, given the dearth of associated cranial and postcranial specimens and the fragmentary nature of the specimens that do exist. Several approaches have been used to address this issue. Often, the more spectacular associated finds, such as 'Lucy' (A.L. 288) or the Nariokotome *H. erectus* skeleton (KNM-WT 15000), are used as reference points. But such an approach limits the observer to a single specimen within a taxon and excludes many taxa all together. For example, *H. habilis* and *H. rudolfensis* both lack associated skeletons of any completeness. It likewise precludes scaling analyses, except across multiple taxa. Another approach uses individual elements to establish species means for body size and brain size: for example, using femoral length for body size or preserved crania for brain size. Such an approach allows the inclusion of more individuals into the analysis, but requires certainty in the species attribution of individual elements. Such attribution is often impossible, at least at a species level, for hominin postcranial elements. Beyond this, using species means also precludes evaluation of intraspecific scaling and interspecific scaling between fewer than three taxa. For these reasons, in these analyses, I emphasize individuals (or chimera) as data points rather than species means. My emphasis on individuals as data points results from considering the issue of taxonomic relationships within early *Homo* unresolved, thus making species means a problematic approach. In my opinion, the variation around the mean is more critical to analyze than the mean

20 Framing the Question: Diet and Evolution in Early *Homo* 449

itself. Likewise, because two main taxa are compared here, intraspecific scaling is of importance. Thus, individual data points not species means are required.

The use of individual data points, while critical to address the question posed here, is problematic given the fossil record because so few associated individuals exist. To address this issue in the scaling analyses of body and brain size to dentognathic size, I supplement the existing fossil individuals with chimeras, or combinations of individual cranial and postcranial data from the same fossil locality and temporal zone. A full listing of specimens used in these chimeras and the species to which they are assigned here is given in the appendix. The chimeras used here expand the numbers of individuals and taxa used in Antón et al. (2007). The criteria used to construct each chimera are relatively straightforward. Specimens from a single locality are grouped only with other specimens from the same locality. Among the potential candidates for a chimera, those that are closest in geological age to one another are grouped together. When there are two or more possible postcranial matches of the same geological age, but there are clues as to cranial size from fragmentary associated remains, the match is made to the closest size match. (For example, the femoral length of KNM-ER 1808 was matched with the cranial capacity of KNM-ER 3733, rather than that of 3883, due to the greater similarity in size between the former and cranial fragments of 1808.) At localities with only a single postcranial or cranial estimate, (such as Dmanisi and Trinil, respectively) that estimate will be grouped with both the smallest and the largest of the estimates for the other variable, in order to consider the possible range of scaling relationships. (For example, at Dmanisi the smallest and largest adult cranial capacities were used to form two chimeras by pairing each with the stature estimate from the metatarsal D2021. At Trinil, the single cranial capacity value of Trinil 2 was paired with both the shortest and the longest of the femoral values from Trinil.) There is no doubt that these chimeras do not reflect actual individuals. However, they have been conservatively constructed and are likely to be a better representation of individual variation than are mean values. Chimeras are also the only means of assessing intraspecific scaling that is feasible from the current state of the fossil record for early *Homo*.

Because previous hypotheses are couched in terms of foraging shifts from *H. habilis* to *H. erectus*, in the text and figures individual points are identified to one of these groups (or in the case of *H. erectus* sometimes to locality). *H. habilis s.l.* includes specimens assigned by others to both *H. habilis. sensu stricto (s.s.)*, *H. rudolfensis,* and *H. sp. nov* (*sensu*, Wood, 1991). *H. erectus s.l.* includes specimens assigned by others to *H. ergaster* and *H. erectus s.s* (see appendix). Despite these identifications, attention is paid to the extent of intermingling between these two large groups and their subgroups in order to address whether morphological changes expected of a foraging shift are apparent.

Due to this virtual absence of associated cranial and postcranial remains in the early *Homo* record, the three data sets used here are generated from different fossil samples (see appendix). Each set of data attempts to sample earliest *Homo* in Africa, early East African *H. erectus*, and non-African *H. erectus*. The samples used from oldest to youngest, are: *H. habilis s.l.* (from Koobi Fora and Olduvai dated between 1.9 and 1.6 Ma; Feibel et al., 1989), Early East African and Georgian *H. erectus* (e.g., Dmanisi and African *H. erectus* from Koobi Fora, West Turkana, and Olduvai; dated from 1.8-1.0 Ma; Feibel et al., 1989; Brown and McDougall, 1993; Vekua et al., 2002), Indonesian *H. erectus* (from Sangiran, Java dated from 1.6-1.2 Ma; Swisher et al., 1994; Antón and Swisher, 2004), middle Pleistocene African *H. erectus* (from Ternifine, Algeria; dated to about 700 ka; Geraads et al., 1986), and middle Pleistocene Chinese *H. erectus* (from Zhoukoudian; dated approximately 400 ka for these specimens; Huang et al., 1991a, b; Shen et al., 1996; Grün et al., 1997). Late Pleistocene *H. erectus* from Ngandong, Indonesia, are used only in the body to brain size scaling analyses as there are no teeth or jaws from this site. Each decision about what to include (and in which category) is critical to the results and should be so recognized. It should be noted that these samples, by necessity, must be considered mixed-sex samples, although this cannot be proven. Adult and subadult samples are treated separately with summary statistics provided only for adult specimens (Table 20.2). Comparative data for a worldwide *Homo sapiens* sample (after Brown, 2005; femur length, mandibular corpus, and buccolingual molar dimensions only) and for *Australopithecus* are provided as necessary.

20.2.2 Variables and Scaling Analyses

Body size and weight are important variables that reflect aspects of diet, home range, and life history in mammals (e.g., McNab, 1963; Milton and May, 1976; Harvey and Clutton-Brock, 1985). However, the fragmentary fossil record often does not provide even intact skeletal elements, let alone the associated elements from a single individual required for robust estimates of body mass. To avoid the issues inherent in estimating body weight (e.g., Ruff, 2002), I focus on individual long bones and stature estimates generated from these (Table 20.3; Antón et al., 2007); these provide a general proxy for overall body size. When necessary, I use long bone lengths predicted from bone fragments by segment-based regression analyses (e.g., Fellmann, 2004). I caution, however, that these predictions assume similar proportions in the target as in the sample population (e.g., Steele and McKern, 1969; Wright and Vásquez, 2003), an assumption that may not hold true for fossil hominins. Because of the fragmentary nature of the fossil record, using estimates of stature often increases sample sizes since stature can be estimated from a number of long bones and then can be used to compare sizes between individuals who do not both preserve the same element. I collect stature estimates from the literature, and estimate stature from isolated skeletal elements where necessary by using regression analyses that predict stature from single elements of the lower limb (e.g., Trotter and Gleser, 1952, 1958), or Feldesman's (Feldesman et al., 1990) femur to stature ratio. Again it should be noted that there are a number of issues with these estimates. Most critically, the accuracy of the estimate is dependent on the similarity of proportions between the known sample and the individual whose stature is to be estimated. For

Table 20.2 Dentognathic summary statistics for *H. habilis* and *H. erectus* (worldwide) and for *H. erectus* by region/locality. Regional columns arranged from geologically oldest to youngest localities. Measurements in mm. Mean/Standard Deviation, (*n*). Data ranges are presented in Figs. 20.2 and 20.3

	Africa	Worldwide	Africa & Georgia	Indonesia	Ternifine	China
	H. habilis s.l.	*H. erectus* s.l.	*H. erectus*	*H. erectus*	*H. erectus*	*H. erectus*
M₁ bucco-lingual	123.8/9.0 (8)	121.3/7.8 (30)	119.2/7.6 (5)	124.7/6.8 (9)	126.7/2.9 (3)	118.5/8.6 (13)
M₁ mesio-distal	138.5/6.1 (8)	129/10 (30)	132.2/6.5 (5)	130.4/10.9 (10)	131.7/8.5 (3)	125.8/0.9 (12)
M₁ Area	1705/181 (8)	1578/204 (30)	1592/131 (6)	1648/206 (9)	1670/141 (3)	1497/225 (12)
Corpus height M₁	32.6/4.3 (11)	32.5/5.6 (20)	29.4/2.9 (6)	36.3/5.9 (7)	36.3/1.5 (3)	27.8/4.2 (4)
Corpus breadth M₁	20.8/3.68 (10)	19.0/2.9 (22)	20.1/0.9 (6)	19.0/3.5 (9)	18.3/1.2 (3)	16/1.2 (4)
Symphyseal height	33.2/6.0 (5)	35.1/4.8 (16)	33.4/2.6 (4)	36.7/7.3 (5)	37.3/2.5 (3)	33.2/4.3 (4)
Symphyseal depth (labiolingual)	21.7/3.2 (5)	17.2/3.2 (17)	18.5/2.4 (4)	19.1/3.5 (5)	18.7/0.6 (3)	13.6/0.5 (5)

Table 20.3 Individuals and chimeras used in body size-brain size scaling relationships

Specimen for cranial estimate	Capacity (cc)	Geological age (Ma) (cranial/postcranial)	Femur length (mm)	Stature estimate (cm)	Specimen for postcranial estimates
A. afarensis					
AL 162-28	380	3.1/3.1	280	106	A.L. 288-1
A. garhi					
Bou-VP 12/130	450	2.5/2.5	335/348	–	Bou-VP 12/1
H. habilis s.l.					
KNM-ER 1813	510	1.9/1.9	350	131	KNM-ER 1503[1]
KNM-ER 1470	740	1.9/1.9	400	150	KNM-ER 1472
KNM-ER 1805	582	1.9/1.9	395	148	KNM-ER 1481
OH 16	638	1.8	395	148	OH 34[2]
OH 24	590	1.8/1.8	280	106	OH 62
H. erectus					
KNM-ER 3733	848	1.8/1.7	480	173	KNM-ER 1808
KNM-ER 3883	804	1.5/1.7	440	164	KNM-ER 737
KNM-WT 15000	909	1.5/1.5	429[3]	160	KNM-WT 15000
OH 9	1067	1.5/1.0	448	167	OH 28
OH 12	727	1.2/1.5	430	160	OH 34[2]
Trinil 2	940	0.9/0.9	447	167	Trinil II
Trinil 2	940	0.9/0.9	433	162	Trinil III
Zhoukoudian (ZKD) Skull III	915	0.58/0.42	413	158	ZKD Femur IV
Zhoukoudian Skull VI	855	0.42/0.42	328	141	ZKD Femur I
Zhoukoudian Skull XII	1225	0.42/0.42	413	158	ZKD Femur IV
Ngandong 7	1013	0.05/0.05	–	158	Ngandong Tibia B
Ngandong 6	1251	0.05/0.05	–	158	Ngandong Tibia B
Dmanisi 2280	780	1.7/1.7	–	146	Dmanisi 2021
Dmanisi 2282	650	1.7/1.7	–	146	Dmanisi 2021

Biological data from:.A. afarensis - A. gahri – Asfaw et al., 1999; Richmond et al., 2002.
H. habilis – Haeusler and McHenry, 2005; McHenry, 1991, 1992; McHenry and Coffing, 2000; Tobias, 1991; Wood, 1991.*H. erectus* – Antón et al., 2007; Gabunia et al., 2000a, b; Holloway, 1980, 1981; Ruff and Walker, 1993; Santa Luca, 1980; Walker and Leakey, 1993; Weidenreich, 1941, 1943; Wood, 1991. *Geological age estimates* as per text and from: Asfaw et al., 1999; Antón and Swisher, 2004; Swisher et al., 1996.

[1] This specimen is assigned alternatively to *H. habilis* or *A. boisei* as are most specimens of this age at Koobi Fora, except those of relatively large size (McHenry, 1991). In order not to unduly skew *H. habilis* results to large size, this specimen, which is near the mean of femoral length values of those specimens classified by McHenry (1991) as either *H. habilis* or *A. boisei*, is used.
[2] The taxonomic status and length of OH 34 is contested. Haeusler and McHenry (2004) estimate the length between 375 and 395 and as a member of earliest *Homo*. However, an estimated length of 430 mm and assignment to *H. erectus* have also been proposed (Day and Molleson, 1976; Howell, 1978). The two estimates and attributions are used here.
[3] Femur length for KNM-WT 15000 used here is the length at the individual's death, not the length that the individual is estimated to have achieved had it reached adulthood.

hominin samples, such as *H. habilis* and the Asian representatives of *H. erectus* for which proportions are entirely unknown, such assumptions may introduce significant and unknown errors.

Increases in body size may indicate a foraging shift (or better nutrition), and brain size may increase as an allometric result of body size increase. Two questions are of interest here. First, the relevant issue from an energetics perspective is the absolute increase in brain tissue, which is more expensive to grow and maintain than most other body tissues (e.g., Leonard and Robertson 1992, 1994); that is, are *H. erectus* brains absolutely larger than *H. habilis* brains? The second relates to the relative scaling of brain size and body size in the two taxa – is the scaling relationship similar and is there any evidence of a grade shift between them? Evaluation of the relationship between brain and body size is assessed using regression analyses based on chimeras of brain and body sizes for different individuals from each species (Table 20.3).

Ideally, variables used to assess dietary adaptations from the jaws would be those correlated with generating bite force or dissipating occlusal loads (e.g., Hylander, 1988; Daegling, 1989; Ravosa, 1990). These include cranial (maxillary breadth and length), dental (occlusal area), and mandibular measures (e.g., corpus and symphyseal height and breadth, mandible length), as well as those that estimate masticatory muscle position and size [e.g., gonion-zygion distance, vertical distance from the temporal line to the temporalis tubercle (temporalis height), masseter insertion angle, etc.; Antón, 1996]. The fossil record, however, precludes the measurement of most of these variables in more than one or a few specimens. Therefore, I aim to maximize sample size while still allowing for some size adjustment between individuals by focusing on frequently preserved areas; the mandibular body, symphysis, and molar teeth.[3] These dimensions are then evaluated between taxa for evidence of the directional changes hypothesized to occur in the case of a foraging shift. One serious drawback of the data at hand is that very few of the mandibles for which molar, corpus, or symphyseal dimensions are available also preserve mandibular length. In addition, many preserve only corpus or only symphyseal dimensions. Thus, scaling individual dimensions to some biomechanically relevant aspect of mandibular size, such as mandibular length, is not possible. The alternative practice of scaling dimensions to some overall size estimate of the mandible (such as the geometric mean of several mandibular dimensions) is similarly impossible. The only feasible scaling is between certain molar dimensions and mandibular dimensions (this sample is composed of actual individuals) and between these and body and brain size proxies (these analyses use mostly chimera).

Here, I consider the scaling relationship between mandible and molar size for evidence of a foraging shift, and the relationship between jaw variables and body size and brain size is explored. The height and width of the mandibular corpus

---

[3] Isolated teeth unassigned to species were avoided in Africa due to the presence of multiple species of *Homo*. This decision severely restricts sample sizes. South African hominins were avoided due to taxonomic uncertainties.

scale isometrically with size (body mass, mandible length) in extant primates (Smith, 1983; Bouvier, 1986; Hylander, 1988). Mandibular height has been used to predict body mass in fossil hominins and to then argue that *H. erectus,* but not *H. habilis,* has human-like mandibular proportions (Wood and Aiello, 1998). The goal of Wood and Aiello's work was to establish how well body-mass-estimates predicted from mandibular measures correlated with the mass estimates predicted from other sources, such as postcranial dimensions, for the same taxon. Their study was not concerned with estimating an individual's body mass, but with estimating the mean body mass for a species. Thus, they did not consider the relationship between mandibular variables and individual body mass, as described here, but tried to establish the extent to which the mandible yielded similar body mass estimates as did other proxies such as femur size. Here, instead, I consider the relationship between corpus dimensions, body size, and brain size using chimeras for jaw and body size as approximations of individuals (see appendix).

Other studies have suggested that most frugivorous primates, particularly anthropoids, have smaller molars for their body size than do folivores (e.g., Kay and Hylander, 1978; Gingerich and Smith, 1985; but see Vinyard and Hanna, 2005). If this difference is related to the mechanical properties of the food items, then the inferred higher-quality, softer diet of *H. erectus* should yield similar predictions relative to *H. habilis*. In strepsirrhines, a positive correlation between brain size and molar area is suggested (when holding body size constant; Vinyard and Hanna, 2005). Since there is a slight trend toward increasing brain size with time across early *Homo* (Henneberg, 1987, 2001; Antón and Swisher, 2001; Rightmire, 2004), I also consider the evidence for a relationship between these variables here.

Finally, because the assemblages span at least 500,000 years, the relationship between each of these variables and geologic time is of interest. Previous work has established that despite a slight trend toward increasing cranial size in early *Homo*, particularly *H. erectus*, (Wolpoff, 1984; Leigh, 1992; Antón and Swisher, 2001), there is not much trend in body size (Antón et al., 2007). It should be noted that the trend toward increasing body size found across the genus with time (e.g., Ruff et al., 1997) is not evident in early *Homo* when chimera of *H. habilis* and *H. erectus* are considered rather than species means (Antón et al., in press). Decreases in some mandibular measures with time have been documented within the Indonesian *H. erectus* sample from Sangiran (Kaifu et al., 2005) by looking at changes in mean values across temporal groups. These changes have not been considered, however, across the taxon for individuals. Here then, I consider whether there is evidence of temporal trends in the dentognathic variables addressed above.

It should be noted that in most cases in which chimeras are required, sample sizes preclude statistical comparisons across groups. This is also the case for some of the analyses that do not employ chimeras. Thus, bivariate plots are provided for the evaluation of trends that will require statistical testing when more robust sample sizes, and definitive individuals (rather than chimeras), are available. In most cases, the absence of trends is clear, whereas the presence of trends may be only a suggestion. At reviewers' request, slope values and regression statistics are provided for preliminary comparison. All plots and statistics are generated from Systat version 10.

20.3 Results

20.3.1 Is Homo erectus *Large-Bodied and Large-Brained?*

Results presented in Fig. 20.1 essentially replicate those of Antón et al. (2007), with the addition of more *Australopithecus* and *H. habilis* data points (Table 20.4). These suggest that average body size may differ between groups of early *Homo*, but that the scaling relationship is essentially the same in all early hominins. As a result, there

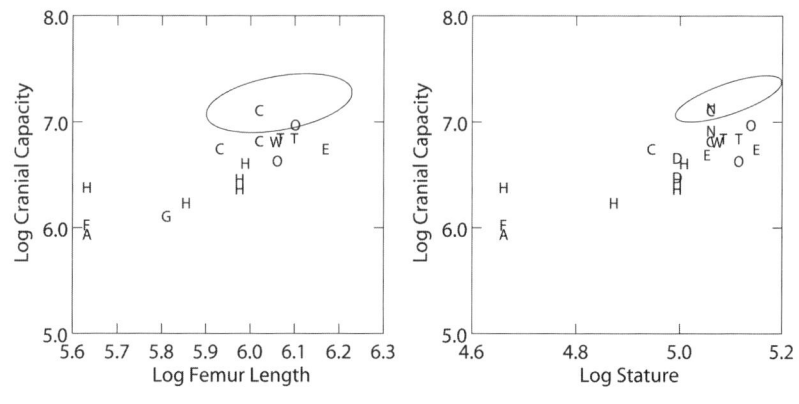

Fig. 20.1 Relationship between body size (femur length and stature) and brain size. Abbreviations are: A = *Australopithecus afarensis*, G = *A. garhi*, F = *H. floresiensis*, H = *H. habilis s.l.*, *H. erectus* subgroups are C = Chinese, D = Dmanisi, E = Koobi Fora, N = Ngandong, Java, O = Olduvai, T = Trinil, Java, W = KNM-WT 15000. Ellipse represents 95% confidence ellipse for *H. sapiens* data. Cranial capacity in cc, Femur length in mm, Stature in cm. *H. sapiens* stature data from Ruff et al., (1997). Regression statistics in Table 20.4

Table 20.4 Regression results for the relationship between body size and cranial capacity

	Femur length					Stature				
	R^2	OLS Slope	Intercept	P	n	R^2	Slope OLS	Intercept	P	n
H. sapiens only	0.14	0.63	3.3	<0.00001	198	0.52	1.5	−0.47	<0.00001	55
H. erectus only	0.03	0.42	4.3	ns	8	0.08	0.89	2.3	ns	13
Early *Homo* (*H. erectus* and *H. habilis*)	0.64	1.36	−1.4	0.006	10	0.47	1.35	−0.03	0.005	15
All fossil hominins	0.78	1.72	−3.6	<0.00001	11	0.67	1.7	−1.8	<0.00001	16

is substantial overlap between larger specimens of *H. habilis* (including KNM-ER 1470 and OH 16) and smaller specimens of *H. erectus* (including Dmanisi and OH 12). The relationship is approximately isometric. One will note that there is no overlap between some individuals (chimera) of the smaller bodied hominins (especially *A. afarensis*) and the early *Homo* data points, and thus extension of the regression between these clouds of data is speculative. *H. sapiens* is transposed above the other hominins due to substantially larger brain sizes, and the slope of the line differs somewhat depending on the body size variable considered.

20.3.2 Do Homo erectus *Jaws and Teeth Suggest a Foraging Shift?*

20.3.2.1 Absolute Size of Teeth and Jaws

Absolute molar and jaw size tend not to differentiate between *H. erectus s.l.* and *H. habilis s.l.* In terms of absolute size, the *H. habilis s.l.* M_1 range falls completely within that of *H. erectus s.l.* (Fig. 20.2; Table 20.2). However, the *H. erectus* sample exhibits a larger range of variation, and for this reason the means for each of the dental measures are slightly lower than the mean values for *H. habilis s.l.* In each case, the mean values of the taxa are within a single standard deviation of one another. The same is true, not surprisingly, of corpus height and breadth in the two groups (Fig. 20.2; Table 20.2), but not of symphyseal dimensions. For symphyseal height, the two groups overlap substantially, but not completely, as some *H. habilis* individuals fall below the range of *H. erectus s.l.* The opposite is true of symphyseal width (labiolingual) for which some *H. habilis* individuals are substantially larger than the *H. erectus s.l.* range. For this dimension, the mean values overlap at not one, but only at two standard deviations. It is interesting to note that subadult *H. erectus* achieve adult dimensions of mandibular corpus height later than mandibular breadth, which mirrors dental development (Fig. 20.4). The same cannot be said of the symphysis. In these plots, the single subadult data point (Dmanisi 2735) falls with the adults for both symphyseal height and width.

Even if we further subdivide the *H. erectus* sample into geographic and time packets, there is still substantial overlap of the ranges of variation from the *H. habilis s.l.* sample and the various *H. erectus* samples (Fig. 20.3). Most importantly, there is substantial overlap for all jaw and tooth dimensions between the nearly coeval East African *H. erectus* sample (also known as *H. ergaster*) and the *H. habilis s.l.* sample.

Not surprisingly, then, if we consider the relationship between molar dimensions (BL vs. MD), corpus dimensions (height vs. width), and symphyseal dimensions (height vs. width), the groups show similar relationships (Table 20.5; Fig. 20.4). The relationship between mesiodistal and buccolingual dimensions of M_1 is similar amongst the groups of early *Homo*, and there is no discernible difference in the placement of purported *H. habilis* (or *H. rudolfensis*) individuals relative to *H. erectus s.l.* (Fig. 20.4a). Likewise, the relationship between mandibular corpus height and breadth dimensions is similar amongst groups of early *Homo* and appears to continue the same relationship seen in *H. sapiens* to larger sizes (Fig. 20.4b). Again, there is no clear differentiation between individuals commonly assigned to

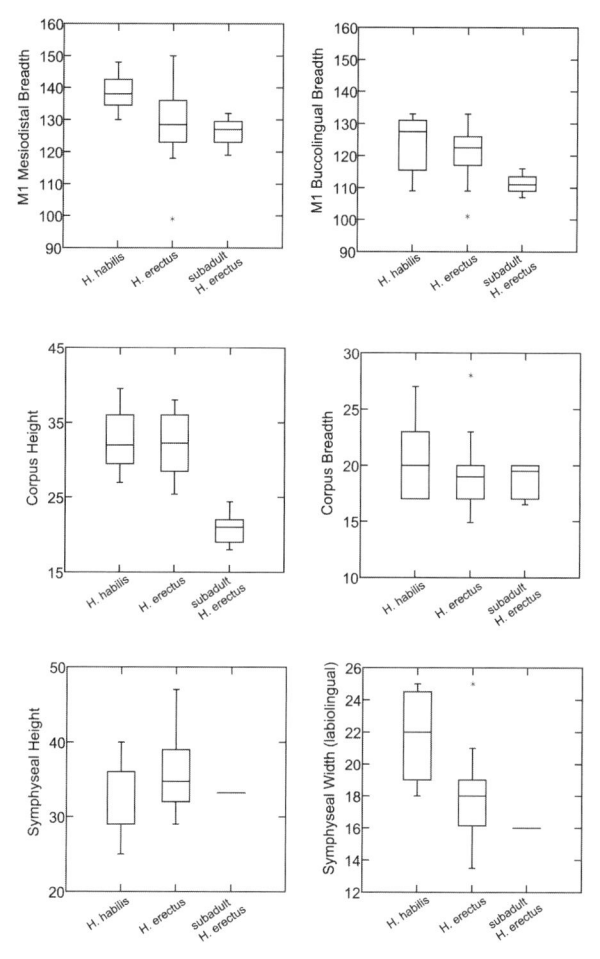

20 Framing the Question: Diet and Evolution in Early *Homo*

Fig. 20.2 Dentognathic summary statistics by species group. Groups are *H. habilis sensu lato*, *H. erectus sensu lato*, and subadult *H. erectus sensu lato*. For each group, middle line represents the median, whiskers represent the range of observed data points except in the case of significant outliers, which are indicated by an asterisk (∗). A single line indicates only one individual in the sample

H. habilis (or *H. rudolfensis*) and those assigned to *H. erectus s.l.*, although the slope is steeper for *H. habilis*. In both the above cases, the largest individuals are those usually assigned to *H. rudolfensis*, along with some Indonesian *H. erectus* that populate the upper right corner of the graphs. Individuals typically classified as *H. habilis s.s.* plot among the *H. erectus* specimens from Africa and elsewhere.

The relationship amongst symphyseal dimensions is somewhat different. Although the relationship amongst all early *Homo* symphyseal widths and heights is not statistically significant, in this instance it is possible to argue that the *H. habilis s.l.* sample is transposed above that of *H. erectus s.l.* (Fig. 20.4c). All individuals classified either as *H. rudolfensis* or *H. habilis* plot above the trajectory for all

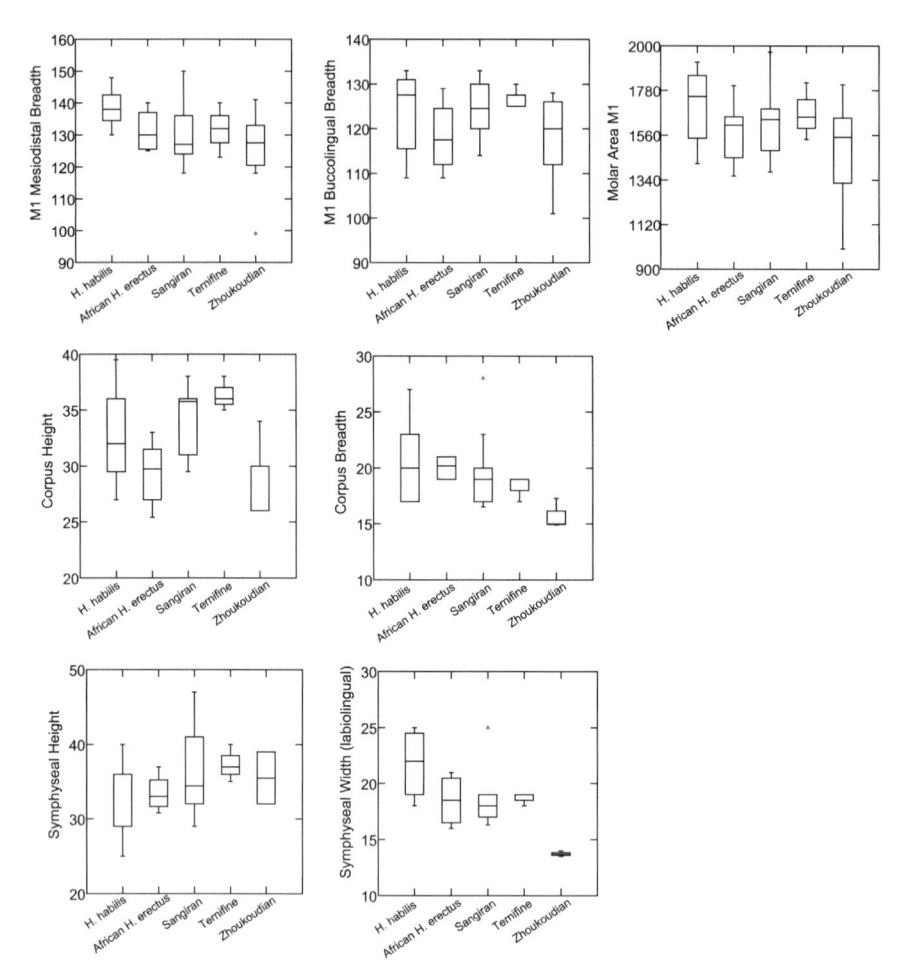

Fig. 20.3 Dentognathic summary statistics for *H. habilis s.l.*, African *H. erectus* (including Dmanisi), Sangiran, Ternifine, and Zhoukoudian *H. erectus*. Sangiran sample is early Pleistocene Indonesia. Ternifine and Zhoukoudian are middle Pleistocene North Africa and China, respectively. For each group, middle line represents the median, whiskers represent the range of observed data points except in the case of significant outliers, which are indicated by an asterisk

individuals classified as *H. erectus s.l.;* a few of the larger *H. erectus* from Sangiran and Olduvai plot close to the *H. habilis* individuals. Due to a lack of data, the relationship between symphyseal height and width dimensions cannot be considered for *H. sapiens*.

20.3.2.2 Scaling of Teeth Relative to Jaw Size

Few significant relationships are evident between tooth and jaw size (Fig. 20.5; Table 20.6). However, when all early *Homo* are grouped, molar area scales with

Table 20.5 Results of ordinary least-squares regression of dentognathic height and breadth dimensions in fossil *Homo*

	N	R^2	OLS slope	P
Molar 1 Mesiodistal Breadth (x)				
M1 buccolingual (early *Homo*)	37	0.50	0.62	<0.00001
M1 buccolingual (*H. habilis*)	8	0.19	0.74	ns
M1 buccolingual (*H. erectus*)	29	0.63	0.66	<0.00001
Corpus Height (x)				
Corpus breadth (early *Homo*)	29	0.43	0.68	<0.00001
Corpus breadth (*H. habilis*)	10	0.85	1.14	<0.00001
Corpus breadth (*H. erectus*)	19	0.32	0.51	0.001
Corpus breadth (*H. sapiens*)	413	0.15	0.44	<0.00001
Symphyseal Height (x)				
Symphyseal depth (early *Homo*)	19	0.17	0.49	ns
Symphyseal depth (*H. habilis*)	5	0.91	0.74	0.01
Symphyseal depth (*H. erectus*)	14	0.27	0.62	ns

corpus breadth (Fig. 20.5). This relationship in buccolingual dimensions follows a similar relationship in *H. sapiens* and *H. erectus*. Similarly, there are no significant relationships between the measures of molar size and the height or width of the mandibular symphysis, regardless of whether all early *Homo* are grouped or the two purported taxa are separated. There are no significant relationships between molar and symphyseal dimensions within species of early *Homo*. Reviewing these plots also suggests no incipient relationship that might prove significant with more robust data. The symphyseal relationship could not be assessed in the recent human sample.

20.3.2.3 Scale of Teeth and Jaws to Body Size

Both estimates of body size (femur length and stature) yield similar results relative to measures of molar size in fossil *Homo* species (Fig. 20.6; Table 20.7). Although none are statistically significant, the plots suggest no correlation between mesiodistal dimensions and body size, but do suggest a more likely relationship between bucco-lingual dimensions and body size. The relationship between buccolingual dimensions and femur length is significant in *H. sapiens* (other variables are not available for *H. sapiens*). In no instance, however, is it feasible to separate *H. habilis s.l.* and *H. erectus s.l.* on the basis of their positions when plotted for tooth and body size.

Femur length and stature also yield similar relationships to jaw dimensions in early *Homo* (Fig. 20.7). Corpus height and breadth in *H. sapiens* are both significantly correlated with femur length. In *H. erectus,* only corpus breadth is correlated with femur length, whereas only corpus height is correlated with stature. *H. sapiens* corpora are thinner than, but as tall as, early *Homo* corpora. Again, it is not feasible to separate groups of early *Homo* based on their plot position.

Symphyseal dimensions lack a significant relationship with the two body size measures, although they show somewhat greater levels of variation than do corporal

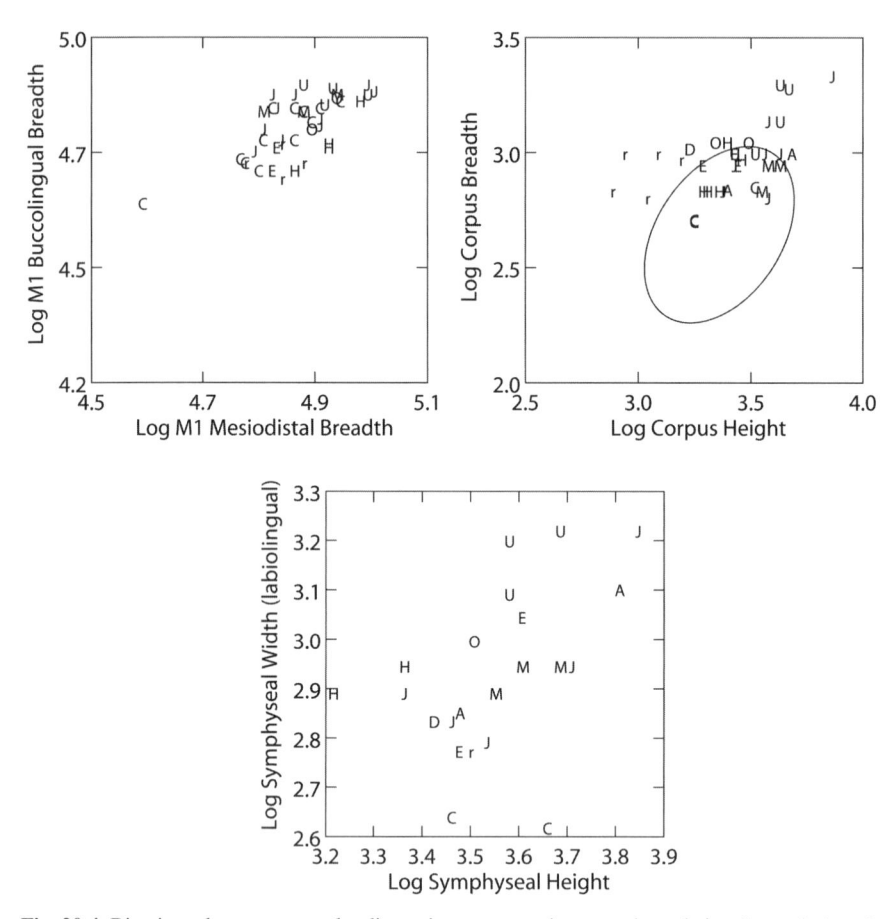

Fig. 20.4 Bivariate plots among molar dimensions, corpus size, symphyseal size. Legend: A = *A. afarensis*, H = *H. habilis*, U = *H. rudolfensis*; *H. erectus* subgroups are C = China, E = Koobi Fora, J = Javan, M = Ternifine, O = Olduvai, r = subadult *H. erectus*. Ellipse represents 95% confidence ellipse for *H. sapiens*. All measures taken in mm and then transformed. Regression statistics reported in Table 20.5

dimensions. Once again, *H. habilis* s.l. and *H. erectus* s.l. are not differentiable on these bases. Chinese *H. erectus* present quite narrow symphyses for their body size, whereas *H. rudolfensis* individuals exhibit quite deep symphyses for their size. *H. habilis* s.l. individuals plot among the other *H. erectus* individuals for these dimensions.

20.3.2.4 Scale of Teeth and Jaws to Cranial Capacity

There is no significant relationship between tooth size and cranial capacity in early *Homo* or its species; however, the relationship between M1 buccolingual dimensions and cranial capacity shows a positive trend in these groups (Fig. 20.8; Table 20.8).

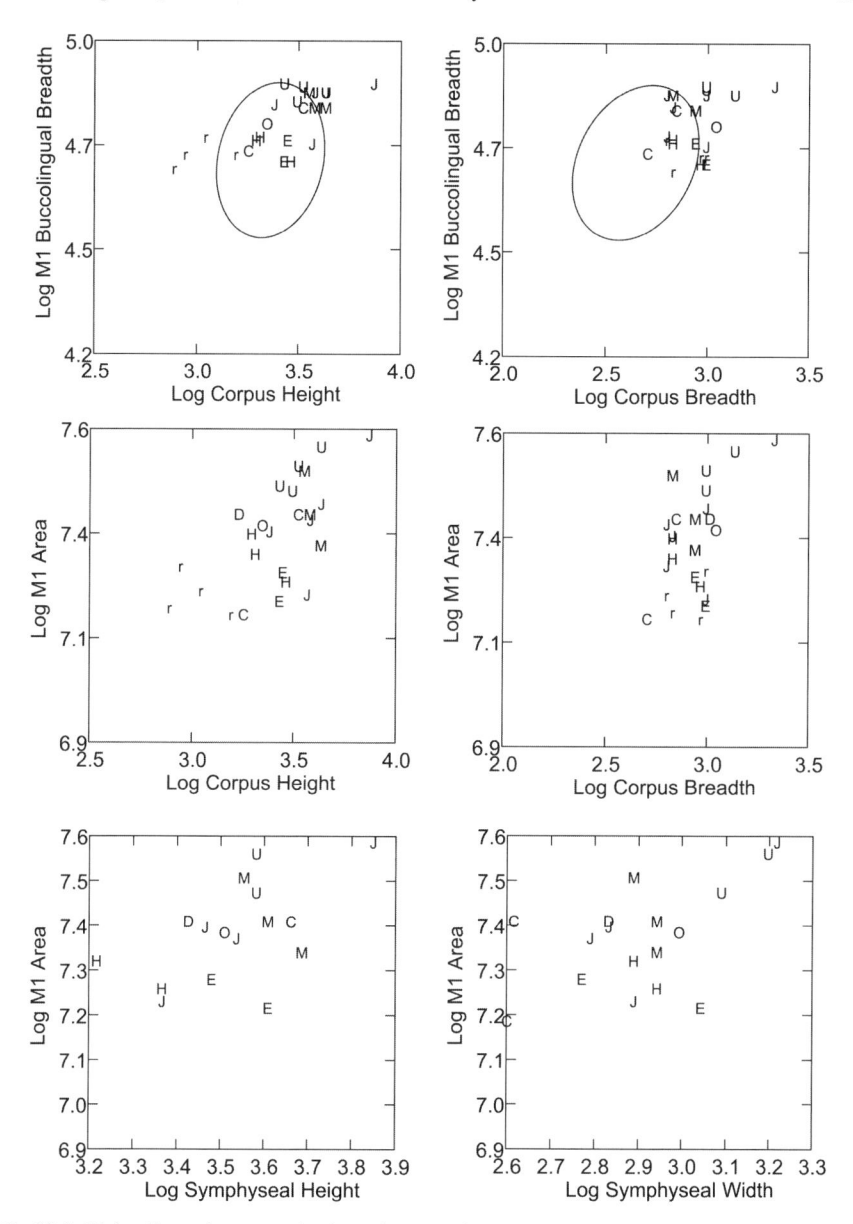

Fig. 20.5 Molar dimensions, symphysis, and corpus size. Legend: A = *A. afarensis*, H= *H. habilis*, U = *H. rudolfensis*; *H. erectus* subgroups are C = China, D = Dmanisi, E = Koobi Fora, O = Olduvai, T = Ternifine, J = Javan, r = subadult *H. erectus*. Ellipse represents 95% confidence ellipse for *H. sapiens* sample. All measures except molar area taken in mm and then transformed. Molar area follows Wood (1991). Regression statistics reported in Table 20.6

Table 20.6 Results of ordinary least-squares regression of dental variables versus jaw size

	n	R^2	OLS slope	P
Corpus Height (x)				
M1 mesiodistal (*H. habilis*)	7	0.20	0.15	ns
M1 buccolingual (*H. habilis*)	7	0.32	0.38	ns
Molar area (*H. habilis*)	7	0.47	0.63	ns
M1 mesiodistal (*H. erectus*)	14	0.25	0.21	ns
M1 buccolingual (*H. erectus*)	13	0.43	0.27	0.02
Molar area (*H. erectus*)	14	0.31	0.37	0.04
M1 buccolingual (*H. sapiens*)	352	0.03	0.12	0.002
Corpus Breadth (x)				
M1 mesiodistal (early *Homo*)	21	0.14	0.17	ns
M1 buccolingual (early *Homo*)	20	0.11	0.16	ns
Molar area (early *Homo*)	21	0.28	0.43	0.01
M1 mesiodistal (*H. habilis*)	6	0.16	0.16	ns
M1 buccolingual (*H. habilis*)	6	0.37	0.45	ns
Molar area (*H. habilis*)	6	0.49	0.73	ns
M1 mesiodistal (*H. erectus*)	15	0.17	0.16	ns
M1 buccolingual (*H. erectus*)	14	0.06	0.10	ns
Molar area (*H. erectus*)	15	0.22	0.35	ns
M1 buccolingual (*H. sapiens*)	361	0.07	0.17	<0.00001
Symphyseal Height (x)				
M1 mesiodistal (*H. habilis*)	4	0.06	0.08	ns
M1 buccolingual (*H. habilis*)	4	0.66	0.43	ns
Molar area (*H. habilis*)	4	0.69	0.64	ns
M1 mesiodistal (*H. erectus*)	12	0.17	0.20	ns
M1 buccolingual (*H. erectus*)	11	0.20	0.22	ns
Molar area (*H. erectus*)	12	0.29	0.43	ns
Symphyseal Depth (x)				
M1 mesiodistal (*H. habilis*)	4	0.31	0.24	ns
M1 buccolingual (*H. habilis*)	4	0.68	0.55	ns
Molar area (*H. habilis*)	4	0.88	0.93	ns
M1 mesiodistal (*H. erectus*)	13	0.23	0.19	ns
M1 buccolingual (*H. erectus*)	12	0.06	0.09	ns
Molar area (*H. erectus*)	13	0.18	0.29	ns

There is an approximately isometric relationship between buccolingual M1 size and cranial capacity in *H. sapiens*. Similarly, corpus height and breadth scale approximately isometrically with cranial capacity in *H. sapiens*. Within early *Homo*, *H. habilis* exhibits significant relationships for all dimensions and cranial capacity, and *H. erectus* shows a relationship between corpus height and cranial capacity. There is, however, no hint of a relationship between breadth and capacity.

20.3.2.5 Jaws and Teeth with Time

Across the early *Homo* assemblage there is no discernible trend in molar or jaw size with time. Molar dimensions (bucco-lingual, mesio-distal, and occlusal areal

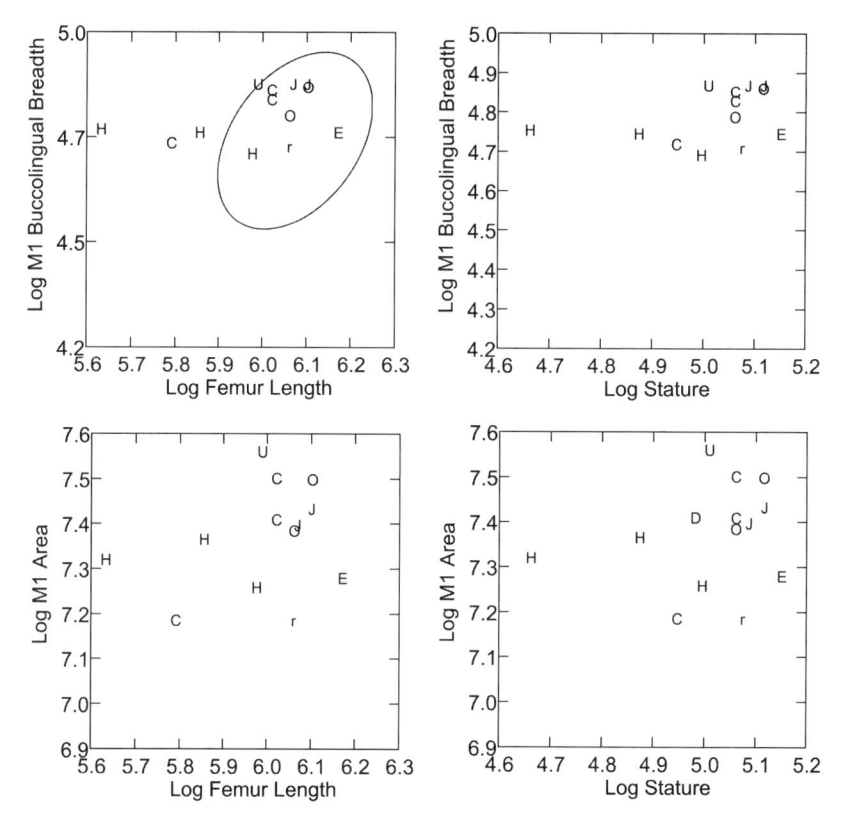

20 Framing the Question: Diet and Evolution in Early *Homo*

Fig. 20.6 Bivariate relationship between molar dimensions and body size (femur length and stature). Legend: A = *A. afarensis*, H= *H. habilis*, U = *H. rudolfensis*; *H. erectus* subgroups are C = China, D = Dmanisi, E = Koobi Fora, O = Olduvai, T = Ternifine, J = Javan, r = subadult *H. erectus*. Ellipse represents 95% confidence ellipse for *H. sapiens* sample. All measures except molar area taken in mm and then transformed. Molar area follows Wood (1991). Regression statistics reported in Table 20.7

measures) show similar levels of variation in the earliest (*H. habilis s.l.*) specimens as in the latest *H. erectus* specimens (Fig. 20.9; Table 20.9). Indeed, some of the largest molars are found in the latest *H. erectus* sampled here, those from Zhoukoudian. Likewise, mandibular corpus height and symphyseal height show no trend with time across early *Homo*. However, both corpus breadth and symphyseal depth appear to diminish somewhat with time due to relatively small sizes in the Chinese *H. erectus* samples, although only the trend in corpus breadth is significant. The pattern of decrease differs, however, between the two dimensions. Corpus breadth appears to show slight diminution with time, starting with large variability from 1.9 to 1.5 Ma and decreased variability after 1.5 Ma; both middle Pleistocene *H. erectus* samples (from Ternifine and Zhoukoudian) are among the narrowest corpora, but only the Zhoukoudian values are (slightly) outside the range of the earlier *H. erectus*. Alternatively, symphyseal depth is markedly smaller in Zhoukoudian

Table 20.7 Results of ordinary least-squares regression of dentognathic variables versus body size estimates

	n	R^2	OLS slope	P
Femur Length (x)				
M1 mesiodistal (*H. habilis*)	4	0.006	0.02	ns
M1 buccolingual (*H. habilis*)	4	0.04	0.09	ns
Molar area (*H. habilis*)	4	0.12	0.28	ns
M1 mesiodistal (*H. erectus*)	8	0.22	0.25	ns
M1 buccolingual (*H. erectus*)	8	0.20	0.23	ns
Molar area (*H. erectus*)	8	0.26	0.49	ns
M1 buccolingual (*H. sapiens*)	94	0.15	0.50	<0.00001
Corpus height (*H. habilis*)	6	0.45	0.76	ns
Corpus breadth (*H. habilis*)	6	0.52	0.89	ns
Symphyseal height (*H. habilis*)	4	0.63	0.95	ns
Symphyseal depth (*H. habilis*)	4	0.48	14.2	ns
Corpus height (*H. erectus*)	8	0.39	0.66	ns
Corpus breadth (*H. erectus*)	8	0.68	0.75	0.01
Symphyseal height (*H. erectus*)	5	0.006	0.05	ns
Symphyseal depth (*H. erectus*)	6	0.35	9.48	ns
Corpus height (*H. sapiens*)	112	0.28	0.86	<0.00001
Corpus breadth (*H. sapiens*)	119	0.02	0.27	<0.00001
Stature (x)				
M1 mesiodistal (*H. habilis*)	4	0.007	0.03	ns
M1 buccolingual (*H. habilis*)	4	0.04	0.09	ns
Molar area (*H. habilis*)	4	0.13	0.29	ns
M1 mesiodistal (*H. erectus*)	8	0.18	0.41	ns
M1 buccolingual (*H. erectus*)	8	0.20	0.43	ns
Molar area (*H. erectus*)	9	0.15	0.60	ns
Corpus height (*H. habilis*)	6	0.47	0.79	ns
Corpus breadth (*H. habilis*)	6	0.53	0.92	ns
Symphyseal height (*H. habilis*)	4	0.63	0.98	ns
Symphyseal width (*H. habilis*)	4	0.50	0.7	ns
Corpus height (*H. erectus*)	9	0.48	1.4	ns
Corpus breadth (*H. erectus*)	9	0.31	0.97	ns
Symphyseal height (*H. erectus*)	6	0.06	0.27	ns
Symphyseal depth (*H. erectus*)	7	0.18	0.79	ns

specimens than in any of the earlier *H. erectus*, whereas the Ternifine remains exhibit average symphyseal depth values for *H. erectus*. Unfortunately, further interpretation of this trend is hampered by the virtual absence of data on symphyseal depth values between about 1.4 and 0.7 Ma.

20.4 Discussion

20.4.1 Evidence for a Foraging Shift

Most of the generalized statements about the body and brain size of *H. erectus* relative to *H. habilis* are not supported by these data (Figs. 20.1 & 20.3; Tables 20.1

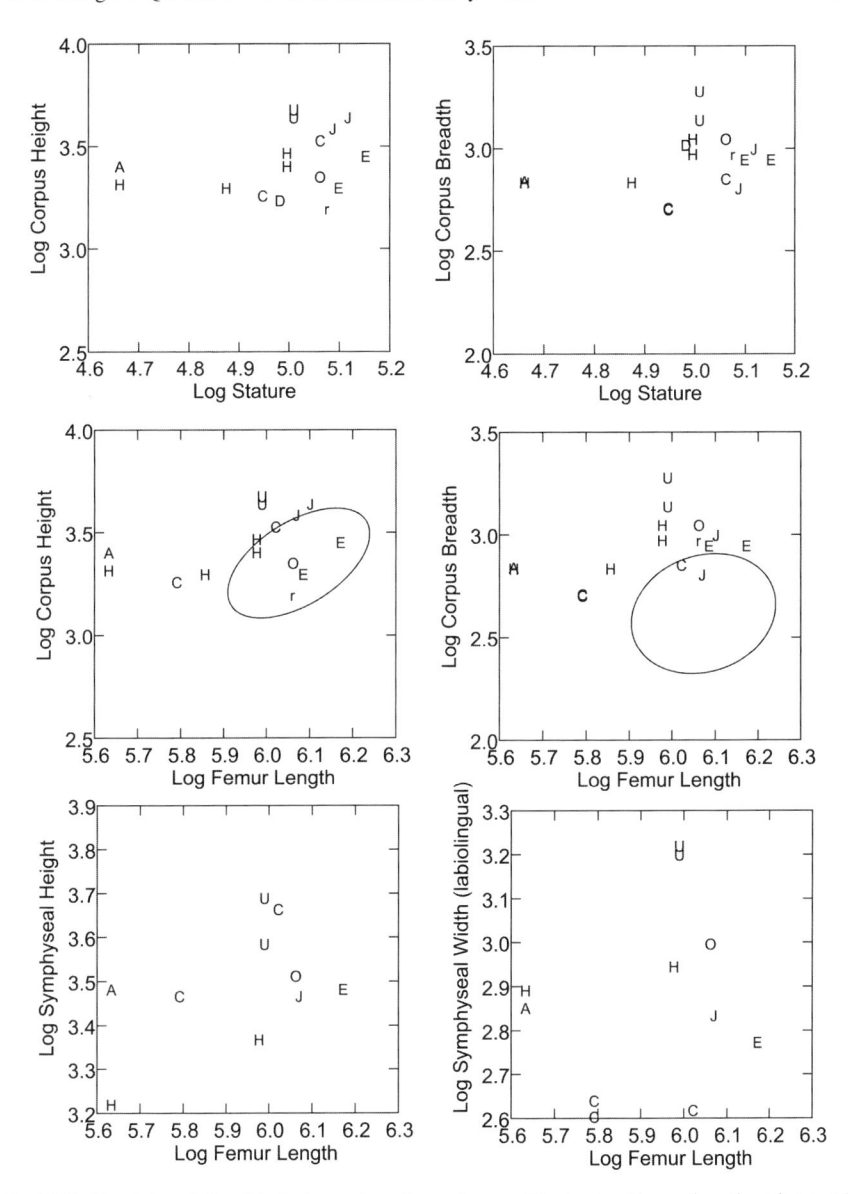

Fig. 20.7 Bivariate relationship between jaw dimensions and body size (femur length and stature). Legend: A = *A. afarensis*, H= *H. habilis*, U = *H. rudolfensis*; *H. erectus* subgroups are C = China, D = Dmanisi, E = Koobi Fora, O = Olduvai, T = Ternifine, J = Javan, r = subadult *H. erectus*. Ellipse represents 95% confidence ellipse for *H. sapiens* sample. Stature in cm, all other dimensions taken in mm and then transformed. Regression statistics reported in Table 20.7

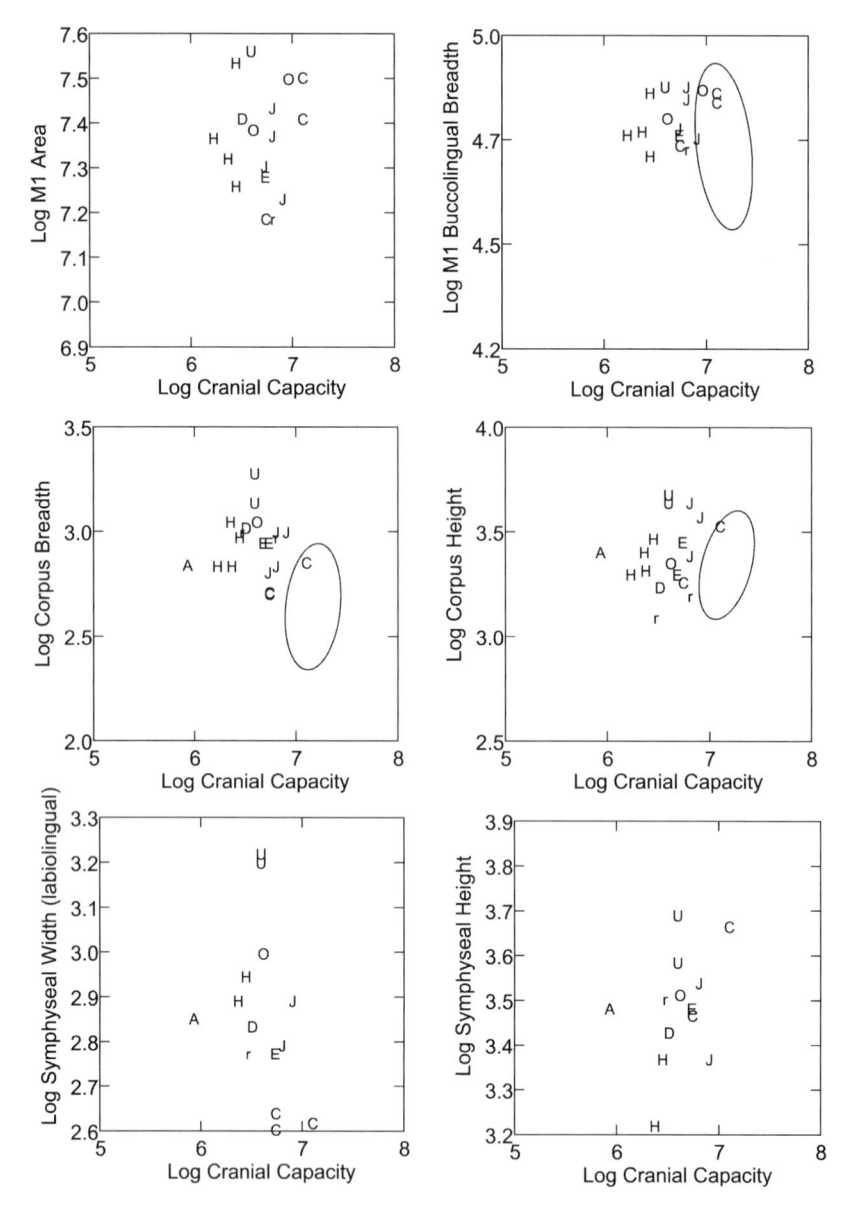

Fig. 20.8 Molar and jaw dimensions relative to cranial capacity. Legend: A = *A. afarensis*, H = *H. habilis*, U = *H. rudolfensis*; *H. erectus* subgroups are C = China, D = Dmanisi, E = Koobi Fora, O = Olduvai, T = Ternifine, J = Javan, r = subadult *H. erectus*. Ellipse represents 95% confidence ellipse for *H. sapiens* sample. Cranial capacity in cc and Molar area after Wood (1991). All other dimensions taken in mm and then transformed. Regression statistics reported in Table 20.8

Table 20.8 Results of ordinary least-squares regression of dentognathic size versus cranial capacity

	n	R^2	OLS slope	P
Cranial Capacity (x)				
M1 mesiodistal (*H. habilis*)	5	0.20	0.17	ns
M1 buccolingual (*H. habilis*)	5	0.32	0.32	ns
Molar area (*H. habilis*)	5	0.29	0.53	ns
M1 mesiodistal (*H. erectus*)	10	0.23	0.17	ns
M1 buccolingual (*H. erectus*)	10	0.25	0.17	ns
Molar area (*H. erectus*)	11	0.12	0.20	ns
M1 buccolingual (*H. sapiens*)	168	0.09	−0.22	<0.00001
Corpus height (*H. habilis*)	6	0.91	1.05	0.003
Corpus breadth (*H. habilis*)	6	0.73	1.02	0.03
Symphyseal height (*H. habilis*)	4	0.95	1.84	0.02
Symphyseal depth (*H. habilis*)	4	0.97	1.48	0.01
Corpus height (*H. erectus*)	10	0.44	0.60	0.04
Corpus breadth (*H. erectus*)	11	0.06	−0.19	ns
Symphyseal height (*H. erectus*)	7	0.28	0.26	ns
Symphyseal depth (*H. erectus*)	8	0.19	−0.34	ns
Corpus height (*H. sapiens*)	201	0.16	0.38	<0.00001
Corpus breadth (*H. sapiens*)	211	0.04	0.20	0.007

& 20.2). Although average body and brain sizes are larger in *H. erectus* than in *H. habilis*, the two overlap substantially in size. This is particularly true if one considers only the nearly coeval African/Georgian *H. erectus* and *H. habilis s.l.* samples (e.g., Vekua et al., 2002). Thus, while average differences between groups may signal something about an increase in the average nutritional needs of a group, it is not clear to me that this requires a wholesale foraging shift between the taxa. Could it not be the case that we are simply sampling different groups with slightly different access to resources rather than completely different foraging strategies related (or not) to taxonomic differences? The data seem particularly equivocal if we then consider that many of the postcranial specimens have previously been assigned to taxa on the basis of size and geological age; thus larger specimens from appropriate time intervals are almost always assigned to *H. erectus* not *H. habilis* (although the two overlap substantially in time, Spoor et al., 2007), and large specimens from early time intervals without definitive cranial remains of *H. erectus* (e.g. femora KNM-ER 1472 and 1481) are considered controversial (e.g., Kennedy, 1983: Trinkaus, 1984). Thus, the aggregation of the samples may be biased to produce a larger body size for *H. erectus*. The discovery of the Dmanisi, Ileret (KNM-ER 42700), and Olorgesailie (KNM-OL 45500) specimens, all of which have very small inferred brain sizes (625–750 cc) but very clear *H. erectus* cranial features (Vekua et al., 2002; Leakey et al., 2003; Potts et al., 2004; Lordkipanidze et al., 2005; Spoor et al., 2005, 2007), indicates that the automatic assignation of large size to *H. erectus* is unwise (Antón et al., 2007). Taken together, additional dentognathic evidence of a foraging shift would be more convincing than are the body and brain size data.

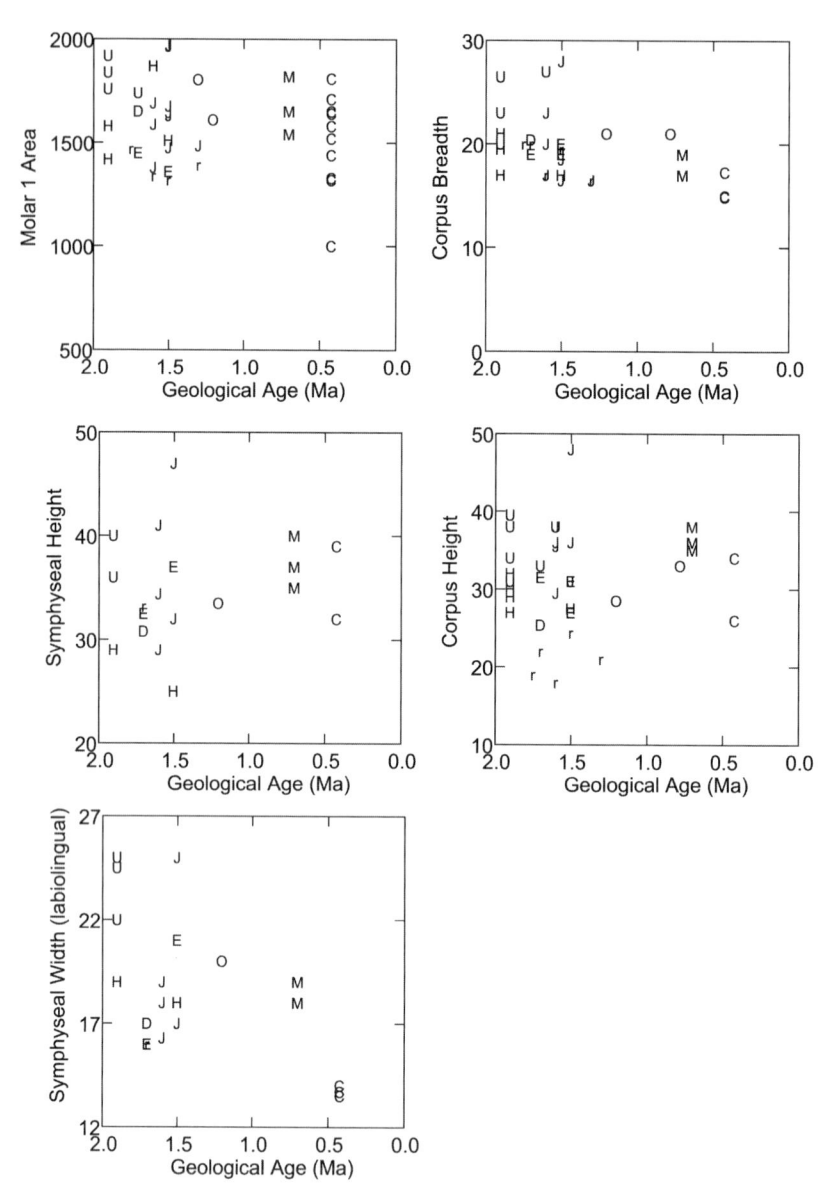

Fig. 20.9 Molar and jaw dimensions through geological time. Legend: A = *A. afarensis*, H= *H. habilis*, U = *H. rudolfensis*; *H. erectus* subgroups are C = China, D = Dmanisi, E = Koobi Fora, O = Olduvai, T = Ternifine, J = Javan, r = subadult *H. erectus*. Ellipse represents 95% confidence ellipse for *H. sapiens* sample. Molar area after Wood (1991). All other dimensions are in mm. Geological age in mega annum (Ma). Regression statistics reported in Table 20.9

Table 20.9 Results of ordinary least-squares regression of dentognathic size and time (in Ma)

	N	R^2	OLS slope	P
Geological Time (x)				
M1 mesiodistal (*H. habilis*)	8	0.06	−9.2	ns
M1 buccolingual (*H. habilis*)	8	0.01	6.2	ns
Molar area (*H. habilis*)	8	0.01	113.9	ns
M1 mesiodistal (*H. erectus*)	30	0.04	3.6	ns
M1 buccolingual (*H. erectus*)	30	0.05	3.1	ns
Molar area (*H. erectus*)	30	0.06	93.9	ns
Corpus height (*H. habilis*)	11	0.005	0.06	ns
Corpus breadth (*H. habilis*)	10	0.001	−0.003	ns
Symphyseal height (*H. habilis*)	5	0.62	0.84	ns
Symphyseal depth (*H. habilis*)	5	0.46	0.56	ns
Corpus height (*H. erectus*)	19	0.01	0.04	ns
Corpus breadth (*H. erectus*)	21	0.22	0.14	0.03
Symphyseal height (*H. erectus*)	14	0.03	−0.05	ns
Symphyseal depth (*H. erectus*)	15	0.23	0.15	ns

However, molar and mandibular dimensions do not differ between groups even on the basis of mean values, although early African *H. erectus* has somewhat shorter and narrower molars, on average, than *H. habilis s.l.* (Table 20.2). Thus, if a dietary shift should be signaled by decreasing molar or mandibular size, then little evidence of such a shift is present between groups of early *Homo*. The one exception to this statement is the diminution of symphyseal depth (labiolingual) and its probable association with decreasing wishboning (Hylander and Johnson, 1994) of the symphysis in *H. erectus*. Given slightly smaller body sizes in *H. habilis*, and the known allometric scaling between mandibular size, wishboning stress, and symphyseal thickness (Vinyard and Ravosa, 1998), the presence of greater symphyseal width in the *smaller* taxon may present a strong dietary signal. Alternatively, it may represent structural differences between the two taxa related to facial and mandibular size in the larger taxa. It should be stressed that even if we divide the sample by time packets and compare between early African *H. erectus* and *H. habilis*, we get similar results (Fig. 20.3). So it is difficult, except possibly in the case of symphyseal depth, to argue for substantial differences between the two samples to support a foraging shift.

There are, however, at least three serious flaws with these data, and thus any interpretations drawn from them. Because of preservation bias, it is impossible to scale any of these dimensions to mandibular length, arguably the most relevant biomechanical size proxy for mastication (e.g., Hylander, 1979, 1985, 1988; Bouvier, 1986; Daegling, 1989, 1992; Ravosa, 1996). Although there is little evidence for substantial length differences between the jaws of *H. habilis* and *H. erectus*, there are also very few jaws – as such, the potential effect on the results is unknown. As has been clearly shown in any number of previous studies, the size proxy to which one scales seriously influences the results of the analyses (e.g., Bouvier, 1986; Smith, 1983). In addition, both mandibular length and jaw breadth influence wishboning stress and thus, symphyseal depth (e.g., Vinyard and Ravosa, 1998); without comparing bicondylar and alveolar breadths between groups, the significance of the decrease in symphyseal depth in *H. erectus* remains obscure. Symphyseal orientation also influences the ability to withstand wishboning stress (e.g., Daegling, 2001). However, due to the fragmentary nature of the specimens, it was impossible to assess orientation differences between the groups.

Second, the characters I consider here represent a limited set of metric information. Although these correlate with diets of particular mechanical quality that influence mandibular loading regimes in extant taxa (e.g., Hylander, 1985) it cannot be certain that the actual diets in early *Homo* so differed. It remains possible, although less likely, that a large shift in diet quality did not involve a shift in mechanical properties. And, as with all measurements, and particularly in this case symphyseal thickness, the external dimensions themselves may obscure morphological differences in cortical thickness or trabecular architecture that may exist between groups (c.f., Daegling, 1989). It should also be noted that, although *H. habilis* s.l. and *H. erectus* s.l. do not seem to differ much in the dimensions considered here, they differ in other dentognathic variables (e.g., Wood, 1991; Gabunia et al., 2000a).

Third, in the scaling analyses relative to body size (femur length and stature) and brain size, the 'individuals' used are chimera. They have been matched for geological age and locality – but the vast majority of them are not truly associated individuals. Arguably, these chimera give us a better appreciation of variation in early *Homo* than does the single associated *H. erectus* skeleton (KNM WT 15000)[4] for which both body size and brain size estimates are available, and they are the only means of assessing intraspecific scaling. However, we must accept that the chimera may be biased, perhaps systematically so, in ways that we do not foresee.

20.4.2 Suggestions About Scaling

Previous studies have suggested that mandibular dimensions should scale with body mass (Smith, 1983; Bouvier, 1986; Hylander, 1988) and cranial capacity (Vinyard and Hanna, 2005) in primates. However, the present data suggest no scaling relationship between most dentognathic variables (symphyseal dimensions, corpus breadth, molar area) and body size. They do, however, suggest a relationship between mandibular height and body size (femur length/stature). The strongest relationships are between jaw dimensions and cranial capacity, which makes intuitive sense as these are likely to track metabolic demands most closely. The difference in results between studies may stem from differences in the body size proxy used. If it is expected that masticatory features, body mass, and cranial capacity all track

---

[4] Although other associated, partial skeletons exist for *H. habilis* (OH 62, KNM-ER 3735) and *H. erectus* (KNM-ER 1808 and 803), these are quite fragmentary and do not preserve both brain and body size estimates. Data for associated Dmanisi specimens were not available when this study was undertaken but were included in Table 20.1 in proof.

metabolic demands, then it is reasonable to expect that they should scale to one another. Alternatively, femur length and stature arguably have a less clear relationship to metabolic demands than does body mass. Femur length and stature may not track metabolic demands, or more likely given secular trends in growth in height that show some relationship between stature and nutrition (e.g., Fogel et al., 1983; Cole, 2000), they may track them less faithfully than does body mass. Thus, lack of scaling to body size in this study may reflect my use of femur length and stature as size proxies rather than body mass. Alternatively, it may reflect my small sample sizes. However, it seems unlikely that the use of chimera would systematically influence postcranial/jaw scaling analyses and not cranial/jaw scaling, although this also remains a possibility.

20.4.3 Trends with Time

Across the early *Homo* lineage, there is little evidence of systematic change in these dentognathic dimensions through time. The geologically youngest specimens included in this analysis, the Zhoukoudian remains, have molars and some jaw dimensions that are nearly as large as all other specimens considered, with the exception of symphyseal height. Only corpus breadth is significantly correlated with geological age.

That said, temporal trends within regional samples may be expected to exist given the different environmental histories of each region. The African sample shows no such tendencies, but is a very small sample. If we only consider African *H. erectus* (the 'Es' of Fig. 20.9), one might even argue for an increase in all dimensions with time, except corpus breadth, which is stable. The similarly aged Indonesian sample from Sangiran, Java, shows a diminution of molar area and corpus breadth with time (cf. Kaifu et al., 2005; Fig. 20.9). However, when using "individual" data points rather than means for time packets (as was the procedure in Kaifu et al., 2005), even these trends are less clear than previously argued. Indonesian trends may speak to regional adaptations or genetic drift through time. Nonetheless, this regional trend is obscured when considering the entire taxon because later *H. erectus*, those at Ternifine and Zhoukoudian, exhibit both larger and smaller sizes than do the Sangiran remains.

Indeed, there may be some justification for considering the middle Pleistocene populations from Ternifine and Zhoudkoudian separately from the older populations and probably from one another. These later populations are likely to represent a different foraging adaptation as the use of fire is more certainly accepted for these individuals than for the early Pleistocene individuals (although this remains contested; Weiner et al., 1998; Goldberg et al., 2001). At least in the case of Ternifine, their stone tool assemblages (Acheulean) are also more advanced (Arambourg and Hofstetter, 1963; Geraads et al., 1986), potentially signaling a different foraging adaptation. Furthermore, the Zhoukoudian cranial sample differs from all other *H. erectus* with regard to posterior vault morphology (Antón, 2002; Kidder and Durband, 2004). This differentiation has been interpreted as the result of some regional isolation (Antón, 2002), and may provide another local reason for expecting a different pattern of variation in masticatory dimensions due either to genetic drift or to different adaptations. Potential differences in foraging pattern are also supported by the extreme symphyseal narrowing, particularly in the Zhoukoudian sample, as well as by their more moderate mandibular height reduction.

Here the importance of framing the right question becomes clear. If we are interested in trends for the entire species – a certain kind of sample needs to be used. If, alternatively, we are concerned with local adaptation and regional trends through time, quite a different sample is needed. Thus, what may seem monolithic, or unchanging, in one sample may be constructed from a fair degree of variation around the mean, variation that represents both idiosyncratic differences among individuals as well as ecological differences between subgroups of that taxon. Furthermore, trends seen in one sample may not be paralleled in others. The implications of such results need to be considered in light of the samples from which they are derived, that is which specimens from which times and regions are included.

20.4.4 Implications for Understanding Early Homo

If we accept that the sample at hand is an adequate first approximation of the processes that differentiate between *H. habilis s.l.* and *H. erectus s.l.*, then these data support the idea that the foraging shift had already substantially occurred in *H. habilis*. An intensification of this shift from *H. habilis s.l.* to *H. erectus s.l.*, may, however, be signaled by the decrease in symphyseal depth in the latter. It is equally as likely that this change relates to other structural changes in the shape or positioning of the face between the taxa, as it remains unclear how big a role wishboning plays in mastication in larger bodied/jawed primates (e.g., Daegling, 2001). That the foraging shift had occurred already in *H. habilis* is supported by those who interpret the fragmentary partial skeletons of *H. habilis* as already showing elongation of the lower limb, which is presumably linked to ranging and foraging strategy (e.g., Haeusler and McHenry, 2004). However, an early foraging shift is contraindicated by those who would prefer to remove *H. habilis* from the genus *Homo* on the basis of foraging adaptation (e.g., Wood and Collard, 1999). It is also supported, at least in part, by the dental topography data (Ungar, 2004). Indeed, the contention of Ungar (2004) that raw underground tubers have mechanical properties consistent with the topography of early *Homo* teeth suggests that such tubers could also be the foods that are relied upon more heavily by *H. habilis* than *H. erectus*. That the difference between *H. habilis s.l.* is one more of degree than of kind is also supported by cranial data that suggest that in many, but not all, cranial variables *H. erectus* and *H. habilis* follow similar scaling relationships and have similar ranges of variation (e.g., Spoor et al., 2005; Antón et al., 2007). *H. erectus* ultimately extends these relationships to absolutely greater sizes overall than does *H. habilis*.

Dietary differences that are not mechanical in nature but that offer different nutritive value may still be implicated between *H. habilis s.l.* and early *H. erectus*.

Experimental evidence suggests that monkeys fed low-protein diets throughout development achieve adult craniofacial shapes, but not sizes, creating, essentially, proportionate dwarves of the neurocranium, but not face (Ramírez-Rozzi et al., 2005). Interestingly, dwarfed animals often have relatively large teeth (Lister, 1996), suggesting that if increased nutrition is responsible for increasing cranial size in *H. erectus* over that seen in *H. habilis*, we should not expect to see an increase in relative tooth size (albeit, absolute tooth size might not be expected to decrease either). The fact that all early *Homo* scale on the same trajectories for most craniofacial and masticatory dimensions, as well as body and brain size relations, suggests that moderate increases in nutritive content could influence adult size dramatically. Such shifts could happen very quickly, between generations, and need not be permanent features of the taxon. We are familiar with similar secular trends in modern humans (e.g., Cole, 2000), including differences in stature that have been documented in *Homo sapiens* over the past 300 years, which are correlated with increases in caloric intake (Fogel et al., 1983; Fogel and Costa, 1997). The reverse argument, local resource stress over several hundred years, could explain the small size of marginal populations, and may be a viable explanation for the small size of the Dmanisi remains.

Such a scenario makes it fundamentally unlikely that a distinct foraging shift between *H. erectus* and *H. habilis* had significant influence on the dispersal ability of *H. erectus* (*contra* Antón et al., 2002). More likely is that a lag between dispersal ability (acquired at the origin of *Homo*) and actual dispersal (undertaken around 1.8 Ma) resulted in the dispersal of *H. erectus* rather than *H. habilis*. Such lags are commonly encountered in mammalian dispersals (Woodburne and Swisher, 1994), both with and without speciation events.

Acknowledgments My very great thanks to Matt, Christine, and Chris for including me in the original AAPA symposium in celebration of Bill. Thanks to previous co-authors and colleagues who have worked on some of these ideas with me: Bill Leonard, Fred Spoor, Connie Fellmann, Carl Swisher, Andrea Taylor. I am grateful to Emma Hite for help with copy-editing. Thanks to the many museums and curators who have facilitated my work around the world. I am grateful to Chris Vinyard for his editorial support and patience. The title of this chapter comes from that thing that Bill tried hard to teach me (not his fault if I never got it right) – "frame the question you want to ask". More importantly, Bill provided me a model for "being" – both in science and in life. At the symposium I came up with my 'Top ten things I learned from Bill', I recap them here, with apologies to ideas I have borrowed. 'Top ten things I learned from Bill': 10. Be careful, there's a file on you somewhere, 9. Squeak! 8. Time flies, 7. Work with people you like, 6. Do not take yourself too seriously, 5. Facial morphology is influenced by the dissipation of masticatory loads, 4. Things always take longer than you think, 3. Everything you know could be wrong, 2. Or conventional wisdom could be wrong, 1. Life is short, enjoy it. Bill Hylander has been a big man in Science; but his influence has been much more profound than this implies. Thanks Bill.

Appendix

Table 20.10 Samples and chimeras used in analyses.[1] Reading down columns yields the fossil samples from which each dimension (variable) was measured. The summary statistics for each dimension (variable) found in the preceding tables were generated from these samples. When a specimen is used more than once in a column, as was sometimes necessitated for the creation of a chimera, the summary statistics include that specimen only once. Reading across rows yields the chimera used in the scaling analyses that consider the relationship between two variables. Thus, when scaling cranial capacity against corpus dimensions, the *A. afarensis* chimera was made by using the capacity estimate for A.L. 162-28 and the corpus dimensions for A.L. 288-1

Specimen for cranial capacity estimate	Specimen for corpus dimensions	Specimen for symphysis dimensions	Specimen for molar dimensions	Specimen for femur length	Specimen for stature estimate
A. afarensis					
A.L. 162-28	A.L. 288-1	A.L. 288-1	A.L. 288-1	A.L. 288-1	A.L. 288-1
—	A.L. 437-1	A.L. 437-1	A.L. 437-1	—	—
A. gahri					
Bou-VP 12/130	—	—	—	Bou-VP 12/1	—
H. habilis s.l. (adult)					
KNM-ER 1470	KNM-ER 1483	KNM-ER 1483	—	KNM-ER 1472	KNM-ER 1472
KNM-ER 1470[2]	KNM-ER 1802	KNM-ER 1802	KNM-ER 1802	KNM-ER 1472	KNM-ER 1472
KNM-ER 1805	KNM-ER 1805	—	—	KNM-ER 1481	KNM-ER 1481
KNM-ER 1813	KNM-ER 1502	—	KNM-ER 1502	KNM-ER 1503	KNM-ER 1503[3]
	KNM-ER 819	—	—	—	—
	KNM-ER 1482	KNM-ER 1482	KNM-ER 1482	—	—
	KNM-ER 1501	—	—	—	—
	KNM-ER 1506	—	KNM-ER 1506	—	—
	KNM-ER 1801	—	KNM-ER 1506	—	—
OH 16	OH 37	OH 37	—	OH 34[3]	OH 34[4]
OH 16	—	—	OH 16	—	—
OH 24	OH 13	OH 13	OH 13	OH 62	OH 62
H. habilis s.l. (juv)					
OH 7					
H. erectus s.l. (adult)					
KNM-ER 3733	KNM-ER 730	KNM-ER 730	KNM-ER 730	KNM-ER 1808	KNM-ER 1808

474   S.C. Antón

20 Framing the Question: Diet and Evolution in Early *Homo*

Table 20.10 (continued)

Specimen for cranial capacity estimate	Specimen for corpus dimensions	Specimen for symphysis dimensions	Specimen for molar dimensions	Specimen for femur length	Specimen for stature estimate
KNM-ER 3883	KNM-ER 731	—	—	KNM-ER 737	KNM-ER 737
—	KNM-ER 992	KNM-ER 992	KNM-ER 992	—	—
OH 9	OH 51	—	OH 51	OH 28	OH 28
OH 12	OH 22	OH 22	OH 22	OH 34[3]	OH 34[3]
—	OH 23	—	—	—	—
Dmanisi 2280	—	—	—	—	Dmanisi 2021
Dmanisi 2282	D211	D211	D211	—	Dmanisi 2021
Sangiran 2[2]	Sangiran SB8103	—	Sangiran SB8103	—	—
Sangiran 4[2]	Sangiran 5	—	Sangiran 5	Trinil II	Trinil II
Sangiran 4[2]	Sangiran 22	Sangiran 22	Sangiran 22	—	—
Sangiran 17[2]	Sangiran 8	Sangiran 8	Sangiran 8	—	—
—	Sangiran 1b	Sangiran 1b	Sangiran 1b	Trinil III	Trinil III
Trinil 2	Sangiran 5	—	Sangiran 5	Trinil II	Trinil II
Trinil 2	—	—	—	Trinil III	Trinil III
—	Sangiran 9	Sangiran 9	Sangiran 9	—	—
—	Sangiran 6	—	Sangiran 6	—	—
—	Sangiran BK 8606	—	Sangiran BK 8606	—	—
—	—	—	Sangiran 7-42	—	—
—	—	—	Sangiran 7-43	—	—
—	—	—	Sangiran 7-61	—	—
—	Sangiran BK 7905	—	Sangiran 7-76	—	—
Zhoukoudian Skull VI	ZKD H1	ZKD H1	—	ZKD Femur I	ZKD Femur I
Zhoukoudian Skull III	—	—	—	ZKD Femur IV	ZKD Femur IV
Zhoukoudian Skull XII	ZKD G1	ZKD G1	ZKD G1	ZKD Femur IV	ZKD Femur IV
Zhoukoudian Skull VI[2]	ZKD A II	ZKD A II	ZKD A2-2	ZKD Femur I	ZKD Femur I
Zhoukoudian Skull XII[2]	—	—	ZKD C3-45	ZKD Femur IV	ZKD Femur IV
—	—	—	ZKD A1-1	—	—
—	—	—	ZKD A3-56	—	—

Table 20.10 (continued)

Specimen for cranial capacity estimate	Specimen for corpus dimensions	Specimen for symphysis dimensions	Specimen for molar dimensions	Specimen for femur length	Specimen for stature estimate
—	—	—	ZKD B1-3	—	—
—	—	—	ZKD B3-9	—	—
—	—	—	ZKD C3-53	—	—
—	ZKD K1	ZKD K1	ZKD F1-5	—	—
—	—	—	ZKD K1-96	—	—
—	—	ZKD M1	ZKD L4-309	—	—
—	—	—	—	—	—
—	—	—	ZKD 02-314	—	—
—	Ternifine 1	Ternifine 1	ZKD PA-87	—	—
—	Ternifine 2	Ternifine 2	Ternifine 1	—	—
—	Ternifine 3	Ternifine 3	Ternifine 2	—	—
—	—	—	Ternifine 3	—	—
Ngandong 6	—	—	—	Ngandong Tibia B	Ngandong Tibia B
Ngandong 7	—	—	—	Ngandong Tibia B	Ngandong Tibia B
H. erectus (subadult)					
KNM-WT 15000	KNM-WT 15000	KNM-WT 15000	KNM-WT 15000	KNM-WT 15000	KNM-WT 15000
—	KNM-ER 820	—	KNM-ER 820	—	—
—	KNM-ER 1507	—	KNM-ER 1507	—	—
Dmanisi 2735	D2735	—	—	—	—
—	Sangiran NG 8503	—	Sangiran NG 8503	—	—

[1] Data from sources in Table 20.3 and Arambourg, 1954; Kaifu et al., 2005; Rightmire, 1993; Weidenreich, 1936, 1937, 1941.
[2] Chimeras not used in establishing capacity/body size relationships.
[3] KNM-ER 1503 is assigned alternatively to *H. habilis* or *A. boisei*, as are most specimens of this age at Koobi Fora, except those of relatively large size (McHenry, 1991). In order not to unduly skew *H. habilis* results to large size, this specimen, which is near the mean of femoral length values of those specimens classified by McHenry (1991) as either *H. habilis* or *A. boisei*, is used.
[4] The taxonomic status and length of OH 34 is contested. Haeusler and McHenry (2004) estimate the length between 375 and 395 and as a member of earliest *Homo*. However, an estimated length of 430 mm and attribution to *H. erectus* have also been proposed (Day and Molleson, 1976; Howell, 1978). The two estimates and attributions are used here.

References

Aiello, L., Wheeler, P. (1995). The expensive tissue hypothesis: the brain and digestive system in human and primate evolution. *Current Anthro.* 36:199–221.
Antón, S.C. (1996). Cranial adaptation to a high attrition diet in Japanese macaques. *Int. J. Primatol.* 17:401–427.
Antón, S.C. (2002). Evolutionary significance of cranial variation in Asian *Homo erectus*. *Am. J. Phys. Anthropo.* 118:301–323.
Antón, S.C., Aziz, F., Zaim, Y. (2001). Dispersal and migration in Plio-Pleistocene *Homo*. In: Tobias, P.V., Raath, M.A., Moggi-Cecchi, J., Doyle, G.A. (eds.), *Humanity from African Naissance to Coming Millennia – Colloquia in Human Biology and Palaeoanthropology*. Florence University Press, Florence, Italy, pp. 97–108.
Antón, S.C., Leonard, W.R., Robertson, M. (2002). An ecomorphological model of the initial hominid dispersal from Africa. *J. Hum. Evol.* 43:773–785.
Antón, S.C., Spoor, F., Fellmann, C., Swisher, C.C. III. (2007). Chapter 11: Defining *Homo erectus*: size considered. In: Henke, R., Tattersall, I. (eds.), *Handbook of Paleoanthropology Volume III*, Springer-Verlag, Berlin, Germany, pp. 1655–1693.
Antón, S.C., Swisher, C.C. III. (2001). Evolution and variation of cranial capacity in Asian *Homo erectus*. In: Indriati, E. (ed.), *A Scientific Life: Papers in Honor of Professor Dr. Teuku Jacob*. Bigraf Publishing, Yogyakarta, Indonesia, pp. 25–39.
Antón, S.C., Swisher, C.C. III. (2004). Early dispersal of *Homo* from Africa. *Ann. Rev. Anthropol.* 33:271–296.
Arambourg, C. (1954). L'hominien fossile de Ternifine (Algérie). *Compte Rendu Hebdom. Séances L'Academie Sci* 239:893–895.
Arambourg, C., Hofstetter, R. (1963). Le Gisement de Ternifine. *Mem. Inst Pal. Hum.* No. 32.
Asfaw, B., White, T., Lovejoy, O., Latimer, B., Simpson, S., Suwa, S. (1999). *Australopithecus garhi*; a new species of early hominid from Ethiopia. *Science* 284:629–635.
Bouvier, M. (1986). A biomechanical analysis of mandibular scaling in Old World monkeys. *Am. J. Phys. Anthropol.* 69:473–482.
Brown, F.H., McDougall, I. (1993). Geological setting and age. In: Walker, A., Leakey, R. (eds.), *The Nariokotome Homo Erectus Skeleton*. Harvard University Press, Cambridge, MA, USA, pp. 9–20.
Brown, P. (2005) http://www-personal.une.edu.au/~pbrown3/resource.htm, accessed February 8, 2005.
Cole, T.J. (2000). Secular trends in growth. *Proc. Nutr. Soc.* 59:317–324.
Daegling, D.J. (1989). Biomechanics of cross-sectional size and shape in the hominoid mandibular corpus. *Am. J. Phys. Anthropol.* 80:91–106.
Daegling, D.J. (1992). Mandibular morphology and diet in the genus *Cebus*. *Int. J. Primatol.* 13:545–570.
Daegling, D.J. (2001). Biomechanical scaling of the hominoid mandibular symphysis. *J. Morphol.* 250:12–23.
Day, M.H., Molleson, T.I. (1976). The puzzle from JK2 – A femur and a tibial fragment (OH 34) from Olduvai Gorge, Tanzania. *J. Hum. Evol.* 5:455–465.
Dobson, S.D., Trinkaus, E. (2002). Cross-sectional geometry and morphology of the mandibular symphysis in Middle and Late Pleistocene *Homo*. *J. Hum. Evol.* 43:67–87.
DuBrul, E.L. (1977). Early hominid feeding mechanisms. *Am. J. Phys. Anthropol.* 47: 305–320.
Feibel, C.S., Brown, F.H., McDougall, I. (1989). Stratigraphic context of fossil hominids from the Omo group deposits: Northern Turkana Basin, Kenya and Ethiopia. *Am. J. Phys. Anthropol.* 78:595–622.
Feldesman, M.R., Kleckner, J.G., Lundy, J.K. (1990). Femur/stature ratio and estimates of Mid- and Late Pleistocene Fossil Hominids. *Am. J. Phys. Anthropol.* 83:359–372.
Fellmann, C.D. (2004). *Estimation of Femoral Length and Stature in Modern and Fossil Hominins: A Test of Methods*. MA Thesis, Rutgers University.

Frisancho, A.R. (1978). Nutritional Influences on Human Growth and Maturation. *Yearbook Phys. Anthropol.* 21:171–191.
Fogel, R.W., Costa, D.L. (1997). A theory of technophysio evolution, with some implications for forecasting population, health care costs, and pension costs. *Demography* 34:49–66.
Fogel, R.W., Engerman, S.L., Floud, R., Friedman, G., Margo, R.A., Sokologg, K., Steckel, R.H., Trussel, T.J., Villaflor, G., Wachter, K.W. (1983). Secular changes in American and British stature and nutrition. *J. Interdiscipl. Hist.* 14:445–481.
Foley, R.A., Lee, P.C. (1991). Ecology and energetics of encephalization in hominid evolution. *Phil. Trans., Soc. London B.* 334:223–232.
Gabunia, L., Vekua, A., Lordkipanidze, D., Swisher, C.C., Ferring, R., Justus, A., Nioradze, M., Tvalchrelidze, M., Antón, S.C., Bosinski, G., Joris, O., de Lumley, M-A., Majsuradze, G., Mouskhelishvili, A. (2000a). Earliest Pleistocene cranial remains from Dmanisi, Republic of Georgia: taxonomy, geological setting, and age. *Science* 288:1019–1025.
Gabunia, L., de Lumley, M-A., Berillon, G. (2000b). Morphologie et fonction du troisième métatarsien de Dmanissi, Géorgie orientale. In: Lordkipanidze, D., Bar-Yosef, O., Otte, M. (eds.), *Early Humans at the Gates of Europe*. L'Etudes et Recherches Archéologiques de l'Université de Liege 92. Liège, Belgium, pp. 29–41.
Geraads, D., Hublin, J-J., Jaeger, J-J., Haiyan, T., Sevketm S., Toubeau, P. (1986). The leistocene hominid site of Ternifine, Algeria: new results on the environment, age, and human industries. *Quat. Res.* 25:380–386.
Gingerich, P.D., Smith, B.H. (1985). Allometric scaling in the dentition of primates and nsectivores. In: Jungers, W.L. (ed.), *Size and Scaling in Primate Biology*, Plenum Press, New York, USA, pp. 257–272.
Gingerich, P.D., Smith, B.H., Rosenberg, K. (1982). Allometric scaling in the dentition of primates and prediction of body weight from tooth size in fossils. *Am. J. Phys. Anthropol.* 58:81–100.
Goldberg, P., Weiner, S., Bar-Yosef, O., Xu, Q., Liu, J. (2001). Site formation processes of Zhoukoudian, China. *J. Hum. Evol.* 41:483–530.
Gould, S.J. (1977). *Ontogeny and Phylogeny*. Harvard University Press, Cambridge, MA USA.
Grün, R., Huang, P.H., Wu, X., Stringer, C.B., Thorne, A.G., McCulloch, M. (1997). ESR analysis of teeth from the paleoanthropological site of Zhoukoudian, China. *J. Hum. Evol.* 32: 83–91.
Harestad, A.S., Bunnell, F.L. (1979). Home range and body weight: a re-evaluation. *Ecology* 60:389–402.
Harvey, P.H., Clutton-Brock, T.H.(1985). Life history variation in primates. *Evolution* 39:559–581.
Hatley, T., Kappelman, J. (1980). Bears, pigs, and Plio-Pleistocene hominids: a case for the exploitation of below-ground food resources. *Hum. Ecol.* 8:371–387.
Haeusler, M., McHenry, H.M. (2004). Body proportions of *H. habilis* reviewed. *J. Hum. Evol.* 46:433–465.
Henneberg, M. (1987). Hominid cranial capacity change through time: a Darwinian process. *Hum. Evol.* 2:213–220.
Henneberg, M. (2001). The gradual eurytopic evolution of humans not from Africa alone. In: Indriati, E., (ed.), *A Scientific Life: Papers in Honor of Professor Dr. Teuku Jacob*. Bigraf Publishing, Yogyakarta, Indonesia, pp. 41–52.
Holloway, R. (1980). Indonesian 'Solo' (Ngandong) endocranial reconstruction: preliminary observations and comparison with Neanderthal and *Homo erectus* groups. *Am. J. Phys. Anthropol.* 53:285–295.
Holloway, R. (1981). The Indonesian *Homo erectus* brain endocasts revisited. *Am. J. Phys. Anthropol.* 55:385–393.
Howell, F.C. (1978). Hominidae. In: Maglio, V.J., Cooke, H.B.S. (eds.), *Evolution of African Mammals*. Harvard University Press, Cambridge, MA USA, pp. 154–248.
Huang, P., Jin, S., Liang, R., Lu, Z., Zheng, L., Yuan, Z., Cai, B., Fang, Z. (1991a). Study of ESR dating for burying age of the first skull of Peking man and chronological scale of the cave deposit in Zhoukoudian site loc. 1. *Acta Anth. Sin.* 10:107–115. (Chinese with English abstract).

Huang, P., Jin, S., Liang, R., Lu, Z., Zheng, L., Yuan, Z., Cai, B., Fang, Z. (1991b). Study of ESR dating for the burial age of the first skull of Peking Man and chronological scale of the site. *Chinese Sci. Bull.* 36:1457–1461. (Chinese).

Hylander, W.L. (1972). *The Adaptive Significance of Eskimo Craniofacial Morphology.* Ph.D. Dissertation, Department of Anthropology, University of Chicago, Chicago, IL.

Hylander, W.L. (1977). Morphological changes in human teeth and jaws in a high-attrition environment. In: Dahlberg, A.A., Graber, T.M. (eds.) *Orofacial Growth and Development.* Mouton Publishers, Paris, France, pp. 301–330.

Hylander, W.L. (1979). The functional significance of primate mandibular form. *J. Morphol.* 160:223–240.

Hylander, W.L. (1985). Mandibular function and biomechanical stress and scaling. *Am. Zool.* 25:315–330.

Hylander, W.L. (1988). Implications of in vivo experiments for interpreting the significance of 'robust' australopithecine jaws. In Grine, F.E. (ed.), *Evolutionary History of the Robust Australopithecines.* Aldine de Gruyter, New York, USA, pp. 55–83.

Hylander, W.L., Johnson, K.R. (1994). Jaw muscle function and wishboning of the mandible during mastication in macaques and baboons. *Am. J. Phys. Anthropol.* 94:523–548.

Kaifu, Y., Baba, H., Aziz, F., Indriati, E., Jacob, T. (2005). Taxonomic affinities and evolutionary history of the early Pleistocene homininds of Java: dentognathic evidence. *Am. J. Phys. Anthopol.* 128:709–726.

Kay, R.F. (1975). The functional adaptations of primate molar teeth. *Am. J. Phys. Anthropol.* 43:195–216.

Kay, R.F. (1985). Dental Evidence for the Diet of *Australopithecus.* Annual Review of Anthropology, 14:315–341.

Kay, R.F., Hylander, W.L. (1978). The dental structure of mammalian folivores with special reference to primates and Phalangeroidea (Marsupialia). In: Montgomery, G.G., (ed.), *The Ecology of Arboreal Folivores.* Smithsonian Institution Press, Washington, DC USA, pp. 173–191.

Kennedy, G.E. (1983). A morphometric and taxonomic assessment of a hominine femur from the lower member Koobi Fora, Lake Turkana. *Am. J. Phys. Anthropol.* 61:429–431.

Kidder, J.H., Durband, A.C. (2004). A re-evaluation of the metric diversity within *Homo erectus.* *J. Hum. Evol.* 46:297–313.

Kinzey, W.G. (1974). Ceboid models for the evolution of hominoid dentition. *J. Hum. Evol.* 3:191–203.

Koganezawa, M. (1975). Food habits of Japanese monkey (*Macaca fuscata*) in the Boso mountains. In: Kondo, S., Kawai, M., Ehara, A. (eds.), *Contemporary Primatology.* 5th International Congress of Primatology, Nagoya 1974. Karger, Basel, pp. 380–383.

Laden, G., Wrangham, R. (2005). The rise of the hominids as an adaptive shift in fallback foods: plant underground storage organs (USOs) and australopith origins *J. Hum. Evol.* 49:482–498.

Leakey, M.G., Spoor, F., Brown, F.H., Gathogo, P.N., Leakey, L.N. (2003). A new hominin calvaria from Ileret (Kenya). *Am. J. Phys. Anthropol. Suppl.* 36:136.

Leigh, S.R. (1992). Cranial capacity evolution in *Homo erectus* and early *Homo sapiens. Am. J. Phys. Anthropol.* 87:1–13.

Leonard, W.R., Robertson, M.L. (1992). Nutritional requirements and human evolution: a bioenergetics model. *Am. J. Hum. Biol.* 4:179–195.

Leonard, W.R., Robertson, M.L. (1994). Evolutionary perspectives on human nutrition: the influence of brain and body size on diet and metabolism. *Am. J. Hum. Biol.* 6:77–88.

Leonard, W.R., Robertson, M.L. (1997). Comparative primate energetics and hominid evolution. *Am. J. Phys. Anthropol.* 102:265–281.

Leonard, W.R., Robertson, M.L., Snodgrass, J.J., Kuzawa, C.W. (2003). Metabolic correlates of hominid brain expansion. *Comp. Biochem. Physiol. A (Mol. Integr. Physiol.)* 136:5–15.

Lister, A.M. (1996). Dwarfing in island elephants: processes in relation to time of isolation. *Symp. Zool. Soc. Lond.* 69:277–292.

Lordkipanidze, D., Vekua, A., Ferring, R., Rightmire, G.P., Agusti, J., Kiladze, G. et al. (2005). The earliest toothless hominin skull. *Nature* 434:717–718.

Lordkipanidze, D., Jashashvili, T., Vekua, A., Ponce de León, M.S., Zollikofer, C.P.E., Rightmire, G.P., Pontzer, H., Ferring, R., Oms, O., Tappen, M., Bukhsianidze, M., Agusti, J., Kahlke, R., Kiladze, G., Martinez-Navarro, B., Mouskhelishvili, A., Nioradze, M. Lorenzo Rook, L. (2007). Postoranial evidence from early Homo from Dmanisi, Georgia, *Nature* 449:305–310.

McHenry, H.M. (1991). Femoral lengths and stature in Plio-Pleistocene hominids. *Am. J. Phys. Anthropol.* 85:149–158.

McHenry, H.M. (1992). Body size and proportions in early hominids.*Am. J. Phys. Anthropol.* 87:407–431.

McHenry, H.M. (1994). Behavioral ecological implications of early hominid body size. *J. Hum. Evol.* 27:77–87.

McHenry, H.M., Coffing, K. (2000). *Australopithecus* to *Homo*: transformations in body and mind. *Ann. Rev. Anthropol.* 29:125–146.

McNab, B.K. (1963). Bioenergetics and the determination of home range size. *Am. Nat.* 97:133–140.

Milton, K., May, M.L. (1976). Body weight, diet and home range in primates. *Nature* 259:459–462.

Nunn, C.L., Barton, R.A. (2000). Allometric slopes and independent contrasts: a comparative test of Kleiber's law in primate ranging patterns. *Am. Nat.* 156:519–533.

O'Connell, J.F., Hawkes, K., Blurton Jones, N.G.B. (1999). Grandmothering and the evolution of *Homo erectus*. *J. Hum. Evol.* 36:461–485.

Potts, R., Behrensmeyer, A.K., Deino, A., Ditchfield, P., Clark, J. (2004). Small mid-Pleistocene hominin associated with East African acheulean technology. *Science* 305:75–78.

Ramírez-Rozzi, F.V., González-José, R., Pucciarelli, H.M. (2005). Cranial growth in normal and low protein fed Saimiri. An environmental heterochrony. *J. Hum. Evol.* 49:515–535.

Ravosa, M.J. (1990). Functional assessment of subfamily variation in maxillomandibular morphology among old world monkeys. *Am. J. Phys. Anthropol.* 82:199–212.

Ravosa, M.J. (1991). Structural allometry of the prosimian mandibular corpus and symphysis. *J. Hum. Evol.* 20:3–20.

Ravosa, M.J. (1992) Allometry and heterochrony in extant and extinct Malagasy primates. *J. Hum. Evol.* 23:197–217.

Ravosa, M.J. (1996). Jaw morphology and function in living and fossil Old World monkeys. *Int. J. Primatol.* 17:909–932.

Richmond, B.G., Aiello, L.C., Wood, B.A. (2002). Early hominin limb proportions. *J. Hum. Evol.* 43:529–548.

Rightmire, G.P. (1993). *The Evolution of* Homo erectus, *Comparative Anatomical Studies of an Extinct Human Species.* Cambridge University Press, New York, USA.

Rightmire, G.P. (2004). Brain size and encephalization in early to Mid-Pleistocene *Homo*. *Am. J. Phys. Anthropol.* 124:109–123.

Robinson, B.W., Wilson, D.S. (1998). Optimal foraging, specialization and a solution to Liem's paradox. *Am. Nat.* 151:223–235.

Ruff, C., Walker, A. (1993). Body size and body shape. In: Walker, A. and Leakey, R. (eds.), *The Nariokotome Homo Erectus Skeleton.* Harvard University Press, Cambridge, MA USA, pp. 234–265.

Ruff, C.B. (2002). Variation in human body size and shape. *Ann. Rev. Anthropol.* 31:211–232.

Ruff, C.B., Trinkaus, E., Holliday, T.W. (1997). Body mass and encephalization in Pleistocene *Homo*. Supplementary information to *Nature* 387:173–176.

Ryan, A.S., Johanson, D.C. (1989). Anterior dental microwear in *Australopithecus afarensis*. *J. Hum. Evol.* 18:235–268.

Santa Luca, A.P. (1980). The Ngandong fossil hominids: a comparative study of a far eastern *Homo erectus* group. *Yale Univ. Publ. Anthropol.* 78:1–175.

Shen, G., Gahleb, B., Yuan, Z. (1996). Preliminary results on U-series dating of Peking man site with high precision TIMS. *Acta Anth. Sin.* 15:211–217. (Chinese with English abstract/tables).

Shipman, P., Walker, A. (1989). The costs of becoming a predator. *J. Hum. Evol.* 18:373–392.
Smith, R.J. (1983). The mandibular corpus of female primates: taxonomic, dietary and allometric correlates of interspecific variations in size and shape. *Am. J. Phys. Anthropol.* 61:315–330.
Sponheimer, M., de Ruiter, D., Lee-Thorp, J-L., Späth, A. (2005). Sr/Ca and early hominin diets revisited: new data from modern and fossil tooth enamel. *J. Hum. Evol.* 48:147–156.
Spoor, F., Leakey, M.G., Leakey, L.N. (2005). A comparative analysis of the KNM-ER 42700 hominin calvaria from Ileret (Kenya). *Am. J. Phys. Anthropol. Suppl.* 40:195–196.
Spoor, F., Leakey, M.G., Gathogo, P.N., Brown, F.H., Antón, S.C., McDougall, I., Kiarie, C., Manthi, F.K., and Leakey, L.G. Implications of new early *Homo* fossils from Ileret, East Lake Turkana (Kenya). *Nature* 448:688–691.
Steele, D.G., McKern, T.M. (1969). A method for assessment of maximum long bone length and living stature from fragmentary long bones. *Am. J. Phys. Anthropol.* 31:215–226.
Suzuki, A. (1965). An ecological study of wild Japanese monkeys in snowy areas – focused on their food habits. *Primates* 6:31–72.
Swisher, C.C. III, Curtis, G.H., Jacob, T., Getty, A.G., Suprijo, A., Widiasmoro. (1994). Age of the earliest known hominids in Java, Indonesia. *Science* 263:118–1121.
Swisher, C.C. III, Rink, W.J., Antón, S.C., Schwarcz, H.P., Curtis, G.H., Suprijo, A., Widiasmoro. (1996). Latest *Homo erectus*, in Java: potential contemporaneity with *Homo sapiens* in Southeast Asia. *Science* 274:1870–1874.
Taylor, A. (2002). Masticatory form and function in the African great apes. *Am. J. Phys. Anthropol.* 117:133–156.
Teaford, M., Ungar, P. (2000). Diet and the evolution of the earliest human ancestors. *Proc. Natl. Acad. Sci.* 97:13506–13511.
Tobias, P.V. (1991). *Olduvai Gorge Volume 4: The Skulls, Endocasts and Teeth of H. habilis*. Cambridge University Press, New York, USA.
Trinkaus, E. (1984). Does KNM-ER 1481 establish *Homo erectus* at 2.0 my B.P.? *Am. J. Phys. Anthropol.* 64:137–139.
Trotter, M., Gleser, G.C. (1952). Estimation of stature from long bones of American and whites and Negroes. *Am. J. Phys. Anthropol.* 10:463–514.
Trotter, M., Gleser, G.C. (1958). A re-evaluation of estimation of stature based on measurements of stature taken during life and of long bones after death. *Am. J. Phys. Anthropol.* 16:79–123.
Ungar, P. (2004). Dental topography and diets of *Australopithecus afarensis* and early *Homo*. *J. Hum. Evol.*, 46:605–622.
Vekua, A., Lordkipanidze, D., Rightmire, G.P., Agusti, J., Ferring, R., Majsuradze, G., Mouskhelishvili, A., Nioradze, M., Ponce de Leon, M., Tappen, M., Tvalchrelidze, M., Zollikofer, C. (2002). A new skull of early *Homo* from Dmanisi Georgia. *Science* 297:85–89.
Vinyard, C.J., Hanna, J. (2005). Molar scaling in strepsirrhine primates. *J. Hum. Evol.* 49:241–269.
Vinyard, C.J., Ravosa, M. (1998). Ontogeny, function, and scaling of the mandibular symphysis in papionin primates. *J. Morph.* 235:157–175.
Walker, A., Leakey, R., (eds.). (1993). *The Nariokotome* Homo erectus *Skeleton*. Harvard University Press, Cambridge, MA USA.
Weidenreich, F. (1936). The mandibles of *Sinanthropus pekinensis*: a comparative study. *Palaeontologia Sin Ser D* 7:1–164.
Weidenreich, F. (1937). The dentition of *Sinanthropus pekinensis*: a comparative odontography of the hominids. *Palaeontologia Sin Ser D* 1:1–180.
Weidenreich, F. (1941). The extremity bones of *Sinanthropus pekinensis*. *Palaeontologia Sin., Ser. D* 5:1–150.
Weidenreich, F. (1943). The skull of *Sinanthropus pekinensis*; a comparative study on a primitive hominid skull. *Palaeont. Sin Ser D* 10:1–298.
Weiner, S., Xu, Q., Goldberg, P., Liu, J., Bar-Yosef, O. (1998). Evidence for the use of fire at Zhoukoudian, China. *Science* 281:251–253.
Wolpoff, M.H. (1984). Evolution in *Homo erectus*: the question of stasis. *Paleobiology* 10:389–406.

Wood, B. (1991). *Koobi Fora Research Project Volume 4, Hominid Cranial Remains.* Clarendon Press, Oxford, UK.
Wood, B., Abbott, S. (1983). Analysis of dental morphology of Plio-Pleistocene homininds. II. Mandibular molars – crown area measurements and morphological traits. *J. Anat.* 136:197–219.
Wood, B., Abbott, S., Graham, S.H. (1983). Analysis of dental morphology of Plio-Pleistocene homininds. II. Mandibular molars – study of cusp areas, fissure pattern and cross sectional shape of the crown. *J. Anat.* 137:287–314.
Wood, B., Aiello, L. (1998). Taxonomic and functional implications of mandibular scaling in early hominins. *Am. J. Phys. Anthropol.* 105:523–538.
Wood, B., Collard, M. (1999). The human genus. *Science* 284:65–71.
Wood, B.A., Xu, Q. (1991). Variation in the Lufeng dental remains. *J. Hum. Evol.* 20:291–311.
Woodburne, M.O., Swisher, C.C. III (1994). Land mammal high-resolution geochronology, intercontinental overland dispersals, sea level, climate and vicariance. In: Berggren, W.A., Kent, D.V., Aubry, M.-P., Hardenbol, J. (eds.), *Geochronology, Time Scales and Global Stratigraphic Correlations*, Society of Economic Paleontologists and Mineralogists, Special Publication, 54, pp. 335–364.
Wrangham, R.W., Jones, J.H., Laden, G., Pilbeam, D., Conklin-Brittain, N. (1999). The raw and the stolen: cooking and the ecology of human origins. *Curr. Anthropol.* 5:567–594.
Wright, L.E., Vásquez, M.A. (2003). Estimating the length of incomplete long bones: forensic standards from Guatemala. Am. J. Phys. Anthropol. 120:233–251.

Index

A

Acheulean, 471
Acoustic microscope, 266, 286
Actin, 243
Adapid, 413
Adapis, 421
Adaptation, 8, 9, 126, 168, 174, 184, 190, 193, 237, 242, 281, 285–287, 294, 297, 321, 352, 378, 379, 419, 421, 422, 425, 447, 471, 472
Adaptive plasticity, 294–297, 302, 305, 308, 313, 318, 319, 320
Adaptive radiation, 244–245, 421
Adductor force, 234, 297, 298, 300, 330, 335
Adductor muscles, 20, 21, 84, 86, 87, 92, 99, 101, 105, 106, 108, 114, 115–116, 218, 222, 225, 227, 230, 231, 234, 236, 330
Adenosine triphosphate (ATP), 254–255
Aegyptopithecus zeuxis, 410–411, 412, 414, 416
Alcelaphus buselaphus, 57
Alisphenoid, 93, 423
Alkaline phosphatase, 296
Allometry, 202, 212, 213, 344, 360, 363, 368, 370
Alpaca, 17, 41, 42, 43, 44, 45, 46, 47, 48, 50, 51, 52, 54, 55, 56, 58, 59, 86, 87, 107
Alveolar process, 55, 131, 133, 136, 137, 138, 141, 142, 144, 145, 160, 165
Alveolar region, 190
Alveolus, 129, 131, 132, 133, 135, 136, 142, 143, 145, 154, 180
American opossum (*Didelphis virginiana*), 84
Amphipithecidae, 425
Analysis of covariance (ANCOVA), 203, 204, 206, 209, 214, 359
Angiosperm, 421–422

Angle of pinnation, 202, 203, 210,220, 224, 229, 244, 248, 250, 251
Anisognathy, 107
Anisotropy, 266, 267, 268, 281, 282, 285–287
Anterior pillar, 126, 174, 175, 190
Anterior temporalis, 18, 66, 84, 93, 100, 113–122, 177, 178, 246, 247, 250, 330, 433
Anthropoid, 20, 35, 40, 41, 67, 76, 79, 214, 264, 407–426
Anthropoidea, 408, 409, 410, 413, 416, 418, 420, 421, 423
Anthropoid origin, 67, 76, 79, 264, 410, 418, 424
Antilocapra americana, 57
Aotus, 66, 67, 77, 79, 205
Apidium, 411, 414, 417, 418, 419, 424
Apidium bowni, 414, 417
Apidium phiomense, 414, 418
Arboreal, 339, 408, 410, 418–420, 421, 422, 423, 425, 426
Arctocebus, 421
Argument from design, 5
Arsinoea, 411, 414, 417, 425
Arsinoea kallimos, 414, 417, 425
Articular cartilage, 295, 296, 297, 300, 301, 302, 305, 306, 307, 308, 310, 311, 312, 315, 316, 317, 318, 319
Articular disc, 91, 94, 295, 301
Articular eminence, 178, 179, 439, 440
Artiodactyl, 57, 105
Artiodactyla, 330
Ascending ramus, 91, 93, 96–97
Australopithecine, 79, 125, 126, 174, 175, 264, 437, 443–444
Australopithecus africanus, 174, 435, 436
Australopithecus boisei, 431, 432, 433, 435, 439, 452, 476
Australopithecus robustus, 431

Autocorrelation, 341, 342, 348
Avahi laniger, 219, 226, 394

B

Baboon, 7, 18, 40, 63, 66, 77, 79, 113–122, 284
Balancing side, 11, 20, 21, 22, 26, 32, 40, 41, 42, 53, 54, 55, 56, 59, 65, 66, 77, 78, 79, 83, 85, 86, 87, 90, 92, 98, 100, 102, 103, 104, 105, 106, 107, 108, 114, 115, 119, 120, 121, 144, 166, 177, 182, 183, 186, 330, 434
Balancing-side deep masseter, 11, 40, 41, 42, 55, 56, 59, 83, 85, 87, 105, 177
Bamboo, 233, 234
Basicranium, 179
Bat, 335, 336, 408
Beam, 6, 17, 40, 41, 52, 55–56, 64, 78, 125, 333, 340
Beam theory, 6, 55–56
Bend, 21, 78
Bending, 19, 21, 22, 23, 31, 32, 35, 40, 41, 43, 44, 45–46, 47, 51–52, 55, 76, 77, 79, 87, 107, 128, 131, 132, 134, 135, 136, 139, 143, 144, 160, 167, 175, 185, 186, 221, 236, 263, 283, 330, 331, 332, 333, 335, 340, 341, 345, 348, 350, 369, 370, 388, 392, 397, 411–412
Beza Mahafaly, 389, 390
Bilobate, 88
Biological role, 294, 320, 379
Biomechanical scaling, 218, 233
Biomechanical standard, 357, 358, 368, 369, 370, 371, 372, 374, 375, 379
Biomechanics, 7, 9, 10, 11, 150, 151, 168, 174, 181, 193, 194–195, 282, 313
Biplanar radiography, 340
Biretia megalopsis, 420
Bite, 1, 126, 191, 192, 193, 201, 202, 212, 213, 214, 217, 221, 295, 330, 331, 332, 390
Bite force, 6, 7, 10, 18, 41, 53, 54, 64, 65, 66, 74, 77, 78, 79, 106, 108, 114, 115, 121, 144, 152, 153, 174, 178, 179, 193, 195, 202, 212, 233, 253, 256, 299, 340, 362, 433, 453
Bite point, 76, 126, 174, 175, 176, 178, 179, 181, 185, 186, 187, 188, 189, 190, 193, 195, 201, 362
Biting, 9, 17, 64, 65, 66, 78, 132, 133, 145, 153, 168, 178, 181, 199, 233, 245, 258, 298, 306, 307, 308, 311, 312, 315, 316, 317, 320, 321, 329, 330, 331, 345, 350, 361, 362, 363, 368, 371, 373, 375, 409, 440

Body mass, 10, 199, 202, 204, 206, 209, 211, 212, 213, 214, 218, 220, 221, 225, 226, 227, 229, 231, 233, 234, 235–236, 238, 299, 335, 337, 338, 339, 369, 370, 450, 454, 470, 471
Body size, 20, 199, 202, 212, 213, 214, 218, 221, 233, 234, 236, 237, 368, 370, 423, 445, 446, 447, 448, 450, 452, 453, 454, 455, 456, 459, 460, 463, 464, 465, 467, 469, 470, 471, 476
Body weight, 202–203, 204, 206, 207, 211, 226, 252, 394, 450
Bolus, 21, 25, 32, 237, 433, 434, 437
Bone mass, 8, 9, 10, 129, 184, 286–287, 298
Bone matrix, 266, 267, 273, 274, 275, 286, 287
Bone microstructure, 263, 266, 272
Bone strain, 7, 8, 9, 11, 40, 44, 63, 64, 65–66, 68, 69, 76, 77, 137, 156, 167, 168, 175–176, 178, 281, 286, 287, 320, 330, 331, 349
Bovid, 334
Braincase, 64, 77, 78, 79, 122, 211
Brain size, 410, 445, 446, 448, 449, 450, 452, 453, 454, 455, 456, 464, 467, 470, 473
Bredt's formula (K), 332, 333, 341
Brittle, 193, 388, 389, 390, 397, 400, 401, 403, 446
Brittleness, 388
Browridge, 10
Burramys parvus, 338, 350

C

Callithrix jacchus, 205, 242, 245, 247, 248, 249, 250, 252, 253, 254, 371, 373, 375
Callitrichid, 6, 199, 218, 241–259, 335, 373, 374, 375
Camel, 41
Canaliculi, 282
Cancellous bone, 269
Canine, 11, 76, 122, 145, 154, 174, 175, 181, 332, 434
Canine/P3 honing complex, 122
Carnivora, 330, 336
Carnivore, 246, 332, 336, 347, 348, 349, 351, 408
Carpolestids, 422
Cartilage, 263, 294, 295, 296, 297, 300–302, 305, 306, 307, 308, 310, 311, 312, 313, 315, 316, 317, 318, 319, 321
Catarrhine, 11, 175, 199, 203, 204, 205, 206, 209, 210, 211, 212, 214, 418, 425
Catopithecus, 411, 414, 415, 416, 419
Cebuella pygmaea, 244, 246, 250, 252, 253, 254, 247, 249

Cebus apella, 11, 193, 194, 335
Central nervous system, 201
Centric occlusion, 90, 178, 437
Cercartetus caudatus, 338
Cercartetus nanus, 338, 350
Cercopithecid, 334
Cercopithecine, 17, 63, 64, 210, 334, 368, 369, 370, 447
Cercopithecoid, 59, 203, 370
Cetartiodactyla, 20
Character, 294, 320, 342, 379, 423, 424, 425
Character polarity, 379
Cheirogaleid, 335, 372, 374, 375
Cheirogaleus major, 375
Cheirogaleus medius, 219, 226
Chewing, 17, 26, 28, 29, 33, 35, 50, 51, 52, 55, 56, 64, 67, 70, 71, 76, 84, 97, 98, 101, 114, 115, 121, 122, 168, 178, 181, 188, 189, 193, 194, 201, 202, 212, 213, 217–240, 245, 264, 297, 306, 307, 308, 311, 312, 315, 317, 320, 321, 329, 330, 331, 341, 345, 362, 368, 370, 409, 410, 431, 437, 440
Chewing frequency, 212, 213
Chewing muscles, 64, 122, 201, 202, 212, 214, 217–240
Chewing rate, 237
Chewing speed, 213–214
Chimera, 264, 448–449, 452, 453, 454, 456, 470, 471, 474, 476
Chimpanzee, 432, 433, 435, 439
Chondroblast, 296, 302, 306, 308, 311
Chondrocytes, 295, 301, 302, 306, 307, 308, 311, 316, 317, 317, 318, 319
Chondroitin sulfate, 301
Circumorbital, 17, 64, 66, 67, 76, 77, 78, 79, 130, 153, 155, 186
Cladistics, 423, 424
Claws, 418, 420
Closed section, 125, 134–135, 144
Closing stroke, 121
Collagen, 150, 283, 285, 287, 296, 297
Collagen crosslinking, 285
Collagen orientation, 287
Colobine, 334, 368, 369, 370, 447
Colobus polykomos, 205, 334
Common wombat, 87, 90, 93, 339
Comparative analyses, 52, 128, 140, 242, 244, 294, 335, 358, 378–380
Comparative anatomy, 3–11, 128
Compliant, 33, 130, 139, 142, 144, 145
Compression, 20, 22, 23, 29, 30, 31, 32, 33, 35, 40, 41, 43, 45, 46, 47, 51, 55, 98, 102, 156, 158, 159, 161, 162, 163, 164, 179, 183, 236, 301, 305, 306, 312, 314, 315, 333, 414
Condylar cartilage, 296, 305, 311, 315
Condylar height, 246, 371, 373, 432, 433, 434, 435
Condylar rotation, 90, 91, 107, 434, 437, 438
Condylar translation, 90, 437, 438, 440
Condyle, 34, 90, 91, 92, 94, 106, 107, 246, 247, 248, 266, 295, 296, 297, 299, 303, 309, 340, 362, 363, 432, 433, 434, 435, 437, 440
Confidence intervals, 203, 204, 208, 212, 213, 214, 225, 228, 233, 235, 236
Confocal microscopy, 263, 272, 281, 283, 285
Connective tissue, 58, 150, 220, 223, 234, 235, 295, 319, 321
Connochaetes gnou, 57
Contractile length, 210
Contraction, 17, 20, 21, 22, 31, 33, 34, 35, 77, 116, 167, 234, 243, 244, 250, 254, 281, 351, 410
Contraction velocity, 243, 244, 254
Convergence, 20, 264, 340, 407–426
Cooking, 223, 445
Corpus, 11, 33, 51, 52, 54, 84, 125, 127–145, 178, 194–195, 263, 264, 282, 294, 295, 299, 303, 309, 313, 329–356, 365, 373, 447, 450, 451, 453, 456, 459, 460, 461, 462, 464, 467, 469, 470, 471, 474, 475, 476
Cortical bone, 41, 130, 134, 140, 143, 144, 145, 160, 168, 169, 176, 179, 263, 265–288, 294, 295, 296, 299, 300, 303, 304, 309, 311, 313, 319, 331, 332, 341
Crack formation, 388, 392, 401, 403
Crack propagation, 388
Crania, 90, 168, 180, 407, 410, 411, 414, 416, 432, 448
Craniobasal length, 203, 204, 205, 206, 207, 209
Cranium, 7, 63, 66, 87, 114, 179, 287, 410, 412, 413, 418, 423
Crepuscular, 425
Cretaceous, 407, 421, 422, 424
Cribriform plate, 128
Criterion of subtraction, 360
Cross-bridge, 243
Cross-section, 274, 333, 436
Cross-sectional area, 10, 93, 106, 121, 177, 178, 199, 201–214, 218, 219, 221, 223, 224, 226, 227, 228, 232, 233–235, 243, 244, 250
Crushing, 7, 11, 107, 193, 411
Curved beam, 40, 51

Cusp, 89, 152, 153, 174, 351, 388, 389, 392, 393, 394, 398, 400, 401, 403, 433, 437, 440
Cusp height, 389, 392, 393, 398
Cyclical loading, 41, 298, 331, 341
Cylinder, 17, 64, 66, 67, 74, 76, 77, 78, 79, 268, 269, 270, 271, 273

D

Daubentonia, 219, 226, 236, 340
Daubentonia madagascarensis, 219, 226
Deep masseter, 20, 42, 44, 55, 87, 93, 94, 95, 96, 100, 177, 222, 223, 228, 232, 234, 236, 237, 249
Deep temporalis, 117, 121, 222, 223, 227, 234
Deformation, 21, 22, 23, 26, 28, 30, 31, 33, 34, 35, 77, 78, 79, 130, 150, 151, 156, 167, 181, 184, 185, 186, 187, 267, 281, 287, 298, 388, 410
Delta rosette strain gage, 68
Denominator, 120, 361, 362, 365, 366, 367, 368, 369, 370–378, 379
Density, 128, 131, 134, 145, 203, 218, 224, 250, 284, 285, 287, 295, 296, 299, 300, 302, 303, 304, 305, 309, 310, 311, 313, 314, 319
Dental arcade, 433, 440
Dentine, 88, 90, 436, 437
Developmental constraint, 175
Diaphysis, 269, 287
Diet, 7, 20, 25, 87, 95, 108, 122, 174, 194, 195, 199, 213, 237, 241, 242, 258, 264, 294, 295, 297, 298, 299, 303, 304, 306, 308, 309, 319, 321, 330–336, 337, 339, 340, 343, 345, 348, 351, 370, 387, 389, 395, 396, 400, 402, 404, 420, 443–476
Digestibility, 331
Dihedral goniometer, 413
Diprodontia, 88
Displacement-limited defense, 388, 403
Diurnal, 390, 419, 420, 425
Divergence, 5, 242, 244, 245, 374, 411
Dmanisi, 449, 450, 452, 455, 456, 458, 461, 463, 465, 466, 467, 468, 473, 475, 476
Dorsal interorbital region, 66, 72, 75, 76, 77, 79
Dorso-ventral shear, 21, 22, 23, 32, 33, 34, 35
Drifting osteon, 284, 287
Durophagy, 331, 349

E

Early *Homo*, 443–476
Ecomorphology, 334, 340, 389
EDTA, 270
Egypt, 411, 412, 414, 424

Elastic orthotropy, 268, 282, 284, 286
Elastic properties, 150, 167, 169, 176, 179–180, 184, 263, 265–288
Elastic (Young's) modulus, 131, 140, 141, 142, 143, 179, 267, 388, 392
Electrodes, 25, 96, 97, 100, 106, 117, 118, 178
Electromyography (EMG), 10–11, 18, 25, 26, 40, 42, 56, 63, 84, 86, 96, 97, 98,101, 114, 115, 117, 118, 119, 121, 175, 178, 181, 218, 258
Elegantulus, 233, 371, 374, 375, 376, 377
Enamel, 88, 89, 90, 94, 246, 339, 436, 437
Endentulous, 135, 136, 184, 286, 287
Endosteal, 287
Energy, 202, 212, 213, 214, 254, 331, 445
Entoglenoid eminence, 91
Environment, 8, 9, 23, 34, 127, 256, 258, 271, 293, 295, 339, 379, 387, 426, 471
Eocene, 218, 407, 411, 413, 414, 422, 425
Eocene adapid, 218, 413
Epigenetics, 296
Eskimo, 7
Ethanol, 203, 219, 221, 223, 269, 301, 422
Eulemur, 11, 17, 65, 66, 67, 68, 69, 70, 71, 72, 73, 74, 76, 77, 78, 79, 80, 205, 233, 393, 401
Eulemur coronatus, 233, 393
Eulemur fulvus, 11, 17, 65, 67, 68, 69, 70, 71, 72, 73, 78, 80, 401
Eulemur rubriventer, 401
Euoticus elegantulus, 371, 374, 375, 376
Euprimates, 408, 422
Eutheria, 332, 407
Eversion, 17, 20, 22, 23, 31, 32, 33, 35
Evolutionary adaptation, 184
Evolutionary morphology, 358, 379
Exocarp, 390, 391, 392, 401
Expensive tissue hypothesis, 445
Experimental approach, 3, 4, 5, 6, 7–11
External auditory canal, 413
External force, 63, 64, 65, 77, 362, 370
Extracellular matrix (ECM), 296, 305, 306, 307, 308, 311, 313, 315, 318, 319
Exudativore, 339
Eye, 64, 408, 409, 410, 411, 418, 419, 421, 422, 423
Eyeball, 407, 408, 409, 410, 411, 417, 420, 423

F

Face, 9, 17, 63–80, 91, 93, 155, 160, 166, 168, 174, 175, 179, 187, 190, 193, 195, 214, 235, 286, 408, 472, 473
Face length, 235

Facial skeleton, 63, 64, 65, 67, 74, 76, 77, 78, 149–169, 179, 184, 190, 409
Failure, 33, 41, 64, 127, 130, 133, 143, 193, 221, 236, 297, 298, 388
Fascicle, 116, 243
Fasciculus length, 210, 248, 250, 251
Fast close phase, 84
Fast-closing, 48
Fatigueable, 116
Fatigue failure, 41, 133
Fatigue-resistant, 116
Fayum, 267, 407–426
Feeding adaptation, 195, 218, 443
Feeding apparatus, 4, 79, 122, 330
Feeding behavior, 1, 9, 63, 97, 173, 194, 200, 234, 241–259, 318, 320, 330, 340, 390
Femoral head, 265
Femur, 265–288, 446, 450, 452, 454, 455, 459, 463, 464, 465, 470, 471, 474, 475, 476
Fiber, 18, 114, 116, 117, 121, 122, 199, 203, 211, 213, 218–220, 221, 223, 225, 226, 229, 231, 232, 235, 236, 237, 242, 243–244, 245, 249, 250, 252, 254–256, 257, 258–259, 287, 299
Fiber architecture, 114, 121, 199, 218–220, 242, 243–244, 245, 250, 252, 254–256, 257, 258–259
Fiber length, 116, 121, 203, 211, 218, 220, 221, 223, 224, 226, 228, 229, 231, 232, 235, 236, 237, 244, 250, 251, 252, 253, 255, 256
Fiber type, 18, 114, 116, 117, 121, 213–214
Fibrocartilage, 42, 297, 301, 305, 308, 315, 318
Fibula, 284
Field work, 380
Finite element analysis (FEA), 132, 144, 149–169, 174, 180, 181, 182, 183, 194, 195
Finite element model (FEM), 125, 130, 133, 135, 167, 175, 176, 177, 185, 187
Fitness, 294, 379, 409
Flower, 234, 339, 389, 390, 391, 394, 395, 396, 397, 402, 403, 421, 422
Flowering plant, 421
Folivore, 41, 226, 227, 228, 233, 234, 237, 331, 332, 335, 339, 345, 347, 349, 350, 389, 402, 454
Folivory, 331, 334, 335, 349, 425
Food, 7, 10, 11, 18, 20, 25, 28, 58, 68, 85, 90, 94, 98, 105, 121, 126, 150, 173, 175, 179, 181, 193, 194, 195, 199, 201, 202, 212, 213, 214, 217, 218, 221, 234, 236, 237, 244, 264, 297, 298, 321, 330, 336, 339, 341, 349, 350, 370, 387–404, 420, 422, 426, 446, 447, 454
Food acquisition, 235
Food material properties, 10, 195, 321, 336, 341
Food properties, 18, 237, 264, 388, 394, 400, 404, 447
Foraging, 264, 339, 410, 420, 444, 445–448, 449, 453, 456–464, 467, 469, 471, 472, 473
Force, 6, 7, 10, 11, 18, 21, 22, 32, 40, 41, 53, 57, 63, 64, 65, 77, 79, 86, 87, 94, 104, 105, 106, 107, 114, 115, 119, 121, 122, 134, 153, 175, 177, 179, 181, 193, 195, 199, 201, 202, 210, 212, 217, 218, 221, 224, 227, 233, 237, 243, 244, 246, 250, 253, 255, 257, 301, 309, 335, 362, 373, 388, 391, 400, 434, 453
Force-displacement graph, 388, 392, 400
Force transducer, 178
Fracture, 130, 199, 202, 212–214, 218, 221, 225, 228, 231, 232, 235, 236–237, 330, 388, 391
Fracture scaling, 199, 202, 212–214, 218, 221, 225, 228, 231, 232, 235, 236–237
Fragmentation, 221, 301, 397
Frankfurt horizontal, 413, 432, 433
Frontal, 55, 64, 65, 76, 77, 78, 79, 150, 153, 154, 155, 165, 166, 177, 180, 185, 187, 411, 414, 423, 433
Frontation, 264, 407–426
Frugivore, 335, 339
Fruit, 193, 219, 339, 390, 391, 392, 394, 395, 396, 397, 400, 401, 402, 403, 404, 410, 421, 425
Function, 1, 5, 7, 8, 9, 11, 17, 31, 39, 50, 85, 109, 114, 116, 117, 121, 125, 128, 129, 131, 133, 143, 145, 150, 168, 201, 217, 233, 241, 242, 243, 245, 256, 258, 263, 264, 266, 269, 286, 287, 294, 296, 297, 301, 302, 311, 320, 329, 330, 332, 334, 358, 367, 378, 410, 411, 434, 437
Functional adaptation, 8, 9, 282, 294, 296, 297, 309, 313, 321
Functional consequence, 358, 359, 360, 361, 362, 363, 365, 366, 368, 370, 378, 379
Functional equivalence, 366
Functional morphology, 3, 4, 166–168, 234, 357, 359, 368, 371, 378, 389, 404
Fusion, symphysis, 19–36, 85, 87, 94, 107, 389

G

Galagid, 335, 372, 374, 375, 376, 377
Galago, 17, 40, 64, 85, 205, 219, 222, 225, 226, 227, 232, 375, 409

Galago demidoff, 227, 232, 233, 375
Galago gallarum, 375
Galagoides alleni, 375
Galagoides demidoff, 219, 226, 227, 232
Galago moholi, 375
Galago senegalensis, 219, 226, 227
Galago zanzibaricus, 375
Gape, 6, 11, 48, 193, 199, 221, 233, 235, 245, 246, 247, 251, 253, 255, 256, 257, 258, 371, 373, 375, 434, 440
Gastrointestinal system, 233
Genetic drift, 471, 472
Genome, 409
Geometric mean, 264, 335, 367, 368, 370, 371, 372
Geometric similarity, 10, 199, 202, 212–214, 218, 221, 361
Glenoid, 91, 94, 108, 258, 371, 439, 440
Glenoid fossa, 439, 440
Glycosaminoglycan (GAG), 295, 296, 301, 305, 306, 312, 314, 315
Goat, 85, 86, 106
Gorilla, 79, 334, 432, 433, 435, 440, 447
Gorilla gorilla beringei, 334
Gorilla gorilla gorilla, 334
Gouging, 6, 199, 241–259, 331, 335, 339, 341, 343, 345, 348, 349, 357, 371, 372, 373, 374, 375, 376, 378, 379
Grinding, 6, 32, 107, 264, 269, 431
Grooming behavior, 235
Guanaco, 58
Gummivory, 233, 241–259

H

Hapalemur, 205, 219, 226, 233, 234, 388, 421
Hapalemur griseus, 219, 226, 233, 234
Hapalemur simus, 233
Haplorhine, 421, 424
Hardness, 140, 141, 142, 298, 332, 351, 388, 392, 397, 400, 401, 402, 404
Hard object feeding, 115, 119, 120, 193, 334
Harversian canal, 267
Herbivore, 18, 84, 87, 94, 105, 106, 109, 387, 388, 389
Herbivory, 94, 388
Hippotragus equinus, 57
Histology, 275, 296, 299, 300–302, 305, 311, 313–316, 321
Home-range, 446
Hominin, 11, 122, 437, 444, 445, 446, 447, 448, 450, 453, 454, 455, 456
Homo, 264, 413, 421, 432, 443–476
Homo erectus, 445, 455–456

Homo habilis, 264, 445
Homo rudolfensis, 446
Homo sapiens, 432, 433, 450, 473
Horse, 86, 94, 408
Human, 4, 7, 21, 31, 79, 116, 122, 128, 132, 134, 136, 142, 150, 151, 173–195, 212, 218, 243, 265–288
Human evolution, 4, 122, 173–195
Hypertrophic, 302, 306, 308, 311, 316
Hypoconid, 393, 398
Hypsodonty, 107, 389
Hyrax, 408

I

IIM isoform, 116
Ileret, 467
Immunohistochemistry, 296, 299, 300–302, 305, 311, 313–316, 321
Incision, 9, 11, 23, 24, 25, 44, 68, 96, 242, 330, 331, 436
Incisor, 20, 24, 26, 32, 33, 35, 57, 59, 84, 90, 107, 135, 153, 154, 156, 157, 158, 159, 160, 165, 166, 244, 245, 298, 330, 363
Indentation, 140, 144, 298, 388, 392, 397
Independent contrasts, 344, 348
Indri, 233, 394, 401
Indriids, 393, 401
Inferior orbital fissure, 423
Infraorbital region, 189
Inion, 69, 412, 413, 414
Insect, 219, 235, 410, 420
Insectivore, 227, 235, 336, 339
In situ hybridization, 299
Interdigitations, 57, 58, 150, 151
Internal load, 242, 370
Interstitial bone, 284, 286
Inversion, 22, 23
In vitro, 43–44, 46, 51–52, 59, 125, 134, 135, 136, 149–169, 341
In vivo, 1, 6, 7, 8, 9, 11, 17–18, 19–36, 63, 76, 78, 79, 125, 127, 128, 130, 131, 133–134, 135, 145, 150, 151, 160, 167, 168, 169, 176, 178, 181, 182, 183, 188, 189, 218, 242, 245, 263, 281, 294, 297, 318, 320, 330, 331, 349, 443
Isoflurane, 96
Isognathy, 20
Isometric biting, 9, 178
Isometry, 199, 203, 204, 212, 220, 221, 228, 231, 232, 233, 236, 346, 361, 363, 366, 367, 394
Isotropy, 129, 140, 266, 267, 268, 281, 282, 284, 285–287

Index

J
Jaw, 17–18, 55–59, 125, 126, 177, 199–200, 241–259, 456–464
Jaw adductor muscles, 113, 114, 115, 121, 214, 217, 218, 233, 237, 296, 299, 320
Jaw depth, 369, 370
Jaw gape, 245, 246, 251, 253, 256, 258, 371, 373
Jaw joint, *see* Temporomandibular joint
Jaw length, 212, 213, 248, 251, 252, 253, 302, 369, 370, 371, 375, 376, 377
Jaw muscles, 17–18, 21, 40, 42, 55, 122, 199–200, 212, 213, 214, 218, 233, 241–259, 263, 299, 362, 363, 440
Jugal, 91, 423

K
Kangaroo(s), 88, 107, 108, 336
Keratan sulfate, 301
Kinematics, 87
Koala(s), 88, 106
Koobi Fora, 445, 450, 455, 460, 461, 463, 465, 466, 468
Kruskal-Wallis test, 394
L-41 locality, 411

L
Laboratory, 4, 5, 6, 7, 23, 116, 150, 242, 245, 256, 258, 264, 284, 287, 298, 380
Lamellae, 285
Lamellar bone, 128, 287
Lamina dura, 128, 135, 139
Lasiorhinus latifrons, 17, 83–109, 339
Lateral pterygoid, 21, 22, 31, 32, 35, 55, 91, 93, 222, 256
Lateral transverse bending, 19, 21, 22, 23, 32, 33, 34, 35, 40, 43, 44, 46, 47, 87, 107
Leaping, 5, 410, 418, 419, 420, 424
Least squares linear regression, 203
Least-squares regression, 363, 459, 462, 464, 467, 469
Leaves, 34, 213, 219, 234, 236, 331, 332, 339, 370, 388, 389, 390, 391, 394, 395, 396, 397, 401, 403, 410, 420, 421
Lemur catta, 219, 226, 233, 389, 390, 393, 395, 401
Lemurids, 388, 389, 393, 394, 401
Lemuroidea, 409
Length-tension relationship, 243
Leontopithecus rosalia, 375
Lepilemur, 219, 226, 233, 234, 421
Lepilemur mustelinus, 219, 226
Leptadapis, 421
Lever, 7, 201, 202, 217, 361, 362, 363

Leverage, 90, 121, 217, 233, 234, 236, 237, 256, 361, 364, 433
Life history, 295, 450
Ligament, 128, 131, 135, 139, 144
Load-arm, 330
Load arm (out-lever), 361
Load cell, 391
Loading, 8, 17, 20, 21, 34, 35, 41, 42, 43, 45, 49–51, 52, 53, 55, 57, 59, 64, 67, 72, 74–78, 125, 132, 152–153, 156, 160, 165, 166, 167, 173–195, 263, 267, 281, 294–297, 299, 302–309, 311, 318, 319, 321, 329, 331, 332, 341, 349, 350
Locomotion, 4, 421
Lophodont, 88
Loris, 421
Lorisid, 233, 236
Lorisiformes, 421
Lucy (AL 288), 448

M
Macaca, 67, 79, 135, 140, 151, 175, 177, 180, 334, 409
Macaca fascicularis, 135, 140, 151, 175
Macaca fuscata, 177, 334
Macaque, 21, 40, 41, 86, 106, 125, 142, 151, 167, 175, 176, 193, 218, 334
Macropodidae, 88
Madagascar, 389, 390, 419
Mammal, 84, 116, 297, 305, 321, 407
Mandible, 7, 11, 20, 21, 22, 23, 24, 28, 29, 31, 32, 34, 41, 43, 44, 46, 52, 63, 85, 87, 90, 91, 92, 96, 101, 102, 105, 106, 113, 114, 127, 131, 132, 134–136, 142, 144, 179, 217, 247, 265–288, 299, 318, 330, 331, 332, 335, 341, 344, 350, 351, 363, 369, 371, 401, 412, 431, 432, 435, 437, 440, 443, 453, 454
Mandibular angle, 84, 87, 91, 93, 94, 96, 106, 432, 433–434
Mandibular condyle, 34, 91, 94, 179, 266, 295, 296, 340, 432, 440
Mandibular corpus, 33, 84, 125, 127–145, 178, 194, 263, 294, 329–356, 373, 450, 453–454, 456, 463
Mandibular corpus depth, 373
Mandibular corpus width, 373
Mandibular length, 10, 334, 335, 340, 341, 342, 344, 345, 346, 347, 348, 349, 371, 373, 435, 437, 453, 469
Mandibular ramus, 24, 102, 104, 117, 264, 431–440

Mandibular symphysis, 11, 20, 30–31, 40, 45, 87, 90, 92, 94, 96, 101, 105, 107, 294, 296, 316, 459

Mann-Whitney U-test, 118, 119, 228, 250, 302, 303, 304, 309, 310, 313, 314, 369, 371, 372, 373, 397, 398

Marmoset, 6, 40, 199, 242, 243, 244, 245, 246, 249, 250, 251, 252, 253, 254–256, 257, 258, 374

Marsupial, 84, 87, 105, 106, 336, 337, 341–342, 343, 349, 350, 418

Marsupial mammal, 84

Masseter, 11, 17, 24, 25, 26, 33, 35, 42, 65, 66, 79, 105, 203, 204, 205, 207, 208, 211, 212, 213, 229, 243, 244, 251, 253, 255, 256–257, 298, 309, 330, 332, 340, 373, 433, 434, 447, 453

Masseter, superficial, 20, 22, 42, 56, 77, 78, 85, 86, 87, 93, 95, 96, 101, 106, 107, 108, 177, 222, 223, 227, 228, 246, 247, 249, 250

Mastication, 4, 6, 8, 9, 10–11, 17–18, 20, 21, 25, 26–33, 40, 42, 44, 46–51, 52–55, 58, 59, 64, 65, 67, 69, 73, 76, 78, 79, 87, 90, 98, 105, 107, 113–122, 129, 133, 144, 150, 167, 174, 177, 184, 190, 193, 199, 201–214, 221, 234, 237, 242, 257, 264, 298, 314, 320, 321, 330, 331, 431, 440, 469, 472

Masticatory apparatus, 10, 11, 40, 87, 94, 122, 125–126, 241, 242, 243, 246, 255, 256, 258, 294, 299, 357–385, 389, 401, 403

Masticatory muscle(s), 52, 150, 167, 173, 179, 201, 202, 213, 214, 221, 222, 256, 257, 281, 330, 351, 373, 453

Masticatory system, 6, 8, 20, 88–94, 108, 212, 234, 237, 294, 302, 319, 431, 434, 435, 436, 440

Material properties, 10, 130, 142, 144, 151, 152, 167, 168, 195, 236, 237, 266, 283, 285, 286, 287, 297, 298, 321, 330, 336, 341, 351

Materials property tester, 286, 298

Maxilla, 78, 154

Maxillo-zygomatic suture, 153, 155, 165, 166

Maximum principal strain, 26, 28, 46, 47, 49, 50, 52, 53, 54, 68, 69, 70, 71, 73, 74, 75, 160, 166, 168, 181, 183, 184, 185, 192

Maximum shear strain, 160, 181, 184, 189, 191

Maximum tetanic tension, 243

Mechanical advantage, 361, 362, 363, 365, 373, 433, 434

Mechanobiology, 294

Mechanotransduction, 296

Medial orbital wall, 64, 66, 154, 180

Medial pterygoid, 11, 20, 26, 65, 84–85, 87, 92, 93, 95, 96, 98, 100, 105, 106, 107, 108, 177, 203, 204, 205, 206, 211, 213, 222, 223, 225, 227, 228, 234, 363, 433

Medial transverse bending, 21, 22, 23, 31, 32, 35, 46, 47, 52

Megaladapis, 410, 421

Mesocarp, 390, 391

Metabolic, 199, 202, 212, 213, 299, 403, 470, 471

Metabolic demands, 202, 470, 471

Metacarpal, 130

Metatarsal, 284, 449

Microcebus murinus, 233, 375

Microcebus rufus, 375

MicroCT, 145, 263, 266, 267, 296, 299–300, 301, 302, 303–305, 308, 309–311, 313–316, 321

Microindentation, 130, 140, 142, 143, 144, 145

Microstrain, 26, 28, 40, 68, 98, 107, 130, 138, 188

Microwear, 7, 193, 194

Midface, 153, 287

Midpalatal suture, 154, 160

Mineralization, 131, 134, 150, 263, 285, 286, 296, 299–300, 301, 302, 303, 304, 309, 310, 311, 313, 314, 318, 319, 320

Mineralizing front, 308, 316

Minimum principal strain, 26, 45, 46, 47, 49, 50, 68, 135, 184, 187

Miocene, 411, 426

Miopithecus talapoin, 210, 211

Mirza coquereli, 219, 226, 375

Modeling, 64, 78, 125–126, 127, 132, 142–145, 168, 176, 179, 181, 217, 242, 266, 287, 294, 331

Modern human (*Homo sapiens*), 432, 473

Modulus, 130, 131, 134, 140, 141, 142, 143, 145, 179, 267, 298, 388, 392

Molar, 6, 22, 32, 88, 90, 97, 106, 107, 125, 128, 152, 154, 160, 163, 166, 173, 175, 213, 233, 267, 282, 307, 317, 329, 340, 351, 388, 389, 392, 393, 398–400, 403, 434, 437, 447, 453, 456, 459, 461, 462, 464, 467, 469, 471, 475, 476

Moment, 10, 21, 41, 54, 55, 65, 131, 142, 178, 203, 204, 206, 217, 233, 332, 333, 335, 340, 341, 343, 361, 362, 363, 371, 373, 433

Moment arm (in-lever), 361

Monochromatic, 425

Morphometric, 218, 263, 264, 302, 309, 348, 357, 358, 359, 366, 368, 371, 378, 379, 380
Motor unit, 116
Multipinnate, 243, 249
Multiple regression, 359
Muscle, 8, 9, 10, 17, 18, 21, 22, 27, 32, 34, 35, 40, 42, 54, 59, 65, 66, 77, 79, 85, 86, 92, 96, 98, 100, 103, 105, 106, 114, 115, 117, 119, 121, 167, 175, 177, 181, 190, 199, 201, 204, 210, 213, 217, 218, 219, 223, 230, 231, 232, 242, 244, 248, 249, 250, 256, 258, 299, 320, 351, 373, 453
Muscle activity, 8, 11, 17, 18, 32, 42, 55, 59, 84, 98–100, 102, 103, 177, 178, 195, 199, 236, 242, 320
Muscle architecture, 121, 199–200, 241–258, 434
Muscle-cross sectional area, 10, 201, 213, 219, 221, 228, 233
Muscle force, 10, 11, 22, 53, 54, 65, 77, 79, 86–87, 118, 175, 177, 178, 181, 201, 202, 210, 218, 221, 233, 234, 236, 237, 245–247, 254, 255, 256
Muscle length, 251, 252
Muscle mass, 113, 199, 202, 203, 214, 218, 220, 221, 223, 224, 225, 226, 227, 228, 230, 231, 232, 233, 234, 235–236, 237, 244, 254, 256, 298, 299, 303, 309
Muscles of mastication, 116, 201–214, 257
Muscle weight, 93, 210, 225, 244
Mus musculus, 297, 298
Myanmar (Burma), 425
Myofibril, 177, 243
Myofilament, 243
Myosin, 114, 116, 117, 243
Myostatin, 294, 297, 298, 299, 309, 310, 311, 319

N

Nanoindentation, 266
Nariokotome (KNM-WT 15000), 448
Nasal aperture, 130
Nasal margin, 189, 191, 192
Nasion, 363, 371, 412, 413, 414
Naso-frontal suture, 150
Natural selection, 379
Nectarivore, 335
Nectary, 391
Neural control, 254, 255
Neurocranium, 64, 473
Neutral axis, 45, 46, 51, 134, 136, 332
Ngandong, 450, 452, 455, 476
Nocturnal, 407, 410, 419, 420, 421, 425

Node, 176, 178, 181, 184, 187, 188, 189, 190, 191, 192, 275, 277, 337, 342
Norms of reaction, 295, 318, 319
Notoryctes typhlops, 338, 350
Numerator, 120, 361, 362, 366, 373
Nutrient, 35, 130, 139, 142, 144, 331, 332, 422
Nutrient foramina, 35, 130, 139, 142, 144
Nycticebus, 219, 226, 421

O

Occipital crest, 122
Occlusal area, 447, 453, 462
Occlusal force, 32, 87, 106, 108, 128, 135, 143, 145, 236, 237, 298, 434, 447
Occlusal load, 21, 22, 32, 144, 190, 433, 453
Occlusal plane, 20, 21, 90, 152, 176, 234, 235, 432, 433, 434, 435, 436
Occlusal surface, 6, 88, 89, 107, 213, 393, 431, 434, 435, 436, 437, 440
Occlusion, 22, 24, 26, 32, 35, 89, 90, 107, 178, 244, 249, 298, 331, 332, 433, 434, 435, 437, 438
Olduvai, 450, 455, 458, 460, 461, 463, 465, 466, 468
Old World monkeys, 370
Olfaction, 420
Olfactory bulb, 420
Oligopithecidae, 411
Olorgesailie, 467
Omnivorous, 174
Ontogeny, 17, 41, 42, 180, 293, 294
Opening stroke, 25, 26, 48, 100
Open section, 127, 131, 134, 135, 136, 144
Opossum, 84, 85, 86, 106, 337, 418
Optimal strain environment, 8, 9
Optimization, 9, 63, 128, 131, 246
Oral cavity, 193
Orbit, 117, 155, 408, 409, 410, 411, 412, 413, 423
Orbital aperture, 413
Orbital rim, 410, 413
Orbital wall, 64, 66, 79, 153, 154, 155, 180
Orthotropy, 268, 282, 284, 286
Oryctolagus cuniculus, 297
Osteoarthritis, 305
Osteoblast, 296
Osteoclast, 150, 296
Osteocyte, 267, 275, 282
Osteocyte lacunae, 275, 282
Osteon, 131, 263, 266, 267, 274, 275, 276, 279, 280, 282, 283, 286
Osteotomy, 24, 28, 32, 33

Otolemur, 66, 67, 74, 76, 78, 79, 205, 219, 226
Ourebia ourebia, 57
Owl monkey, 40, 64

P
Palate, 77, 153, 180, 371
Paleocene, 407, 422
Paleogene, 411
Papio, 67, 79, 114, 115, 117, 119, 120
Papio anubis, 114, 115, 117
Papionins, 369
Paramasticatory activity, 443
Paranthropus boisei, 175
Paranthropus robustus, 174, 435
Parapithecidae, 411, 416
Parapithecus, 411, 412, 413, 414, 419
Parapithecus grangeri, 412, 413, 414
Parasaggital bending, 369
Partial correlation, 359
Peccaries, 20
Pedicel, 391
Performance, 6, 128, 144, 242, 256, 258, 294, 296, 318, 320, 321, 362, 363, 368, 379
Periodontal ligament, 128, 131, 135, 139, 144
Periosteum, 44, 96, 153
Petal, 391
Phalanger gymnotis, 338, 350
Phaner, 233, 371, 374, 375
Phaner furcifer, 371, 374, 375
Phascolarctidae, 88, 338
Phase I, 7, 437
Phase II, 6, 7, 437
Phenophase, 389, 396, 403
Photoelastic, 132
Phylogenetic autocorrelation, 341, 342
Phylogeny, 336, 341–342, 343, 345, 349, 409
Physiological cross-sectional area (PCSA), 199, 213, 221, 223–224, 225, 226, 227, 228, 229, 230, 231, 232, 233–235, 237, 243, 244, 250, 251, 252, 253, 255, 256, 257, 258
Pig, 19–36, 244, 246
Pinnae, 391
Pinnation, 121, 202, 203, 210, 211, 218, 219, 220, 222, 223, 224–225, 226, 227, 228, 229, 233, 234, 235, 243, 244, 248, 250, 251, 252
Pinnation angle, 234, 243, 244, 250, 252
Pinnule, 391
Pithecines, 194

Placental mammal, 84, 87, 94, 330, 339, 352, 407
Platyrrhine, 203, 204, 205, 206, 209, 210, 211, 212, 213, 214, 334, 423, 425
Pleiotropy, 360
Plesiadapoidea, 407
Poisson's ratio, 179
Pondaungidae, 425
Pongo, 334, 413
Pongo abelii, 334
Pongo pygmaeus, 334
Porosity, 141, 145, 266, 286
Positive allometry, 199, 202, 204, 211, 212, 213, 228, 232, 237, 320, 334, 361, 363, 370
Postcanine teeth, 264, 435, 436, 437, 440
Postorbital bar, 17, 64, 66, 69, 70, 71–72, 74, 76, 77, 78, 79, 176, 179, 180, 183, 407, 408, 409, 410
Postorbital closure, 408, 423
Postorbital septum, 423
Potto, 205, 219, 226, 421
Power stroke, 6, 19, 21, 23, 26, 34, 35, 40, 41, 42, 45, 49, 50, 52, 54, 55, 59, 69, 70, 73, 84, 87, 90, 94, 98, 99, 100, 105, 106, 107, 114, 121, 178, 234, 235, 264, 431, 434, 440
Predator, 334, 402, 410, 419, 420, 426
Preglenoid plane, 439, 440
Premaxilla, 34, 154, 155, 180
Pre-maxillo-maxillary suture, 150
Premolar, 20, 125, 126, 132, 152, 153, 154, 160, 161, 162, 165, 166, 173–195, 401, 434
Primate(s), 1, 4, 5, 6, 7, 8, 10, 11, 17, 21, 33, 40, 63, 64, 66, 68, 76, 125, 127, 150, 151, 168, 188, 195, 199, 201, 204, 211, 212, 214, 218, 221, 237, 241, 243, 258, 264, 287, 329, 357, 358, 359, 361, 362, 366, 378, 408, 410, 411, 413, 414, 415, 418, 420, 421, 422, 424, 432, 444
Principal strain, 26, 28, 45, 46, 47, 49, 50, 53, 55, 68, 70, 71, 73, 75, 104, 135, 166, 181, 182, 183, 184, 185, 187, 189, 192
Principal stress, 283
Procolobus badius, 145, 334
Product-moment correlation, 142, 203, 204, 206, 343
Propalinal, 87
Propithecus verreauxi, 11, 219, 226
Propliopithecidae, 411, 422
Proportion, 115, 184, 210, 233, 234, 255, 271, 272, 280, 320, 344, 345, 346, 347, 348, 349, 359, 404
Prosimians, 199, 214, 218, 220, 221, 225, 232, 233, 237, 423

Prosthetic, 151
Prosthion, 69, 248, 371
Protein, 213, 302, 403, 445, 473
Proteinase, 296
Proteoglycan, 296, 297, 301, 302, 305, 306, 308, 311, 312, 314, 315, 318, 319
Proteopithecidae, 411, 416
Proteopithecus, 411, 414, 415, 416, 417, 419
Protocone, 393, 398, 399, 400, 403, 404
Protrusion, 21
Pterygoid hamulus, 93
Puncture-crushing, 11
Puncturing, 194, 236
Pyriform aperature, 174

Q

Quadritubercular, 389, 393

R

Rachis, 391
Radiography, 340
Radius, 287, 332, 389, 398, 400
Rat, 301
Ratio, 50, 54, 98, 100, 105, 115, 118, 119, 120, 129, 131, 135, 156, 179, 182, 183, 184, 189, 191, 225, 246, 250, 254, 298, 361, 365, 367, 371, 372, 376, 398, 401, 432, 435, 450
Reaction force, 7, 53, 54, 179, 362
Reduced major axis regression (RMA), 203, 204, 205, 207, 208, 211, 212, 213, 21, 225, 228, 230, 231, 342, 344
Reduced physiological cross-sectional area (RPCSA), 224, 225, 227, 228, 229, 233–236
Reflex microscope, 392, 393
Regression, 203, 204, 206, 207, 209, 211, 213, 214, 220, 225, 228, 232, 340, 344, 345, 346, 348, 359, 360, 363, 393, 450, 453, 455, 456, 459, 460, 462, 464, 467, 469
Regression-least squares, 363, 393, 459, 462, 464, 467, 469
Relative size, 358–366, 374, 375
Remodeling, 8, 41, 184, 266, 286, 287, 294, 295, 296, 331
Residual, 156, 157, 167, 359, 360, 342, 346
Resting length, 244, 246, 249, 256
Reverse wishboning, medial transverse bending, 47
Rhinarium, 421
Ring-tailed lemurs, 40, 85, 264, 402
"Robust" australopithecine, 125, 126, 264, 437

Robust australopiths, 125, 126, 173, 190, 193, 194, 195, 264, 431–440
Rodent, 87, 212, 244, 298, 309, 418, 419
Root mean square (RMS), 118, 178
Rosette strain gage, 23, 68
Rostrum, 78, 174, 175, 180, 184, 185, 186, 189, 190, 191, 192, 193, 194, 421

S

Safranin O, 301, 305, 306, 311, 312, 313, 314, 315
Sagittal crest, 122
Sagittal suture, 34
Saguinus fuscicollis, 375
Saguinus Oedipus, 242, 245, 247, 249, 250, 252, 253, 254, 255, 256, 257
Saimiri sciureus, 11
Sangiran, 450, 454, 458, 471, 475, 476
Sarcomere, 203, 210, 218, 224, 243, 244
Sarcomere length, 203, 210, 224
Scaling, 8, 10, 41, 144, 199, 201, 202, 203, 210– 214, 218, 226, 228, 231, 232, 233, 235, 236, 237, 264, 361, 363, 448, 449, 450–455, 455, 469, 470, 471, 472, 474, 221, 225
Scissors cutting test, 391, 392
Scotopic vision, 425
Secondary compounds, 331
Second moment of inertia, 41, 332, 333, 341
Seed predation, 402
Selection, 99, 236, 237, 351, 379, 389, 400, 403, 409, 419, 422, 424, 425
Selenodont artiodactyls, 39–62
Sepal, 390
Shape, 8, 9, 10, 52, 64, 78, 130, 132, 174, 179, 184, 201, 219, 251, 273, 285, 330, 335, 336, 340, 358, 359, 360, 361, 362, 365, 366, 368, 370, 372, 373, 375, 378, 379, 388, 421, 445, 446, 472
Shape ratio, 361, 362, 363, 365, 366, 367, 376
Shear, 21, 22, 23, 26, 28, 30, 32, 33, 34, 35, 45, 48, 52, 53, 54, 57, 76, 90, 107, 134, 135, 138, 139, 155, 156, 160, 166, 179, 181, 182, 183, 184, 187, 188, 189, 191, 301, 316, 388, 401
Shear, anteroposterior, 52, 53, 54
Shear, dorsoventral, 53, 57
Shear flow, 134
Shearing, 53, 77, 89, 107, 128, 131, 133, 134
Shear modulus, 179
Shear strain, 26, 28, 30, 33, 45, 135, 138, 160, 183, 184, 188, 189, 191
Shear strength, 388, 401

Sifaka, 264, 389, 390, 398, 400, 401, 403, 404
SINE marker, 409, 421
Size, 8, 9, 10, 19, 20, 24, 25, 28, 52, 57, 59, 64, 95, 121, 129, 130, 133, 135, 174, 185, 187, 195, 199, 201, 202, 210, 212, 221, 233, 237, 249, 264, 270, 275, 279, 283, 287, 299, 303, 309, 330, 335, 340, 357, 359, 361, 367, 373, 388, 393, 401, 418, 423, 425, 432, 445, 447, 449, 450, 455, 459, 467, 469, 471, 473, 476
Size adjustment, 264, 335, 357, 358, 360, 366, 453
Size estimate, 359, 361, 364, 367, 368, 453
Skeletal biomechanics, 282
Skeletal muscle, 243, 298
Skull, 1, 3, 4, 6, 8, 9, 11, 17, 24, 34, 65, 67, 69, 70, 72, 73, 76, 78, 88, 90, 96, 107, 125, 150, 151, 153, 155, 160, 167, 175, 176, 179, 180, 201, 203, 204, 211, 212, 213, 242, 246, 258, 263, 294, 297, 299, 302, 335, 357, 363, 373, 394, 411, 413, 423, 433, 443, 447
Sloth lemurs, 421
Sonomicrometry, 23, 24, 25, 28, 30, 31
Specific density, 203, 224, 250
Specific gravity, 210
Split-line pattern, 266
Squamosal, 34, 91, 94
Squirrel, 418
Statistical control, 360, 361
Stature, 446, 449, 450, 452, 455, 459, 463, 465, 470, 473, 474, 475
Stereology, 266
Stereoscopic vision, 410, 418, 423
Sterkfontein, 436
Stiffness, 131, 135, 136, 140, 144, 151, 167, 179, 180, 266, 267, 268, 269, 271, 273, 275, 278, 279, 280, 282, 283, 284, 286, 287, 298, 388
Stone tool, 471
Strain, 7, 8, 9, 11, 17, 20, 21, 23, 25, 27, 28, 30, 32, 34, 35, 41, 43, 45, 46, 47, 49, 50, 55, 58, 59, 63, 64, 66, 68, 73, 78, 79, 84, 92, 95, 97, 98, 101, 104, 125, 127, 131, 133, 135, 136, 139, 140, 144, 150, 151, 153, 155, 157, 160, 165, 166, 167, 169, 179, 183, 189, 287, 320, 349, 388
Strain dampener, 150
Strain gage, 19, 21, 23, 24, 31, 63, 68, 69, 70, 71, 72, 73, 132, 136, 139, 153–154, 168, 169, 178, 281
Strain gauge, rosette, 43, 96
Strain gauge, single-element, 43

Strain orientation, 32, 67, 69, 70, 71, 72, 73, 76, 77, 78, 79, 92, 155, 160, 189
Strength, 9, 63, 64, 130, 131, 134, 135, 136, 143, 144, 150, 151, 167, 266, 287, 297, 302, 321, 332, 333, 334, 335, 336, 339, 340, 348, 351, 388, 401
Strepsirrhine(s), 66, 67, 74, 76, 79, 421
Stress, 10, 21, 45, 59, 63, 121, 125, 127, 129, 130, 133, 136, 139, 140, 143, 144, 168, 179, 201, 202, 212, 213, 214, 266, 267, 281, 294, 298, 307, 318, 320, 331, 335, 388, 397, 400, 403, 409, 421, 469, 470, 473
Stress concentration, 129, 130, 131, 132, 133, 136, 139, 142
Stress concentration factor, 129, 130
Stress-limited defense, 388, 397, 403
Stress-strain curve, 214
Subchondral bone, 295, 301, 304, 306, 308, 309, 310, 312, 314, 318, 319
Subfossil lemur, 218
Sugar, 403, 422
Suidae, 20
Sulfuric acid, 223
Suoidea, 20
Superficial masseter, 20, 22, 42, 56, 77, 78, 85, 86, 87, 93, 95, 96, 101, 105, 106, 107, 108, 177, 222, 223, 227, 228, 246, 247, 249, 250
Superficial temporalis, 18, 117, 121, 122, 222, 223
Supraorbital, 64, 66, 79, 154, 155, 176, 179, 184
Supraorbital torus, 154, 155, 176, 179, 184
Surface area, 57, 190, 201, 202, 213
Surface lamellae, 286
Suture, 26, 30, 31, 34, 36, 90, 96, 101, 102, 150, 153, 154, 155, 156, 157, 160, 165, 166, 168, 180
Swallow, 201
Symphyseal classes (I-IV), 57
Symphyseal curvature, 41
Symphyseal fusion, 17, 20, 21, 30, 35, 39–59, 86–87, 297, 320
Symphysis, 11, 17, 20, 21, 22, 24, 25, 26, 28, 29, 30, 31, 34, 35, 40, 41, 43, 44, 45, 46, 47, 50, 51, 52, 53, 54, 55, 57, 58, 59, 84, 85, 87, 90, 92, 94, 95, 96, 97, 11, 103, 105, 107, 108, 140, 194, 294, 296, 297, 299, 300, 301, 303, 309, 311, 313, 314, 315, 316, 318, 319, 320, 389, 447, 453, 459, 469, 474, 476
Syndesmoses, 297, 320
Synovial fluid, 296
Systematics, 411, 424

T

Talonid, 388, 393, 398, 399, 400, 401, 403
Tamarin, 40, 199, 200, 242, 245, 246, 250, 251, 252, 257, 258, 389, 391, 396
Tamarindus indica, 389, 391, 396
Taphonomy, 433
Tarsier, 205, 409, 418, 420, 421, 423, 424
Tarsioidea, 409
Tarsius bancanus, 235
Tarsius syrichta, 219, 226
Tayassuidae, 20
Tearing, 236
Teeth, 6, 7, 8, 9, 21, 32, 87, 94, 125, 127, 131, 135, 136, 139, 151, 160, 167, 173, 179, 181, 184, 187, 189, 190, 193, 194, 201, 256, 264, 330, 351, 389, 409, 431, 433, 435, 438, 440, 450, 453, 473
Temporalis, 10, 11, 18, 26, 53, 65, 66, 77, 79, 84–85, 93, 95, 96, 100, 105, 113, 114, 115, 117, 119, 120, 121, 122, 151, 177, 178, 203, 204, 205, 206, 222, 223, 227, 228, 234, 236, 243, 250, 251, 255, 257, 258, 259, 298, 330, 363, 410, 433, 453
Temporalis, anterior, 18, 66, 84, 93, 100, 113, 114, 115, 116, 117, 119, 120, 121, 177, 178, 246, 250, 330, 433
Temporalis, deep, 117, 121, 222, 223, 227, 234
Temporalis, posterior, 93, 95, 96, 114, 115, 151
Temporalis, superficial, 18, 117, 121, 122, 222, 223
Temporomandibular joint disorder (TMD), 7
Temporomandibular joint (TMJ), 7, 20, 91, 107, 108, 167, 181, 263, 294, 295, 296, 297, 298, 299, 300, 301, 302, 303, 304, 305, 306, 307, 308, 309, 310, 311, 312, 313, 314, 315, 316, 317, 318, 319, 320, 321, 332, 362
Tendon, 91, 113, 117, 210, 220, 224, 227, 235, 248, 249, 250, 251, 252, 254, 255, 256
Tenrec, 85
Tension, 17, 22, 23, 28, 29, 31, 32, 33, 40, 41, 43, 45, 47, 51, 55, 98, 102, 107, 116, 129, 132, 158, 159, 161, 162, 163, 164, 183, 236, 243, 244, 254, 255, 301, 313, 333
Terminal branch feeding, 339, 421
Ternifine, 450, 451, 458, 460, 461, 463, 464, 465, 466, 468, 471, 476
Terrestrial, 389, 407, 408, 419
Tertiary, 277, 411, 422
Third-class lever, 7, 217
Three-point bending, 221, 236, 392
Tooth area, 202, 212, 213, 389, 392
Tooth basin area, 389, 393, 401

Tooth crest length, 388, 389, 392, 393, 398, 400
Tooth root, 125, 128, 131, 134, 135, 136, 139, 143, 181
Tooth size, 174, 201, 40, 445, 448, 460, 473
Tooth-tooth contact, 11, 440
Torque, 21, 32, 63–80, 104, 107, 108
Torsion, 71, 72, 73, 76, 78, 132, 134, 135, 136, 139, 185, 186, 190, 330, 331, 332, 333, 341, 347, 349, 350
Total morphological pattern, 418
Tough, 20, 91, 108, 118, 150, 193, 194, 221, 233, 295, 297, 330, 331, 341, 345, 346, 347, 348, 370, 388, 389, 396, 397, 404, 447
Tough, herbaceous vegetation (THV), 447
Toughness, 131, 214, 236, 298, 332, 351, 388, 391, 392, 394, 396, 397, 398, 401, 402, 403, 404
Trabecular, 128, 176, 179, 265, 266, 316, 341, 470
Trabecular bone, 128, 176, 179, 266, 316, 341
Trabecular orientation, 265
Transducer, 23, 178, 266, 269
Transpalatal suture, 153, 154
Transverse isotropy, 140, 268, 282, 284, 286
Tree gouging, 6, 199, 241–259, 331, 335, 339, 341, 348, 357, 371, 373, 376, 378, 379
Treeshrew, 85
Trigon(id), 388, 393, 398
Trinil, 449, 452, 455, 475
Triplet I, 85, 86, 94, 105
Triplet II, 85, 86, 87, 94, 104
Twisting, 17, 21, 31, 33, 52, 53, 54, 57, 58, 64, 66, 67, 72, 74, 76, 77, 78, 79, 87, 107, 128, 132, 133, 135, 136, 143, 160, 167, 194, 263, 332, 440
Twisting, transverse, 33, 52, 53, 54, 58
Type I collagen, 313
Type I error, 336, 341
Type I fibers, 116
Type II collagen, 296, 301, 306, 307, 308, 311, 312, 313, 315, 316, 317, 318, 319
Type II fibers, 115, 116, 117, 121, 122
Typology, 4

U

Ulna, 287
Ultimate failure, 133
Ultrasonic, 142, 263, 266, 267, 268, 269, 272, 273, 282, 284, 286
Ungulate, 85, 87, 91, 105, 107
Unipinnate, 223, 243
Unloading, 40, 45, 49, 50, 51, 52, 54

V

Varecia, 219, 226, 393, 421
Variance, 140, 156, 181, 279, 284, 351, 359–361, 366, 367, 394, 401
Vicuña, 41, 58
Video, 7, 44, 97, 105, 392, 393
Vision, 410, 418, 419, 422, 423
Visual acuity, 408, 409
Visual predation, 410, 420–421
Volksman's canal, 275
Vombatus ursinus, 339

W

Wallabies, 88
Wear pattern, 90, 107, 435–437, 440
Wedge test, 392
West Turkana, 450
Wheatstone bridge, 68, 97
Wishboning, *see* Lateral transverse bending
Wombat, 17, 83–109, 339
Work of fracture, *see* Toughness
Work of friction, 391
Working/balancing EMG ratio, 86

Working side, 11, 21, 22, 26, 27, 33, 35, 42, 53, 54, 56, 65, 77, 79, 84, 87, 89

X

Xenarthran, 408

Y

Yield stress, 267
Young's modulus, 298, 388

Z

Zhoukoudian, 450, 452, 458, 463, 471, 472, 475
Zygoma, 9, 63, 66, 77, 79, 84, 93, 108, 126, 174, 175, 176, 180
Zygomatic arch, 63, 64, 68, 71, 90, 91, 117, 154, 155, 156, 157, 167, 176, 179, 180, 186, 189, 190, 249, 434
Zygomatico-frontal suture, 154, 155, 165, 166
Zygomaticomandibularis, 20, 84, 117, 222
Zygomatico-temporal suture, 150, 154, 155, 156, 157, 165, 166
Zygomatic temporalis, 222, 223, 228, 232, 234, 235, 236, 237